BACTERIA IN NATURE

Volume 2

*Methods and Special Applications
in Bacterial Ecology*

BACTERIA IN NATURE

Volume 1   Bacterial Activities in Perspective
Volume 2   Methods and Special Applications in Bacterial Ecology

Forthcoming
Volume 3   Foundations of Bacterial Existence

BACTERIA IN NATURE

Volume 2

Methods and Special Applications in Bacterial Ecology

Edited by

Jeanne S. Poindexter

Long Island University
Brooklyn Campus
Brooklyn, New York

and

Edward R. Leadbetter

The University of Connecticut
Storrs, Connecticut

PLENUM PRESS • NEW YORK AND LONDON

Library of Congress Cataloging in Publication Data

(Revised for vol. 2)

Bacteria in nature.

Includes bibliographies and index.
Contents: v. 1. Bacterial activities in perspective — v. 2. Methods and special applications in bacterial ecology.
1. Bacteria — Ecology — Collected works. 2. Bacteria — Collected works. I. Leadbetter, Edward R. II. Poindexter, Jeanne S. (Jeanne Stove).
QR100.B33   1986   589.9′05   85-3433
ISBN 0-306-42346-4

Cover illustration: Stalked bacteria *(Caulobacter)* in their natural superior position
within the prokaryote community attached to the diatom *Melosira*. Scanning electron micrograph by J. S. Poindexter, taken at the Marine Biological Laboratory,
Woods Hole, Massachusetts.

© 1986 Plenum Press, New York
A Division of Plenum Publishing Corporation
233 Spring Street, New York, N.Y. 10013

Printed in the United States of America

PREFACE TO THE TREATISE

The effects of bacteria on their environments were known and variously explained by human societies long before these microorganisms were recognized. Even after they had been detected microscopically, nearly two centuries elapsed before it was demonstrated that bacteria were causes, rather than effects, of fermentations, infectious diseases, and transformations of both organic and inorganic materials in soils, waters, and sediments. It was these demonstrations of the ecological roles of bacteria that gave birth to bacteriology as an experimental science. The applications of the understanding of ecological activities of bacteria have in no small part been responsible for this century's revolution in human health and longevity through changes in agricultural, medical, and sanitation practices.

However, the ecology of bacteria has only relatively recently emerged as a science in itself, having as its goal the elucidation of the interactions of bacteria and their habitats whether or not those activities appear immediately relevant to human affairs. In this ten-volume treatise, it is our intention to present this broadened view of bacterial existence, that the work may serve as a synthesis of current ideas and information that will be valuable both to basic scientists and to those directly engaged in applications of science to specific problems of human existence. Our hope is that the completed project will expose and explore the diversity of bacterial capabilities and culpabilities, limitations and sensitivities, and will imply the equally diverse ways in which they can be exploited. We hope, especially, that investigators trained in other disciplines—clinicians, oceanographers, molecular biologists, engineers—who may not expect that their disciplines are interrelated with bacterial ecology, will find this treatise both stimulating and valuable.

The introductory volume traced main points in the history of bacteriology that have led to the present state of bacterial ecology, to the awareness that bacteria constitute distinctive populations that separately and in concert affect the physiochemical conditions of the biosphere and interact, sometimes intimately, with other organisms. This second volume reviews and evaluates the technical and philosophical tools presently available to the student of bacteria in nature. It is intended to provide evaluations of methods and not to serve as a procedural manual.

Although the initial stimulus to interest in bacteria arose from attempts to understand natural phenomena and to distinguish abiotic from biotic causes of these phenomena, most of the progress in managing and learning about bacteria has been accomplished in the laboratory, very largely through the study of pure cultures. While some devoted naturalists eschew the study of monotypic populations as artificialties, it would not be possible to understand the activities of a bacterial community in ignorance of the separate, respective potential activities of the members of the community. For this reason, two volumes will comprise information regarding structure, composition, genetics, physiology, and biochemistry of bacteria, obtained predominately from pure culture studies, that is essential to unraveling the interactions of bacteria with their environment and with each other. The goal of those volumes is not simply to review the information, but to demonstrate its importance to inferences based on studies of natural, polytypic populations.

The remaining six volumes in this series will explore bacterial habitats. Because a bacterial activity of any type—polymer solubilization, oxygen consumption, toxin production, or other—is not confined to one type of habitat, we anticipate that some groups of bacteria and some bacterial activities will be mentioned in more than one ecological context. However, since the emphasis throughout the series will be on interactions, the role of the habitat in influencing the extent and consequence of bacterial activities will vary with its own inherent stability, its resilience to the effects of bacteria, and its capacity for supporting and restricting those activities. An additional reason for organizing the treatise principally around habitats reflects the fact that human problems and advantages that arise from bacterial activities are most often met within the context of a particular kind of site. Similarly, the general ecological significance of a baterial activity is proportional to the rate of activity allowed by a habitat, the geographical extent of the habitat, and the degree of dependence of other forms of life on the condition of the habitat affected by that activity. Accordingly, for all conceptual and practical purposes other than classification, the study of bacterial ecology is, we believe, most usefully presented by grouping the information into volumes that reflect the manner in which bacterial communities are gathered and interact in nature.

The treatise will conclude with a consideration of the frontiers and the relic habitats of the biosphere, those environments inhabited almost solely by bacteria. Modern biology recognizes bacteria as the pioneering populations— demonstrably so in the past and predictably so in the future—of this and possibly also of other worlds. It is the humble yet confident hope of the editors that the insights and experimental results of today's bacterial ecologists, compiled in these volumes, will contribute significantly to the continuing elucidation of the roles and potentials of our bacterial cohabitants—so long a major influence on this earth, yet only so recently appreciated.

This series was conceived partly as a result of our participation as instructors in the summer program in microbial ecology at the Marine Biological Laboratory, Woods Hole, Massachusetts. Our interests in bacterial ecology antedate that participation by many years, having been earlier stimulated and guided by R. A. Slepecky and J. W. Foster (E.R.L.), by W. A. Konetzka and R. Y. Stanier (J.S.P.), and by one of the greatest appreciators of microorganisms, C. B. van Niel. To our colleagues—students and faculty—in the M.B.L. course, and to our several mentors, we dedicate this series.

The Editors

CONTRIBUTORS

RONALD M. ATLAS, *Department of Biology, University of Louisville, Louisville, Kentucky 40292*

J. W. COSTERTON, *Department of Biology, University of Calgary, Calgary, Alberta, Canada T2N IN4*

ROBERT D. FALLON, *University of Georgia Marine Institute, Sapelo Island, Georgia 31327*

ROLF FRETER, *Department of Microbiology and Immunology, The University of Michigan Medical School, Ann Arbor, Michigan 48109*

JAN C. GOTTSCHAL, *Department of Microbiology, University of Groningen, 9751 NN Haren, The Netherlands*

DAVID M. KARL, *Department of Oceanography, University of Hawaii, Honolulu, Hawaii 96822*

T. I. LADD, *Department of Biology, St. Mary's University, Halifax, Nova Scotia, Canada B3H 3C3*

E. R. LEADBETTER, *Department of Molecular and Cell Biology, The University of Connecticut, Storrs, Connecticut 06268*

MICHAEL J. McINERNEY, *Department of Botany and Microbiology, University of Oklahoma, Norman, Oklahoma 73019*

STEVEN Y. NEWELL, *University of Georgia Marine Institute, Sapelo Island, Georgia 31327*

J. C. NICKEL, *Department of Urology, Queens University, Kingston, Ontario, Canada K7L 2V7*

J. S. POINDEXTER, *Science Division, Long Island University, Brooklyn Campus, Brooklyn, New York 11201*

PAUL S. TABOR, *Biotechnology Section, Naval Research Laboratory, Washington, D.C. 20375; present address: Department of Life Sciences, Indiana State University, Terre Haute, Indiana 47809*

DAVID C. WHITE, *Center for Biomedical and Toxicological Research, Florida State University, Tallahassee, Florida 32306*

PREFACE

The very small size of bacteria permits their cultivation on scales ranging from depressions in microscope slides through fermentation vats two stories high. This is indisputably an asset in their use for laboratory experiments and for many commercial and agricultural processes. Their small size, as well as their haploid condition and their sex-independent reproduction, have also proved invaluable in genetic studies; astronomical numbers of individuals can be produced overnight in hand-held vessels requiring minute expenditures of time, space, and materials. On the other hand, the extremely small size of bacteria and the correspondingly microscopic dimensions of their habitats present the major technical problem in bacterial ecology: how to conduct reproducible, quantitative, and unobtrusive ecological investigations of invisible organisms.

For the second volume of *Bacteria in Nature,* contributors were invited to evaluate methods currently in use or being developed for bacterial ecology. The evaluations comprise recognition of both the suitability of a given method for particular investigative purposes and its limitations and potential for misleading interpretations. The critical discussions are accompanied by descriptions of methods and sample investigations that serve as illustrations; they are not intended as procedural instructions, which are available in the references cited.

Each of the four chapters (1 through 4) concerned with immediate analysis of natural samples emphasizes the importance of employing more than one method in assessing bacterial numbers and activities; in each case, a microscopical method is included as one of those regarded as useful in interpreting the results of other, parallel, studies. This is not surprising—with the notable and promising exception of the employment of microelectrodes, microscopical investigations are the only direct method of study that correspond in practice to the dimensions of bacterial populations in natural situations. Nevertheless, these methods involve varying degrees of subjectivity, not only at the level of discerning their implications, but also at the more immediately practical level of discerning bacteria and adopting criteria for inferring their viability or metabolic activity. Among the evaluations presented in this volume, there is some difference of opinion regarding the value and interpretability of microscopical observations; we have chosen to include the different views. There is general agreement, however, that microscopical investigations are indispensable for bacterial ecology, and efforts to expand their capabilities and dependability continue along with controversies regarding their import.

The next four chapters (5 through 8) are concerned with bacterial behavior under conditions intentionally different from their natural situation, imposed in an effort to study the interactions of the bacteria and the changed environment, the bacteria and an animal host, or the bacteria among themselves. Unincubated samples cannot provide the kinds of information available from such studies, which are included here as examples of the ecological methods available uniquely with microorganisms.

The final chapter examines principles recognized in general ecology for their relevance to the study of the distribution, interactions, and fate of microbes in nature. To our regret, a second relatively theoretical chapter, on mathematical modeling in bacterial ecology, was not available in time for inclusion in this volume.

This examination of methods appears as an early volume in this treatise for two principal reasons. First, it is intended to illustrate that the problems of understanding microorganisms as they exist in nature, typically as minute yet heterogeneous populations, are being resolved by technological and procedural innovations, as well as by use of methods long familiar in microbiology and applicable to ecological studies. Second, most of the methods discussed, as well as their principles, are or can be employed in studies of various microbial habitats. Accordingly, this evaluation should prove useful to a variety of microbial ecologists and their students, to readers attempting to judge the dependability of approaches and techniques in bacterial ecology, and as an advance critique of assertions presented in later volumes of this treatise.

The Editors

CONTENTS

CHAPTER 1

Direct Microscopy of Natural Assemblages

    STEVEN Y. NEWELL, ROBERT D. FALLON, AND PAUL S. TABOR

Introduction 1
Light Microscopic Determination of Standing Crop 2
    Transmitted Light 2
    Epifluorescence 3
Light Microscopic Determination of Productivity 10
    Without Incubation 10
    With Incubation 14
Light Microscopic Determination of Individual Cell Activity 19
    Fluorochromes in Combination with Direct Activity Measurements ... 19
    A Comparative Study: Numbers of Synthetically Active Bacteria,
        Cells with Electron-Transport Activity, and Substrate Uptake-
        Active Cells 22
    Fluorochromes Alone as Active Cell Indicators 30
Immunolabeling in Bacterial Ecology 31
    Introduction 31
    Development and Use of Immunolabeling Reagents 32
    Recent Applications and Current Status 35
References 37

CHAPTER 2

Suitable Methods for the Comparative Study of Free-Living and
Surface-Associated Bacterial Populations

    J. W. COSTERTON, J. C. NICKEL, AND T. I. LADD

Introduction 49
Direct Examination by Transmission Electron Microscopy 50
Direct Examination by Scanning Electron Microscopy 57
Sampling of Biofilm Bacteria 64
Cultivation of Bacteria in Biofilms 70
Detection of Activity within Biofilms 75

Summary 81
References 81

CHAPTER 3

Determination of *in situ* Microbial Biomass, Viability, Metabolism,
and Growth

DAVID M. KARL

Introduction 85
Habitat and Community Description 86
  Diversity and Scale 86
  Microbial Communities in Nature 87
Scope of Existing Methods 88
  Sampling, Subsampling, and Incubation Considerations 92
  Use of Isotopic Tracers 93
  Attribution of Total Biomass, Metabolic Activity, or Growth to
    Taxonomic Compartments 99
Biomass 100
  Observational Methods 100
  Biochemical Methods 106
Viability 116
  Vital Fluorogenic Dyes 117
  Dye Exclusion Test 118
  Microautoradiography 118
  INT Labeling 121
  Yeast Extract/Nalidixic Acid 123
Physiological Potential/Nutritional State 123
  Enzymic Activity or Potential 124
  Nucleotide Ratios, Nucleosides, and Related Intracellular
    Compounds 127
  Synthesis and Catabolism of Storage or Reserve Polymers 128
  Photosynthetic Potential 129
  Chemical Composition and Growth State 130
  Patterns of Autotrophic ^{14}C Assimilation 131
Metabolic Activity 132
  O_2 and CO_2 Fluxes 132
  Uptake and Assimilation of ^{14}C- and ^{3}H-Labeled Organic
    Compounds 134
  Microcalorimetry 138
Growth, Production, and Cell Division 139
  Primary Production of Organic Matter 140
  Dark CO_2 Assimilation 143
  Mitotic Index/Frequency of Dividing Cells 144
  Assimilatory Sulfur Metabolism 146

Rates of Nucleic Acid Synthesis 148
ATP and Adenine Nucleotide Pool Turnover 155
Increase in Cell Numbers/Biomass during Timed Incubations 156
Summary and Prospects for the Future 158
References 159

CHAPTER 4

Quantitative Physicochemical Characterization of Bacterial Habitats

DAVID C. WHITE

Introduction 177
Consequences of Small Size 178
Microbial Modification of Microenvironments 179
Microbial Modification of the Biosphere 182
Direct Measurement of the Microbial World 182
Microelectrodes 182
Epifluorescence Microscopy 183
Electron Microscopy 185
Surface Analysis 187
Fourier Transforming Infrared Spectrometry 188
Indirect Inferences from Microbes 189
Biomass and Community Structure 190
Microbial Metabolic Activity 193
Nutritional Status 194
Effects of Surfaces 196
Conclusions 196
References 197

CHAPTER 5

Gnotobiotic and Germfree Animal Systems

ROLF FRETER

Introduction 205
Definitions 206
Germfree Animals 206
Techniques and Tests 206
The Function of a Complete Indigenous Microflora: Comparison
of GF and CV Animals 209
Gnotobiotic Animals 213
Etiology and Pathogenic Mechanisms (Models of Disease) 213
Physiologic Effects 216
Interactions among Microbes of the Indigenous Microflora 218
Summary and Conclusions 222
References 223

CHAPTER 6

Enrichment Cultures in Bacterial Ecology

J. S. POINDEXTER AND E. R. LEADBETTER

Introduction 229
Purposes and Principles 230
Bacterial Diversity 233
Conditions for Elective Cultivation 240
   Closed-System Liquid Enrichment Cultures 240
   Open-System Elective Cultivation 246
   Two-Phase Elective Cultivation 248
Elective Cultures as Microbial Successions 249
Limitations and Prospects for Elective Cultures in Bacterial Ecology .. 250
   Enrichment Cultures as Enumeration Techniques: The MPN
      Method 251
   Eukaryotes as Living Media 252
   Stable Prokaryote Associations 253
   Unexpected Organisms 254
   Prospects 256
References 257

CHAPTER 7

Mixed Substrate Utilization by Mixed Cultures

JAN C. GOTTSCHAL

Introduction 261
Theory of Growth in Continuous Culture 262
   Substrate-Limited Growth 263
   Multiple Substrate Limitation 265
   Mixed Cultures 267
Pure Cultures, Mixed Substrates 273
Mixed Cultures, Mixed Substrates 277
Chemostat Enrichments with Mixed Substrates 285
Conclusion 287
References 288

CHAPTER 8

Transient and Persistent Associations among Prokaryotes

MICHAEL J. McINERNEY

Introduction 293
Classification of Bacterial Interactions 294
   Neutralism 295
   Commensalism 296

Protocooperation and Mutualism 297
Competition 299
Amensalism (Antagonism) 304
Predation and Parasitism 305
Syntrophic Associations in Methanogenic Ecosystems 306
Microbial Interactions in the Rumen 306
Microbial Interactions in Other Methanogenic Environments 314
Associations with Phototrophs 322
Syntrophic Associations 322
Symbiotic Associations 323
Competition between Sulfate Reducers and Methanogens 324
Degradation of Xenobiotic Compounds 326
Intraperiplasmic Growth of Bdellovibrios 327
Summary 328
References 329

CHAPTER 9

Applicability of General Ecological Principles to Microbial Ecology

RONALD M. ATLAS

Ecosystems and Microbial Populations 339
Energy Flow through Ecosystems 340
Photoautotrophy 341
Chemolithotrophy 343
Heterotrophy 343
Food Webs 344
Biogeochemical Cycling within Ecosystems 346
Cycling of Carbon 346
Cycling of Nitrogen 347
Cycling of Sulfur 350
Other Cycling Activities 350
Interactions of Populations with their Abiotic Surroundings 351
Factors Controlling Populations 351
Adaptation in Microbial Populations 353
Interactions between Diverse Populations 354
Competitive Exclusion 355
Cooperative Relationships 357
Community Structure 358
Diversity, Succession, and Stability 358
Summary 361
References 361

INDEX 371

DIRECT MICROSCOPY OF NATURAL ASSEMBLAGES

Steven Y. Newell, Robert D. Fallon, and Paul S. Tabor

INTRODUCTION

A recent trend in microbial ecology is for biochemical methods to be adapted to the study of bacteria in nature (e.g., see Chapter 3). This type of approach will continue to provide invaluable information regarding the activities and chemical impact of bacteria within ecosystems. Yet indirect biochemical methods cannot supplant direct-observational methods as long as it is of interest to know the dynamics of biovolume and biomass of the intact bacterial organism(s) within ecosystems. It is not reasonable to expect that insertion of procedural steps (components of chemical assay techniques, each subject to error) in the attainment of values for biomass or biovolume will result in more accurate values. Nor is it to be expected that community structure (the relationships between species constituting communities or assemblages) will be more readily detectable biochemically than by appropriately specific direct-observational methods. As a famous American sports figure once succinctly put it, "You can see a lot by lookin'."

As two recent examples of the power of direct microscopic methods in bacterial ecology, consider the following two reports. Parsons *et al.* (1981) discovered, partially by monitoring the dynamics of bacterial standing crop by direct microscopy in marine habitats, that saprotrophic production could support substantial animal secondary and tertiary production. Wright *et al.* (1982) used direct microscopic determination of bacterial cell concentration upstream and downstream from three species of bivalve mollusks. Sharp

Steven Y. Newell and Robert D. Fallon • University of Georgia Marine Institute, Sapelo Island, Georgia 31327. *Paul S. Tabor* • Biotechnology Section, Naval Research Laboratory, Washington, D. C. 20375; *present address:* Department of Life Sciences, Indiana State University, Terre Haute, Indiana 47809.

differences were revealed in the abilities of the mollusks to remove bacteria from the water column, indicating that only one of the species was capable of taking advantage of saprotrophic production based on material from salt-marsh vascular plants.

This chapter reviews the methods that have been used for direct microscopic determination of distributions of bacterial cells and bacterial productivity in time and space and for the detection of activities of individual bacterial cells, both of total assemblages and within single morphological or immunological types. For the most part, this chapter, completed in January 1985, deals with light microscopic methods. For corresponding electron microscopic information, see Chapter 2, this volume.

LIGHT MICROSCOPIC DETERMINATION OF STANDING CROP

Transmitted Light

Many of the earlier attempts to obtain direct counts of natural bacterial assemblages involved the use of transmitted light microscopy. Samples of water, soil, or sediment were collected, homogenized or ground (soil and sediment), diluted or concentrated as necessary, deposited on membrane filters, glass slides, in agar films, or in counting chambers, stained (usually erythrosine) or not stained, and examined at × 1000 or greater, in most cases by phase-contrast microscopy (e.g., Jannasch and Jones, 1959; Oláh, 1974; Fog, 1977; Straškrabová and Komárková, 1979; Seki and Nakano, 1981; Allison and Sutherland, 1984; see also Hossell and Baker, 1979). As these references reveal, some fine direct-count work has been done in this way, although early workers were concerned that they were not recognizing all bacterial cells in their preparations due to the almost inevitable presence of noncellular debris that obscured cells. It was also likely that small bits of nonbiomass material would be confused with bacterial cells (Joint and Morris, 1982). Comparison of erythrosine/phase-contrast direct counts of bacteria with epifluorescence direct counts has shown the latter to yield higher, and probably more accurate, values (Coveney et al., 1977; Salonen, 1977; Schmidt and Paul, 1982). A further problem is that the erythrosine method yields inaccurate values for cell sizes that were directly measured, as a result of cell shrinkage (Straškrabová and Komárková, 1979). Mitskevich and Kriss (1982) surmounted the problem of cell shrinkage by concentrating natural cell suspensions and then counting erythrosine-stained cells within plane-parallel capillaries; however, they avoided counting bacteria associated with particles because the erythrosine method failed to permit clear distinction of cells from noncellular debris. Bacterial ecologists are using epifluorescence microscopy almost exclusively in their direct microscopic work; therefore, we shall not discuss transmitted light analyses further.

Epifluorescence

The epifluorescence microscopic technique most commonly used for direct counting of bacteria in natural samples is the acridine orange direct count (AODC) method (Daley, 1979; Rublee, 1982). Although introduced for use in soil microbiology (Strugger, 1948), AODC has now been used repeatedly in a wide variety of terrestrial, freshwater, and marine environments. The major advantage of AODC over transmitted light methods lies in the greater ability permitted the observer to discriminate between bacterial cells and non-bacterial entities. This is because acridine orange (AO) is a nucleic acid fluorochrome, fluorescing green to red when combined with DNA or RNA (Kasten, 1981). Since bacterial cells have loosely centralized "nuclear regions" and small (70 S) densely packed ribosomes throughout the cytosol (Ingraham *et al.*, 1983) rather than nuclear and extranuclear DNA packaged into organelles and large (80 S) less densely packed ribosomes, they are relatively easily distinguished by means of AODC from eukaryotic organelle-containing cells and from abiotic particles. A wide variety of other direct and indirect techniques have been used to cross-check AODC as used in aquatic ecology (see discussion and references cited in Daley, 1979; Fuhrman, 1981; Newell and Christian, 1981; Moriarty and Hayward, 1982; Paul and Jeffrey, 1984); the result in each case has been that AODC values for bacterial direct counts and cell volumes have proved accurate. However, Domsch *et al.* (1979) found evidence that AODC yields inaccurate low values when used with terrestrial soil samples (see also Casida, 1971; Jenkinson *et al.*, 1976; Nishio, 1983); this may well be largely a problem of inefficiency of extraction.

Three commonly cited formulas for AODC are those suggested by Trolldenier (1973), Zimmerman and Meyer-Reil (1974), and Hobbie *et al.* (1977). There are potentially important differences among the three sets of instructions, with respect to the comparability of results. Components of the methods are listed in Table I, along with recent variants. Daley (1979) expressed the opinion that establishment of a standard procedure should soon be realized. This will probably not be the case, however, unless one or more workers carry out a concerted, strictly methodological research project, the goal of which would be to test components of AODC methodology and obtain sound, convincing, comparative results. Otherwise, each AODC devotee will continue to use the most familiar method.

In spite of the several existing experimental verifications of the accuracy of AODC, there are several potential problems associated with obtaining accurate AODC values. One of these is the subjectivity involved in identifying green-to-red fluorescing objects as bacterial cells. Although this is less of a problem than with some other fluorescent microscopic techniques (cf. Fry and Humphrey, 1978; Newell and Hicks, 1982), practice with microbial assemblages of known construct, familiarity with the range of potential bacterial morphologies, and training with experienced observers are highly desirable as preludes to the use of AODC. Regarding potential bacterial morphologies,

TABLE I

Some Combinations of Components of Methods for Acridine Orange Direct Counting of Bacteria in Natural Samples[a,b]

Reference (sample)	Preservation[c]	Extraction (solid samples)	AO% (solvent); and time in AO	Viewing support[d]	Final treatment of preparation	Reference
Fresh/seawater	2% formaldehyde	—	0.01% (distilled H_2O); 1–2 min	Polycarbonate filter, 0.2-μm pore size, irgalan black stained	No rinsing or drying	Hobbie *et al.* (1977)
Seawater	2% formaldehyde	—	0.01% (phosphate buffer); 3 min	Polycarbonate filter, 0.2-μm pore size	Destaining with isopropyl alcohol + xylene; drying; clearing with cinnamaldehyde + eugenol	Zimmerman and Meyer-Reil (1974)
Soils	Storage at 2°C	Shaking in phosphate buffer + dilute agar	0.007% (not given); 2 min	Agar film on engraved area of glass slide	Rinsing (water) and drying	Trolldenier (1973) (cited in Domsch *et al.,* 1979)

Sample	Fixation	Homogenization	Stain (concentration); time	Filter/surface	Rinsing/drying	Reference
Sediments	3.5% formaldehyde, storage at 4°C	Homogenization, 2×10^4 rpm, 30 sec	0.0004% (seawater); 2 min	Black cellulose acetate filter, 0.2-μm pore size	No rinsing or drying	Moriarty (1980)
Seawater	1.5% formaldehyde + organic buffer	—	0.01% (buffered seawater); hours	Polycarbonate filter, 0.2-μm pore size	Rinsing (water) and clearing with cinnamaldehyde + eugenol	Larsson and Hagström (1982)
Freshwater	Glutaraldehyde vapor following filtration	—	0.01% (phosphate buffer); several minutes	Polycarbonate filter, 0.2-μm pore size	Destaining with isopropyl alcohol; drying	Geesey and Costerton (1979)
Plant surfaces	0.0002% formaldehyde, during fluorochroming	Not performed	0.0001% (solution of Triton X + 0.005% fluorescent brightener); 15–30 sec	Natural surfaces	Rinsing (water); no drying	Paton (1982); see also Pettipher and Rodriguez (1982)

[a] See also reviews by Daley (1979), Jones (1979), Rublee (1982), and Schmidt and Paul (1982).
[b] AO, acridine orange.
[c] Daley (1979) contends that rapid air drying of filtered samples without aldehydes works well (see Ramsay, 1978; Schmidt and Paul, 1982).
[d] See Brock (1983).

a problem that has not been satisfactorily resolved for AODC is the question of the low end of the range of sizes of bacterial cells. When only pinpoints of AO fluorescence are seen, it cannot be determined whether the tiny fluorescent object is amorphous or membrane bounded. This problem can be particularly severe in inshore marine environments (Wilson and Stevenson, 1980; S. Newell, unpublished observations). Are these dots tiny functional bacteria, bits of nonliving or nonbacterial nucleic acid polymers (bacteriovore egesta?) or some form of nonmicrobial fluorescent material? Sieburth and Johnson have established via electron microscopy that marine planktonic prokaryotes as small as 1 to $2 \times 10^{-2} \, \mu m^3$ cell volume are membrane bounded and exhibit internal ultrastructure typical of bacterial cells (Sieburth, 1979; see his plate 14-5). Krambeck et al. (1981) and Fuhrman (1981) found, by electron microscopic examination, that the smallest planktobacteria had cell volumes of about 1 to $2 \times 10^{-2} \, \mu m^3$. The smallest starved bacteria studied by Novitsky and Morita (1978) were about $3 \times 10^{-2} \, \mu m^3$ in cell volume. A practical solution to this problem may be to use magnification sufficiently high (Fuhrman, 1981; S. Y. Newell, 1984) to permit exclusion of fluorescent bodies that appear to be smaller than $1 \times 10^{-2} \, \mu m^3$ cell^{-1} during AODC.

Other potential problems with respect to accuracy are patchy distribution of bacteria (in nature and in filtered concentrates), presence of cells attached to undersurfaces of or intermingled with opaque suspended material, and failure of preservation techniques to prevent decreases in sample counts during storage. Coping with or resolution of these difficulties is discussed by Daley (1979); see also Rublee and Dornseif (1978), Kirchman and Mitchell (1982) on opaque particles, Montagna (1982) on heterogeneous distribution in sediments, Pomroy (1984) on an alternative to aldehyde fixation, and Roser et al. (1984) on "plotless" counting methods. Kirchman et al. (1982b) discussed the problem of uneven distribution of bacteria on filters prepared for AODC; these workers concluded that microscopic fields on filters must be chosen randomly for counting. Indeed, this would be ideal, but unless one is to choose the fields by reference to a table of random numbers (identifying fields as points on a superimposed grid), the fields are not likely to be random with respect to the filter area. Looking away from the microscope and moving the stage controls (Kirchman et al., 1982b) could certainly lead to unintentional nonrandom sampling of the filter area. It may be more practical to sample the filter area evenly by using a standard set of locations on the filter and to strive to achieve as even a distribution of bacteria as possible on filters. There are two further problems: (1) the potential presence of very small cyanobacterial and eukaryotic algal cells (many less than $1–2 \, \mu m^3$ per cell) that could be mistaken for heterotrophic bacteria; and (2) in the case of solid samples, incomplete extraction of bacterial cells from their attachment to the substrate.

If the goal in using AODC is to enumerate total bacteria, or total nonphotoautotrophic bacteria, one must be aware that small cyanobacterial and eukaryotic algal cells may be present in the samples, at least in samples from photic marine and probably freshwater environments (Krempin and Sullivan,

1981; Cole *et al.*, 1982; Johnson and Sieburth, 1982). The cyanobacterial and algal cells can be enumerated in preparations parallel to those for AODC by omitting AO treatment; the photosynthetic pigments of cyanobacteria and eukaryotic algal cells will autofluoresce orange to red under the same epi-illumination conditions used for AODC (Davis and Sieburth, 1982). Current protocols, e.g., that suggested by Krempin and Sullivan (1981), involve counting only orange-autofluorescing (phycoerythrin-containing) bacteria-sized cells as cyanobacteria; it may be that nanoeukaryotic algae can be separately enumerated as red-autofluorescing bacteria-sized cells (blue light excitation) and phycoerythrin-negative cyanobacteria distinguished by brightness with green light excitation (Gantt, 1975; Table I of Wilde and Fliermans, 1979). Counts of autofluorescing cells must be done within 48 hr of aldehyde preservation, as autofluorescence of chlorophyll can be lost more rapidly than can AO-nucleic acid fluorescence (cf. Daley, 1979; Sieburth and Davis, 1982).

Samples that include particulate matter too large for practical inclusion in the final AODC preparation (e.g., sandy sediments, leaves of vascular plants) must be treated in such a way that removal of bacterial cells from the bulky particles is maximized. Mechanical homogenization has been accepted as the best means among those tested for providing maximal removal of bacteria from sediment sand grains, soil particles, and plant materials (Babiuk and Paul, 1970; Newell, 1981; Montagna, 1982). Newell (1981) found homogenization of pieces of seagrass leaves to result in release of greater than 99% of attached, epiphytic bacterial cells (see also Fry and Humphrey, 1978; but see Hossell and Baker, 1979). However, it is clear from the findings of Rublee (1982), Moriarty (1980), and Newell and Fallon (1982, and unpublished results) that homogenization leaves a substantial fraction of bacterial cells behind when sediments containing sand are processed. These investigators found, respectively, that 26–40%, 33–40%, and 15–50% (excluding two high values of 67–80%) of cells (Rublee and Newell–Fallon) or cell-wall components (Moriarty) were not detached from heavy particles and brought into suspension. The percentages are likely to vary with specific type of homogenization and sample; therefore, unless one conducts a carefully designed experiment to determine cell-extraction efficiency for the particular type of sample examined, the direct-count values for bacteria from sediment and soil samples are likely to be inaccurate. Ultrasonication techniques with surfactants or deflocculants were recently found more efficient in extraction of intact bacterial cells from soil or sediments than such techniques as agitation or homogenization (Ellery and Schleyer, 1984; Ramsay, 1984; Velji and Albright, 1985), although only Ellery and Schleyer made a thorough experimental comparison of ultrasonication versus homogenization. These investigators recognized that ultrasonication could destroy some bacterial cells; Ellery and Schleyer (1984) recommend determination of a correction factor, just as for homogenization.

If one wishes to obtain values for biovolume and biomass via AODC, additional problems with respect to accuracy must be faced. Measurements of lengths and widths (or diameters) of bacterial cells by light microscopic examination may yield systematic errors in calculation of cell volumes, as these dimensions approach the limits of light microscopic resolution (Quesnel, 1971). Fuhrman (1981), however, compared measurement of bacterial cell volumes by both light microscopy (after photographic enlargement) and electron microscopy, and concluded that, although imprecision is clearly a potential problem, inaccuracy probably is not. Newell (1984) described a means of enlarging fluorescent microscopic images to ×4000; this can be used in place of photographic enlargement, as described by Fuhrman (1981), or television image analysis (Fry and Davies, 1984). Problems of inaccuracy are more likely to be associated with conversion of biovolume values to biomass. It is common practice to arrive at a dry mass density (g dry · cm^{-3} fresh) value for bacteria using literature values (including references earlier than 1960) for percent water and specific gravity (e.g., Robertson et al., 1982); this dry mass density is then used with measured biovolume to calculate biomass. Recent experimental work (van Veen and Paul, 1979; Schiemer, 1982; Robinson et al., 1982; Bakken and Olsen, 1983; Bratbak and Dundas, 1984) indicates that values commonly used for bacterial dry mass density (e.g., 100 mg C · cm^{-3} cell; Newell and Fallon, 1982) may be low by factors of ×2 to ×4 (the value suggested by Bratbak and Dundas is 220 mg C · cm^{-3} cell). These findings indicate that current estimates of biomass of natural bacterial assemblages may be substantially inaccurate; further careful experimental analysis of bacterial dry mass and carbon densities is urgently needed because of the implications of this inaccuracy for models of carbon flow involving the use of AODC values (e.g., see Cammen, 1980).

The degree of precision with AODC, aside from the problems associated with cell sizing pointed out in the previous paragraph, is a function of the heterogeneity of distribution of bacterial concentrations at particular sampling sites, quantity of sample filtered, and the amount of replication of subsampling. Two papers that all AODC users should consult are those by Kirchman et al. (1982b) and Montagna (1982) (see also J.G. Jones, 1979). As an example of the precision that can be obtained, Kirchman et al. (1982b) found an average coefficient of variation for replicate water column subsamples at three separate sites of 6.9%. Sensitivity of AODC is limited only by the concentration of fine suspended particulate matter that could block excitation or fluorescent light from the bacterial cells (see Kirchman and Mitchell, 1982), as even very rarefied bacterial suspensions can be concentrated to detectable levels by filtration. In practice, this has rarely been found to be an obstacle to effective use of AODC (S. Y. Newell, unpublished observations). Even in and beneath the turbid waters of coastal salt marshes (common depth of 75% reduction of sunlight intensity, 0.5 m), samples containing too few bacteria for effective counting have been encountered only in rare, very dense clay layers of deeper ($\geq$20 cm) sediments.

In addition to AO, a variety of fluorochromes have been proposed for direct bacterial counting. Babiuk and Paul (1970) found fluorescein isothiocyanate (FITC) to yield higher counts of bacteria than AO when used for terrestrial soil analysis, but Lundgren (1981) obtained the opposite result.

Marsh and Odum (1979) used a europium-chelate/fluorescent brightener technique for counting bacteria on suspended particles in the water of a salt marsh. Roser (1980) discussed the merits of the use of ethidium bromide with soil bacterial assemblages, relative to FITC and AO. Porter and Feig (1980) and Robarts and Sephton (1981) recommended the use of 4'6-diamidino-2-phenylindole (DAPI) for counting of aquatic bacteria when concentrations of inorganic matter are high, and Coleman (1980) recommended DAPI for enumerating bacteria on the surfaces of autofluorescent substrates (e.g., diatom cells). Paul (1982) introduced the use of Hoechst dye (33258) for counting bacteria from marine environments. Haas (1982) suggested that proflavine is better than AO if a range of microbes from bacteria to autotrophic and heterotrophic microflagellates are to be enumerated. However, unless it is clearly shown that use of one or more of these substances yields statistically significantly more accurate values than AO in a broad range of sample types, the use of AO is advisable for meaningful comparison from a wide variety of types of sites. To those who choose to use a fluorochrome other than AO, caution is suggested. Although the use of DAPI has been reported to yield higher bacterial counts than AODC when large amounts of inorganic matter are present in freshwater samples (Porter and Feig, 1980), AODC yields higher bacterial counts than DAPI (Table II) in similar situations with coastal seawater, even though concentrations of DAPI up to 50 times greater than that suggested by Porter and Feig were tested (Newell, unpublished). Furthermore, Tabor and Neihof (1982a) and Pomeroy et al. (1983) found that DAPI may yield lower values than AO with seawater samples.

TABLE II

Mean Counts of Bacterial Cells per Microscope Field (± 95% Confidence Interval) for Parallel Samples of Equal Volume, Using AO and DAPI Fluorescence Detection[a-c]

Site[d]	AO[a]	DAPI[a]	AO : DAPI	PC : FC[e]
Nearshore	14.2[B]	15.3[B]	0.91	0.05
	(±1.2)	(±2.8)		
Beach	20.0[C]	10.3[A]	1.94	0.76
	(±2.3)	(±2.2)		
Marsh 1	27.7[E]	10.9[A]	2.54	0.51
	(±3.8)	(±3.0)		
Marsh 2	23.9[D]	22.5[CD]	1.06	0.12
	(±3.1)	(±2.7)		

[a] Final concentrations: AO (acridine orange), 0.004%; DAPI (4'6-diamidino-2-phenylindole), 0.0005%. Light source: HBO 50-Hg lamp; otherwise microscopic conditions as described by Newell and Christian (1981) for AO and by Porter and Feig (1980) for DAPI.

[b] The ratio of the two counts (AO : DAPI) and the ratio of cells on/in particles to those free of particles (PC : FC) for each site was used.

[c] Mean counts bearing the same superscript capital letter (A–E) could not be statistically differentiated (ANOVA, $p > 0.05$). To convert counts to cells $\times ml^{-1}$ of original sample, multiply by 1.55×10^5.

[d] Nearshore, 9 km from the beach; beach, in the surf; marsh 1, tidal creek, high ebbing tide; marsh 2, tidal river, high flooding tide. All samples taken at the surface.

[e] As determined in AO preparations.

LIGHT MICROSCOPIC DETERMINATION OF PRODUCTIVITY

Without Incubation

Net Change in Standing Crop

One means of determining bacterial productivity via AODC is to take samples periodically from within a well-defined system and calculate production using the sums of the positive net differences observed (see also Chapter 3, this volume). By well-defined system we mean one that is not mixed regularly with other ecosystems, so that input of bacterial cells from outside the system is negligible or zero. Thus, any positive change in concentrations of cells must be a consequence of cell production. This yields only minimum values, since negative contributions to the net positive changes (e.g., predation, lysis, sinking in aquatic systems) are not measured. These minimum values can be very useful for comparison with the lowest values found by alternative methods. Examples of the use of this method can be found in Coveney *et al.* (1977), Clarholm and Rosswall (1980), and Jordan and Likens (1980). Hobbie and Rublee (1975) used this method and added estimates of grazing output and bacterial respiratory carbon output in order to estimate bacterial carbon flow for an arctic pond. Using AODC in conjunction with measured flow and sinking rates in a relatively open estuarine system and comparing bacterial productivities derived via a radiotracer method, Ducklow (1983) employed the net-change technique to obtain values for rates of grazing by bacteriovores.

Frequency of Dividing Cells (FDC)

Although only recently adapted for bacteria (Hagström *et al.*, 1979), this method is based on an idea introduced long ago for use with microalgae and protozoa (see references in Weiler and Eppley, 1979; Coats and Heinbokel, 1982). As described by Hagström *et al.* (1979; Larsson and Hagström, 1982), the method involves determination by AODC of the percentage of cells within natural assemblages that are in the process of dividing. The percentages of dividing cells are then used to calculate the growth rates that were extant in the field, based on the mathematical relationship between FDC and growth rate. Because bacteria are not expected to exhibit synchronous or phased division in the field, and because duration of the paired division stage is not readily determinable, the FDC is not used to calculate growth rate directly (McDuff and Chisholm, 1982; Trueba *et al.*, 1982*a*). Rather, the mathematical relationship between growth rate and FDC is determined experimentally, using predator-free bacterial assemblages from the environment in question (Hagström *et al.*, 1979; Newell and Christian, 1981; Hanson *et al.*, 1983; Davis and Sieburth, 1984). A range of directly measurable (AODC) growth rates is induced, corresponding FDCs are determined, and regression techniques are used to determine the mathematical formula(s) required to calculate *in situ* growth rates from *in situ* FDC. The calculated growth rates are then used with AODC standing-crop values to calculate bacterial productivity.

The FDC method is an attractive one because natural samples do not require manipulation (e.g., incubation) for determination of growth rate (μ) and productivity. Those who choose to use it, however, should be aware that there are many unresolved questions regarding its validity. Several of these are pointed out by S. Y. Newell and Christian (1981). McDuff and Chisholm (1982) should be consulted regarding the mathematical questions involved [see their equation (8)]. The major question is: How realistic are the types of calibration experiments (Fig. 1) performed by Hagström *et al.* (1979), Newell and Christian (1981), and Hanson *et al.* (1983)? Although bacterial assemblages from the natural environments were used in the laboratory incubations, it is quite likely that during incubation, particular types of bacteria (high K_t-high V_{max} bacteria; Azam and Hodson, 1981) were selected (Ferguson *et al.*, 1984). This may lead to recording of an FDC–μ relationship that is unrepresentative of that in the field. It may be that incubation techniques such as those discussed in the section, With Incubations, Prevention of Grazing, may be more appropriate for calibration of the FDC technique.

The diversity of the relationships found for μ–FDC (Fig. 1) points up another problem. If one is not in a position to perform calibration experiments, which of the available formulas should be used (e.g., see Pedrós-Alió and Brock, 1982)? Obviously these formulas could yield widely differing results for growth rate predicted from FDC. Very little is known about the cause

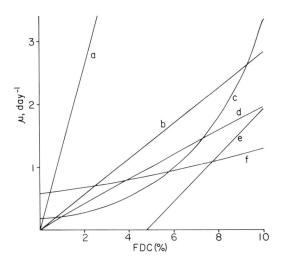

FIGURE 1. Suggested mathematical relationships between frequency of dividing cells (FDC, %) and instantaneous growth rate (μ, day^{-1}). Formulas and references: a, $\mu = 131$ ln $(1 + $ FDC/100) (McDuff and Chisholm, 1982), using $t_d = 11$ min (from Grover *et al.*, 1977); b, $\mu = 24$ FDC/85, for temperatures $>15°C$ (Larsson and Hagström, 1982); c, $\mu = 24$ $e^{0.299 \text{ FDC} - 4.961}$ (Newell and Christian, 1981); d, $\mu = 20.5$ ln $(1 + $ FDC/100) (McDuff and Chisholm, 1982), using a strictly hypothetical value of $\mu = 1$ day^{-1} (see Hagström *et al.*, 1984; Williams, 1984) at a commonly recorded FDC of 5% (Newell and Fallon, 1982); e, $\mu = 24$ (FDC $-$ 4.8)/65.2, for temperatures near 5°C (Larsson and Hagström, 1982); f, $\mu = 24$ $e^{0.081 \text{ FDC} - 3.73}$ (Hanson *et al.*, 1983). t_d, duration of paired division stage (McDuff and Chisholm, 1982).

and extent of variability in the formulas; Larsson and Hagström (1982) and Hanson *et al.* (1983) examined the effect of differing temperatures and have obtained apparently contradictory results. Clearly much more experimental work is needed in analysis of the relationship between FDC and bacterial growth rate before bacterial ecologists can feel comfortable in using this method. In doing this work, future investigators should consider the basic relationship to be expected between the duration of the paired division stage (t_d) and FDC (McDuff and Chisholm, 1982; Trueba *et al.*, 1982a). If t_d were constant, the relationship would be

$$\mu = (1/t_d)(\ln[1 + 0.01 \text{ FDC}])$$

Since t_d is probably not constant, but declining, especially at higher growth rates (Trueba *et al.*, 1982a), it will probably not be the case that a simple linear relationship will exist between FDC and μ.

Another problem with respect to accuracy in application of the FDC method is the dependence of its validity upon the existence of continuous exponential growth of bacteria in the field (S. Y. Newell and Christian, 1981). If bacterial assemblages in a particular environment fluctuate rather rapidly between exponential growth and stationary or senescent phases (see, e.g., Clarholm and Rosswall, 1980; Burney *et al.*, 1981; Moriarty and Pollard, 1982), FDC may be a poor predictor of growth rate. S. Y. Newell and R. R. Christian (unpublished observations) found that FDC did not fall off rapidly to zero as their bacterial assemblages entered stationary and senescent phases, wherein net growth was zero or negative (Fig. 2) (see Christian *et al.*, 1982). Newell and Fallon (1982) suspected that this factor was a partial cause of the spurious

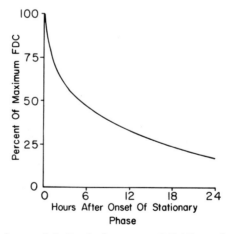

FIGURE 2. Approximate image of decline in frequency of dividing cells after stationary phase was reached in batch culture of a marine bacterial assemblage (Newell and Christian, 1981; also unpublished observations). Between hours 0 and 6 of this graph, net growth was zero; between 6 and 24 hr, net growth was negative.

values for bacterial production, which they found in their attempt to apply the FDC method to marine sediment bacterial assemblages (see Fallon *et al.*, 1983).

The type of regression model (model I) that has been used for determination of the mathematical relationship between FDC and μ may be inappropriate; perhaps model II regressions would be more correct. Those who plan to experiment with calibration of the FDC method should consult Laws and Archie (1981) and Sokal and Rohlf (1981) in this regard. When the data of Newell and Christian (1981) were processed via Bartlett's three-group model II regression analysis, the resultant equation was virtually identical to the model I result:

$$Model\ I: \quad \ln \mu = 0.299\ FDC - 4.961$$

$$Model\ II: \quad \ln \mu = 0.305\ FDC - 5.000$$

Two technical factors that have considerable potential for affecting accuracy of FDC results are subjectivity and interference from suspended particulate matter. Dividing pairs of bacterial cells are easily and unequivocally detectable when they are near the point of separation, but subjectivity enters into the analysis when paired cells in the early stages of constriction are counted (e.g., examine Fig. 14-2A of Sieburth, 1979). A minimum magnification of ×1250 should be used to optimize ability to make correct decisions (see Newell, 1984, p. 874). When particulate debris is present in AODC preparations, it is usually not so opaque that bacterial cells within it cannot be recognized; however, the debris exacerbates the subjectivity problem when dividing cells are to be counted. Particulate material could conceivably also prevent divided cells from separating; this may have been another factor in the improbably high FDC-derived productivity values found for sediments by Newell and Fallon (1982).

Although there have been only a few tests of the FDC method versus other measures of bacterial productivity (Newell and Fallon, 1982; S. Y. Newell *et al.*, 1983; Marcussen *et al.*, 1986; Riemann *et al.*, 1984), it appears generally to be the case that FDC results are higher than parallel results for the radioisotopic method, e.g., [³H]thymidine incorporation into macromolecules or into DNA, which is presently most favored (see Chapter 3, this volume). In oligotrophic marine water and in marine sediments, very substantial disparities have been found. For sediments, it appears that the FDC method is inapplicable (Newell and Fallon, 1982; Fallon *et al.*, 1983); for the water column, which method yields more accurate results is unknown.

Neither precision nor sensitivity can be seen as strong points of the FDC method. In their calibration experimentation, Newell and Christian (1981) found ±95% confidence limits for predicted growth rate, which were about ±40–50% of the predicted rates. These workers suggested that multiple regression analysis, including FDC, temperature, and ambient nutrient status, might improve the predictability of μ (see Larsson and Hagström, 1982; Trueba *et al.*, 1982*a*). The sensitivity problem for counting dividing cells is more severe than that for AODC, since the frequencies of dividing cells in natural assemblages are usually less than 10%. Samples with low FDC values cannot be concentrated sufficiently for precise measurement of FDC without excessive crowding of cells within microscope fields. Davis and Sieburth (1984) suggested that frequencies of recently divided cells be added to those of dividing cells in order to alleviate this sensitivity problem.

Mean Cell Volume

Mean cell volume of bacterial populations, and perhaps assemblages as well, may be related to growth rate (Jannasch, 1979; Kirchman *et al.*, 1982*a*; Larsson and Hagström, 1982); Jannasch pointed out that this possibility has not been explored as a means of detecting *in situ* growth rates. Jannasch (1979), however, also points out that Shehata and Marr (1971) found that mean bacterial cell volume may be related to growth rate in such a complex fashion that it would be difficult to use it as a predictor of growth rate under natural conditions (see also Trueba *et al.*, 1982*b*). There is also the possibility that, under natural conditions, predators of bacteria could control mean cell size (Fenchel, 1980; Ryther and Sanders, 1980; Ammerman *et al.*, 1984) such that it would not change with changing growth rates; it is also possible that populations of nongrowing bacteria could decrease in mean cell volume without substantial change in total cell biovolume (Tabor *et al.*, 1981). Van Verseveld *et al.* (1984) found that for bacteria having instantaneous generation times greater than about 24 hr, proportionality between specific growth rate and cell size can be absent. These factors weaken the potential value of using measured cell volume to estimate growth rate and productivity.

With Incubation

General

Valid use of all of the following methods for estimating bacterial growth rates and productivity depends on the assumption that the bacterial phenomena that occur within incubation vessels are accurate reflections of naturally occurring bacterial phenomena. Although little understood, distortions caused by separation of samples from the field, placement within small boundaries, and incubation have been recognized for many years (Zobell, 1946; see also Taylor, 1978; Eppley, 1981; Oviatt, 1981; Ferguson and Sunda, 1984); nevertheless, the complexities of natural ecosystems will inevitably tempt investigators to continue to conduct experiments of this kind in an effort to clarify results by reduction of interactive effects. Ferguson *et al.* (1984) and Fuhrman and Azam (1980) conducted investigations of extents of aspects of the bottling-incubation effects. Ferguson and co-workers found profound changes in characteristics of marine bacterial assemblages (e.g., increases in percent culturable on solid media, increases per cell in velocity of uptake of amino acids, changes in distribution of total biovolume among morphological types), which they attributed to confinement for 16–32-hr periods. Fuhrman and Azam (1980) found negligible potential adverse effects caused by growth on the walls of incubation vessels before 15 hr of confinement of marine bacterial assemblages, but their incubations were performed at 18°C; applicability of their results to other temperatures (and other water types) is uncertain. Continued research into potential effects of incubation in itself, especially for brief incubations, would be useful in improving interpretability of results of the incubation techniques discussed herein.

In using the following techniques, bacterial growth rates are calculated from AODC-measured changes with time in bacterial numbers and biovolume. In so doing, one should be aware that simple assumption of logarithmic growth may not yield accurate calculated results, largely because some fraction of the assemblage may not be growing or may be growing much more slowly than the remainder. Kirchman *et al.* (1982*a*) presented a discussion of this problem, with suggestions for dealing with it.

Isolates at Natural Substrate Concentration

Two separate tacks have been taken here: use of batch volumes (Carlucci and Williams, 1978), and continuous culture (Jannasch, 1979). Although in neither case was direct microscopy used to monitor change in standing crop, the potential for its use is obvious. The major question with respect to ecological interpretation of results is: How representative of the activities of natural heterotrophic bacterial assemblages as a whole are the axenic responses of the particular strains of bacteria used? Since we know very little about the species level of most naturally occurring bacterial cells (see, e.g., Sieburth, 1979), we cannot pass judgment on the ecological value of this method in this respect. R. C. Newell *et al.* (1982, and references cited therein) introduced modifications of the Carlucci and Williams (1978) approach by adding natural substrates to batch culture setups containing natural bacterial assemblages and by using AODC to monitor bacterial production prior to development of significant activity of bacterial predators (soil analogue; see references cited in Jenkinson and Ladd, 1981). This leads to two important questions. First, is protozoan grazing during the early stages of growth of the bacterial assemblage truly insignificant?, Second, does pretreatment of substrates (disruption of phytoplankton cells, freeze-drying and grinding of algal thalli) result in artifactitious bacterial responses? Ammerman *et al.* (1984) and Hagström *et al.* (1984) recently introduced variants of the axenic batch and continuous-culture techniques in which bacterial-assemblage inocula and AODC were used with bacteria-free unsupplemented seawater. Application of these techniques should yield important insights into aquatic bacterial ecology, yet a question remains: Does the fraction of the total bacterial assemblage used act in the same way as the whole assemblage in nature?

Prevention of Grazing Output

If it can be fairly assumed that simple death and autolysis are negligible factors in output of bacterial cells from particular ecosystems, one way to measure growth rates and production is to prevent output of cells to predators in systems manipulated such that the only input source is growth. Under these circumstances, AODC can be used to measure increase in numbers and biovolume directly. In this method, grazing pressure has been eliminated by three basic methods: filtration, dilution, and chemical inhibition. Filtration and dilution are applicable only to aquatic systems, but chemical inhibition can be applied to both terrestrial and aquatic samples.

Filtration has been used in two ways for creation of predation-free microecosystems: (1) collection of bacterial cells upon filters, and incubation of the wet filters floating on or containing bacteria-free water from the environment in question (Meyer-Reil, 1977; Straškrabová and Fuksa, 1982); and (2) filtration of water samples through 1.0–3.0-μm sieves and incubation of the filtrate (DeLattre *et al.*, 1979; Fuhrman and Azam, 1980; Vyshkvartsev, 1980; Wright and Coffin, 1984). The success of the dilution approach (Kirchman *et al.*, 1982a; Landry *et al.*, 1984) depends on the extent of dilution being great enough to reduce concentrations of bacterial cells below levels at which bacteriovores can or will feed upon them (Berk *et al.*, 1976; Fenchel, 1980; Swift *et al.*, 1982; Davis and Sieburth, 1984); the diluent used is bacteria-free water from the environment in question. The chemicals used in the chemical inhibition method are chosen for their ability to selectively prevent bacteriovores from feeding, without affecting bacterial growth (Alexander, 1981; Newell *et al.*, 1983). In the case of each of these methods, following manipulation, samples are incubated at conditions as close to natural as practicable, and subsamples periodically taken for AODC. In the case of the filtrational removal, dilution, and perhaps the chemical inhibition methods, the *in situ* membrane chamber may be the most ecologically realistic incubation technique for aquatic samples (Olàh, 1974; DeLattre *et al.*, 1979; see also Fliermans and Gorden, 1977; Sieburth, 1979; Crumpton and Wetzel, 1982).

The obvious primary question associated with the accuracy of this method is: Do the manipulations involved truly prevent predatory bacterial output? Since a film of water is present on or in the filters in the wet-filter method, it cannot be safely assumed, without experimentation, that heterotrophic microprotozoa are prevented from ingesting bacteria on the incubated filter pad or membrane. The use of 3-μm sieves to remove predators is likely to be an inefficient technique, as small zooflagellates will pass these readily (Newell and Fallon, 1982; Wright and Coffin, 1984; see Fenchel, 1982). S. Y. Newell (unpublished observations) also finds that glass-fiber filters of small nominal porosity (1.2 μm) will pass heterotrophic flagellates as well as bacteria; membrane filters of 0.8 μm or less pore size are required to yield only bacteria in filtrates of marine water samples.

Other technical problems potentially affect the accuracy of these techniques as well:

1. The fact that the bacteria in the wet-filter method are brought from suspension onto a solid surface for incubation leads one to suspect that artifactitious results may be obtained (see Fenchel and Jørgensen, 1977; Marshall, 1979).

2. Filtration may have subtle but important effects on the physiology of filtered bacteria as well as on the saprotrophic nutritional characteristics of the liquid filtrate (Karl, 1980; Newell and Fallon, 1982; Hewes and Holm-Hansen, 1983; Ferguson *et al.*, 1984; Fuhrman, 1984; Laanbroek and Verplanke, 1986) that could cause inaccuracy in various assessments. These effects include rapid change in adenylate ratios, depression of division rates, enhancement of oxygen uptake, increase in concentrations of labile organics, and perhaps disruption of natural extracellular structures (e.g., damage to pili) and natural microenvironmental structure as well (see Azam and Ammerman, 1984; J. C. Goldman, 1984). This problem also applies to the dilution approaches, since 0.2-μm filtration is used to produce the diluent.

3. Chemicals used to selectively inhibit predation on bacteria could cause (a) output of cellular carbon by damaged eukaryotes, leading to artifactitious bacterial response; (b) disruption of eukaryotic photosynthesis and consequent cessation of the normal outflow of photosynthetically fixed carbon to bacteria (Newell *et al.*, 1983); and (c) disruption of normal rates of recycling of nitrogen and other nutrient elements provided by protozoan feeding and excretory activity (Güde, 1985). Furthermore, in using the chemical inhibition method, one should be careful to establish that the predator poisons prevent feeding, and not just growth, by the predators. The possibility of substantial predation upon bacteria by predacious bacteria (e.g., McCambridge and McMeekin, 1980; Germida and Casida, 1983) should also be considered.

These methods are relatively untried, so it is too early to judge their precision. They will be limited in their sensitivity by the precision of AODC: Because incubations must be kept short for accuracy, slow growth rates (μ < approximately 0.03 hr^{-1}?: Newell *et al.*, 1983) will not be detectable. This sensitivity problem is probably the most serious shortcoming of the direct-count incubation-productivity methods.

Estimation of Grazing Output

For these methods, the same basic assumptions are required as discussed for prevention of grazing output, except that if total bacterial production is to be estimated, it must be assumed that predatory control of bacterial standing crop is complete (i.e., bacterial growth rate = predator ingestion rate) and in steady state. Although the first two of the following approaches have not been used to devise methods for AODC determination of bacterial productivity, the potential for this will be evident. The methods are (1) selective chemical inhibition of bacterial growth, without rapid induction of bacterial death and lysis (Habte and Alexander, 1978; Fuhrman and McManus, 1984); (2) addition to samples and dispersal of known quantities of nongrowing, palatable bacterial cells that would be easily recognizable among cells originally present (mycological analogue: Frankland, 1975); and (3) determination of ingestion rates of major bacteriovores in laboratory experiments by AODC and the use of these rates along with concentrations of these bacteriovores in the environment to calculate potential bacterial output (Peterson *et al.*, 1978; Fenchel, 1982).

Selective chemical inhibition of bacterial growth as a method for determining total bacterial output was attempted in a preliminary trial by Newell *et al.* (1983). Although these investigators did not succeed in finding appropriately selective inhibitors, the potential for success of the method remains, especially considering the results of Habte and Alexander (1978) and of Fuhrman and McManus (1984), who used the method to investigate predation upon a *Rhizobium* species and upon marine bacterial assemblages, respectively. Researchers who would use this method must be aware that below threshold concentrations of prey bacteria, predators may cease feeding (e.g., Fenchel, 1982).

Monitoring disappearance rates of recognizable "tracer" bacterial cells is a method suggested by the work of McCambridge and McMeekin (1980) and of Borsheim (1984). Using plate counts, McCambridge and McMeekin monitored the disappearance rates from seawater of cells of an individual bacterial species, comparing results with and without inhibition of bacterial predators. Borsheim (1984) used bacteria-sized, monodispersant, fluorescent Latex particles in suspension to measure clearance rates by freshwater ciliates. Brock's (1971) ideas regarding marking of cells and monitoring their fate via autoradiography also prompt the proposal that cells distinct from those usually seen by AODC in aquatic environments could be used as tracers. For example, cells grown to stationary phase in dilute yeast extract solutions, and then concentrated, washed, and chilled, could be added to marine water samples and incubated. These cells are larger (about ×3–5) and red to red-orange, as seen in AODC; virtually none of the naturally occurring saprotrophic cells is so characterized. Incubations would be kept short enough to prevent the tracer cells from moving out of stationary phase, as monitored in protozoa-free controls. If tracer cells were equally palatable and ingestible to bacteriovores as naturally occurring cells, if their addition did not in itself cause a change in predation rates, and if size selectivity by predators over the range of prey sizes present was not a confounding factor (see Fenchel, 1982), the rate of disappearance of the tracer bacterial cells as determined by AODC would be equal to the total bacterial output rate.

Finally, if the species of bacteriovores responsible for the major portion of the predation on bacteria are known for a particular environment, one can use measured concentrations of bacteriovores, in conjunction with their known rates of ingestion of bacteria, to calculate *in situ* output of bacterial cells. In so doing, one would need to know how large a range of feeding rates could be expected and to what extent these rates were affected by changes in environmental variables. The rates can be determined using AODC in laboratory incubations of particular bacteriovores and nongrowing bacteria. Fenchel's (1982) work with marine heterotrophic microflagellates provides a good example of the potential for applicability of this method. Bassøe (1984) describes an automation (flow cytometry) of this type of analysis for the special situation of human leukocytes phagocytizing bacteria; it may well be that this type of automation is also feasible for some aquatic ecological situations.

Further questions of accuracy, and of precision and sensitivity, will remain open until these grazing assay methods are put into practice.

LIGHT MICROSCOPIC DETERMINATION OF INDIVIDUAL CELL ACTIVITY

Fluorochromes in Combination with Direct Activity Measurements

Active metabolic functions and productivity of composite microbial assemblages are quantitated by the use of methods that are nonselective and that cause minimum perturbation of environmental conditions (see also Chapter 3, this volume). Among these are methods for determination of active substrate uptake, incorporation and respiration of specific substrates, synthesis by microheterotrophs (e.g., Wright and Hobbie, 1966; Hobbie and Crawford, 1969; Fuhrman and Azam, 1982; Karl, 1982), and determination of total respiration in terms of electron-transport-system (ETS) activity measurements (e.g., Christensen *et al.*, 1980; Packard, 1971; Kenner and Ahmed, 1975). These methods yield information on the microbial system as a whole. No information is obtained, however, on the structure of the microbial assemblage in terms of the specific fraction of the total number of bacteria engaged in the measured function or identification of specific types of bacteria that are active. Autecological methods are available to relate active metabolic function to specific types of individual bacterial cells. Direct determination of ETS activity, active substrate uptake, and active cellular synthesis can be made for individual bacterial cells.

ETS activity is linked to many metabolic pathways in virtually all bacteria; thus, measurements of ETS activity are useful in recognition of actively respiring bacteria. ETS activity is conveniently measured by the use of artificial electron acceptors such as tetrazolium salts (Packard, 1971; Altman, 1976; Trevors, 1984). ETS activity of whole aquatic assemblages is determined spectrophotometrically after extraction of the reduced insoluble tetrazolium, which is deposited within microorganisms. A correlation between respiration rates of phytoplankton and ETS activity has been demonstrated; for marine bacteria, however, the species of bacteria as well as the physiological state of cultures can cause significant variation in the relationship between ETS activity and oxygen consumption. Ratios of respiration ($\mu l\ O_2\ hr^{-1}\ cell^{-1}$): ETS activity ($\mu l\ O_2$-equivalents $hr^{-1}\ cell^{-1}$) ranging from 2.9 to 8.2 for growing cultures of different species and from 0.1 to 5.9 for starving and growing cultures, respectively, were observed with a marine vibrio (Christensen *et al.*, 1980). Organisms in mixed microbial assemblages in nature cannot be assumed to be in a relatively homogeneous physiological state. No information on the structure of an assemblage in terms of numbers or types of organisms engaged in respiration is gained.

Direct microscopic determinations of bacteria in natural samples with ETS activity can be made by a modification of the method. The electron acceptor 2-(*p*-iodophenyl)-3-(*p*-nitrophenyl)-5-phenyl tetrazolium chloride (INT) is commonly used in determining ETS activity. INT accumulates in respiring bacteria, and the reduced form (INT-formazan) is a water-insoluble precipitate observed microscopically as opaque intracellular deposits. A number of investigations to determine numbers of respiring bacteria in natural aquatic systems have been made. Iturriaga (1979) measured the total ETS activity of the composite population of particulates collected in sediment traps in the Baltic Sea and, for the same samples, determined the number of ETS-active bacteria by microscopy. He found 17–574×10^6 ETS-active bacteria per mg particulate organic matter, and respiration rates interpreted from ETS activity of 10 to >700 µl O_2-equivalents hr^{-1} per mg particulate organic matter. No correlation between total ETS activity and ETS-active bacteria was observed. Zimmerman *et al.* (1978) combined determination of ETS activity of individual bacteria and AODC. The fraction of the AODC exhibiting ETS activity for microbial assemblages collected in the winter in Kiel Bight ranged from 6 to 7%. For freshwater samples collected in July (Zimmerman *et al.*, 1978), ETS-active bacteria were 36% of the AODC. Results of microscopic determination of ETS-active bacteria in aquatic samples from a salt marsh creek (Harvey and Young, 1980) showed, strikingly, that most (62–99%) of the ETS-active organisms were associated with particulates (see Macdonald, 1980, for results with soil samples). Recently, technical aspects of the method have been improved (see INT-Reduction Method) and bacteria recognized as ETS active were observed to be 60% of the total AODC of aquatic assemblages of Chesapeake Bay (Tabor and Neihof, 1982*a*). Additional applications of the direct-microscopic INT-reduction method are those of Kurath and Morita (1983), who used it to monitor decrease in the proportion of respiring cells among starving bacterial cells, and Baker and Mills (1982), who combined the INT-reduction method with a fluorescent-antibody technique.

Quantitative measurements of substrate uptake of composite assemblages are used in the determination of kinetic uptake parameters (Christian and Hall, 1977). Indirect relationships between metabolic activity and a structural component of the assemblage, i.e., number of bacteria, are made by expressing uptake parameters relative to the AODC of the total microbial assemblage as specific-activity indices (Wright, 1978). A specific-activity index is an approximation of the relationship between the activity of the microbial assemblage and the population structure. Analogous to the direct determination of ETS-active bacteria, individual bacteria engaged in active substrate uptake are determined by microautoradiography, as opposed to active substrate uptake, for composite assemblages determined by measurement of the accumulation of radiolabeled substrate by the total assemblage. Individual microhetero-trophs with active specific uptake systems can be recognized using radiolabeled substrates and microautoradiography. Substrate uptake is a more specifically defined measurement of metabolic function than the determination of ETS activity in that only organisms with an active transport system corresponding to the labeled substrate and having sufficient substrate affinity for the labeled substrate to produce a microautoradiogram are determined as active.

Microautoradiography is useful in determining the specific individual microorganisms active in metabolic functions (Brock and Brock, 1968; Munro and Brock, 1968; Bright and Fletcher, 1983). This autecological technique has generally been used only in qualitative studies for determination of active substrate uptake by individual organisms. Stanley and Staley (1977) calculated the *in situ* specific activity of added [^{3}H]acetate in aeration lagoon samples and determined the quantity of substrate uptake required to produce an autoradiogram. These workers were able to calculate kinetic parameters of *in situ* acetate uptake for several morphological types of bacteria using microautoradiography. Horstmann and Hoppe (1981) used a microautoradiographic method, [^{3}H]methylamine, to detect competitive uptake of ammonia by bacteria and phytoplankton. The seasonal distribution of microorganisms active in substrate uptake as determined by microautoradiography has been reported for areas of the Baltic Sea (Hoppe, 1977, 1978). A correlation was observed for the number of bacteria active in glucose uptake and the glucose uptake rate for the total assemblage (Hoppe, 1977; Meyer-Reil, 1978). However, AODC, numbers of colony-forming bacteria, and biomass values were not correlated with the microautoradiographic activity measurements. Combined recognition of microorganisms by immunofluorescence (see section on Immunolabeling in Bacterial Ecology) or fluorochrome staining and determination of substrate uptake-active organisms improves the utility of the microautoradiographic method (Fliermans and Schmidt, 1975) and presents the possibility of combining direct identification of organisms and determination of *in situ* metabolic function. Results of an improved method combining microautoradiography and epifluorescence microscopy show that up to 94% of the AODC of Chesapeake Bay water samples are active in amino acid uptake at a final activity of 0.1 μCi $\cdot$ ml^{-1} (Tabor and Neihof, 1982b).

DNA synthesis by bacteria from aquatic assemblages is specifically inhibited when natural water samples are amended with sublethal concentrations of nalidixic acid. Organisms sensitive to nalidixic acid at these concentrations (principally gram-negative bacteria) continue to synthesize cellular components, and cells become enlarged because cell division does not occur. Such bacteria in open ocean and near-coastal samples increase in size from <0.3 × <0.3 μm to >0.8 × >1.0 μm on incubation of samples with nalidixic acid and yeast extract (Kogure *et al.*, 1979, 1980). The method was proposed for use in direct identification and enumeration of viable bacteria, especially in oligotrophic marine waters. Organisms identified as viable by this method are 10–300-fold the number of colony-forming bacteria determined for the same samples (Kogure et al., 1979). An investigation of the method for use in oceanic samples has shown that larger numbers of synthetically active bacteria are found for samples amended with yeast extract and tryptone than substrates that are single carbon or nitrogen sources. The concentration of substrate used to amend the sample influences the number of organisms recognized as active by as much as fivefold (Peele and Colwell, 1981). In addition, the efficacy of nalidixic acid is influenced by the concentration of dissolved organics in unpolluted estuarine samples to the extent that it is necessary to amend samples with concentrations of nalidixic acid fivefold higher than those used by Kogure et al. (1979) to achieve effective inhibition of cell division resulting in determination of a maximum number of synthetically active bacteria (P. Tabor, unpublished data). Synthetically active bacteria have been used as a basis for a specific-activity index expressing the inhibitory effects of a pesticide on natural aquatic bacterial assemblages (Orndorff and Colwell, 1980). The index, total utilization of ^{14}C-amino acids/direct viable count (equivalent to synthetically active bacteria), proved useful in describing the effects of various concentrations of insecticide on the in situ microbial activity of Chesapeake Bay bacterial assemblages. The nalidixic acid method has also proved useful as a means of detecting viability of potentially pathogenic bacteria when introduced into seawater (Xu et al., 1982).

Direct determinations of ETS activity, active substrate uptake and synthetic activity are made in conjunction with observation of the total microbial assemblage by epifluorescence microscopy of the samples. Thus, these distinct metabolic functions can be contrasted for a single assemblage by determination of the fraction of the total number of bacteria recognized as active by each method (Tabor and Neihof, 1984). A direct comparison of these three methods using parallel Chesapeake Bay water samples over a period of 1 year (Tabor, unpublished observations) is described briefly in the following section.

A Comparative Study: Numbers of Synthetically Active Bacteria, Cells with Electron-Transport Activity, and Substrate Uptake-Active Cells

Methods

Water samples were collected from 1 m below the surface on 11 occasions during 1980–1981 at a single station in the Chesapeake Bay 4 km from Chesapeake Beach, Maryland (38° 41.4' N, 76° 30.0' W; station depth 10 m). Numbers of synthetically active bacteria (SAB) were determined by the nalidixic acid method (Kogure et al., 1979; Peele and Colwell, 1981). A range of nalidixic acid concentrations (0.002–0.01%) was used, with and without addition of 0.025% yeast extract, in 6-hr incubations of samples. The bacteria counted had morphological characteristics comprising the operational definition of SAB: cells of >0.5 μm in the shorter dimension (diameter), elongated cells (>2.0 μm in length), cells distorted in growth in the long axis (resulting in a bent rod-shaped or club-shaped cell), and cells with mottled staining (a combination of red, yellow, and green fluorescence). Most commonly, cells with mottled fluorescence had a green or yellow fluorescing area within the red-fluorescing cell. Preliminary work with pure cultures has shown that the entire bacterial cell originally fluoresces either green because of high DNA content relative to RNA or yellow due to a combination of green fluorescence of DNA and red fluorescence of RNA. Bacteria inhibited in DNA synthesis by nalidixic acid and still active in RNA synthesis increase in size. This is indicated by the red fluorescence of the RNA in the area of the cell surrounding the outline of the original, yellow-fluorescing cell shape.

Numbers of cells with electron-transport activity (INT-reducing microorganisms, INTRM) were determined by the method of Tabor and Neihof (1982a). This method is an improvement on the method of Zimmerman et $al.$ (1978) in that cells are observed within a gelatin matrix, rather than upon a membrane filter, thereby eliminating dissolution of INT-formazan by immersion oil and permitting increased detectability of INT-reducing cells. Newell (1984) has described a modification on the Tabor–Neihof method that does not include gelatin sandwiching. Substrate uptake-active cells were enumerated using the microautoradiographic method described by Tabor and Neihof (1982b). A mixture of tritiated amino acids was used as the substrate for uptake (final concentration 2.8 $\mu g \cdot liter^{-1}$; specific activity 0.1 $\mu Ci \cdot ml^{-1}$). Isolated, individual, developed silver grains associated with single fluorescing organisms were equated with clusters of silver grains associated with single organisms, and both were accepted as microautoradiographic counts (MC). Silver grains recognized under transmitted light were observed closely under epifluorescence illumination for small cells displaying low fluorescence.

Direct counts (AODC) of total number of fluorescing bacterial cells were made for each sample by the method of Hobbie et $al.$ (1977) (Table I). No significant difference was found in the AODC among corresponding samples prepared for enumeration of SAB, INTRM, or MC, with one exception (see the section, INT-Reduction Method).

Results

Nalidixic Acid Method. In Chesapeake Bay samples, the number of SAB represented from 7% (March) to >77% (August) of the total AODC determined from the same sample preparation (Table III). In Table III are given the largest numbers of SAB found from enumeration of the various samples amended with yeast extract and nalidixic acid or nalidixic acid only. The differences in numbers of SAB for the various treatments were significant. In April, May, and June, numbers of SAB determined for samples amended with 0.002% nalidixic acid with or without yeast extract were <30% of the numbers for samples amended with 0.006% nalidixic acid. These samples were observed to have high concentrations of phytoplankton. Filtrates of these samples, when added in a 1 : 1 concentration to samples otherwise affected by a concentration of 0.002% nalidixic acid, inhibited the action of nalidixic

TABLE III

Total Direct Counts and Counts of Synthetically Active Bacteria, Respiring INTRM,
and Bacteria Active in Amino Acid Uptake[a] for Chesapeake Bay Surface Water
Samples[b]

Date	DC	SAB	INTRM	MC	°C[c]
1980					
3/26	4.1[b]	0.3	1.0	0.7	7.0
4/21	5.0	1.5	1.2	3.1	13.4
5/1	3.9	1.2	1.4	2.7	12.9
5/29	7.4	1.3	2.2	3.4	19.5
6/17	9.9	4.1	7.7	8.5	20.8
7/9	10.7	5.9	5.8	8.8	23.3
8/5	11.0	8.5	10.4	9.1	28.3
9/21	13.6	5.4	12.8	10.1	24.8
10/31	10.2	2.4	5.9	6.4	12.6
12/10	6.4	2.3	3.6	2.9	7.3
1981					
2/9	3.3	0.6	1.2	1.3	0.0

[a] Microautoradiographic counts.
[b] DC, direct counts; SAB, synthetically active bacteria; INTRM, INT-reducing microorganisms; MC, microautoradiographic counts.
[c] In situ and incubation temperature.
[d] Counts $\times 10^6$ ml^{-1}.

acid. The addition of 0.01% nalidixic acid resulted in numbers of SAB equivalent to or less than numbers for corresponding samples amended with 0.006%
nalidixic acid. In preliminary studies using samples collected from areas of
Chesapeake Bay with salinities of <5–30%, samples amended with 0.006%
nalidixic acid were found to have a significantly larger number of SAB than
did samples amended with lower concentrations. For July and August samples
of this study, numbers of SAB were equivalent when samples were amended
with 0.006% nalidixic acid, with and without yeast extract. Nalidixic acid
concentrations of 0.002% were as effective as 0.006% nalidixic acid for samples
from December through March. These samples amended with yeast extract
yielded significantly larger numbers of SAB than did samples with no substrate
addition.

In samples fixed with formaldehyde before the addition of nalidixic acid
and yeast extract, as well as in samples prepared for AODC, bacteria scored
as SAB represented <1–75% of the SAB enumerated for the samples incubated with nalidixic acid and yeast extract. Samples made in March had the
lowest percentages (<1%) of SAB in fixed controls, whereas the number of
organisms determined as SAB for the August control sample was 75% of the
number found for the incubated sample. In most cases, the bacteria recognized as SAB in the fixed controls were 10–20% of the SAB enumerated in
the incubated samples. No correction for the number of SAB recognized in
the fixed sample control was made in determination of SAB for incubated
samples.

The largest numbers of SAB and the highest SAB percentages of AODC were seen in months with water temperatures of >20°C (Table III). SAB determined for samples from February and March (0°C and 7.0°C, respectively) were significantly less than the numbers of SAB determined for the remaining samples with water temperatures <20°C. A set of duplicate samples collected in February and May 1980, and in February 1981 from 8.5 m as well as 1 m were incubated at 22°C to determine whether a loss in sensitivity in detection of SAB in samples was due to low temperatures. For the February 1981 samples, the number of SAB was 7% larger (i.e., SAB were 24% of AODC) than that determined for the samples incubated at *in situ* temperature (0.0°C). For the February and May 1980 samples incubated at 22°C, the numbers of SAB were not significantly different from those for samples incubated at *in situ* temperature (4.0°C and 12.9°C, respectively).

INT-Reduction Method. The number of INTRM (INT-reducing microorganisms) determined for Chesapeake Bay samples is presented in Table III. The samples were not amended with ETS activators because preliminary results had not shown a significant increase in reduction of INT by cells with the addition of NaCN or succinate (Zimmerman *et al.,* 1978; Tabor and Neihof, 1982*a*). Numbers of INTRM in fixed-sample controls were <2% of the DC in all samples. The number of INTRM represented from 24% (March and April) to 94% (August and September) of the total AODC (Table III). INT-formazan deposits recognized by the use of transmitted-light illumination were examined for associated small AO-fluorescing cells, and additional organisms with faint fluorescence were recognized. Because additional cells were recognized by this procedure, AODC for August and September samples prepared for INTRM determination were significantly greater than AODC of the corresponding controls.

The number of INTRM and the portion of the total population recognized as INTRM were higher when water temperatures were >20°C. The number of INTRM determined for the March, September, and February samples was not different with incubation times of 10 min up to 3 hr (Tabor and Neihof, 1982*a*). Thus, our observations indicated that the sensitivity of the method for detection of microorganisms with ETS activity was not dependent on the length of incubation when Chesapeake Bay temperatures were low.

Microautoradiography. The number of bacteria active in amino acid uptake (MC) determined for surface Chesapeake Bay samples is presented in Table III. MC was 17% (March) to >82% (June, July, and August) of the AODC determined for the same samples. Incubation times lengthened to 4 hr for duplicate samples collected in September and October resulted in labeling of microorganisms to the extent that clusters of silver grains extended to adjacent cells. When temperatures were <20°C, samples incubated 2 hr had significantly lower MC than did samples incubated for 2.5 hr. Thus, it was critical to determine the incubation time that permitted labeling of the maximum number of organisms without developing autoradiograms with grain accumulations so large that they could not be discerned as associated with individual cells.

Results reported in Table III are for samples amended with ^{3}H-amino acids to give a final activity of 0.1 μCi · ml^{-1} and final added concentration of 2.8 μg · liter^{-1}. Samples were also amended with tritiated amino acids to give a final activity of 0.05 μCi · ml^{-1}. Numbers of MC were significantly less for all samples with the lower initial specific activity, ranging from 8% to 40% less than corresponding samples. The magnitude of the difference between the numbers of MC for samples amended with different activities of substrate seemed unrelated to season or water temperature.

Samples collected in March, April, and May 1981 were amended with [^{3}H]methionine at six activities ranging from 0.05 to 0.3 μCi · ml^{-1} (1–6 nM added methionine). For March samples, an increase in the activity of the sample resulted in a proportional increase in the number of MC. MC increased from 20% (0.05 μCi · ml^{-1}) to 50% (0.3 μCi · ml^{-1}) of the AODC. No significant difference in the number of MC (representing 35% of the AODC) was found for the May samples amended with various activities of [^{3}H]methionine. In April, the increase in the number of MC with increasing activity was observed to be nonlinear, and the biphasic relationship indicated two distinct assemblages of bacteria defined by different substrate affinity constants.

In addition to the use of amino acids, acetate and thymidine at equivalent final activities of radioactive label were used as substrates in replicate samples in order to determine the uptake activity of individual organisms in the population over the 1-year period. Results of these determinations were compared. In summary, numbers of MC determined for samples amended with three substrates increased significantly from March to April and showed a significant decrease from September to October. These changes corresponded with a large increase and a large decrease, respectively, in water temperature (Table III). With the exception of these cases, the change in the number of MC for one substrate from month to month did not correspond to changes observed for the others.

Discussion

Nalidixic Acid Method.   Enumeration of SAB requires recognition of bacterial cell types that are enlarged. However, questions arose when a set of standard morphological criteria was used to define SAB. For example, large portions (up to 75%) of the Chesapeake Bay bacteria meeting the definition of SAB, including red or yellow fluorescence, were recognized in the formaldehyde-fixed controls (see photomicrographs in Orndorff and Colwell, 1980). Correction of SAB enumerated in incubated samples for the morphologically SAB-like cells in fixed-sample controls may not be appropriate, as many large bacteria in duplicate samples were consistently observed to be active in amino acid and thymidine uptake, as determined in parallel by microautoradiography, and active in respiration, as determined by the INT-reduction method. Thus, a correction would presumably result in underestimation of SAB, especially in samples collected in June through September (Table III).

The increased number of SAB observed for some, but not all, of the samples incubated at higher than *in situ* temperature suggested an induction of cellular synthetic activity for a fraction of the bacterial population which is dormant at *in situ* temperature. For summer samples with temperatures of >20°C, higher rates of cellular biosynthesis were expected relative to the rates of samples with temperatures of 0–13.4°C. Indeed, the highest percentages of the AODC active in synthesis were observed in the months June through September. Thus, organisms may be synthetically active, but the method is not so sensitive when used at low *in situ* temperatures. Longer incubation times may increase the sensitivity of the method. However, organisms not affected by nalidixic acid, including gram-positive bacteria, will proliferate and change the AODC of the sample on extended incubation. In the determination of SAB for freshwater lakes, an increase in total counts was observed for samples amended with nalidixic acid and yeast extract and incubated at 20°C (Maki and Remsen, 1981). Grazing of the bacterial population in samples incubated for extended periods must also be considered. It must be pointed out that the method is limited to detection of SAB that are sublethally inhibited by nalidixic acid, i.e., in which cell division, but not growth, is arrested. Generally, DNA synthesis of gram-negative bacteria is inhibited by nalidixic acid, whereas gram-positive bacteria and eukaryotes are unaffected.

INT-Reduction Method. Fluorescing bacteria of filtered aquatic samples were viewed in a structureless background, as the Nuclepore filter had been removed in the preparation of the samples (Tabor and Neihof, 1982a). The pore structure of the filter imprinted in the initial gelatin film was eliminated by embedding the samples in a gelatin matrix. Viewing the samples through the cover glass (i.e., the embedded sample is below the cover glass when viewed) prevented dissolution of the intracellular INT-formazan deposits by immersion oil used in microscopy (Tabor and Neihof, 1982a). With these modifications, the number of INTRM were at least threefold the number determined by using the method of Zimmerman *et al.* (1978). Total DC determined by this method was not significantly different from DC determined by a commonly used AO technique (Table I, item 1). Thus, organisms are quantitatively retained on the gelatin-coated slide when the filter is removed (Tabor and Neihof, 1982a).

Extended incubation periods (up to 3 hr at *in situ* temperature) did not result in increased numbers of INTRM for February, March, or September samples. This indicated that few organisms had active ETS at levels below the sensitivity of the method. Results of the determination of amino acid uptake-active bacteria by microautoradiography also supported the seasonal distribution of numbers of INTRM (Table III).

ETS activity was thought to be a more general measurement of meta-bolically active microorganisms than was uptake of amino acids; thus, if the methods were accurate and equally sensitive, the number of INTRM should have been larger than the number of MC. This was not true in many cases (Table III). The addition of ETS activators and substrates of the ETS may increase electron transport activity. Stimulating electron transport would in-crease the reduction of the electron acceptor, INT, and detectability of ETS-active bacteria would be increased. Zimmerman et al. (1978) and Tabor and Neihof (1982a) amended samples with NaCN and with succinate. No clear effect on INT reduction by bacteria in natural water samples was observed. However, detailed studies using ETS activators, NADH, and other substrates of ETS enzymes, similar to that by Owens and King (1975) for marine zoo-plankton, are needed to determine the limiting factors in reactions of the ETS for microorganisms in natural samples. Studies of sensitive methods for detection of ETS activity may be expected to lead to the development of an improved method for determining individual metabolically active organisms. An accurate measurement such as ETS activity is needed for generalizations concerning viability or active (versus dormant) states of metabolism.

The low number of INTRM in comparison with MC of corresponding samples (Table III) suggested that amino acids were used by a large majority of the metabolically active free-living assemblage of microheterotrophs in Chesapeake Bay and that the INT reduction method was underestimating numbers of respiring bacteria. This was consistent with the additional finding that ETS activity could not be detected for certain distinct morphological types of bacteria in any of the year's samples (Tabor, unpublished observa-tions). Although numbers of INTRM and MC were similar for several samples (Table III), and the numbers of bacteria as well as the fraction of the total assemblage recognized as active were large in comparison with previous re-ports, it was not possible to conclude unambiguously that either or both meth-ods were accurate in detecting all bacteria metabolically active in situ. Numbers of SAB were lower than numbers of INTRM and MC in most samples (Table III). This was expected because the detection of SAB was limited to gram-negative bacteria and, in addition, a relatively large amount of energy and numerous metabolic steps are necessary to synthesize cellular components sufficient to be recognized by the nalidixic acid method.

Microautoradiography.   In the microautoradiograms prepared by the method of Tabor and Neihof (1982b) fluorescing microorganisms were superimposed on opaque clusters of silver grains. Small cells active in amino acid uptake were not masked by silver grains as in a previous method (Meyer-Reil, 1978; Tabor and Neihof, 1982b). Masking can cause a significant loss of counts; samples prepared by the method of Meyer-Reil (1978) yielded DCs 50–80% of the DC determined for samples prepared for microautoradiography by the method of Tabor and Neihof (1982b). The Nuclepore filter, which causes interference in recognition of silver grains (the same size as the pores), was removed before microscopic observation, resulting in a structureless background similar to samples prepared for determination of INTRM.

Technical difficulty, standardization problems, and variability of results from one preparation to another are criticisms leveled against microautoradiography. For these reasons, microautoradiography is not considered a quantitative tool (Wright, 1978; Fuhrman and Azam, 1982). However, the method of Tabor and Neihof (1982b) was not technically difficult to use (60 microautoradiograms could be prepared by two people in one day) and results were reproducible.

Fuhrman and Azam (1982) point out a need for using equivalent activity levels of substrates (used at tracer concentrations) in samples when comparisons are made of the ability of individual organisms in a population to transport different substrates actively. Tabor and Neihof (1982*b*) proposed the use of microautoradiography as a quantitative method and outlined the use of various specific activities of [^{3}H]methionine, all at tracer concentrations, to recognize the distribution of microheterotrophs in Chesapeake Bay with various minimum uptake rates. Prerequisites for the use of microautoradiography as a method of quantitating uptake rates of individual organisms of a population include (1) a sensitive autoradiographic emulsion, (2) a high-efficiency microautoradiographic method, (3) calculation of the minimum quantity of accumulated radioactivity required to produce an autoradiogram per organism (which can be accomplished by the use of standard radioactive sources of a size comparable to microbial cells found in natural samples), and (4) measurement of the endogenous concentration of the substrate of interest for determination of the specific activity of labeled substrate in the amended sample. An alternative to quantitation of label accumulated by an organism by grain-density autoradiography, a method that is technically difficult and that poses many theoretical problems (Knoechel and Kalff, 1976), is the determination of minimum uptake rates. This requires the assumption that the accumulation of label needed to produce an autoradiogram is known and determinations are made simply by observing whether an organism was active (i.e., produced an autoradiogram of one silver grain or a cluster of silver grains) at a given specific activity of substrate. Enumeration of bacterial types with minimum uptake rates at a number of specific activities can aid in determining the distribution of bacteria within microbial assemblages active in uptake of a specific substrate over a range of minimum rates. Unique information on the distribution of uptake rates of bacteria within an assemblage can be obtained by use of this type of quantitative microautoradiographic method. With determination of actual minimum uptake rates for bacteria, direct relationships can be made among numbers or types of bacteria, their minimum uptake rates, and the rate of uptake of the composite population.

Summary. The nalidixic acid method is useful for providing an index of synthetically active bacteria in a population, but the results are subject to marked influence by differences in type of sample (e.g., quantity of organic matter present). In addition, because of the presence in natural assemblages of bacteria resistant to nalidixic acid, the method yields an underestimate of bacteria active in cellular synthesis. The INT-reduction method is a valuable and practical means of determining the percentage of cells within an assemblage that are respiring (have functional electron-transport systems); there is, however, need for improvement in the method, as there are indications that respiring cells of particular morphological groups are not detected as active. The microautoradiographic method as described here is not only an efficient and practical means of detecting cells active in substrate uptake, but it may also be useful as a method of determining minimum rates of uptake within assemblages of bacteria.

Fluorochromes Alone as Active Cell Indicators

Fluorochromes have also been suggested and tested for use without additional substances (i.e., INT, tritiated substrates, nalidixic acid) in assays of percentages of living and dead cells within natural assemblages. Acridine orange is one of the substances that has been used (Pugsley and Evison, 1974; Delattre *et al.*, 1979; Kasten, 1981); the AO is used with samples before fixation, and cells that fluoresce green are counted as living and cells fluorescing orange to red as dead. For bacteria, this usage of AO has been tested only in extreme situations, such as exposure of assemblages to high chlorine concentrations (Delattre *et al.*, 1979); thus, its accuracy requires evaluation, especially because it has also been suggested that red-fluorescing bacteria in formaldehyde-preserved samples may represent fast-growing cells (Daley, 1979).

A second fluorochrome method for determining the proportion of living bacterial cells is the combined use of fluorescein diacetate (FDA) and ethidium bromide (EB) (Jarnagin and Luchinger, 1980; see Calich *et. al.*, 1978; Roser, 1980; Paton, 1982; Roser *et. al.*, 1982; for other, similar methods, see Johnen, 1978; J. G. Jones, 1979; Kasten, 1981; Lundgren, 1981, 1982). Nonpreserved samples are incubated with FDA (usually for about 1 hr). During this period, the esterases of living cells enzymatically produce fluorescein within cells, where it is mostly retained. The preparation is then incubated for a few minutes with an EB solution; EB does not readily penetrate into living cells but rapidly enters and combines with components of dead or damaged cells (Calich *et. al.*, 1978; Roser, 1980). The bacteria are concentrated and examined by epifluorescence microscopy (UV blue excitation, green-red return). Cells that fluoresce green (fluorescein fluorescence) are assumed to have been living; cells that fluoresce red (ethidium bromide fluorescence) are taken to have been dead. A major potential problem lies in the fact that many bacterial strains are unable to take up FDA and/or convert it to fluorescein (Jones, 1979; Jarnagin and Luchsinger, 1980; Lundgren, 1981; Paton, 1982; Chrzanowski *et al.*, 1984); thus, whether the method yields accurage percentages for living cells within natural assemblages is uncertain. In this connection, Domsch *et al.* (1979) found an FDA, percent-active technique for fungal mycelium in soils to give values unreasonably small relative to those of an alternative technique for determination of living fungal biomass.

IMMUNOLABELING IN BACTERIAL ECOLOGY

Introduction

The following discussion addresses some critical factors in immunolabeling (IL) methods used in bacterial ecology, especially fluorescent antibody (FA) techniques. In addition, some recent applications that demonstrate the usefulness of such approaches are reviewed. This information is intended to convey an impression of the current status and directions of IL in bacterial ecology. It is not an exhaustive literature review; instead, a representative cross section of bacterial ecological studies published since Bohlool and Schmidt (1980) is cited.

Coons *et al.* (1942) first demonstrated the feasibility of fluorescein-labeled antibodies to identify bacterial antigens in tissue specimens. However, not until the mid-1950s did Hobson and Mann (1957) use the FA technique in a study of rumen bacteria. Commercial availability of fluorescein isothiocyanate (FITC) and fluorescent microscopes greatly facilitated adoption of the FA method. Because most bacteria are morphologically indistinguishable from each other, IL is the only means of directly identifying a specific nondistinctive bacterial type in natural, diverse bacterial assemblages. Over the past two decades, habitats ranging from the cockroach gut (Bracke and Markovetz, 1978) to hot springs (Bohlool and Brock, 1974*b*) and bacteria from the common *E. coli* to the exotic *Sulfolobus acidocalcarius* have been studied using immunolabeling techniques. These techniques depend on antibodies that specifically adhere to target organisms and that, when appropriately tagged, permit microscopic recognition of the target organism. These techniques are limited to studies of organisms whose properties are sufficiently characterized to allow their cultivation as pure populations.

Many bacterial ecological studies have benefited from immunolabeling techniques, but as with any method, there are certain disadvantages. Development and trials of the reagents can be both time consuming and frustrating. When used for enumeration by direct observation, interferences common to any direct observational approach may make quantitative studies difficult (e.g., see Kingsley and Bohlool, 1981; see also the section, Epifluorescence). However, IL remains an excellent means of identifying a specific bacterial type in the presence of a complex visual matrix. IL techniques are of great value in qualitative observational studies (i.e., simple detection, relative numbers, relative position); this fact is reflected in the observation that more than 50% of the reports on immunolabeling reviewed for this chapter are qualitative in nature. In addition, for organisms that are difficult to grow, that grow slowly, or that are present at a low frequency in the natural assemblage, IL techniques represent a more efficient and sensitive enumeration method than previously used most probable number or plate count methods (Chadwick, 1966; Rennie et al., 1977; Fliermans et al., 1981a).

In ecological studies, the FA technique has been used more often than other IL approaches. Bohlool and Schmidt (1980) recently reviewed this technique and its applications. Because of ready availability of epifluorescence microscopes, now commonly used for nonspecific direct bacterial counts, and the relatively straightforward FA procedures, this approach is likely to continue to be widely used in bacterial ecology. Moreover, IL techniques for transmission electron microscopy (TEM) and scanning electron microscopy (SEM) have found occasional application in ecological studies. For example, Lalonde and Quispel (1977) and Lalonde et al. (1975) used FA labeling for epifluorescence in combination with immunoferritin labeling for TEM in studies of the Alnus root nodule endophyte. Bracke and Markovetz (1978) demonstrated an immuno-Latex reagent for SEM studies of Fusobacterium sp. The Latex particles (0.23 μm) are easily recognized in the SEM fields presented, and the authors claim better resolution than with SEM procedures, which depend on fluorochrome cathodoluminescence (Hough et al., 1976). However, the microspheres might be more difficult to recognize in topographically more complex preparations.

Development and Use of Immunolabeling Reagents

As FAs are, almost exclusively, the immunolabel used in environmental bacterial ecology, this discussion concentrates on this method. However, much of the material in this section is broadly applicable to any IL technique. For specific protocols, it is best to consult previous reports on similar bacterial species. Alternatively, such reviews as M. Goldman (1968), Garvey et al. (1977), Kawanamura (1977), Jones et al. (1978), and Hunn and Chantler (1980) may provide enough general information to guide the successful development of FA reagents.

When the desired FA reagent must be prepared de novo, purified immunoglobulin specific for the target organism is usually produced from hyperimmune rabbit serum. Injections for serum production require 10–1000 μg dry weight of target cells/injection and 10–100 injections. Schemes using minimal amounts of antigen/injection generally produce antibodies of higher specificity (Hunn and Chantler, 1980). Even so, since good growth in pure culture can sometimes be difficult or impossible to obtain (Cherry et al., 1978), production of sufficient target cell mass can be a deterrent to FA production. The use of nonaxenic cultures can be successful but may sometimes require absorbance of the contaminant-specific antibodies. For example, Strayer and Tiedje (1978) developed an FA for a methanogen using a slightly contaminated (<1% cell volume) culture. However, nonaxenic cultures should be avoided. In addition, to ensure specificity, clones used for culture should represent recent isolates from the habitats of interest; it may be desirable to employ culture conditions that mimic growth conditions in situ. Occasional reports have noted that strains from culture collections did not react well with

FA developed against recent local isolates (Farrah and Unz, 1975; Fliermans and Hazen, 1980). Numerous studies have also shown that culture conditions can influence surface FA-reactive antigens (Hill and Gray, 1967; Vestal *et al.*, 1973; Garibaldi and Gibbons, 1975; Apel *et al.*, 1976). Nevertheless, recent environmental isolates often produce useful FA reagents even when grown on rich laboratory media.

After culturing, target cells are generally washed and resuspended in a balanced salts solution for injection. Cells are often formalin killed to prevent active infections in the rabbit (especially important with suspected pathogens). Other specific postcultural treatments may sometimes be used, such as isolation of cell walls or outer membranes (Bohlool and Brock, 1974*a,b*) or removal of sources of surface antigens, such as flagella, known to be nonspecies specific (especially with Enterobacteriaceae) (Cherry *et al.*, 1975; Fliermans and Hazen, 1980).

Test bleeding permits determination of when optimal antibody development has occurred; the antibody is then purified from the serum. In ecological studies, purification has generally been accomplished by repeated (2–4×) ammonium sulfate precipitation (G. L. Jones *et al.*, 1978). However, other techniques such as ion exchange (Kawamura, 1977) or affinity chromatography (Swaminathan *et al.*, 1978) could prove more efficient.

With purified antibody, two FA approaches that permit microscopic recognition of target-adsorbed antibody are then possible. For the direct method, a marker recognizable microscopically is attached directly to the antibody. FITC has most often been used as the marker (Bohlool and Schmidt, 1980). In the *indirect* method, a second antibody, which will bind to the target cell antibody, is attached to the marker—usually FITC. Such second antibodies, often FITC conjugated, are available commercially. Other indirect approaches are also possible, although they have not received much attention in ecological studies. For example, Dowdle and Hansen (1961) used a phage–antiphage FA system to label *Bacillus anthracis*. The phage, specifically attached to the bacterium, binds the FA, marking the bacterium. Intermediate binding agents such as phage or lectins can be less strain specific than antibodies, hence may offer the possibility of developing more broadly specific FA reagents. Direct or indirect FA reagents have been developed for the simultaneous labeling of two biogeochemically important species (Reed and Dugan, 1979), using two differently colored fluorochromes (e.g., FITC (green) and rhodamine (or tetramethylrhodamine) isothiocyanate (RITC or tRITC) (red-orange). However, this method has not been commonly used with environmental samples. In spite of the variety of alternatives, single-label direct FA methods dominate ecological practice.

Newly produced immunolabeling reagents must be characterized with respect to their specificity before they can be used in natural samples. Proper controls, described for the FA procedure by Schmidt (1973), must be performed to ensure that the IL reagent is performing correctly. The main concern of such tests is to ensure the specificity of the reagent.

Too much or too little specificity can hamper IL studies. Upon micro-

scopic evaluations, FA stains are scored qualitatively for brightness by assign-
ing an integer value of $1+$ to $4+$. Generally, cells reacting $3+$ or $4+$ are
considered positive. The qualitative nature of such judgments is a disadvantage of the FA approach. The FA reagent is subsequently used at the maximum dilution that still gives a $4+$ reaction with the target strain. This reduces
the frequency of cross-reactions. Staining specificity is checked against related
strains and species to define the serotype specificity. Also, unrelated organisms, especially isolates from the habitat to be studied, are checked for possible,
but unexpected, cross-reactions. Some investigators have been especially careful to demonstrate the absence of cross-reactions with unrelated species. For
example, Fliermans *et al.* (1974) showed negative reactions for *Nitrobacter* FA
against 668 soil isolates. Usually, however, 50 or fewer isolates are tested.
Specificity can often be increased by adsorbing out undesired antibodies (e.g.,
Farrah and Unz, 1975; Fliermans *et al.*, 1974), and some investigators have
routinely included adsorption against species carrying commonly encountered
surface antigens (Apel *et al.*, 1976; Reed and Dugan, 1978). Usually, some
loss of sensitivity accompanies the increased specificity. In ecological studies,
it is often desirable to have a less specific IL reagent—one that could identify
a species or genus responsible for a particular physiological process. However,
such reagents have rarely, if ever, been found. Serotypes abound in nature
(Farrah and Unz, 1975; Belser and Schmidt, 1978; Fliermans *et al.*, 1981*a*),
often resulting in immunolabels with more specificity than is desirable. In
such cases, the usual recourse is to develop IL reagents specific for each type
present and to combine them in a polyvalent reagent.

In spite of proper testing and control procedures, the problem of unidentified fluorescent objects (UFOs) can still arise when the FA reagent is
used with field samples. Different types of UFO phenomena have been observed in various studies.

First, nonspecific staining occurs because the FA binds to surfaces via
nonspecific electrostatic or hydrogen-bonding interactions. Bohlool and Schmidt
(1968) introduced the rhodamine isothiocyanate (RITC)-hydrolyzed gelatin
counterstain technique to overcome this problem in soil studies. The use of
this or similar techniques has significantly reduced the problem of nonspecific
staining in ecological studies. The essential feature is the use of a peptide
solution (hydrolyzed gelatin) that occupies nonspecific adsorption sites and
prevents binding of the antibody protein. Similar effects can be achieved with
bovine serum albumin (BSA) (Clarke and Naylor, 1978; Baker and Mills,
1982) or normal rabbit serum (Guthrie and Reeder, 1969).

Second, autofluorescence (fluorescence in the absence of added fluorochrome) is another occasionally troublesome problem in FA procedures. Many
autofluorescent objects show distinguishable colors or shapes and thus can be
differentiated from target cells. In some cases, however, the problem is one
of distinguishing target cells against a bright background of a similar color.
With planktonic samples, the use of Irgalan black-stained filters, as used in
AODC methods (Hobbie *et al.*, 1977; see Table I) has eliminated the problem

of filter fluorescence. Irgalan black generally produces a lower autofluorescence than do many of the commercially available black filters (Schmidt *et al.*, 1968; Kingsley and Bohlool, 1981). For autofluorescent objects appearing in the sample, some investigators have reported that dyes may be used to quench fluorescence. This approach has generally been unsatisfactory, however, (Hill and Gray, 1967; Diem *et al.*, 1978; Swaminathan *et al.*, 1978). Autofluorescence is often noted as being particularly troublesome in samples containing root tissue, in which RITC-protein counterstaining cannot block the intense fluorescence (Lalonde and Quispel, 1977; Diem *et al.*, 1978; Schank *et al.*, 1979). Lalonde and Quispel (1977) report that pectinase treatment can remove such autofluorescence. Hughes *et al.* (1979) used rhodamine conjugates (which contrast with green autofluorescence) to avoid the problem, and Brlansky *et al.* (1982) used tRITC-conjugated markers for immunofluorescent detection of xylem-limited bacteria *in situ*. It appears that these red-orange fluorescent markers are becoming the standard for IL studies of bacteria on plant surfaces lacking chlorophyll fluorescence.

The third and perhaps most troublesome UFO is what Bohlool and Schmidt (1980) refer to as "universal acceptors," which bind the antibody specifically but not at the immune-specific site. These objects bind both preimmune and hyperimmune globulins and are not blocked by nonspecific protein blockers. This phenomenon has been observed in association with fungal spores (Schank *et al.*, 1979; Bohlool and Schmidt, 1980) and actinomycetes (Bohlool and Brock, 1974*a*). The occasional appearance of such phenomena is to be expected, as immunoglobulin-binding proteins have long been recognized in some pathogens; for example, *Staphylococcus aureus* protein A binds IgG (Bjork *et al.*, 1972). Differences in size or morphology often enable such universal acceptors to be differentiated from the target organism. The use of proper controls, including checks with preimmune serum, is important to ensure that universal acceptors are recognized.

Recent Applications and Current Status

With several of the pitfalls, complications, and limitations of IL procedures having been outlined, our discussion now turns to the value of immunolabeling, especially with FAs, in microbial ecology.

In the summer of 1976, an explosive outbreak of pneumonia in Philadelphia began a continuing case of epidemiological sleuthing in which the FA technique played a central role. In preliminary studies, Cherry *et al.* (1978) noted that organisms serologically related to legionnaire's bacterium (*Legionella pneumophila*) were found in soils from several locations. Soon thereafter, Fliermans *et al.* (1979, 1981*a*), using direct FA, demonstrated that *Legionella* serotypes were common in aquatic habitats not associated with epidemics. Further studies showed *L. pneumophila* to be metabolically active in aquatic habitats at temperatures up to 60°C and able to metabolize algal extracellular products (Tison *et al.*, 1980; Fliermans *et al.*, 1981*b*). In urban areas, cooling towers appeared to be an important habitat for the maintenance of *L. pneumophila* (Orrison *et al.*, 1981; Fliermans *et al.*, 1982). Because no selective enrichment medium had been developed for *L. pneumophila*, direct FA in combination with sample concentration (Fliermans *et al.*, 1979; Orrison *et al.*, 1981) was the only rapid enumeration technique available. Culturing in guinea pigs was used for confirmation in many habitats, but only the FA procedure permitted rapid, precise detection.

The inability to differentiate active from inactive cells has been a disadvantage of IL techniques (Bohlool and Schmidt, 1980). However, several recent studies have shown that FA can be combined with indicators of cell activity, permitting total and viable counts. Fliermans *et al.* (1981*b*), C. B. Fliermans (unpublished observations,), and Baker and Mills (1982) combined FA with 2-*p*-iodophenyl-3-*p*-nitrophenyl-5-phenyl tetrazolium chloride (INT), as an indicator of cell electron-transport activity. Using this approach, Fliermans *et al.* (1981*b*) demonstrated that *L. pneumophila* is active up to 60°C in aquatic systems. More recently, the FAINT technique (FA + INT) (Baker and Mills, 1982) has been used to study natural samples containing *Thiobacillus ferroxidans*. FAINT-positive counts were 10–1000 times greater than MPN estimates, although only 20–80% of the FA-positive cells were INT active. This demonstrates that direct-count techniques can be more efficient at detecting active cells than cultivation techniques. These data also appear to contradict the argument of Bohlool and Schmidt (1980) that because biomass turnover is rapid, most FA-positive cells are active. The difference in turnover rates between soils (Bohlool and Schmidt, 1980) and acid mine drainage waters (Baker and Mills, 1982) probably accounts for the contradiction. In addition to INT, Fliermans and Schmidt (1975) demonstrated the use of FA–autoradiography in enumerating active cells. The procedure is more tedious than the FAINT technique. However, both approaches ameliorate a major disadvantage of FA techniques and hold great promise with further development. They would be especially useful in studies of population turnover or correlations between cell numbers and physiological activity, exemplified by recent studies in marine ecosystems by Ward (1982), Ward *et al.* (1982), and Dahle and Laake (1982).

Finally, potential improvements in equipment may soon increase the applicability of IL procedures, especially FA techniques, in ecological studies. Recent developments in quantitative light-measuring systems, image analysis, and automated cell counting may provide the bacterial ecologist with a less tedious means of analysis than manual counting. Fliermans and Hazen (1980) reported the use of photometric techniques that could potentially replace the usual manner of scoring cell brightness in the microscopic system using qualitative integer values. One problem with such systems is that fading of the fluorochrome makes photometric measurement difficult. However, possible solutions to this problem are being examined (Kasatya *et al.*, 1974). The use of more sensitive digital detectors, such as charge-coupled devices now used in astronomy, or of high-speed film systems (Breener, 1983), in combination with image-analyzing systems (e.g., Gregory, 1983), may lead to more rapid and precise cell count determinations than are currently possible with FA analysis.

ACKNOWLEDGMENT. This work is contribution no. 488 from the University of Georgia Marine Institute.

REFERENCES

Alexander, M., 1981, Why microbial predators and parasites do not eliminate their prey and hosts, *Annu. Rev. Microbiol.* **35**:113–133.

Allison, D. G., and Sutherland, I. W., 1984, A staining technique for attached bacteria and its correlation to extracellular carbohydrate production, *J. Microbiol. Meth.* **2**:93–99.

Altman, F. P., 1976, Tetrazolium salts and formazans, *Prog. Histochem. Cytochem.* **9**:1–56.

Ammerman, J. W., Fuhrman, J. A., Hagström, Å, and Azam, F., 1984, Bacterioplankton growth in seawater: I. Growth kinetics and cellular characteristics in seawater cultures, *Mar. Ecol. Prog. Ser.* **18**:31–39.

Apel, W. A., Dugan, P. R., Filppi, J. A., and Rheins, M. S., 1976, Detection of *Thiobacillus ferroxidans* in acid mine environments by indirect fluorescent antibody staining, *Appl. Environ. Microbiol.* **32**:159–165.

Azam, F., and Ammerman, J. W., 1984, The cycling of organic matter by bacterioplankton in pelagic marine ecosystems: Microenvironmental considerations, in: *Flows of Energy and Materials in Marine Ecosystems* (M. J. R. Fasham, ed.), pp. 345–360, Plenum Press, New York.

Azam F., and Hodson, R. E., 1981, Multiphasic kinetics for D-glucose uptake by assemblages of natural marine bacteria, *Mar. Ecol. Prog. Ser.* **6**:213–222.

Babiuk, L. A., and Paul, E. A., 1970, The use of fluorescein isothiocyanate in the determination of the bacterial biomass of grassland soil, *Can. J. Microbiol.* **16**:57–62.

Baker, K. H., and Mills, A. L., 1982, Determination of the number of respiring *Thiobacillus ferroxidans* cells in water samples by using combined fluorescent antibody-2-(*p*-iodophenyl)-3-(*p*-nitrophenyl)-5-phenyltetrazolium chloride staining, *Appl. Environ. Microbiol.* **43**:338–344.

Bakken, L. R., and Olsen, R. A., 1983, Buoyant densities and dry-matter contents of microorganisms: Conversion of a measured biovolume into biomass, *Appl. Environ. Microbiol.* **45**:1188–1195.

Bassøe, C. -F., 1984, Processing of *Staphylococcus aureus* and zymosan particles by human leukocytes measured by flow cytometry, *Cytometry* **5**:86–91.

Belser, L. W., and Schmidt, E. L., 1978, Serological diversity within a terrestrial ammonia oxidizing population, *Appl. Environ. Microbiol.* **36**:589–593.

Berk, S. G., Colwell, R. R., and Small, E. B., 1976, A study of feeding responses to bacterial prey by estuarine ciliates, *Trans. Am. Microsc. Soc.* **95**:514–520.

Bjork, I., Peterson, B. A., and Sjoquist, J., 1972, Some physicochemical properties of protein A from *Staphylococcus aureaus*, *Eur. J. Biochem.* **29**:579–584.

Bohlool, B. B., and Brock, T. D., 1974a, Immunofluorescence approach to the study of the ecology of *Thermoplasma acidophilum* in coal refuse material, *Appl. Microbiol.* **28**:11–16.

Bohlool, B. B., and Brock, T. D., 1974b, Population ecology of *Sulfolobus acidocaldarius*. II. Immunological studies, *Arch. Miocrobiol.* **97**:181–194.

Bohlool, B. B., and Schmidt, E. L., 1968, Nonspecific staining: Its control in immunofluorescence examination of soil, *Science* **162**:1012–1014.

Bohlool, B. B., and Schmidt, E. L., 1980, The immunofluorescence approach in microbiol ecology, in: *Advances in Microbiol Ecology*, Vol. 4 (M. Alexander, ed.), pp. 203–241, Plenum Press, New York.

Borsheim, K. Y., 1984, Clearance rates of bacteria-sized particles by freshwater ciliates, measured with monodisperse fluorescent latex beads, *Oecologia (Berl.)* **63**:286–288.

Bracke, J. W., and Markovetz, A. J., 1978, Immunolatex localization by scanning electron microscopy of intestinal bacteria from cockroaches, *Appl. Environ. Microbiol.* **35**:166–170.

Bratbak, G., and Dundas, I., 1984, Bacterial dry matter content and biomass estimations, *Appl. Environ. Microbiol.* **48**:755–757.

Brenner, M., 1983, Low light microscopy and high-speed photography, Am. Lab. **15**:51–55.

Bright, J. J., and Fletcher, M., 1983, Amino acid assimilation and electron transport system activity in attached and free-living marine bacteria, *Appl. Environ. Microbiol.* **45**:818–825.

Brlansky, R. H., Lee, R. F., Timmer, L. W., Percifull, D. E., and Raju, B. C., 1982, Immunofluorescent detection of xylem-limited bacteria *in situ, Phytopathology* **11**:1444–1448.

Brock, T. D., 1971, Microbial growth rates in nature, *Bacteriol. Rev.* **35**:39–58.

Brock, T. D., 1983, *Membrane Filtration: A User's Guide and Reference Manual,* Science Technology, Madison, Wisconsin.

Brock, M. L., and Brock, T. D., 1968, The application of micro-autoradiographic techniques to ecological studies, *Mitt. Int. Ver. Limnol.* **15**:1–29.

Burney, C. M., Davis, P. G., Johnson, K. M., and Sieburth, J. McN., 1981, Dependence of dissolved carbohydrate concentrations upon small scale nanoplankton and bacterioplankton distributions in the western Sargasso Sea, *Mar. Biol.* **65**:289–296.

Calich, V. L. G., Purchio, A., and Paula, C. R., 1978, A new fluorescent viability test for fungi cells, *Mycopatholoqia* **66**:175–177.

Cammen, L. M., 1980, The significance of microbial carbon in the nutrition of the deposit feeding polychaete *Nereis succinea, Mar. Biol.* **61**:9–20.

Carlucci, A. F., and Williams, P. M., 1978, Simulated *in situ* growth rates of pelagic marine bacteria, *Naturwissenschaften* **65**:541–542.

Casida, L. E., 1971, Microorganisms in unamended soil as observed by various forms of microscopy and staining, *Appl. Environ. Microbiol.* **21**:1040–1045.

Chadwick, P., 1966, The relative sensitivity of fluorescent antibody and cultural methods in detection of small numbers of pathogenic serotypes of *Escherichia coli, Am. J. Epidemiol.* **84**:150–155.

Cherry, W. B., Thomason, B. M., Gladden, J. B., Holsing, N., and Murlin, A. M., 1975, Detection of *Salmonellae* in foodstuffs, feces, and water by immunofluorescence, in: *Fifth International Conference on Immunofluorescence and Related Techniques* (W. Hijams and M. Schaeffer, eds.), *Ann. N. Y. Acad. Sci.* **254**:350–369.

Cherry, W. B., Pittman, B., Harris, P. P., Herbert, G. A., Thomason, B. M., Thacker, L., and Weaver, R. E., 1978, Detection of legionnaire's disease bacteria by direct immunofluorescent staining, *J. Clin. Microbiol.* **8**:329–338.

Christensen, J. P., Owens, T. G., Devol, A. H., and Packard, T. T., 1980, Respiration and physiological state in marine bacteria, *Mar. Biol.* **55**:267–276.

Christian, R. R., and Hall, J. R., 1977, Experimental trends in sediment microbial heterotrophy: Radioisotopic techniques and analysis, in: *Ecology of Marine Benthos* (B. C., Coull, ed.), pp. 67–88, University of South Carolina Press, Columbia.

Christian, R. R., Hanson, R. B., and Newell, S. Y., 1982, Comparison of methods for measurement of bacterial growth rates in mixed batch cultures, *Appl. Environ. Microbiol.* **43**:1160–1165.

Chrzanowski, T. H., Crotty, R. D., Hubbard, J. G., and Welch, R. P., 1984, Applicability of the fluorescein diacetate method of detecting active bacteria in freshwater, *Microb. Ecol.* **10**:179–185.

Clarholm, M., and Rosswall, T., 1980, Biomass and turnover of bacteria in a forest soil and a peat, *Soil Biol. Biochem.* **12**:49–57.

Clarke, R. T. J., and Naylor, G. E., 1978, Fluorescence microscopy of gut microbes, in: *Microbial Ecology* (M. W. Loutit and J. M. R. Miles eds.), pp. 244–245, Springer-Verlag, Berlin.

Coats, D. W., and Heinbokel, J. F., 1982, A study of reproduction and other life cycle phenomena in planktonic protists using an acridine orange fluorescence technique, *Mar. Biol.* **67**:71–79.

Cole, J. J., Likens, G. E., and Strayer, D. L., 1982, Photosynthetically produced dissolved organic carbon: An important carbon source for planktonic bacteria, *Limnol. Oceanogr.* **27**:1080–1090.

Coleman, A. W., 1980, Enhanced detection of bacteria in natural environments by fluorochrome staining of DNA, *Limnol. Oceanogr.* **25:**948–951.

Coons, A. H., Creech, H. J., Jones, R. N., and Berliner, E., 1942, The demonstration of penumococcal antigen in tissues by use of fluorescent antibody, *J. Immunol.* **45:**159–170.

Coveney, M. F., Cronberg, G., Enell, M., Larsson, K., and Olofsson, L., 1977, Phytoplankton, zooplankton and bacteria—standing crop and production relationships in a eutrophic lake, *Oikos* 29:5–21.

Crumpton, W. G., and Wetzel, R. G., 1982, Effects of differential growth and mortality in the seasonal succession of phytoplankton populations in Lawrence Lake, Michigan, *Ecology* **63:**1729–1739.

Dahle, A. B., and Laake, M., 1982, Diversity dynamics of marine bacteria studies by immunofluorescent staining on membrane filters, *Appl. Environ. Microbiol.* **43:**169–176.

Daley, R. J., 1979, Direct epifluorescence enumeration of native aquatic bacteria: Uses, limitations, and comparative accuracy, in: *Native Aquatic Bacteria: Enumeration, Activity, and Ecology* (J. W., Costerton and R. R. Colwell, eds.), pp. 29–45, American Society for Testing and Materials, Philadelphia.

Davis, P. G., and Sieburth, J. McN., 1982, Differentiation of the photosynthetic and heterotrophic nanoplankton populations in marine waters by epifluorescence microscopy, *Ann. Inst. Oceanogr. (Suppl.)* **58:**249–260.

Davis, P. G., and Sieburth, J. McN., 1984, Estuarine and oceanic microflagellate predation of actively growing bacteria: Estimation by frequency of dividing-divided bacteria, *Mar. Ecol. Prog. Ser.* **19:**237–246.

Delattre, J. M., Delesmont, R., Clabaux, M., Oger, C., and Leclarc, H., 1979, Bacterial biomass, production and heterotrophic activity of the coastal seawater at Gravelines (France), *Oceanol. Acta* **2:**317–324.

Diem, H. G., Schmidt, E. L., and Dommergues, J. R., 1978, the use of fluorescent-antibody technique to study the behavior of a *Beijerinckia* isolate in the rhizosphere and spermosphere of rice, *Ecol. Bull.* (Stockholm) **26:**312–318.

Domsch, K. H., Beck, T., Anderson, J. P. E., Söderström, B., Parkinson, D., and Trolldenier, G., 1979, A comparison of methods for soil microbial population and biomass studies, *Z. Pflanzenernaehr. Bodenkd.* **142:**520–533.

Dowdle, W. R., and Hanse, P. A., 1961, A phage-fluorescent antiphage staining system for *Bacillus anthracis*, *J. Infect. Dis.* **108:**125–135.

Ducklow, H. W., 1983, The production and fate of bacteria in the oceans, *BioScience* **33:**494–501.

Ellery, W. N., and Schleyer, M. H., 1984, Comparison of homogenization and ultrasonication as techniques in extracting attached sedimentary bacteria, *Mar. Ecol. Prog. Ser.* **15:**247–250.

Eppley, R. W., 1981, Relations between nutrient assimilation and growth in phytoplankton with a brief review of estimates of growth rate in the ocean, *Can. Bull. Fish. Aquat. Sci.* **210:**251–263.

Fallon, R. D., Newell, S. Y., and Hopkinson, C. S., 1983, Bacterial production in marine sediments: Will cell-specific measures agree with whole-system metabolism?, *Mar. Ecol. Prog. Ser.* **11:**119–127.

Farrah, S. R., and Unz, R. F., 1975, Fluorescent antibody study of natural finger-like zoogloeae, *Appl. Microbiol.* **30:**132–139.

Fenchel, T., 1980, Suspension feeding in ciliated protozoa: Feeding rates and their ecological significance, *Microb. Ecol.* **6:**13–25.

Fenchel, T., 1982, Ecology of heterotrophic microflagellates. IV. Quantitative occurrence and importance as bacterial consumers, *Mar. Ecol. Prog. Ser.* **9:**35–42.

Fenchel, T., and Jørgensen, B. B., 1977, Detritus food chains of aquatic ecosystems: The role of bacteria, in: *Advances in Microbial Ecology*, Vol. 1 (M. Alexander, ed.), pp. 1–58, Plenum Press, New York.

Ferguson, R. L., and Sunda, W. G., 1984, Utilization of amino acids by planktonic marine bacteria: Importance of clean technique and low substrate additions, *Limnol. Oceanogr.* **29:**258–274.

Ferguson, R. L., Buckley, E. N., and Palumbo, A. V., 1984, Response of marine bacterioplankton to differential filtration and confinement, *Appl. Environ. Microbiol.* **47:**49–55.

Fliermans, C. B., and Gorden, R. W., 1977, Modification of membrane diffusion chambers for deep-water studies, *Appl. Environ. Microbiol.* **33:**207–210.

Fliermans, C. B., and Hazen, L. D., 1980, Immunofluorescence of *Aeromonas hydrophila* as measured by photometric microscopy, *Can. J. Microbiol.* **26:**161–168.

Fliermans, C. B., and Schmidt, E. L., 1975, Autoradiography and immunofluorescence combined for autecological study of a single cell activity with *Nitrobacter* as a model system, *Appl. Microbiol.* **30:**676–684.

Fliermans, C. B., Bohlool, B. B., and Schmidt, E. L., 1974, Autecological study of the chemoautotroph *Nitrobacter* by immunofluorescence, *Appl. Microbiol.* **27:**124–129.

Fliermans, C. B., Cherry, W. B., Orrison, L. H., and Thacker, L., 1979, Isolation of *Legionella pneumophila* from nonepidemic related aquatic habitats, *Appl. Environ. Microbiol.* **37:**1239.

Fliermans, C. B., Cherry, W. B., Tison, D. L., Smith, R. B., and Pope, D. H., 1981*a*, Distribution of *Legionella pneumophila*, *Appl. Environ. Microbiol.* **41:**9–16.

Fliermans, C. B., Soracco, R. J., and Pope, D. H., 1981*b*, Measure of *Legionella pneumophila* activity *in situ*, *Current Microbiol.* **6:**89–95.

Fliermans, G. E., Bettinger, G. E., and Fynsk, A. W., 1982, Treatment of cooling systems containing high levels of *Legionella pneumophila*, *Water Res.* **16:**903–906.

Fog, K., 1977, Studies on decomposing wooden stumps. I. The microflora of hardwood stumps, *Pedobiologia* **17:**240–261.

Frankland, J. C., 1975, Fungal decomposition of leaf litter in a deciduous woodland, in: *Biodégradation et humificacion* (G. Kilbertus *et al.*, eds.), pp. 33–40, Pierron, Sarreguemines.

Fry, J. C., and Davies, A. R., 1984, An assessment of methods for measuring volumes of planktonic bacteria, with particular reference to television image analysis, *J. Appl. Bacteriol.* **58:**105–112.

Fry, J. C., and Humphrey, N. C. B., 1978, Techniques for the study of bacteria epiphytic on aquatic macrophytes, *Soc. Appl. Bacteriol Tech. Ser.* **11:**1–29.

Fuhrman, J. A., 1981, Influence of method on the apparent size distribution of bacterioplankton cells: Epifluorescence microscopy compared to scanning electron microscopy, *Mar. Ecol. Prog. Ser.* **5:**103–106.

Fuhrman, J. A., 1984, Biological considerations in the measurement of dissolved free amino acids in seawater: Implications for chemical and microbiological studies, *EOS* **65:**926.

Fuhrman, J. A., and Azam, F., 1980, Bacterioplankton secondary production estimates for coastal waters of British Columbia, Antarctica, and California, *Appl. Environ. Microbiol.* **39:**1085–1095.

Fuhrman, J. A., and Azam, F., 1982, Thymidine incorporation as a measure of heterotrophic bacterioplankton production in marine surface waters: Evaluation and field results, *Mar. Biol.* **66:**109–120.

Fuhrman, J. A., and McManus, G. B., 1984, Do bacteria-sized marine eukaryotes consume significant bacterial production? *Science* **224:**1257–1260.

Gantt, E., 1975, Phycobilisomes: Light-harvesting pigments, *BioScience* **25:**781–788.

Garibaldi, A., and Gibbins, L. N., 1975, Induction of avirulent variants in *Erwinia stewartii* by incubation at supraoptimal temperatures, *Can. J. Microbiol.* **21:**1282–1287.

Garvey, J. S., Cremer, N. E., and Susdorf, D. H., 1977, *Methods in Immunology*, 3rd ed., W. A. Benjamin, Reading, Massachusetts.

Geesey, G. G., and Costerton, J. W., 1979, Bacterial biomass determinations in a silt-laden river: Comparison of direct count epifluorescence microscopy and extractable adenosine triphosphate techniques, in: *Native Aquatic Bacteria: Enumeration, Activity, and Ecology* (J. W. Costerton and R. R. Colwell, eds.), pp. 117–127, American Society for Testing and Materials, Philadelphia.

Germida, J. J., and Casida, L. E., 1983, *Ensifer adhaerens* predatory activity against other bacteria in soil, as monitored by indirect phage analysis, *Appl. Environ. Microbiol.* **45:**1380–1388.

Goldman, J. C., 1984, Oceanic nutrient cycles, in: *Flows of Energy and Materials in Marine Ecosystems* (M. J. R. Fasham, ed.), pp. 137–170, Plenum Press, New York.

Goldman, M., 1968, *Fluorescent Antibody Methods*, Academic Press, New York.

Gregory, P., 1983, Advances in automatic image analysis, *Am. Lab.* **15:**29–37.

Grover, N. B., Woldringh, C. L., Zaritsky, A., and Rosenberger, R. F., 1977, Elongation of rod-shaped bacteria, *J. Theor. Biol.* **67**:181–193.

Güde, H., 1985, Influence of phagotrophic processes on the regeneration of nutrients in two-stage continuous culture systems, *Microb. Ecol.* **11**:193–204.

Guthrie, R. K., and Reeder, D. J., 1969, Membrane filter-fluorescent antibody method for the detection and enumeration of bacteria in water, *Appl. Microbiol.* **17**:399–401.

Haas, L. W., 1982, Improved epifluorescent microscopic technique for observing planktonic micro-organisms, *Ann. Inst. Oceanogr. (Suppl.)* **58**:261–266.

Habte, M., and Alexander, M., 1978, Mechanisms of persistence of low numbers of bacteria preyed upon by protozoa, *Soil Biol. Biochem.* **10**:1–6.

Hagström, A., Larsson, U., Hörstedt, P., and Normark, S., 1979, Frequency of dividing cells, a new approach to the determination of bacterial growth rates in aquatic environments, *Appl. Environ. Microbiol.* **37**:805–812.

Hagström, Å, Ammerman, J. W., Henrichs, S., and Azam, F., 1984, Bacterioplankton growth in seawater: II. Organic matter utilization during steady-state growth in seawater cultures, *Mar. Ecol. Prog. Ser.* **18**:41–48.

Hanson, R. B., Shafer, D., Ryan, T., Pope, D. H., and Lowery, H. K., 1983, Bacterioplankton in Antarctic Ocean waters during the late austral winter: Abundance, frequency of dividing cells and estimates of production, *Appl. Environ. Microbiol.* **45**:1622–1632.

Harvey, R. W., and Young, L. Y., 1980, Enumeration of particle-bound and unattached respiring bacteria in the salt marsh environment, *Appl. Environ. Microbiol.* **40**:156–160.

Hewes, C. D., and Holm-Hansen, O., 1983, A method for recovering nanoplankton from filters for identification with the microscope: The filter-transfer-freeze (FTF) technique, *Limnol. Oceanogr.* **28**:389–394.

Hill, I. R., and Gray, T. R. G., 1967, Application of the fluorescent-antibody technique to an ecological study of bacteria in soil, *J. Bacteriol.* **93**:1888–1896.

Hobbie, J. E., and Crawford, C. C., 1969, Respiration corrections for bacterial uptake of dissolved organic compounds in natural waters, *Limnol. Oceanogr.* **14**:528–532.

Hobbie, J. E., and Rublee, P., 1975, Bacterial production in an arctic pond, *Verh. Int. Verein. Limnol.* **19**:466–471.

Hobbie, J. E., Daley, R. J., and Jasper, S., 1977, Use of Nuclepore filters for counting bacteria by fluorescence microscopy, *Appl. Environ. Microbiol.* **3**:1225–1228.

Hobson, P. N., and Mann, S. O., 1957, Some studies on the identification of rumen bacteria with fluorescent antibodies, *J. Gen. Microbiol.* **16**:463–471.

Hoppe, H. -G., 1977, Analysis of actively metabolizing bacterial populations with the autoradiographic method, in: *Microbial Ecology of a Brackish Environment. Ecological Studies*, Vol. 25 (G. Rheinheimer, ed.), pp. 179–197, Springer-Verlag, New York.

Hoppe, H. -G., 1978, Relations between active bacteria and hererotrophic potential in the sea, *Neth. J. Sea Res.* **12**:78–98.

Horstmann, U., and Hoppe, H. -G., 1981, Competition in the uptake of methylamine/ammonium by phytoplankton and bacteria, *Kiel. Meeresforsch. Sonderh.* **5**:110–116.

Hossell, J C , and Baker, J. H., 1979, A note on the enumeration of epiphytic bacteria by micrcopic method siwth particular reference to two freshwater plants, *J. Appl. Bacteriol.* **46**:87–92.

Hough, P. U. C., McKinney, W. R., Ledbetter, M. C., Pollack, R. E., and Moos, H. W., 1976, Identification of biological molecules *in situ* at high resolution via the fluorescence excited by a scanning electron beam, *Proc. Natl. Acad. Sci. U.S.A.* **73**:317–321.

Hughes, T. A., Lecce, J. G., and Ellsan, G. H., 1979, Modified fluorescent technique, using rhodamine, for studies of *Rhizobium japonicum*-soybean symbiosis, *Appl. Environ. Microbiol.* **37**:1243–1244.

Hurn, B. A. L., and Chantler, S. M., 1980, Production of reagent antibodies, *Methods Enzymol.* **70**:104–142.

Ingraham, J. L. Maaløe, O., and Neidhardt, F. C., 1983, *Growth of the Bacterial Cell*, Sinauer Associates, Sunderland, Massachusetts.

Iturriaga, R., 1979, Bacterial activity related to sedimenting particulate matter, *Mar. Biol.* **55:**157–169.

Jannasch, H. W., 1979, Microbial ecology of aquatic low nutrient habitats, in: *Strategies of Microbial Life in Extreme Environments* (M. Shilo, ed.), pp. 243–260, Verlag Chemie, New York.

Jannasch, H. W., and Jones, G. E., 1959, Bacterial populations in seawater as determined by different methods of enumeration, *Limnol. Oceanogr.* **4:**128–139.

Jarnagin, J. L., and Luchsinger, D. W., 1980, The use of fluorescein diacetate and ethidium bromide as a stain for evaluating viability of mycobacteria, *Stain Technol.* **55:**253–258.

Jenkinson, D. S., and Ladd, J. N., 1981, Microbial biomass in soil; measurement and turnover, in: *Soil Biochemistry*, Vol. 5 (E. A. Paul and J. N. Ladd, eds.), pp. 415–472, Marcel Dekker, New York.

Jenkinson, D. S., Powlson, D. S., and Wedderburn, R. W. M., 1976, The effects of biocidal treatments on metabolism in soil—III. The relationship between soil biovolume, measured by optical microscopy, and the flush of decomposition caused by fumigation, *Soil Biol. Biochem.* **8:**189–202.

Johnen, B. G., 1978, Rhizosphere microorganisms and roots stained with europium chelate and fluorescent brightener, *Soil Biol. Biochem.* **10:**495–502.

Johnson, P. W., and Sieburth, J. McN., 1982, *In-situ* morphology and occurrence of eucaryotic phototrophs of bacterial size in the picoplankton of estuarine and oceanic waters, *J. Phycol.* **18:**318–327.

Joint, I. R., and Morris, R. J., 1982, The role of bacteria in the turnover of organic matter in the sea, *Oceanogr. Mar. Biol. Annu. Rev.* **20:**65–118.

Jones, G. L., Hebert, G. A., and Cherry, W. B., 1978, Fluorescent antibody techniques and bacterial applications, *USDHEW CDC No. 788-8364.*

Jones, J. G., 1979, A guide to methods for estimating microbial numbers and biomass in freshwater, *Sci. Publ. Freshwater Biol. Assoc.* **39:**1–112.

Jordan, M. J., and Likens, G. E., 1980, Measurement of planktonic bacterial production in an oligotrophic lake, *Limnol. Oceanogr.* **25:**719–732.

Karl, D. M., 1980, Cellular nucleotide measurements and applications in microbial ecology, *Microbiol. Rev.* **44:**739–796.

Karl, D. M., 1982, Selected nucleic acid precursors in studies of aquatic microbial ecology, *Appl. Environ. Microbiol.* **44:**891–902.

Kasatya, S. S., Lambert, N. G., and Lawrence, R. A., 1974, Use of tunable pulsed dye laser for quantitative fluorescence in syphilis serology (FTA-ABS test), *Appl. Microbiol.* **27:**838–843.

Kasten, F. H., 1981, Methods for fluorescence microscopy, in: *Staining Procedures*, 4th ed. (G. Clark, ed.), pp. 39–103, Williams & Wilkins, Baltimore.

Kawanamura, A., 1977, *Fluorescent Antibody Techniques and Their Applications*, 2nd ed., University of Tokyo Press, Tokyo.

Kenner, R. A., and Ahmed, S. I., 1975, Correlation between oxygen utilization and electron transport activity in marine phyotplankton, *Mar. Biol.* **33:**129–133.

Kingsley, M. T., and Bohlool, B. B., 1981, Release of *Rhizobium* spp. from tropical soils and recovery for immunofluorescence enumeration, *Appl. Environ. Microbiol.* **42:**241–248.

Kirchman, D., and Mitchell, R., 1982, Contribution of particle-bound bacteria to total microheterotrophic activity in five ponds and two marshes, *Appl. Environ. Microbiol.* **43:**200–209.

Kirchman, D., Ducklow, H. W., and Mitchell, R., 1982a, Estimates of microbial growth from changes in uptake rates and biomass, *Appl. Environ. Microbiol.* **44:**1296–1307.

Kirchman, D., Sigda, J., Kapuscinski, R., and Mitchell, R., 1982b, Statistical analysis of the direct count method for enumerating bacteria, *Appl. Environ. Microbiol.* **44:**376–382.

Knoechel, R., and Kalff, J., 1976, The applicability of grain density autoradiography to the quantitative determination of algal species production, *Limnol. Oceanogr.* **21:**583–590.

Kogure, K., Simidu, U., and Taga, N., 1979, A tentative direct microscopic method for counting living marine bacteria, *Can. J. Microbiol.* **25:**415–420.

Kogure, K., Simidu, U., and Taga, N., 1980, Distribution of viable marine bacteria in neritic seawater around Japan, *Can. J. Microbiol.* **26:**318–323.

Krambeck, C., Krambeck, H. -J., and Overbeck, J., 1981, Microcomputer-assisted biomass determination of plankton bacteria on scanning electron micrographs, *Appl. Environ. Microbiol.* **42:**142–149.

Krempin, D. W., and Sullivan, C. W., 1981, The seasonal abundance, vertical distribution, and relative microbial biomass of chroococcoid cyanobacteria at a station in southern California coastal waters, *Can. J. Microbiol.* **27:**1341–1344.

Kurath, G., and Morita, R. Y., 1983, Starvation-survival physiological studies of a marine *Pseudomonas* sp., *Appl. Environ. Microbiol.* **45:**1206–1211.

Laanbroek, H. J., and Verplanke, J. C., 1986, On the use of size fractioning for determining *in situ* bacterial oxygen consumption rates, *Int. Colloq. Mar. Bacteriol.* **2** (in press).

Lalonde, M., Knowles, R., and Fortin, J. -A., 1975, Demonstration of the isolation of non-infective *Alnus crispa* var. *mollis* Fern. nodule endophyte by morphological immunolabelling and whole cell composition studies, *Can. J. Microbiol.* **21:**1901–1920.

Lalonde, M., and Quisepl, A., 1977, Ultrastructural and immunological demonstration of the nodulation of the European *Alnus glutinosa* (L.) Gaertn. host plant by the North American *Alnus crispa* var. *mollis* Fern. root nodule endophyte, *Can. J. Microbiol.* **23:**1529–1547.

Landry, M. R., Haas, L. W., and Fagerness, V. L., 1984, Dynamics of microbial plankton communities: Experiments in Kaneohe Bay, Hawaii, *Mar. Ecol. Prog. Ser.* **16:**127–133.

Larsson, U., and Hagström, A., 1982, Fractionated phytoplankton primary production, exudate release and bacterial production in a Baltic eutrophication gradient, *Mar. Biol.* **67:**57–70.

Laws, E. A., and Archie, J. W., 1981, Appropriate use of regression analysis in marine biology, *Mar. Biol.* **65:**13–16.

Lundgren, B., 1981, Fluorescein diacetate as a stain of metabolically active bacteria in soil, *Oikos* **36:**17–22.

Lundgren, B., 1982, Bacteria in a pine forest soil as affected by clear-cutting, *Soil Biol. Biochem.* **14:**537–542.

Macdonald, R. M., 1980, Cytochemical demonstration of catabolism in soil microorganisms, *Soil Biol. Biochem.* **12:**419–423.

Maki, J. S., and Remsen, C. C., 1981, Comparison of two direct-count methods for determining metabolizing bacteria in freshwater, *Appl. Environ. Microbiol.* **41:**1132–1138.

Marcussen, B., Nielsen, P., and Jeppesen, M., 1986, Diel changes in bacterial activity determined by means of microautoradiography, *Verh. Int. Verein. Limnol.* (in press).

Marsh, D. H., and Odum, W. E., 1979, Effect of suspension and sedimentation on the amount of microbial colonization of salt marsh microdetritus, *Estuaries* **2:**184–188.

Marshall, K. C., 1979, Growth at interfaces, in: *Strategies of Microbial Life in Extreme Environments* (M. Shilo, ed.), pp. 281–290, Verlag Chemie, New York.

McCambridge, J., and McMeekin, T. A., 1980, Relative effects of bacterial and protozoan predators on survival of *Escherichia coli* in estuarine water sample, *Appl. Environ. Microbiol.* **40:**907–911.

McDuff, R. E., and Chisholm, S. W., 1982, The calculation of *in situ* growth rates of phytoplankton populations from fractions of cells undergoing mitosis: A clarification, *Limnol. Oceanogr.* **27:**783–788.

Meyer-Reil, L. -A., 1977, Bacterial growth rates and biomass production, in: *Microbial Ecology of a Brackish Water Environment* (G. Rheinheimer, ed.), pp. 223–236, Springer-Verlag, New York.

Meyer-Reil, L. -A., 1978, Autoradiography and epifluorescence microscopy combined for the determination of number and spectrum of actively metabolizing bacteria in natural waters, *Appl. Environ. Microbiol.* **36:**506–512.

Mitskevich, I. N., and Kriss, A. E., 1982, Distribution of the number, biomass and production of microorganisms in the World Ocean, *Int. Rev. Ges. Hydrobiol.* **67:**433–458.

Montagna, P. A., 1982, Sampling design and enumeration statistics for bacteria extracted from marine sediments, *Appl. Environ. Microbiol.* **43:**1366–1372.

Moriarty, D. J. W., 1980, measurement of bacterial biomass in sandy sediments, in: *Biogeochemistry of Ancient and Modern Environments* (P. A. Trudinger, M. R. Walter, and B. J. Ralph, eds.), pp. 131–138, Australian Academy of Science, Canberra.

Moriarty, D. J. W., and Hayward, A. C., 1982, Ultrastructure of bacteria and the proportion of Gram-negative bacteria in marine sediments, *Microb. Ecol.* **8:**1–14.

Moriarty, D. J. W., and Pollard, P. C., 1982, Diel variation in bacterial productivity in seagrass beds, *Mar. Biol.* **72:**165–173.

Munro, A. L., and Brock, T. D., 1968, Distinction between bacterial and algal utilization of soluble substances in the sea, *J. Gen. Microbiol.* **51:**35–42.

Newell, R. C., Field, J. G., and Griffiths, C. L., 1982, Energy balance and significance of micro-organisms in a kelp bed community, *Mar. Ecol. Prog. Ser.* **8:**103–113.

Newell, S. Y., 1981, Fungi and bacteria in or on leaves of eelgrass (*Zostera marina* L.) from Chesapeake Bay, *Appl. Environ. Microbiol.* **41:**1219–1224.

Newell, S. Y., 1984, Modification of the gelatin–matrix method for enumeration of respiring bacterial cells, for use with salt-marsh water samples, *Appl. Environ. Microbiol.* **47:**873–875.

Newell, S. Y., and Christian, R. R., 1981, Frequency of dividing cells as an estimator of bacterial productivity, *Appl. Environ. Microbiol.* **42:**23–31.

Newell, S. Y., and Fallon, R. D., 1982, Bacterial productivity in the water column and sediments of the Georgia (USA) coastal zone: Estimates via direct counting and parallel measurement of thymidine incorporation, *Microb. Ecol.* **8:**33–46.

Newell, S. Y., and Hicks, R. E., 1982, Direct-count estimates of fungal and bacterial biovolume in dead leaves of smooth cordgrass (*Spartina alterniflora* Loisel.), *Estuaries* **5:**246–260.

Newell, S. Y., Sherr, B. F., Sherr, E. B., and Fallon, R. D., 1983, Bacterial response to presence of eukaryote inhibitors in water from a coastal marine environment, *Mar. Environ. Res.* **10:**147–157.

Nishio, M., 1983, Direct-count estimation of microbial biomass in soil applied with compost, *Biol. Agric. Hort.* **1:**109–125.

Novitsky, J. A., and Morita, R. Y., 1978, Possible strategy for the survival of marine bacteria under starvation conditions, *Mar. Biol.* **48:**289–295.

Oláh, J., 1974, Number, biomass and production of planktonic bacteria in the shallow Lake Balaton, *Arch. Hydrobiol.* **73:**193–217.

Orndorff, S. A., and Colwell, R. R., 1980, Effect of Kepone on estuarine microbiol activity, *Microb. Ecol.* **6:**357–368.

Orrison, L. H., Cherry, W. B., and Milan, D., 1981, Isolation of *Legionella pneumophila* from cooling tower by filtration, *Appl. Environ. Microbiol.* **41:**1202–1205.

Oviatt, C. C., 1981, Effects of different mixing schedules on phytoplankton, zooplankton and nutrients in marine microcosms, *Mar. Ecol. Progr. Ser.* **4:**57–67.

Owens, T. G., and King, F. D., 1975, The measurement of respiratory electron-transport system activity in marine zooplankton, *Mar. Biol.* **30:**27–36.

Packard, T. T., 1971, The measurement of respiratory electron transport activity in marine phytoplankton, *J. Mar. Res.* **29:**235–244.

Parsons, T. R., Albright, L. J., Whitney, F., Wong, C. S., and Williams. P. J. LeB., 1981, The effect of glucose on the productivity of seawater: An experimental approach using controlled aquatic ecosystems, *Mar. Environ. Res.* **4:**229–242.

Paton, A. M., 1982, Light-microscopic techniques for the microbiological examination of plant materials, in: *Bacteria and Plants* (M. E. Rhodes-Roberts and F. A. Skinner, eds.), pp. 235–243, Academic Press, New York.

Paul, J. H., 1982, Use of Hoechst dyes 33258 and 33342 for enumeration of attached and planktonic bacteria, *Appl. Environ. Microbiol.* **43:**939–944.

Paul, J. H., and Jeffrey, W. H., 1984, Measurement of diameters of estuarine bacteria and particulates in natural water samples by use of a submicron particle analyzer, *Curr. Microbiol.* **10:**7–12.

Pedrós-Alió, C., and Brock, T. D., 1982, Assessing biomass and production of bacteria in eutrophic Lake Mendota, Wisconsin, *Appl. Environ. Microbiol.* **44:**203–218.

Peele, E. E., and Colwell, R. R., 1981, Application of a direct microscopic method for enumeration of substrate-responsive marine bacteria, *Can. J. Microbiol.* **27:**1071–1075.

Peterson, B. J., Hobbie, J. E., and Haney, J. F., 1978, *Daphnia* grazing on natural bacteria, *Limnol. Oceanogr.* **23:**1039–1044.

Pettipher, G. L., and Rodrigues, U. M., 1982, Rapid enumeration of microorganisms in foods by the direct epifluorescent filter technique, *Appl. Environ. Microbiol.* **44:**809–813.

Pomeroy, L. R., Atkinson, L. P., Blanton, J. O., Campbell, W. B., Jacobsen, T. R., Kerrick, K. H., and Wood, A. M., 1983, Microbial distribution and abundance in response to physical and biological processes on the continental shelf of southeastern U.S.A., *Cont. Shelf Res.* **2:**1–20.

Pomroy, A. J., 1984, Direct counting of bacteria preserved with Lugol iodine solution, *Appl. Environ. Microbiol.* **47:**1191–1192.

Porter, K. G., and Feig, Y. S., 1980, The use of DAPI for identifying and counting aquatic microflora, *Limnol. Oceanogr.* **25:**943–948.

Pugsley, A. P., and Evison, L. M., 1974, A membrane filtration staining technique for detection of viable bacteria in water, *Water Treat. Exam.* **23:**205–214.

Quesnel, L. B., 1971, Microscopy and micrometry, in: *Methods in Microbiology,* Vol. 5A (J. R. Norris and D. W. Ribbons, eds.), pp. 1–103, Academic Press, New York.

Ramsay, A. J., 1978, Direct counts of bacteria by a modified acridine orange method in relation to their heterotrophic activity, *N.Z. J. Mar. Freshwater Res.* **12:**265–269.

Ramsay, A. J., 1984, Extraction of bacteria from soil: Efficiency of shaking or ultrasonication as indicated by direct counts and autoradiography, *Soil Biol. Biochem.* **16:**475–481.

Reed, W. M., and Dugan, P. R., 1978, Distribution of *Methylomonas methanica* and *Methylosinus trichosporium* in Cleveland Harbor as determined by an indirect fluorescent antibody membrane filter technique, *Appl. Environ. Microbiol.* **35:**422–430.

Reed, W. M., and Dugan, P. R., 1979, Study of developmental stages of *Methylosinus trichosporium* with the aid of fluorescent-antibody staining techniques, *Appl. Environ. Microbiol.* **38:**1179–1183.

Rennie, R. J., Reyes, U. G., and Schmidt, E. L., 1977, Immunofluorescence detection of the effects of wheat and soybean roots on *Nitrobacter* in soil, *Soil Sci.* **124:**10–15.

Riemann, B., Nielsen, P., Jeppesen, M., Marcussen, B., and Fuhrman, J. A., 1984, Diel changes of bacterial biomass and growth rates in coastal environments determined by means of thymidine incorporation into DNA, frequency of dividing cells (FDC), and microautoradiography, *Mar. Ecol. Prog. Ser.* **17:**227–235.

Robarts, R. D., and Sephton, L. M., 1981, The enumeration of aquatic bacteria using DAPI, *J. Limnol. Soc. S. Afr.* **7:**72–74.

Robertson, M. L., Mills, A. L., and Zieman, J. C., 1982, Microbial synthesis of detritus-like particulates from dissolved organic carbon released by tropical seagrasses, *Mar. Ecol. Prog. Ser.* **7:**279–285.

Robinson, J. D., Mann, K. H., and Novitsky, J. A., 1982, Conversion of the particulate fraction of seaweed detritus to bacterial biomass, *Limnol. Oceanogr.* **27:**1072–1079.

Roser, D. J., 1980, Ethidium bromide: A general purpose fluorescent stain for nucleic acid in bacteria and eucaryotes and its use in microbial ecology studies, *Soil Biol. Biochem.* **12:**329–336.

Roser, D. J., Keane, P. J., and Pittaway, P. A., 1982, Fluorescent staining of fungi from soil and plant tissues with ethidium bromide, *Trans. Br. Mycol. Soc.* **79:**321–329.

Roser, D., Nedwell, D. B., and Gordon, A., 1984, A note on "plotless" methods for estimating bacterial cell densities, *J. Appl. Bacteriol.* **56:**343–347.

Rublee, P. A., 1982, Bacteria and microbial distribution in estuarine sediments, in: *Estuarine Comparisons* (V. S. Kennedy, ed.), pp. 159–182, Academic Press, New York.

Rublee, P. A., and Dornseif, B. E., 1978, Direct counts of bacteria in the sediments of a North Carolina salt marsh, *Estuaries* **1:**188–191.

Ryther, J. H., and Sanders, J. G., 1980, Experimental evidence of zooplankton control of the species composition and size distribution of marine phytoplankton, *Mar. Ecol. Prog. Ser.* **3:**279–283.

Salonen, K., 1977, The estimation of bacterioplankton numbers and biomass by phase contrast microscopy, *Ann. Bot. Fenn.* **14:**25–28.

Schank, S. C., Smith, R. L., Weiser, G. C., Zuberer, D. A., Bouton, J. H., Quesenberry, K. H., Tyler, M. E., Milano, J. R., and Littell, R. C., 1979, Fluorescent antibody technique to identify *Azospirillum brasiliense* associated with roots of grasses, *Soil Biol. Biochem.* **11:**287–295.

Schiemer, F., 1982, Food dependence and energetics of freeliving nematodes. II. Life history parameters of *Caenorhabditis briggsae* (Nematoda) at different levels of food supply, *Oecologia* **54:**122–128.

Schmidt, E. L., 1973, Fluorescent antibody techniques for the study of microbial ecology, *Bull. Ecol. Res. Comm.* **17:**67–76.

Schmidt, E. L., and Paul, E. A., 1982, Microscopic methods for soil organisms, in: *Methods of Soil Analysis*. Part 2. *Chemical and Microbiological Properties*, 2nd ed. (A. L. Page, ed.), pp. 803–814, American Society of Agronomy/Soil Sciences Society of America, Madison, Wisconsin.

Schmidt, E. L., Bankole, R. O., and Bohlool, B. B., 1968, Fluorescent-antibody approach to study of rhizobia in soil, *J. Bacteriol.* **95:**1987–1992.

Seki, H., and Nakano, H., 1981, Production of bacterioplankton with special reference to dynamics of dissolved organic matter in a hypereutrophic lake, *Kiel. Meeresforsch. Sonderh.* **5:**408–415.

Shehata, T. E., and Marr, A. G., 1971, Effect of nutrient concentration on the growth of *Escherichia coli, J. Bacteriol.* **107:**210–216.

Sieburth, J. McN., 1979, *Sea Microbes*, Oxford University Press, New York.

Sieburth, J. McN., and Davis, P. G., 1982, The role of heterotrophic nanoplankton in the grazing and nuturing of planktonic bacteria in the Sargasso and Caribbean Sea, *Ann. Inst. Océanogr. (Suppl.)* **58:**285–296.

Sokal, R. R., and Rohlf, R. J., 1981, *Biometry. The Principles and Practice of Statistics in Biological Research*, 2nd ed., W. H. Freeman, San Francisco.

Stanley, P. M., and Staley, J. T., 1977, Acetate uptake by aquatic bacterial communities measured by autoradiography and filterable radioactivity, *Limnol. Oceanogr.* **22:**26–37.

Straškrabová, V., and Fuksa, J., 1982, Diel changes in numbers and activities of bacterioplankton in a reservoir in relation to algal production, *Limnol. Oceanogr.* **27:**660–672.

Straškrabová, V., and Komárková, J., 1979, Seasonal changes of bacterioplankton in a reservoir related to algae. I. Numbers and biomass, *Int. Rev. Ges. Hydrobiol.* **64:**285–302.

Strayer, R. I., and Tiedje, J. M., 1978, Application of the fluorescent-antibody technique to the study of a methanoginic bacterium in lake sediments, *Appl. Environ. Microbiol.* **35:**192–198.

Strugger, S., 1948, Fluorescence microscope examinations of bacteria in soil, *Can. J. Res. Ser. C* **26:**188–193.

Swaminathan, B., Ayres, J. C., and Williams, J. E., 1978, Control of nonspecific staining in the fluorescent antibody technique for the detection of *Salmonellae* in foods, *Appl. Environ. Microbiol.* **35:**911–919.

Swift, S. T., Najita, I. Y., Ohtaguchi, K., and Fredrickson, A. G., 1982, Some physiological aspects of the autecology of the suspension-feeding protozoan *Tetrahymena pyriformis, Microb. Ecol.* **8:**201–215.

Tabor, P. S., and Neihof, R. A., 1982a, Improved method for determination of respiring individual microorganisms in natural waters, *Appl. Environ. Microbiol.* **43:**1249–1255.

Tabor, P. S., and Neihof, R. A., 1982b, Improved microautoradiographic method to determine individual microorganisms active in substrate uptake in natural waters, *Appl. Environ. Microbiol.* **44:**945–953.

Tabor, P. S., and Neihof, R. A., 1984, Direct determination of activities for microorganisms of Cheasapeake Bay populations, *Appl. Environ. Microbiol.* **48:**1012–1019.

Tabor, P. S., Ohwada, K., and Colwell, R. R., 1981, Filterable marine bacteria found in the deep sea: Distribution, taxonomy, and response to starvation, *Microb. Ecol.* **7:**67–83.

Taylor, W. D., 1978, Growth responses of ciliate protozoa to the abundance of their bacterial prey, *Microb. Ecol.* **4**:207–214.

Tison, D. L., Pope, D. H., Cherry, W. B., and Fliermans, C. B., 1980, Growth of *Legionella pneumophila* in association with blue-green algae (cyanobacteria), *Appl. Environ. Microbiol.* **39**:456–459.

Trevors, J. T., 1984, Electron transport system activity in soil, sediment, and pure cultures, *CRC Crit. Rev. Microbiol.* **11**:83–100.

Trolldenier, G., 1973, the use of fluorescence microscopy for counting soil organisms, *Bull. Ecol. Res. Comm. (Stockh.)* **17**:53–59.

Trueba, F. J., Neijssel, and Woldringh, C. L., 1982*a*, Generality of the growth kinetics of the average individual cell in different bacterial populations, *J. Bacteriol.* **150**:1048–1055.

Trueba, F. J., van Spronsen, E. A., Traas, J., and Woldringh, C. L., 1982*b*, Effects of temperature on the size and shape of *Escherichia coli* cells, *Arch. Microbiol.* **131**:235–240.

Van Veen, J. A., and Paul, E. A., 1979, Conversion of biovolume measurements of soil organisms, grown under various moisture tensions, to biomass, and their nutrient content, *Appl. Environ. Microbiol.* **37**:686–692.

Van Verseveld, H. W., Chesbro, W. E., Braster, M., and Stouthamer, A. H., 1984, Eubacteria have 3 growth modes keyed to nutrient flow. Consequences for the concept of maintenance and maximal growth yield, *Arch. Microbiol.* **137**:176–184.

Velji, M. I., and Albright, L. J., 1986, Microscopic enumeration of attached marine bacteria of seawater, marine sediment, fecal matter, and kelp blade samples following pyrophosphate and ultrasound treatments, *Can. J. Microbiol.* **32**:121–126.

Vestal, J. R., Lundgren, D. G., and Milner, K. C., 1973, Toxic and immunological differences among lipopolysaccharides from *Thiobacillus ferroxidans* grown autotrophically and heterotrophically, *Can. J. Microbiol.* **19**:1335–1339.

Vyshkvartsev, D. I., 1980, Bacterioplankton in shallow inlets of Posyeta Bay, *Microbiology (USSR)* **48**:603–609.

Ward, B. B., 1982, Oceanic distribution of ammonium oxidizing bacteria determined by immunofluorescent assay, *J. Mar. Res.* **40**:1155–1172.

Ward, B. B., Olson, R. J., and Perry, M. J., 1982, Microbial nitrification rates in the primary nitrite maximum off southern California, *Deep-Sea Res.* **29**: 247–255.

Weiler, C. S., and Eppley, R. W., 1979, Temporal pattern of division in the dinoflagellate genus *Ceratium* and its application to the determination of growth rate, *J. Exp. Mar. Biol. Ecol.* **39**:1–24.

Wilde, E. W., and Fliermans, C. B., 1979, Fluorescence microscopy for algal studies, *Trans. Am. Microsc. Soc.* **98**: 96–102.

Williams, P. J. LeB., 1984, Bacterial production in the marine food chain: the emperor's new suit of clothes?, in: *Flows of Energy and Materials in Marine Ecosystems*, Theory and Practice (M. J., Fasham, ed.), pp. 271–299, Plenum Press, New York.

Wilson, C. A., and Stevenson, L. H., 1980, The dynamics of the bacterial population associated with a salt marsh, *J. Exp. Mar. Biol. Ecol.* **48**:123–138.

Wright, R. T., 1978, Measurement and significance of specific activity in the heterotrophic bacteria of natural waters, *Appl. Environ. Microbiol.* **36**:297–305.

Wright, R. T., and Coffin, R. B., 1984, Measuring microzooplankton grazing on planktonic marine bacteria by its impact on bacterial production, *Microb. Ecol.* **10**:137–149.

Wright, R. T., and Hobbie, J. E., 1966, Use of glucose and acetate by bacteria and algae in aquatic ecosystems, *Ecology* **47**:447–464.

Wright, R. T., Coffin, R. B., Ersing, C. P., and Pearson, D., 1982, Field and laboratory measurements of bivalve filtration of natural marine bacterioplankton, *Limnol. Oceanogr.* **27**:91–98.

Xu, H. -S., Roberts, N., Singleton, F. L., Attwell, R. W., Grimes, D. J., and Colwell, R. R., 1982, Survival and viability of nonculturable *Escherichia coli* and *Vibrio cholerae* in the estuarine and marine environment, *Microb. Ecol.* **8**:313–323.

Zimmerman, R., and Meyer-Reil, L. -A., 1974, A new method for fluorescence staining of bacterial populations on membrane filters, *Kiel. Meeresforsch.* **30:**24–27.

Zimmerman, R., Iturriaga, R., and Becker-Birck, J., 1978, Simultaneous determination of the total number of aquatic bacteria and the number thereof involved in respiration, *Appl. Environ. Microbiol.* **36:**926–935.

Zobell, C. E., 1946, *Marine Microbiology,* Chronica Botanica Press, Waltham, Massachusetts.

SUITABLE METHODS FOR THE COMPARATIVE STUDY OF FREE-LIVING AND SURFACE-ASSOCIATED BACTERIAL POPULATIONS

J. W. Costerton, J. C. Nickel, and T. I. Ladd

INTRODUCTION

Direct observation of bacteria in a very large number of natural and pathogenic ecosystems has clearly demonstrated that most of these organisms grow in sessile matrix-enclosed biofilms on available surfaces (Costerton *et al.*, 1981*a*). A much smaller number of planktonic bacterial cells are periodically released from these biofilms; these are the swimming and floating forms that we capture in fluid "grab" samples (Geesey *et al.*, 1978). When the organisms in these fluid samples are separated by plating techniques and inoculated into media to produce monospecies cultures, we obtain the conventional derived "pure cultures" that have been used successfully since Pasteur's time to identify and study the bacteria present in the original sampled system. Within these cultures, the bacteria are provided with optimal nutrition and are not subjected to challenge by any antibacterial factors, so that the fast-growing planktonic growth phase of each organism is heavily favored and soon predominates (Chan *et al.*, 1984). These pure cultures are ideally suited to species identification and for metabolic and genetic studies because free-living cells are randomly distributed in the fluid, subculture and quantitation are facilitated,

J. W. Costerton • Department of Biology, University of Calgary. Calgary, Alberta, Canada T2N 1N4.   *J. C. Nickel* • Department of Urology, Queens University, Kingston, Ontario, Canada K7L 2V7.   *T. I. Ladd* • Department of Biology, St. Mary's University, Halifax, Nova Scotia, Canada B3H 3C3.

and extrapolation from aliquots data is valid. Data generated from pure cultures are assuredly accurate and very useful, but we must examine what it means in relationship to the microbial ecology of a whole system, because planktonic bacteria represent <0.1% of the total bacterial population of most of the ecosystems that have been quantitatively examined (Costerton *et al.*, 1981*b*).

The first problem encountered in the "pure culture" approach to microbial ecology is that of sampling. The planktonic population consists of those cells that have been released from the larger sessile population at a moment in time. No one has suggested that this release is representative of all species in the established consortia of a natural biofilm or that it is quantitatively proportional to the numerical size of the sessile population. Many very avidly adherent microorganisms are not even removed from colonized surfaces by vigorous swabbing (McCowan *et al.*, 1979), and certainly not by spontaneous detachment into the fluid phase.

The second and greatest problem concerns major differences between the planktonic and sessile growth phases of the same organism. Whittenbury and Dow (1977) showed that planktonic cells differ profoundly in their metabolism from sessile cells of the same species. However, the major practical differences concern the relative accessibility to nutrients and to antibacterial agents of free-living planktonic cells in fluids versus sessile cells encased in their polyanionic extracellular matrices and organized into complex functional consortia. Clearly, the menstruum immediately surrounding a biofilm cell is radically different from that surrounding a free-living cell; these differences affect most of the important parameters of bacterial survival in an ecosystem: availability of nutrients, permissive chemical and physical conditions, and protection from antibacterial agents. Thus, whereas pure culture data may suggest that a toxin A-producing organism that is susceptible to tobramycin is present in a given system, the organism may be only a minor transient species in the ecosystem in question; and it may be highly protected from tobramycin by its growth in a biofilm, as well as severely limited in its release of toxin A. Clearly, pure culture data are often correct and valuable, as long as we remember what they mean. However, having determined that sessile biofilm bacteria predominate in natural and pathogenic ecosystems, it is incumbent upon us to develop new methods for the study of sessile bacteria actually growing in biofilms.

DIRECT EXAMINATION BY TRANSMISSION ELECTRON MICROSCOPY

Transmission electron microscopy (TEM) of sections of bacterial biofilms is very informative because it yields data on biofilm thickness; it also yields morphological data on the biofilm organisms themselves (Fig. 1) and on the structural details of their juxtaposition in consortia. If the colonized substra-

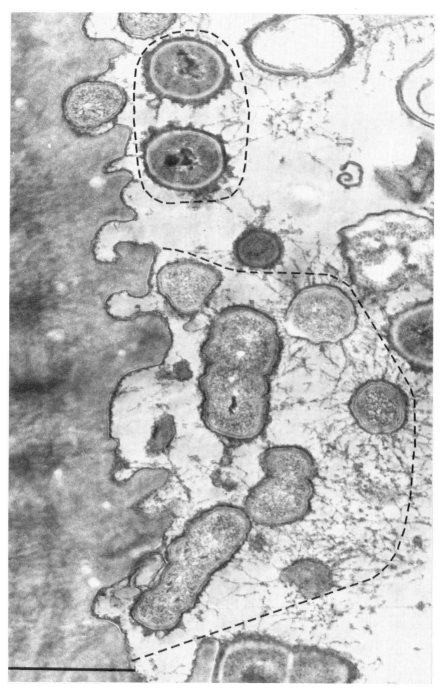

FIGURE 1.   TEM of a section of a ruthenium red-stained preparation of the bovine rumen. This squamous epithelium is colonized by adhering bacteria that form microcolonies (----) and consortia on the tissue surface. The fibrous dehydration-condensed glycocalyces of these cells are clearly seen, and their cell wall type can be determined morphologically. Top microcolony, gram-positive; thick peptidoglycan, no outer membrane; lower microcolony, gram-negative; thin peptidoglycan plus an outer membrane. Bar: 1.0 μm.

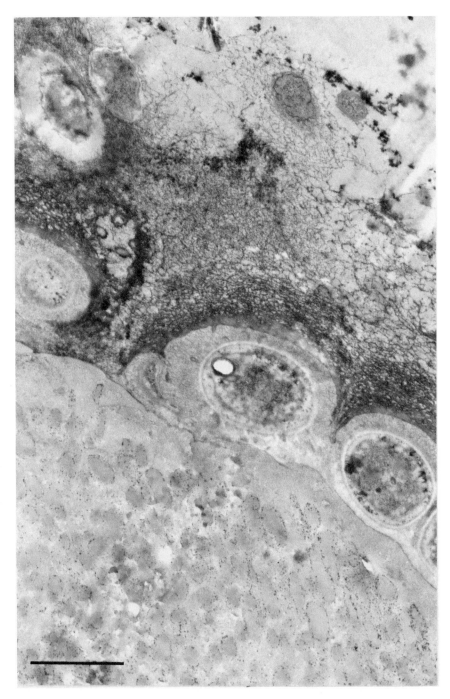

FIGURE 2. TEM of a section of a ruthenium red-stained preparation of a plastic substratum that had been immersed in the Athabasca River for 3 weeks. This plastic substratum has been colonized by blue-green bacteria that adhere by means of dense, highly structured glycocalyces, and a biofilm has developed that is composed of bacterial cells and their extensive fibrous glycocalyces. Bar: 1.0 μm.

tum is soft and amenable to sectioning (e.g., tissue), as in Fig. 1, the relationship of the biofilm to this substratum is easily preserved. In cases in which the substratum is hard, two alternatives are useful: (1) a substitute substratum can be immersed in the system and the biofilm formed thereon (Fig. 2) assumed to be equivalent to that formed on the hard surface, or (2) a plasticene "dam" can be built on the colonized surface and fixation, dehydration, and plastic embedding carried out within this dam to produce a plastic aggregate that can later be removed from the surface by cold shock (liquid nitrogen).

Two major technical difficulties limited our ability to visualize the bacterial glycocalyx matrix of bacterial biofilms throughout the first three decades of biological TEM (Costerton, 1979). The first was the lack of affinity of conventional electron microscopy stains for the anionic glycocalyx polysaccharides that constitute the matrix of the biofilm; the second problem was the radical condensation of this extracellular glycocalyx material, which is 99% water (Sutherland, 1977), during the dehydration necessary in electron microscopy. This unfortunate combination of difficulties produced an "empty" space surrounding bacterial cells (Fig. 3), where we now know that their surface components should have been visualized.

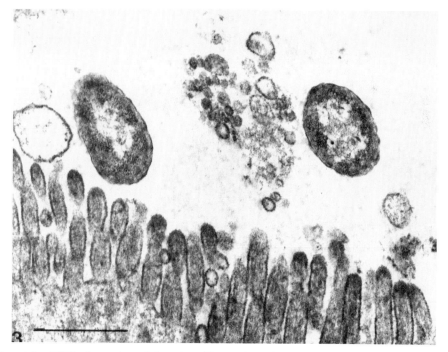

FIGURE 3. TEM of a section of the microvillar surface of the ileum of a calf infected with an enterotoxigenic strain of *E. coli* (ETEC) to cause experimental "scours." Because neither ruthenium red nor any of the modern "stabilization" techniques was used, no extracellular bacterial structures are seen and the bacteria appear to be surrounded by an "empty" space. Bar: 1 μm.

The first problem was solved by the use of ruthenium red (Luft, 1971), which is electron dense and has a marked specificity for polyanions such as glycocalyx polysaccharides. The use of this stain enabled us to see the electron-dense residues of the dehydration-condensed bacterial glycocalyces of adhering bacteria (Fig. 4), and we now know that this radical condensation deposits these residues on available surfaces, such as the bacterial cell surface and the pili, which may be serendipitously outlined (Chan *et al.*, 1983) by condensed electron-dense glycocalyx material (Fig. 4). The second problem was solved by the use of bivalent specific antibodies (Bayer and Thurow, 1977) and lectins (Birdsell *et al.*, 1975) that cross-link the fibrous elements of the bacterial glycocalyx, when it is in its natural hydrated state, and hold the biofilm matrix in approximately its natural configuration (Fig. 5) against the condensation forces of dehydration (Chan *et al.*, 1982). This antibody stabilization technique has been used to demonstrate the real spatial extent of the glycocalyces of cells in pure culture (Mackie *et al.*, 1979; Mayberry-Carson *et al.*, 1984), but it is especially useful when employed to stabilize the matrices of mixed biofilms of unknown cellular composition because positive stabilization constitutes a tentative immunological identification of the organisms growing therein (Fig. 6). A modification of this technique can be used to

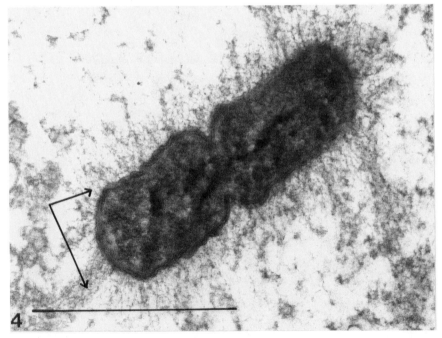

FIGURE 4. TEM of a section of a ruthenium red-stained preparation of the ileum of an ETEC-infected calf. The bacterial glycocalyx, which is radically condensed by dehydration, has collapsed onto the bacterial cell surface and onto its numerous K99 pili (arrows) so that the pili are clearly visualized as a consequence of this accretion. Bar: 1 μm.

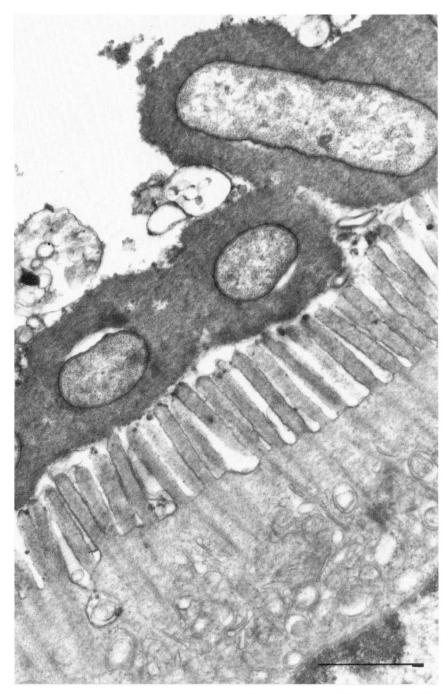

FIGURE 5. TEM of a section of a preparation of the ileum of an ETEC-infected calf "stabilized" with anticapsular (K30) antiserum and stained with ruthenium red. The thick coherent capsular glycocalyces of the infecting cells are clearly seen to make direct contact with the microvillar surface of the infected tissue. Bar: 1 μm.

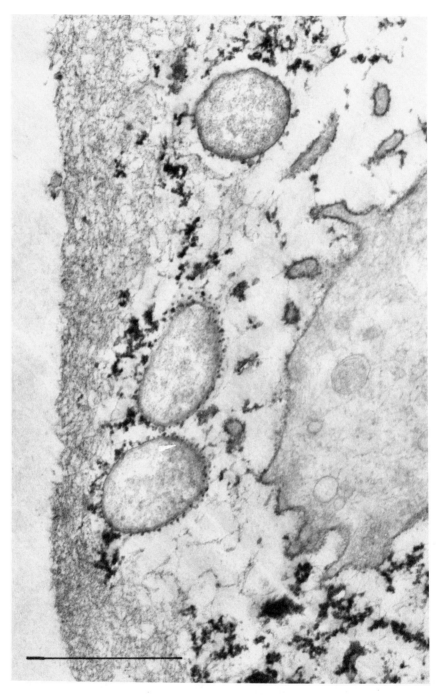

FIGURE 6.   TEM of a section of a preparation of sediment from an infected bladder (UTI) that had been stabilized with the patient's own serum and stained with ruthenium red. Gram-negative bacilli are seen within a biofilm on the surface of a sloughed epithelial cell, and the outer elements of their fibrous glycocalyces are stabilized; however, the antibodies have failed to penetrate and stabilize the inner elements of this structure during dehydration. Bar: 1 μm.

visualize pili, which normally are too thin to be resolved in TEM of sectioned material, but which can be thickened to facilitate their visualization (Fig. 7) by reaction with monoclonal antipilus antibodies (Chan *et al.*, 1982).

These new TEM preparative techniques permit definition of surface structures associated with bacteria on colonized substrata and on infected tissues, and they should be employed before extrapolation from "pure culture" data is attempted. It is now especially apparent that bacterial surface components produced in pure cultures may not even be present when the bacteria are growing under nutrient stress (e.g., iron limitation) and under attack by environmental antibacterial factors (Brown *et al.*, 1984). Accordingly, before a particular bacterial surface component is assigned a role in pathogenesis or designated as a potential antigen for vaccine production, it is imperative that its actual production *in situ* be established by direct biochemical study of recovered bacterial cells (Brown *et al.*, 1984) or by its demonstration *in situ* using specific antibodies or lectins (Figs. 5–7).

DIRECT EXAMINATION BY SCANNING ELECTRON MICROSCOPY

Scanning electron microscopy (SEM) produces its image by the differential analysis of backscattered and secondarily emitted electrons from the specimen. It provides a very useful topographic view of large areas of the surface of a specimen, but it requires that the specimen be conductive (essentially "metalized"), and the electron beam penetrates only 1–3 μm and provides very little information regarding subsurface structures. Once the surface of a bacterial biofilm has been made conductive, that exopolysaccharide surface constitutes the specimen surface; bacterial cells are seen only when they deform the biofilm or protrude when the polysaccharide matrix has been radically condensed during dehydration (Figs. 8–10). Even with these limitations, SEM is a very useful technique for the examination of the colonization of surfaces by bacteria and for the examination of biofilm development by these adhering microorganisms.

The two major problems in the use of SEM to examine bacterial biofilms are dehydration damage during preparation and retention of the biofilm on the colonized substratum. The dehydration necessary for SEM is equivalent to that necessary for TEM, so that hydrated glycocalyx structures are condensed to about 1% of their original volume. This condensation is not usually prevented in SEM because it enables us to see bacterial cells within the biofilm when the matrix collapses (Fig. 11). To avoid severe dehydration damage that distorts bacterial cell shape, simple air drying is usually replaced by the critical point drying technique (Cohen *et al.*, 1968), which exerts much less drying force. The conventional method of rendering the specimen conductive is to spray it with fine metal particles in a vacuum until a conductive metal layer covers the entire specimen. However, this metal coating technique has an inevitable "snow drift" effect that obscures fine detail, and we now obtain

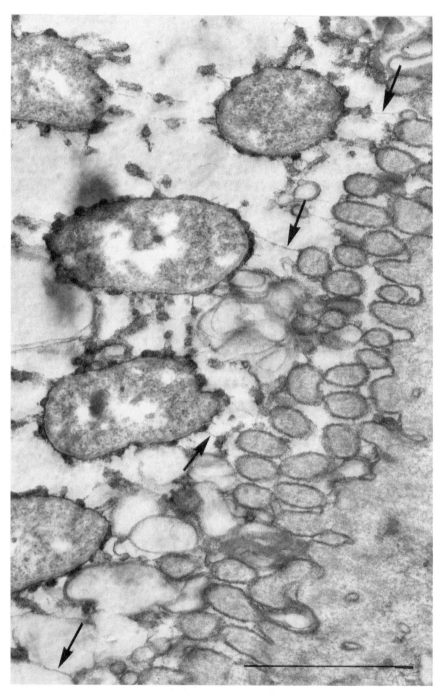

FIGURE 7. TEM of a section of a preparation of ETEC-infected calf ileum (as in Figs. 3–5) that had been reacted with monoclonal anti-K99 antibodies and stained with ruthenium red. The stained glycocalyx is condensed onto the bacterial cell surface, and the very thin K99 pili are sufficiently thickened by antibody accretion to make them clearly visible (arrows) in this sectioned material. Bar: 1 μm.

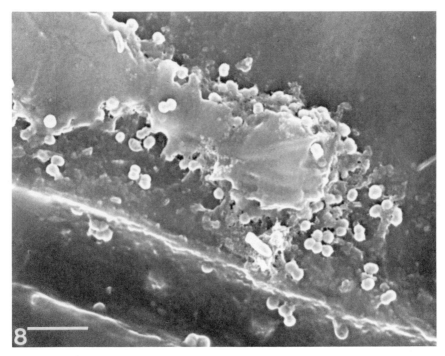

FIGURE 8.   SEM of a TCH-metalized, critical point-dried plastic surface bearing a bacterial biofilm in the early stages of its formation. Coccoid cells at the periphery of the colonized area are clearly visible, but some cells are already buried in their accumulated glycocalyx material in the center of the developing biofilm. Bar: 5 μm.

much better specimen resolution by internally metalizing the specimen by reacting it with OsO_4 and then by linking more osmium into the specimen by the use of a thiocarbohydrazide (TCH) compound (Malick and Wilson, 1975). This internal metalization technique allows exquisite resolution of biological structures (Figs. 14–19, below), but it can only be used on continuous biofilms and on tissues and cannot be used to visualize intermittent microcolonies on a nonconductive substratum (Fig. 11) because severe local "charging" occurs.

The most serious problem in SEM examination of bacterial biofilms is the loss of these structures from the substratum during the many washes and manipulations involved in the complex preparative procedures. Obviously, any bacterial cell, microcolony, or biolfilm still adhering to the substratum after the 20–24 washes typically used during preparation for SEM was truly adherent at the outset. The coherence and adherence of bacterial biofilms is proportional to the shear forces operative during their formation (Bryers and Characklis, 1981), and we can expect high-shear biofilms, such as dental plaque and pipeline biofilms, to be very resistant to loss during processing. However, biofilms formed on substrate in stationary culture vessels and biofilms formed on the surface of mucus-covered tissues are notoriously subject to mechanical

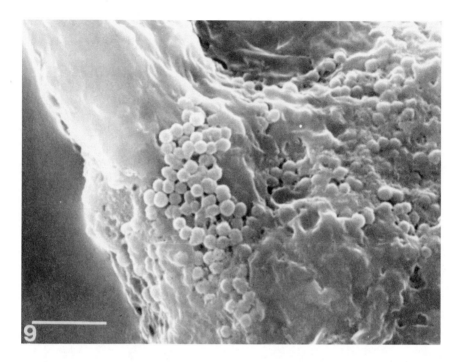

FIGURE 9.   SEM of the same surface shown in Fig. 8 at a later stage of biofilm formation, showing that most bacterial cells are now "buried" in their accumulated glycocalyces forming the amorphous surface of the metalized biofilm. Bar: 5 μm.

removal during processing for SEM. This loss of the biofilm became apparent when we examined the surface of vaginal tissues, known from light microscopy and culture data to be very heavily colonized and found only the pristine surface of a squamous epithelium (Fig. 12). We therefore developed (Rozee *et al.*, 1982) methods for the retention of the mucus layer on tissues during processing for SEM, using an antibody stabilization technique and very gentle handling; elements of these techniques can also be used to retain soft, weakly adherent biofilms developed in low-shear systems. The essential element in monitoring this procedure is continuous observation by low-power light microscopy because the loss of segments of the biofilm is a dramatic and unmistakable event.

In the intestine, a small proportion of the indigenous microorganisms are actually embedded in or firmly adherent to the microvillar surface of the tissue (Davis and Savage, 1974), and these organisms are very clearly seen after the removal of the mucus layer (Figs. 15 and 18, below). However, the great majority of autochthonous intestinal organisms (Cheng *et al.*, 1981) live in the 400-μm thick mucus layer (Rozee *et al.*, 1982) where they demonstrate a pronounced chemotactic attraction to mucus components (Freter *et al.*, 1981). When the mucus is retained intact on the intestinal surface (Fig. 13), the villi

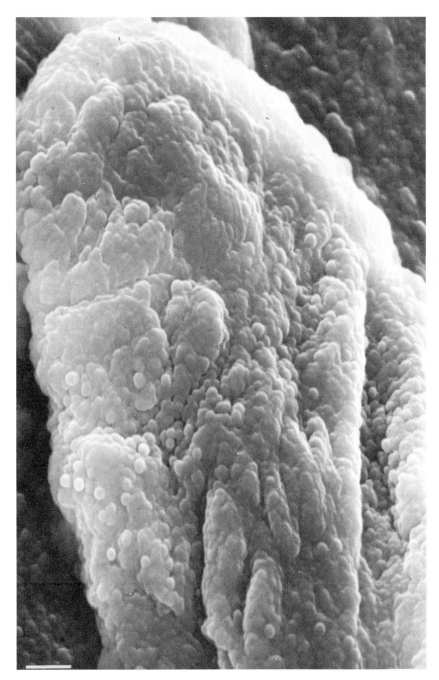

FIGURE 10.   SEM of a bacterial biofilm that developed on the luminal surface of an endotracheal tube. Bacterial cell profiles (largely coccoid) can be distinguished where biofilm cells have deformed the surface of the biofilm, and it is clear that this large pendulous mass is composed largely of bacterial cells and of their extracellular products. Bar: 5 μm.

FIGURE 11. SEM of a critical point-dried preparation of a plastic portion of an endocardial pacemaker colonized by cells of *Staphylococcus aureus*. The coccoid cells of this adherent microcolony are clearly demonstrated by the dehydration collapse of their glycocalyces to form amorphous masses on the substratum around the microcolony. Bar: 5 μm.

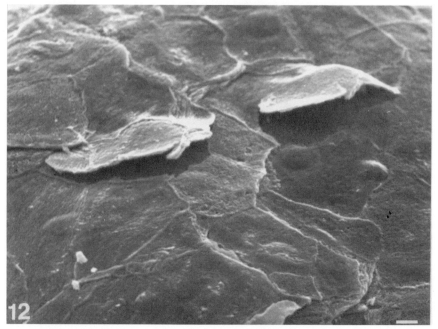

FIGURE 12. SEM of the surface of the squamous epithelial tissue of the mouse vagina. This tissue is known to be covered by mucus and to be heavily colonized by bacteria; however, because of insufficiently careful handling, the loss of the mucus layer in this preparation resulted in the loss of the bacterial population. Bar: 5 μm.

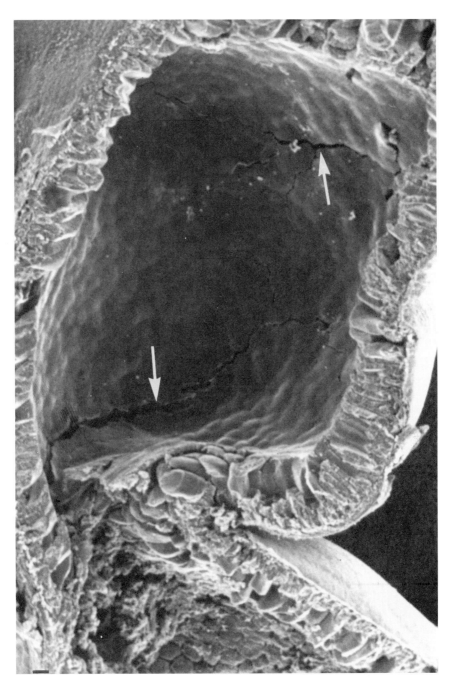

FIGURE 13. SEM of the surface of a preparation of mouse ileum that had been stabilized with antimucus antibodies and handled very carefully in order to retain this mucus layer (Rozee *et al.*, 1982). This layer was retained in most areas of the tissue (but not retained in the lower section of the photograph), where it is seen to be partially deformed by the tips of the villi and to be occasionally cracked (arrows) by dehydration. Bar: 50 μm.

are visible only because they deform the thick mucus layer; detailed examination of the surface of the mucus layer reveals only a few bacteria embedded in an amorphous matrix of mucus and glycocalyx components. However, cracks formed within the mucus layer (Fig. 13, arrows) during dehydration permit examination of the rich and varied microbial population that lives within the mucus "blanket" (Fig. 14). Where the mucus has been lost during processing, firmly tissue-adhering bacteria and protozoa (Fig. 15) are seen. However, the vast majority of intestinal bacteria and protozoa live in the mucus layer (Fig. 16), and most of the microvillar surface of the intestinal tissue of healthy animals is not colonized by bacteria. Exceptions are seen in pathogenic states, and intestinal disease can be induced by using lectins (Banwell *et al.,* 1985) to bind mucus-dwelling autochthonous intestinal bacteria (Fig. 17) and protozoa (Fig. 18) to specific regions of the microvillar surface.

SEM can be useful in the direct examination of bacterial biofilms on colonized substrata, but we must guard against biofilm loss during processing, using methods detailed in Rozee *et al.,* (1982) and Banwell *et al.* (1985), and rationalize the image according to the preparative method. For example, the "pits" surrounding the adhering cells seen in Fig. 19 would seem to indicate digestion of the substratum, but in reality, the cell is surrounded by the craterlike condensed residue of its previously hydrated glycocalyx, and the substratum (glass) is actually entirely unaffected. Similarly, the projections linking the adherent cells in Fig. 20 to the substratum might seem to be pili but this organism is not piliated; these structures are actually thin linear residues of the large mass of hydrated glycocalyx that linked the cell to the substratum in its natural hydrated state.

SAMPLING OF BIOFILM BACTERIA

Bacterial biofilms shed planktonic cells into surrounding fluids, but this shedding is neither sufficiently regular nor sufficiently representative of the species present in the biofilm to allow us to extrapolate from planktonic cell-count data to quantitative data on biofilm populations. However, if many planktonic and sessile data are obtained from a particular system (Geesey *et al.,* 1978), a parallel relationship between these populations can often be detected.

There are two major problems that affect the direct sampling and quantitation of biofilm populations because the biofilm must be both completely removed from the colonized surface and completely disrupted, without killing significant numbers of sessile bacteria (Gessey *et al.,* 1978). Initially, the colonized surface should be rinsed lightly to remove nonadhering planktonic bacteria; a designated area (usually 4 cm^2) should then be scraped with a sterile blade and irrigated with sterile buffer to remove the adherent biofilm. The coherent flakes of biofilm recovered by this technique are then subjected to mechanical shear forces, using a vortex mixer and a low-output ultrasonic

FIGURE 14. SEM showing the detail of the mucus blanket of the mouse ileum as seen in a crack (Fig. 13, arrows) caused by dehydration. A rich variety of bacteria and protozoa are seen to occupy the mucus blanket from its outer surface (B) to the microvillar surface of the ileal tissue (T). Bar: 5 μm.

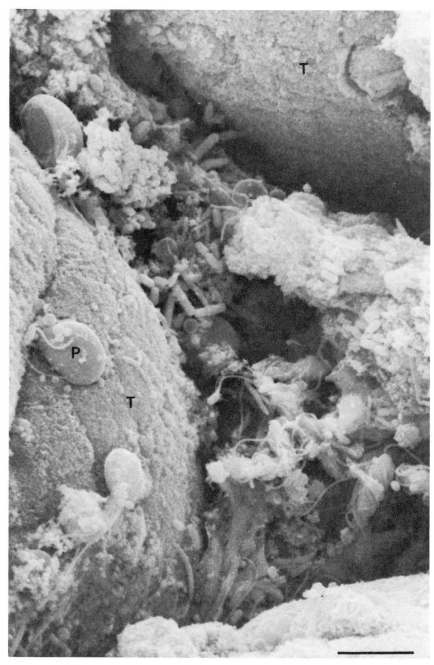

FIGURE 15.   SEM of the mouse ileum after the removal of most of the mucus blanket. Protozoa (P) adhere to the tissue (T), on which they leave perceptible "scars," but most of the autochthonous protozoa and bacteria are actually seen in remnants of the mucus layer between the villi. Bar: 5 μm.

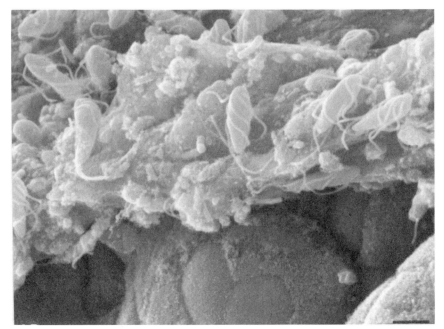

FIGURE 16. In the normal mouse ileum, the microvillar surface of the tissue is only very sparsely colonized by bacteria, such as the elongated filamentous forms seen in Fig. 18; most protozoa and bacterial actually live within the mucus blanket. Bar: 5 μm.

FIGURE 17. When intestinal disease is induced in the rat by feeding phytohemagglutinin (PHA) autochthonous *Escherichia coli* no longer lives exclusively in the mucus layer but colonizes the microvillar surface of the jejunum, where its cells appear to be surrounded by an amorphous matrix. Bar: 5 μm.

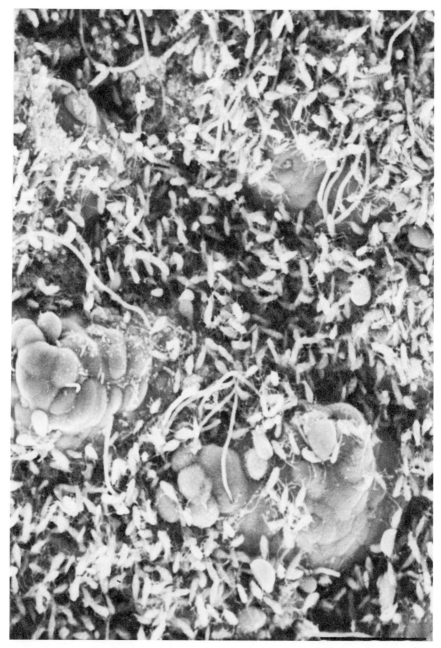

FIGURE 18.   When phytohemagglutinin (PHA) lectin is fed to rats, the microvillar surface of the jejunum becomes colonized by bacteria and the ileum becomes "overgrown" by protozoa. Here, a phenomenal number of protozoa of a species that normally lives autochthonously in the mucus layer have occupied the surfaces of the villi. All other autochthonous microorganisms have disappeared except the filamentous bacteria often embedded in the ileal tissue of rodents. Bar: 50 μm.

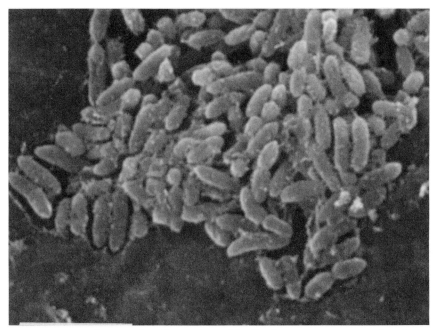

FIGURE 19. In SEM photographs, adherent bacterial cells often seem to penetrate the substratum because they are seen within "pits." These very common artifacts are caused by the dehydration and retraction of the bacterial glycocalyx from the bacterial cell to form a pitlike structure. Bar: 5 μm.

cleaner, so that they are broken up to yield single cells or small bacterial aggregates that will each constitute one colony-forming unit (CFU) in quantitation by plating or by dilution techniques. The obvious errors inherent in these techniques—that some cells remain on the colonized substrate and some cells are killed during the disruption of the aggregates—are both subtractive. We should therefore bear in mind that bacterial population data obtained by this means are "minimum estimates." In all natural aquatic systems studied to date with these methods, the sessile bacterial count per cm^2 has been 500–1000-fold higher than the planktonic population per cm^3. In all systems studied to date, sessile bacterial counts obtained by these methods have agreed well with direct cell counts by epifluorescence microscopy (see Chapter 1, this volume) of intact or detached biofilms except that the culture data are consistently 10–100-fold lower. We attribute this difference to the fact that all bacterial cells seen by direct microscopy are not capable of growth on the media used, and to the imperfect disruption of the biofilm into single cells.

Scraping and disruption methods work well in the quantitation of biofilm bacteria on mucus-covered tissue surfaces, and these disruption methods appear appropriate for the quantitative analysis of clinical specimens from chronic infections in which the causative agents are known to live in biofilms. Many

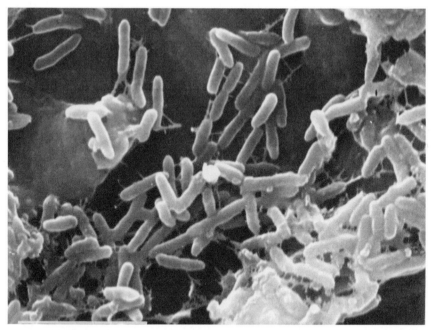

FIGURE 20.   The bacterial glycocalyx is 99% water and condenses radically upon dehydration. The cells in this biofilm were surrounded by masses of hydrated glycocalyx, in their native state, but the condensed residua of this extracellular structure tend to persist where the bacteria were attached to the substratum. These thin linear residua were often mistaken for pili. Bar: 5 μm.

chronic bacterial infections, such as those that are associated with Tenckhoff peritoneal catheters (Marrie *et al.*, 1983) in chronic ambulatory peritoneal dialysis patients, yield negative bacterial cultures even though the patient shows unequivocal clinical signs of acute peritoneal infection. The bacteria present in peritoneal fluid from these patients may be present in very large aggregates that escape detection because they produce only a single colony upon plating, and disruption is advisable. Some bacteria are so firmly adherent to tissue surfaces that they cannot be quantitatively recovered by swabbing or by scraping (McCowan *et al.*, 1979) and, where possible, we quantitate that these bacteria by diluting and plating the "brie" that results from the homogenization of pieces of the colonized tissue using a tissue homogenizer.

CULTIVATION OF BACTERIA IN BIOFILMS

Removing the bacteria from the surface disrupts the spatial arrangement of the microorganisms within the biofilm and lessens or eliminates the influence that the polysaccharide matrix would have on the flux of solutes and macromolecules into the biofilm (Williamson and McCarty, 1976; Ladd *et al.*,

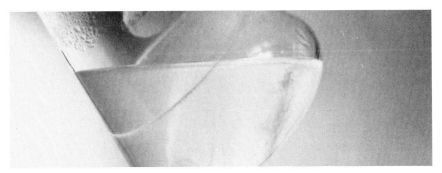

FIGURE 21. When an aerobic bacterial species such as *Pseudomonas aeruginosa* is grown in broth on an orbital shaker, a bacterial biofilm forms at the air–medium interface; this sessile form of growth predominates in these shaken cultures, where it is often seen as a diaphanous veil-like structure within the medium.

1979). These potential problems can be overcome by using techniques that permit the use of intact biofilm populations.

Substrata can be removed from a stream bed, cut into precisely measured squares (1.0 cm × 1.0 cm × 0.1 cm) and then resubmerged in the same stream (Ladd *et al.*, 1979). This procedure provides a uniform surface area that can be easily viewed and counted by scanning electron and epifluorescence microscopy, respectively; it also allows the investigator to follow the development of natural algalbacterial biofilm consortia (Gessey *et al.*, 1978) under ambient conditions on a surface which has physical and chemical characteristics identical with those of the parent substratum (Ladd *et al.*, 1979; Mills and Maubrey, 1981).

Perhaps the simplest method of developing bacterial biofilms *in vitro* is that developed by John Govan (1975); an aerobic organism is agitated in liquid culture, forming a vortex at the air–medium interface (Chan *et al.*, 1984), and a bacterial biofilm develops on available glass surfaces (Fig. 21). A more complex and more useful *in vitro* system involves the passage of cells of a pure culture of bacteria, in a nutrient medium, through a Robbins device (McCoy *et al.*, 1981) (Fig. 22) or a modified Robbins device (Nickel *et al.*, 1985*a*) (Fig. 23). These devices expose the face of a metal or plastic "stud" to the flowing culture while the sides and back of the stud are kept sterile. After the development of a bacterial biofilm on the face, the stud can be aseptically removed from the walls of the Robbins device for quantitation or examination.

In a recent study using the modified Robbins device, we developed biofilms of a uropathogenic strain of *Pseudomonas aeruginosa* on the surface of catheter Latex disks (Figs. 24 and 25) and showed that, while planktonic cells of this strain had a minimum bactericidal concentration (MBC) of 40 µg/ml of tobramycin, biofilm cells were not killed by 1000 µg/ml of this antibiotic (Nickel *et al.*, 1985*b*). This explains the generally poor record of conventional

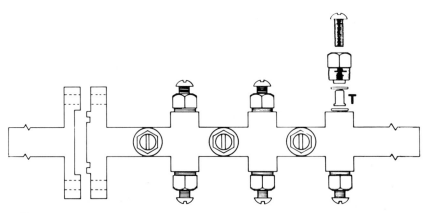

FIGURE 22.   The Robbins device consists of a series of metal or plastic "studs" (T) the sides and back of which are kept sterile by a sealed housing with their "faces" exposed to bacteria-containing fluids passing through a 1-inch pipe. Once biofilms have developed on the faces, the studs can be aseptically removed, and the biofilm bacteria can be examined or quantitated.

antibiotics in resolving biomaterial-related infections (Dougherty and Simmons, 1982) and suggests that this biofilm MBC method might be successfully employed to determine antibiotic concentrations that will actually kill all of the bacterial cells in the biofilms that are characteristic (Costerton, 1984) of so many chronic bacterial diseases. Bacteria within biofilms have also been shown to be resistant to antiseptics used routinely in hospitals (Marrie and Costerton, 1981).

The first bacteria that adhere to the stud face form a biofilm composed of cells embedded in an amorphous intercellular glycocalyx matrix (Fig. 26), and this film has little effect (McCoy *et al.*, 1981) on frictional resistance (f_F).

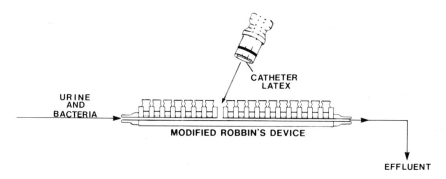

FIGURE 23.   The modified Robbins device (MRD) is made of Perspex; the aseptically removable studs contain a disk of catheter Latex or other material that is exposed to flowing fluids on its "face" only. Media or body fluids containing bacteria pass through this device, and "studs" are examined to monitor biofilm development. Once bacterial biofilms are established on these stud surfaces, the sessile bacteria therein can be challenged with antibiotics to establish a sessile MBC value.

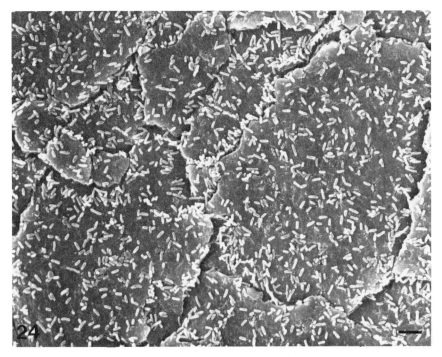

FIGURE 24. SEM of the surface of a disk of catheter Latex exposed for 1 hr to flowing urine containing exponential phase cells of a uropathogenic strain of *Pseudomonas aeruginosa*. A few bacteria have adhered to the platelike Latex surface, and very small amounts of glycocalyx have been produced. Bar: 5 μm.

However, cells within the bacterial biofilm elongate (Fig. 27) so that tufts are formed in the biofilm that oscillate in the flowing fluid and sharply increase frictional resistance (f_F) to fluid flow. In some recent work, conventional bacterial quantitation has been abandoned, and the development of bacterial biofilms has been monitored accurately using a double manometer reading expressed as frictional resistance (f_F) (McCoy *et al.*, 1981).

The Robbins device, in both its low-pressure (<700 psi) and high-pressure (<6000 psi) designs, is now widely used as a biofilm sampler in industrial systems (Watkins and Costerton, 1984). It enables the operator to monitor the development of bacterial biofilms on surfaces, such as those of transmission lines and heat exchangers, and to detect biofilm organisms of particular interest, such as the corrosion-causing sulfate-reducing bacteria. Laboratory tests using the Robbins device to grow sessile biofilms of mixed natural bacterial populations from industrial waters have shown that cells within these biofilms are much more resistant to industrial biocides (Ruseska *et al.*, 1982) than are their planktonic counterparts. Therefore, when deleterious bacteria are detected in an industrial system, the chosen biocide must be used at a combination of concentration and contact time that will completely kill the

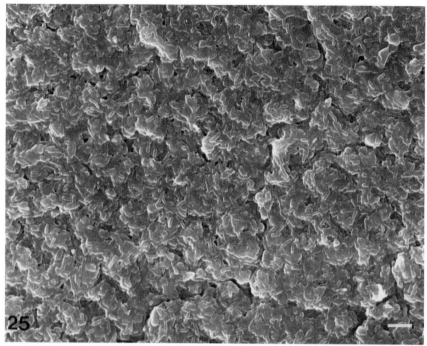

FIGURE 25.   SEM of the surface of a disk of catheter Latex exposed for 8 hr to flowing urine containing exponential phase cells of a uropathogenic strain of *Pseudomonas aeruginosa.* The surface of the Latex disk is entirely occluded by a thick bacterial biofilm within which the cells are heavily enveloped in amorphous glycocalyx components. Cells within this established biofilm withstood 1000 μg/ml of tobramycin, whereas equivalent planktonic cells of the same strain were killed by 40 μg/ml of the same agent. Bar: 5 μm.

biofilm population of the organism in question, as detected by the Robbins device.

Another form of the bacterial biofilm that is especially troublesome to the oil and gas industry is the plugging film that develops on the formation face in water-injection wells (Shaw *et al.,* 1985). Bacteria in these systems produce large amounts of exopolysaccharide "slime" that reduces flow and redirects injection water (Clementz *et al.,* 1982); they also produce H_2S, which can "sour" the whole oil-bearing formation. We have developed a model system (Shaw *et al.,* 1985) in which bacterial cultures or actual injection waters are passed through a core of sandstone or an artificial core of sintered glass beads; the development of bacterial biofilms on surfaces within these solid matrices can be monitored. These model systems have been very useful in testing water-injection well-cleaning strategies; they attest to the universality of the tendency of bacteria to adhere to surfaces and to form biofilms in all natural ecosystems studied to date.

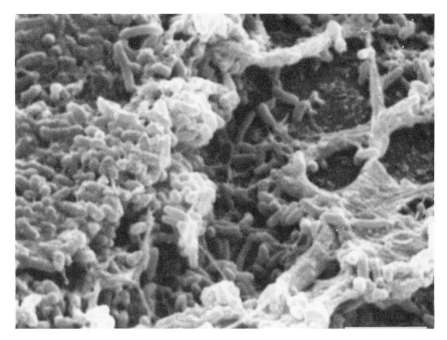

FIGURE 26.   SEM of the face of a metal Robbins device stud showing the bacterial cells and their extracellular products that constitute the early adherent biofilm. These smooth bacterial biofilms create little turbulence and do not increase frictional resistance (f_F) significantly. Bar: 5 μm.

DETECTION OF ACTIVITY WITHIN BIOFILMS

Although viable plate counts can be used to estimate the number of viable cells in a biofilm population, this technique is widely recognized as being highly selective and frequently underestimates total cell numbers by several orders of magnitude (Peele and Colwell, 1984; Quinn, 1984). The widely used acridine orange direct count (AODC) (see Chapter 1, this volume) provides a more accurate estimate of bacterial numbers, but the method does not distinguish between active and inactive microorganisms (Daley, 1979); therefore, it gives no information concerning activity within a biofilm. Recently, several direct microscopic methods have been proposed for determining the percentage of actively metabolizing microorganisms within a bacterial population. The techniques of Zimmermann *et al.* (1978) and Kogure *et al.* (1979) use tetrazolium and acridine orange with nalidixic acid, equating activity with intracellular formazan deposition and nalidixic acid sensitivity (elongation), respectively (see Chapter 1, this volume). Microautoradiographic counts of uptake-active microorganisms can be determined by incubating the cells with an appropriate labeled substrate and viewing the acridine orange-stained au-

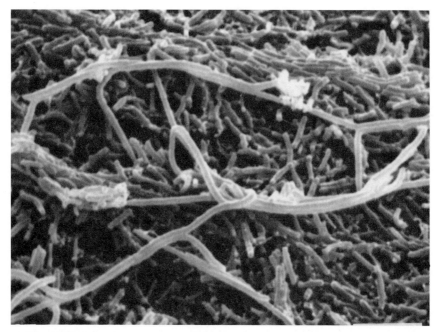

FIGURE 27.   SEM of the face of a metal Robbins device stud showing the elongation of bacterial cells that precedes the development of "tufts" protruding from the bacterial biofilm. These tufts oscillate in the flowing fluid and exert a measurable frictional resistance (f_F) to flow that can be detected manometrically. Bar: 5 μm.

toradiograms via epifluorescence microscopy (Meyer-Reil, 1978; Tabor and Neihof, 1984). These methods provide a rapid means of enumerating metabolically active microorganisms in free-living (planktonic) and sediment populations (Meyer-Reil, 1978; Zimmermann *et al.*, 1978; Jones, 1979; Tabor and Neihof, 1984) and can be used with dispersed biofilm populations, but none of them works well with intact biofilms (T. I. Ladd, unpublished results). We have found that these methods generally give poor resolution or display high background activity or surface autofluorescence when used in conjunction with intact biofilms.

Recent studies have also shown that results obtained with a given method may fluctuate in response to alterations in incubation procedures (Peele and Colwell, 1981) and that there may be significant differences in the estimates of metabolically active microorganisms when several methods are used simultaneously on the same sample (Tabor and Neihof, 1984; Quinn, 1984). Thus, the results from a single method must be interpreted with caution, and the use of combined techniques is highly recommended.

We have had good success detecting activity within intact biofilms by measuring the uptake and/or mineralization of uniformly labeled ^{14}C-labeled compounds. This measurement can be made without distinguishing between active and inactive bacteria and employs radioisotopic techniques commonly used to assess the microheterotrophic activity of planktonic and sediment bacterial populations in aquatic ecosystems (Wright and Burneson, 1979; Wyndham and Costerton, 1981; Ladd et al., 1982).

The rates of substrate utilization by bacteria in biofilms associated with waste water reactors and rotating drum contactors have been studied extensively by engineers for the past two decades and numerous mathematical models have been proposed to describe biofilm kinetics (LaMotta, 1976; Williamson and McCarty, 1976; Rittmann and McCarty, 1980). Substrate removal by biofilms is a complex combination of mass transfer and metabolic processes (Eighmy and Bishop, 1984) and the derived biological rate equations generally describe substrate removal as a process of molecular diffusion with simultaneous biological reaction (LaMotta, 1976; Williamson and McCarty, 1976). The model systems used in these studies are often technically complex and the actual measurement of substrate removal may be both time consuming and cumbersome (LaMotta, 1976). These limitations preclude their use for routine biofilm monitoring in industry and medicine.

An important component of substrate removal by biofilm bacteria that has been largely overlooked until recently is the actual transport or uptake of specific substrates. Ladd et al. (1979) employed the heterotrophic activity technique of Hobbie and Crawford (1969) to measure the uptake and subsequent mineralization of [U-14C]glutamic acid by intact mountain stream biofilms attached to natural cobble substrata. More recently, Eighmy and Bishop (1984) used a similar heterotrophic activity technique to demonstrate aspartate transport in thin wastewater biofilms attached to clear polyvinyl chloride (PVC) stubs mounted in a tubular flow system. One of the principal advantages of this technique is that a biofilm can be developed on any specified substratum (surface) material and the activity can be subsequently measured under appropriate in situ conditions. Flow velocities over the biofilm surface are generally difficult to duplicate with this method, but they can be simulated by gently agitating the reaction vials during incubation.

The results shown in Table I were obtained using the single time-multiple substrate concentration or kinetic assay described by Wright and Hobbie (1965) and Hobbie and Crawford (1969). Briefly, colonized rock squares were retrieved from the stream at appropriate intervals, gently rinsed to dislodge reversible sorbed cells, added in triplicate to individual vials containing filter sterilized stream water and [U-14C]phenylalanine over a range of concentrations (0.16–32.0 µg/liter), and incubated for 4 hr at ambient stream temperature. Mineralization and assimilation reactions were terminated by acidification and filtration, respectively, as described by Ladd et al. (1979). Radioactivity associated with the biofilm bacteria was determined by autooxidation (Packard model 306 Autooxidizer), and both mineralization and assimilated uptake were corrected for duplicate acid-killed controls at each concentration. The uptake by biofilm bacteria was compared to that of planktonic bacteria collected at the same time. The results in Table 1 indicate the total bacterial

TABLE I

Kinetic Analysis of Phenylalanine Uptake by Biofilm and Planktonic Bacteria from a Mountain Stream in Marmot Basin, Alberta, Canada

Days	$AODC^a$	$V_{max}{}^b$	T_t (hours)	Specific activity index[c] (μg/cell per hr $\times 10^{-12}$)	Percentage respiration[d]
			Biofilm Bacteria		
42	2.20×10^6	0.01×10^{-1}	925.00	0.45	8.23
56	5.31×10^7	1.21×10^{-1}	355.00	2.28	9.72
			Planktonic Bacteria		
42	0.89×10^5	1.77×10^{-3}	1233.30	19.89	11.47
56	1.03×10^5	2.00×10^{-3}	1801.10	19.42	9.00

[a] Number of cells per cm^2 (biofilm) or per ml (planktonic).
[b] Expressed in ng · cm^2 (biofilm) or per ml (planktonic) per hour. Each sample represents a triplicate analysis at six concentrations at a single sampling site. Linear regression analysis gave a correlation coefficient of $r \leq 0.89$ for each analysis.
[c] This value was obtained by dividing V_{max} by the AODC (Wright, 1978).
[d] Percentage total uptake as $^{14}CO_2$.

count per cm^2 is approximately 10–100-fold greater than that found per ml of overlying water. The maximum rates of uptake V_{max} for phenylalanine were between 0.5 and 60 times greater for the biofilm bacteria than for the planktonic bacteria. Turnover times T_t for the phenylalanine were between 1.3 and 5 times lower in the biofilm samples than in the planktonic samples.

The differences in activity exhibited in these experiments undoubtedly reflected the variations in bacterial densities of the respective samples. Activity per bacterium or the specific activity index (Wright, 1978) can be calculated for each population by dividing the V_{max} value by the AODC, and this index can be used to compare the activity of the population samples. Minor fluctuations in the specific activity index over time may indicate slight changes in the fraction of metabolically active cells in the population. Major fluctuations in the specific activity index that cannot be accounted for by a potential change in the percentage of actively metabolizing cells may reflect an actual change in the activity per bacterium, possibly due to a change in the physiological state of the cells. Large changes in the specific activity index may indicate shifts in species composition or in preference for a specific substrate (Spencer and Ramsay, 1978; Wright, 1978). As is illustrated in Table I, the specific activity index for the planktonic bacteria showed little variation. The activity per bacterium within the biofilm showed a 5-fold increase, accompanying a 24-fold increase in the bacterial density. The observed change in the specific activity index could have been accounted for by a major shift in the proportion of active cells in the biofilm. It seems more probable, however, that other transitional events occurring in the biofilm, including an increase in biofilm thickness and a concomitant change in diffusional resistances, contributed to the observed fluctuation in the activity index.

The multiple substrate concentration technique provides useful kinetic data about the biofilm bacteria that can be replotted to indicate the type and degree of diffusional resistance (external or internal) influencing the transport assay (Eighmy and Bishop, 1984). By contrast, corrosion engineers and clinicians are frequently more interested in simply knowing whether or not there is activity within the biofilm, and perhaps more importantly, determining whether the biofilm bacteria are sensitive to a biocide or antibiotic challenge. A technically easier radioisotopic assay can be used to address these concerns.

The results shown in Table II were obtained using the single substrate concentration–single time radioisotopic assay (Kadota *et al.*, 1966). The assay is essentially a modified ^{14}C-radiorespirometric technique (Lehmicke *et al.*, 1979) in which $^{14}CO_2$ is collected at the termination of the incubation period and radioactivity is recorded in disintegrations per minute (dpm). Counts that were three times the acid-killed control were scored as positive and interpreted as indicative of the presence of metabolically active bacteria within the biofilm (Lehmicke *et al.*, 1979).

In this experiment, a pseudomonad biofilm was developed on Silkolatex (Raush, West Germany) catheter material mounted in a modified Robbins device as described in Nickel *et al.* (1985a). The surfaces were colonized for 24 hr at a flow rate of 60 ml/hr, retrieved, rinsed, and then added to vials containing either phosphate-buffered saline or phosphate-buffered saline plus 500 μg/ml tobramycin sulfate (Eli Lilly). Both the control and treatment samples were incubated at 37°C for 48 hr. At the end of this period, 3.2 μg/liter of [^{14}C]glutamic acid was added to each vial; the vials were incubated for an additional 30 min. All samples, including the acid-killed controls, were collected in quadruplicate. The evolved $^{14}CO_2$ was collected in a closed-loop circulation system following acidification (Wyndham and Costerton, 1981).

As shown in Table II, bacterial numbers increased four orders of magnitude between the 2- and 24-hr sampling periods. This increase can be

TABLE II

Short Term (1-hr) Mineralization of [U-^{14}C] Glutamic Acid (3.2 μg/liter) by Pseudomonad Biofilms Attached to Silkolatex Foley Catheter Studs Mounted in a Modified Robbins Device

Time colonized (hr)	AODC cells/cm^2 or per ml	dpm $^{14}CO_2$ evolved ± SD[a]		
		Acid killed	No tobramycin	500 μg/ml[c] tobramycin
2	$8.7 \pm 3.2 \times 10^2$	27.5 ± 9.5	36.5 ± 6.3	35.2 ± 5.5
2	$6.9 \pm 1.8 \times 10^2$	36.2 ± 3.5	527.2 ± 10.2	N.D.
10	$8.3 \pm 1.2 \times 10^4$	87.5 ± 17.2	$9,895.2 \pm 1,130.7$	$1,222.3 \pm 91.5^d$
24	$11.2 \pm 3.4 \times 10^6$	111.2 ± 27.3	$17,548.4 \pm 3,306.8$	$1,469.8 \pm 179.9^d$
Planktonic sample	$5.8 \pm 2.3 \times 10^5$	37.8 ± 9.3	$8,211.3 \pm 102.9$	101.7 ± 23.6

[a] Quadruplicate samples were incubated for 30 min at 37°C.
[b] A 24-hr incubation period was used for this radiorespirometric assay.
[c] Established biofilms were exposed to 500 μg/ml tobramycin sulfate for 48 hr prior to the addition of ^{14}C-glutamic acid to the reaction vessel.
[d] No viable cells were isolated on brain heart infusion agar.

attributed to growth and division of biofilm bacteria and the continued sorption of bacteria from the overlying fluid to the surface. The mineralization of the glutamic acid was readily detected when the bacterial density approached 10^5 cells/cm^2, but no activity was observed on studs with approximately 10^3 cells/cm^2. This finding suggested the potential detection limits of the short-term (1-hr) radiorespirometric assay. The observed detection limit for this technique, however, is a function of the incubation time (Table II). Moreover, the potential detection limit varies with the substrate used in the assay and reflects the overall metabolism of the substrate by the biofilm bacteria (T. I. Ladd, unpublished results). For example, a 4- to 8-hr incubation period is generally required to detect the activity of 10^5 cells/cm^2 when [^{14}C]phenylalanine is used as the substrate. The need for the longer incubation time may be partially attributed to the relatively low respiration observed in many of our [^{14}C]phenylalanine uptake experiments (Table I). Because the technique is dependent on the evolution of $^{14}CO_2$, care must be taken to select a substrate which is rapidly mineralized by the biofilm population. [^{14}C]glutamic acid has provided satisfactory results in our environmental samples (Ladd *et al.*, 1979; Wyndham and Costerton, 1981) and met our needs in the current studies but, depending on the system under study, other labeled amino acids, a mixture of amino acids, or simple sugars may be useful as well.

The sensitivity of the biofilm bacteria to an antibiotic was measured by exposing established biofilms to 500 μg/ml tobramycin sulfate for 48 hr before adding the labeled substrate. As illustrated in Table II, the biofilm bacteria were indeed sensitive to the antibiotic challenge and displayed a significantly lower ($p < 0.05$) level of mineralization activity in the treatment group. A similar response was shown in the planktonic sample used for comparison. Despite the apparent sensitivity of the biofilm bacteria to the tobramycin sulfate, the amount of $^{14}CO_2$ evolved by the challenged biofilm was 14 times that of the corresponding acid-killed controls. By contrast, there was no significant activity in the planktonic sample following the antibiotic challenge. These findings generally support the contention that bacteria within a biofilm are more resistant to antibiotics than are free-living cells, even at high antibiotic concentrations (Nickel *et al.*, 1985*b*).

In summary, we have found that the presence of metabolically active bacteria in a biofilm can more easily be detected by means of radioisotopic assays than by the direct microscopic methods. Like the direct microscopic methods, however, a number of important factors affect the final interpretation of the radioisotopic assay, such as affinity and preference for a particular substrate, degree to which a substrate is mineralized, and incubation time. Before using such assays, it is necessary that several preliminary experiments be conducted on representative samples in order to establish the appropriate experimental conditions. For routine sampling of biofilm populations, the simple, rapid radiorespirometric technique is useful. The method permits processing of many samples within a short period of time, thus providing adequate data for statistical analysis.

SUMMARY

Direct observations of environmental, industrial, and medical systems have shown the development of bacterial biofilms on available surfaces. Within these biofilms, bacteria are seen to be embedded in a polymeric matrix of extracellular products that traps molecules and particles from the surrounding fluid until the biofilm consists of bacterial cells and a wide variety of accretions.

Conventional microbiological techniques for the morphological examination, cultivation, and physiological examination of bacteria are all predicated on the notion that these organisms are essentially planktonic (floating or swimming). In spite of increasingly frequent reports of important biofilm populations following Henrici's (1933) and Zobell's (1943) original reports, this divergence of technique from reality seemed to be largely an academic matter until it was demonstrated that sessile cells differ physiologically from planktonic cells of the same organism (Whittenbury and Dow, 1977) and until we showed that biofilm bacteria are much more resistant to antiseptics (Marrie and Costerton, 1981), biocides (Ruseska *et al.*, 1982), and antibiotics (Nickel *et al.*, 1985*b*). Thus, whereas *in vitro* studies of planktonic bacteria continue to yield very useful data on the physiologic and genetic potential of individual bacterial species, it now seems dangerous and unwise to extrapolate from these data to explain the physiologic activities of biofilm bacteria or to predict their susceptibility to antimicrobial agents.

This chapter has outlined some preliminary attempts to devise new techniques for the examination of biofilm bacteria that should be considered by workers in the field; thus, far more definitive techniques can eventually be developed for the study of these important biofilm populations.

REFERENCES

Banwell, J. G., Howard, R., Cooper, D., and Costerton, J. W., 1985, Intestinal microbial flora after feeding phytohemagglutinin lectins (*Phaseolus vulgaris*) to rats, *Appl. Environ. Microbiol.* **50:**68–80.

Bayer, M. E., and Thurow, H., 1977, Polysaccharide capsule of *Escherichia coli:* Microscopic study of its size, structure and sites of synthesis, *J. Bacteriol.* **130:**911–936.

Birdsell, D. C., Doyle, R. J., and Morgenstern, M., 1975, Organization of teichoic acid in the cell of *Bacillus subtilis, J. Bacteriol.* **121:**726–734.

Brown, M. R. W., Anwar, H., and Lambert, P. A., 1984, Evidence that mucoid *Pseudomonas aeruginosa* in the cystic fibrosis lung grows under iron-restricted conditions, *FEMS Microbiol. Lett.* **21:**113–117.

Bryers, J. D., and Characklis, W. G., 1981, Early fouling biofilm formation in a turbulent flow system: Overall kinetics, *Water Res.* **15:**483–491.

Chan, R., Acres, S. D., and Costerton, J. W., 1982, The use of specific antibody to demonstrate glycocalyx, K99 pili, and the spatial relationships of K99+ enterotoxigenic *E. coli* in the ileum of colostrum-fed calves, *Infect. Immun.* **37:**1170–1180.

Chan, R., Lian, C. J., Costerton, J. W., and Acres, S. D., 1983, The use of specific antibodies to demonstrate the glycocalyx and spatial relationships of a K99−, F41− enterotoxigenic strain of Escherichia coli colonizing the ileum of colostrum-deprived calves, *Can. J. Comp. Med.* **47:**150–156.

Chan, R., Lam, J. S., Lam, K., and Costerton, J. W., 1984, Influence of culture conditions on expression of the mucoid mode of growth of *Pseudomonas aeruginosa, J. Clin. Microbiol.* **190:**8–16.

Cheng, K.-J., Irvin, R. T., and Costerton, J. W., 1981, Autochthonous and pathogenic colonization of animal tissues by bacteria, *Can. J. Microbiol.* **27:**461–490.

Clementz, D. B., Patterson, D. E., Aseltine, R. J., and Young, R. E., 1982, Stimulation of water injection wells in the Los Angeles basin by using sodium hypochlorite and mineral acids, *J. Petrol. Tech.* **34:**2087–2096.

Cohen, A. L., Marlow, D. P., and Garner, G. E., 1968, A rapid critical point method using fluorocarbons ("freons") as intermediate and transitional fluids, *J. Microsc. (Paris)* **7:**331–342.

Costerton, J. W., 1979, The role of electron microscopy in the elucidation of bacterial structure and function, *Annu. Rev. Microbiol.* **33:**459–479.

Costerton, J. W., 1984, The etiology and persistence of cryptic bacterial infections: A hypothesis, *Rev. Infect. Dis.* (Suppl 3) **6:**S608–S616.

Costerton, J. W., Irvin, R. T., and Cheng, K.-J., 1981a, The bacterial glycocalyx in nature and disease, *Annu. Rev. Microbiol.* **35:**299–324.

Costerton, J. W., Irvin, R. T., and Cheng, K.-J., 1981b, Role of bacterial surface structures in pathogenesis, *Crit. Rev. Microbiol. (CRC)* **8:**303–338.

Daley, R. J., 1979, Direct epifluorescence enumeration of native aquatic bacteria: Uses, limitations, and comparative accuracy, In: *Native Aquatic Bacteria: Enumeration, Activity, and Ecology,* ASTM Technical publication No. 695 (J. W. Costerton and R. R. Colwell, eds.), pp. 29–45, American Society of Testing and materials, Philadelphia.

Davis, C. P., and Savage, D. C., 1974, Habitat, succession, attachment and morphology of segmented, filamentous microbes indigenous to the murine gastrointestinal tract, *Infect. Immun.* **10:**948–956.

Dougherty, S. H., and Simmons, R. L., 1982, Infections in bionic man: The pathology of infections in prosthetic devices. Part II, *Curr. Prob. Surg.* **19:**269–318.

Eighmy, T. T., and Bishop, P. L., 1984, Multiplicity of aspartate transport in thin wasterwater biofilms, *Appl. Environ. Microbiol.* **48:**1151–1158.

Freter, R., O'Brien, P. C. M., and Mascai, M. S., 1981, Role of chemotaxis in the association of motile bacteria with intestinal mucosa: In vivo studies, *Infect. Immun.* **34:**234.

Geesey, G. G., Mutch, R., Costerton, J. W., and Green, R. B., 1978, Sessile bacteria: An important component of the microbial population in small mountain streams, *Limnol. Oceanogr.* **23:**1214–1223.

Govan, J. R. W., 1975, Mucoid strains of *Pseudomonas aeruginosa:* The influence of culture medium on the stability of mucus production, *J. Med. Microbiol.* **8:**513–522.

Henrici, A. T., 1933, Studies of freshwater bacteria. I. A direct microscopic technique, *J. Bacteriol.* **25:**277–286.

Hobbie, J. E., and Crawford, C. C., 1969, Respiration corrections for bacterial uptake of dissolved organic compounds in natural waters, *Limnol. Oceanogr.* **14:**528–532.

Jones, J. G., 1979, *A Guide to Methods for Estimating Microbial Numbers and Biomass in Freshwater,* Freshwater Biological Association Scientific Publications No. 39, Freshwater Biological Association, Windermere, U.K.

Kadota, H., Hata, Y., and Miyoshi, H., 1966, A new method for estimating the mineralization activity of lake water and sediment, *Mem. Res. Inst. Food. Sci. Kyoto Univ.* **27:**28–30.

Ladd, T. I., Costerton, J. W., and Geesey, G. G., 1979, Determination of the heterotrophic activity of epilithic microbial populations, In: *Native Aquatic Bacteria: Enumeration, Activity, and Ecology,* ASTM Technical Publications No. 695 (J. W. Costerton and R. R. Colwell eds.), pp. 180–195, American Society of Testing and Materials, Philadelphia.

Ladd, T. I., Ventullo, R. M., Wallis, P. M., and Costerton, J. W., 1982, Heterotrophic activity and biodegradation of labile and refractory compounds by groundwater and stream microbial populations, *Appl. Environ. Microbiol.* **44:**321–329.

LaMotta, E. J., 1976, Kinetics of growth and substrate uptake in a biological film system, *Appl. Environ. Microbiol.* **31:**286–293.

Lehmicke, L. G., Williams, R. T., and Crawford, R. L., 1979, ^{14}C most probable number method for enumeration of active heterotrophic microorganisms in natural waters, *Appl. Environ. Microbiol.* **12**:644–649.

Luft, J. H., 1971, Ruthenium red and ruthenium violet. I. Chemistry, purification, methods for electron microscopy and mechanism of action, *Anat. Rec.* **171**:347–368.

Mackie, E. B., Brown, K. N., Lam, J., and Costerton, J. W., 1979, Morphological stabilization of capsules of group B streptococci types Ia, Ib, II and III, with specific antibody, *J. Bacteriol.* **138**:609–617.

Malick, L. E., and Wilson, B. W., 1975, Modified thiocarbohydrazide procedure for scanning electron microscopy: Routine use for normal, pathological or experimental tissues, *Strain Technol.* **50**:265–269.

Marrie, T. J., and Costerton, J. W., 1981, An electron microscopic study of the prolonged survival of *Serratia marcesiens* in chlorhexidine, *Appl. Environ. Microbiol.* **42**:1093–1102.

Marrie, T. J., Noble, M. A., and Costerton, J. W., 1983, Examination of the morphology of bacteria adhering to peritoneal dialysis catheters by scanning and transmission electron microscopy, *J. Clin. Microbiol.* **18**:1388–1398.

Mayberry-Carson, K. J., Tober-Meyer, B., Smith, J. K., Lambe, D. W., and Costerton, J. W., 1984, Bacterial adherence and glycocalyx formation in osteomyelitis experimentally induced with *Staphylococcus aureus, Infect. Immun.* **43**:825–833.

McCowan, R. P., Cheng, K.-J., and Costerton, J. W., 1979, Colonization of a portion of the bovine tongue by unusual filamentous bacteria, *Appl. Environ. Microbiol.* **37**:1224–1229.

McCoy, W. F., Bryers, J. D., Robbins, J., and Costerton, J. W., 1981, Observations on fouling biofilm formation, *Can. J. Microbiol.* **27**:910–917.

Meyer-Reil, L. A., 1978, Autoradiography and epifluorescence microscopy combined for the determination of number and spectrum of actively metabolizing bacteria in natural waters, *Appl. Environ. Microbiol.* **36**:506–512.

Mills, A. L., and Maubrey, R., 1981, Effect of mineral composition on bacterial attachment to submerged surfaces, *Microbiol. Ecol.* **7**:315–322.

Nickel, J. C., Ruseska, I., Marrie, T. J., Whitfield, C., and Costerton, J. W., 1985a, Antibiotic resistance of *Pseudomonas aeruginosa* colonizing a urinary catheter in vitro, *Eur. J. Clin. Microbiol.* **4**:213–218.

Nickel, J. C., Ruseska, I., and Costerton, J. W., 1985b, Tobramycin resistance of cells of *Pseudomonas aeruginosa* cells as a biofilm on urinary catheter material, *Antimicrob. Agents Chemother.* **27**:619–624.

Peele, E. R., and Colwell, R. R., 1981, Applications of a direct microscopic method for enumeration of substrate-responsive marine bacteria, *Can. J. Microbiol.* **27**:1071–1075.

Quinn, J. P., 1984, The modification and evaluation of some cytochemical techniques for the enumeration of metabolically active heterotrophic bacteria in the aquatic environment, *J. Appl. Bacteriol.* **57**:51–57.

Rittmann, B. E., and McCarty, P. L., 1980, A model of steady state biofilm kinetics, *Biotechnol. Bioeng.* **22**:2343–2357.

Rozee, K. R., Cooper, D., Lam, K., and Costerton, J. W., 1982, Microbial flora of the mouse ileum mucous layer and epithelial surface, *Appl. Environ. Microbiol.* **43**:1451–1463.

Ruseska, I., Robbins, J., Lashen, E. S., and Costerton, J. W., 1982, Biocide testing against oilfield bacteria helps control plugging, *Oil Gas J.* **March 8**:253–264.

Shaw, J. C., Bramhill, B., Wardlaw, N. C., and Costerton, J. W., 1985, Bacterial fouling of a model core system, *Appl. Environ. Microbiol* **49**:693–701.

Spencer, M. J., and Ramsay, A. J., 1978, Bacterial populations, heterotrophic potentials and water quality in three New Zealand rivers, *N. Z. J. Mar. Freshwater Res.* **12**:415–427.

Sutherland, I. W., 1977, Bacterial polysaccharides—their nature and production, In: *Surface Carbohydrates of the Prokaryotic Cell* (I. W. Sutherland, ed.), pp. 27–96, Academic Press, London.

Tabor, P. S., and Neihof, R. A., 1984, Direct determination of activities for microorganisms of Chesapeake Bay populations, *Appl. Environ. Microbiol.* **48**:1012–1019.

Watkins, L., and Costerton, J. W., 1984, Growth and biocide resistance of bacterial biofilms in industrial systems, *Chem. Times Trends* **7**:35–40.

Whittenbury, R., and Dow, C. S., 1977, Morphogenesis and differentiation in Rhodomicrobium vannielii and other budding and prosthecate bacteria, *Bacteriol. Rev.* **41**:754–808.

Williamson, K., and McCarty, P. L., 1976, A model of substrate utilization by bacterial biofilms, *J. Water Pollut. Control Fed.* **48**:9–24.

Wright, R. T., 1978, Measurement and significance of specific activity in the heterotrophic bacteria of natural waters, *Appl. Environ. Microbiol.* **36**:297–305.

Wright, R. T., and Burnison, B. K., 1979, Heterotrophic activity measured with radio-labeled organic substrates, In: *Native Aquatic Bacteria: Enumeration, Activity and Ecology*, ASTM Technical Publication No. 695 (J. W. Costerton and R. R. Colwell, eds.), pp. 140–155, American Society of Testing and Materials, Philadelphia.

Wright, R. T., and Hobbie, J. E., 1965, The uptake of organic solutes in lake water, *Limnol. Oceanogr.* **10**:22–28.

Wyndham, R. C., and Costerton, J. W., 1981, Heterotrophic potentials and hydrocarbon biodegradation potentials of sediment microorganisms within the Athabasca oil sands deposit, *Appl. Environ. Microbiol.* **41**:783–790.

Zimmermann, R., Iturreaga, A., and Becker-Birk, J., 1978, Simultaneous determination of the total number of aquatic bacteria and the number thereof involved in respiration, *Appl. Environ. Microbiol.* **36**:926–935.

Zobell, C. E., 1943, The effect of solid surfaces on bacterial activity, *J. Bacteriol.* **46**:39–56.

DETERMINATION OF IN SITU MICROBIAL BIOMASS, VIABILITY, METABOLISM, AND GROWTH

David M. Karl

> The study of the growth of bacteria does not constitute a specialized subject or a branch of research; it is the basic method of microbiology.
>
> *J. Monod (1949)*

INTRODUCTION

The observation and preliminary description of microorganisms in their natural habitats date back to the seventeenth century with the initial microscopic studies of van Leeuwenhoek (1677); however, the founding principles of microbial ecology as a specialized branch of science were not established until the late nineteenth century. The study of microorganisms in nature represents one of the least developed and most poorly quantitated areas of microbiological research. Classical microbiological methods, devised for estimating cell numbers and biomass in laboratory cultures, are generally unsatisfactory when applied to the sparse and physiologically diverse populations present in most natural habitats. Furthermore, many ecosystems are dominated by nonliving particulate organic matter which makes indirect assessments of biomass based on total carbon, nitrogen, protein or dry weight virtually impossible. In addition to the analytical limitations imposed by low standing stocks (or biomass), rates of microbial production and cell division in most natural habitats may be lower than maximum physiological potentials. Consequently, the methods employed to assess metabolic activities must be sensitive enough to detect these anticipated low rates. Unfortunately, the overwhelming need for new methods was neither fully appreciated nor adequately satisfied during the initial stages of research in microbial ecology. Even today, our understanding

David M. Karl • Department of Oceanography, University of Hawaii, Honolulu, Hawaii 96822.

of the integrated functioning of microorganisms in nature is methods-limited, as evidenced by periods of rapid advance following successful development of new experimental approaches. Field-oriented microbiologists are still striving to develop, evaluate, and calibrate the basic methods of microbial ecology.

Van Niel (1949) noted that "growth is the expression *par excellence* of the dynamic nature of living organisms." It is well known from laboratory studies that regulatory mechanisms operate to ensure that cells function to obtain the maximum yield of biomass from the environment in the minimum time (Mandelstam and McQuillen, 1976). However, in nature, additional constraints may be placed on microbial populations, such that the demonstration of growth, per se, may not always yield the most reliable and meaningful ecological information. Measures of microbial biomass (i.e., standing stock of living organisms), metabolic activity (which may be uncoupled from growth and cell division), viability, dormancy, and death may also yield information on the effect of the environment on the microbiota and on microbiologically mediated environmental changes.

It is sometimes stated, and often implied, that many of the currently used methods are useless since they can neither be verified *in situ* nor conveniently compared with absolute standards. This resigned attitude is unjustified given the present level of research effort and analytical sophistication in the field of microbial ecology. This chapter summarizes and evaluates available techniques for assessing microbial biomass and rates of various metabolic processes in nature. Many of the methods discussed are contemporary and, therefore, lack an extensive data base. This chapter reviews (1) the theoretical principles, (2) assay assumptions and evaluation thereof, (3) demonstrated attempts at laboratory and field calibration, and (4) shortcomings or limitations of each experimental approach. Detailed methodologies are not presented except as they relate to the above-mentioned concerns. The reader should consult the references cited for access to detailed protocols and historical roots of each technique. In preparing this chapter, I have tried to avoid duplication of the excellent presentations of Brock (1971), Rosswall (1973), Costerton and Colwell (1979), Litchfield and Seyfried (1979), van Es and Meyer-Reil (1982), White (1983), and Leftley *et al.* (1983) concerning related methods for evaluating microbial processes in nature.

HABITAT AND COMMUNITY DESCRIPTION

Diversity and Scale

Microbial habitats in nature are varied and diverse, ranging from soil, sediment, air, water, and ice to epi- and endobiotic associations with other microbes, higher plants, and animals to infection, parasitism and pathogenesis, and infection. The constraints of this review, which limit a detailed discussion of specific requirements for the investigation of individual or selected habitats, should not be taken to mean that unique requirements do not exist. Arbitrary or noncritical use of the so-called standard methods may sometimes yield data of questionable accuracy. Consequently, the approaches discussed herein should be carefully evaluated for reliable application to each individual or unique habitat.

The diversity of microbial habitats is even greater than already implied. An important consideration in microbial ecology is the fact that microorganisms live in microenvironments defined on scales of millimeters or less. As a result, the environment sensed by individual cells may be distinct from the chemical composition of the surrounding bulk environment usually measured in ecological studies. Consequently, the measurement of pH, redox potential, dissolved oxygen or nutrient concentrations may have little direct bearing on the physicochemical characteristics of the microenvironments. This macroscale description may, in fact, mislead the investigator with regard to predominant microbial processes.

Microbial Communities in Nature

There are several notable distinctions between the bacterial cultures studied under defined laboratory conditions and the microbial communities observed in nature. Monospecific bacterial populations are only infrequently encountered in natural habitats, partly due to the high diversity in microscale habitats and to the occurrence of mutualistic relationships among microbial populations.

In the past, the terms bacteriology and microbiology were generally used interchangeably, perhaps based upon the erroneous assumption that bacteria were the only microorganisms worth studying. However, microscopic members from all three kingdoms (Fox *et al.*, 1980) should be discussed collectively as components of natural microbial assemblages. For example, the journal *Applied and Environmental Microbiology*, published by the American Society for Microbiology (Washington, D. C.), actively solicits manuscripts on "all aspects of applied and basic ecological research on bacteria and other microorganisms, including fungi, protozoa and other simple eucaryotic organisms." However, many of the techniques for assessing microbial biomass, viability, and growth in nature fail to distinguish bacterial from nonbacterial and, depending on the objectives of the particular ecological study, this may be either an advantage or a disadvantage. The existence of bacterial ecology as a separate subdiscipline of microbial ecology probably does not exist except in specialized habitats. It is therefore essential to study all microscopic organisms within the discipline of microbial ecology. The subsequent discussion of selected experimental methods makes the important distinction between bacterial and microbial standing stock and production measurements.

No single approach to the study of microbial ecology is universally accepted or acceptable. One option is to investigate the occurrence of a single taxon or metabolic process through an analytical breakdown of the natural ecosystem. Unfortunately, this approach ignores the complexity and biological interaction that most ecologists intend to investigate and understand. The mere presence of a particular microorganism does not necessarily imply ecological importance. Furthermore, measurements of metabolic potential, derived from pure culture studies, may not always be correlated with *in situ* processes. In spite of such criticisms, enrichment and pure culture studies do have an important role in field-oriented research; the relevance of these techniques to the study of microbial ecology is reviewed in Chapter 6, this volume. An alternative approach is to study the ecosystem *in toto*. Since most microbial communities cannot easily be separated into gross trophic categories (see Chapter 9, this volume), much less into refined taxonomic groups prior to most *in situ* measurements, the resultant synecological data are generally more difficult to interpret. However, *both* autecological and synecological investigations are ultimately required for a complete resolution of microbial processes in nature.

SCOPE OF EXISTING METHODS

Many of the earliest ecological studies relied on enumeration by cultural or direct microscopic methods as the basic metric for assessing the role of microorganisms in nature. The assumption that cell numbers were positively correlated with biomass, metabolic activity, growth, and reproductive rates was often implied in these works. Standing stock assessments alone may be misleading, however, for the purpose of estimating the *in situ* activity of the indigenous microbial community. In natural ecosystems, the rates of biogeochemical transformations are dependent on microbial metabolic activity rather than on cell concentration per se.

Quite often, the terms physiological potential, metabolic activity, growth, and cell division are used interchangeably (hence incorrectly) in discussing ecological rate data. There are important distinctions to be made among these various cellular processes, although available techniques may not always adequately distinguish among them. Physiological potentials are generally derived from *in vitro* measurements of specific enzymes or other rate processes and can be only indirectly, if at all, related to *in vivo* activities. Metabolic activity is a vague term used to describe transformations of matter, energy, or both. Typically measured by respirometry or microcalorimetry, metabolic activity is a more direct measure of the total rate of energy transformation occurring under *in situ* conditions. Nevertheless, these activity measurements fail to provide information regarding the extent to which catabolism is coupled to growth or cell division. Metabolically active but inefficient microorganisms, i.e., low ATP yield, (Y_{ATP}), and low biomass yield, may have a greater impact on their environment than do rapidly growing cells (on a per-unit carbon production basis) due to excessive product formation and the possible effects of thermogenesis. An extremely important but sometimes overlooked distinction between growth (defined as increase in cell mass) and cell division should be made. Cell division requires DNA synthesis. In nature, growth can frequently occur without cell division, and under certain environmental conditions, cell division may proceed without cell growth or elevated metabolic activity, e.g., under starvation conditions (see Novitsky and Morita, 1977, 1978). It is therefore imperative that the proper method or suite of methods be employed in order to obtain the data most appropriate and informative for a particular study. In this regard, it is advantageous to use several complementary measurements in order to describe adequately the role of microorganisms in their natural habitats.

Carbon is traditionally used in ecological studies as the basic unit of living biomass, production, and flux among trophic levels. Unfortunately, few methods in microbial ecology measure carbon pools or rates of carbon turnover directly. Consequently, one must rely on laboratory-derived conversion factors to translate the various indirect measurements of microbial biomass and metabolism to a more universal ecological metric. The conversion factors used in ecological studies are not analogous to those commonly used in chemistry and physics and should be more correctly termed extrapolation factors. The reliability and accuracy of most extrapolation factors is far less than the parameters actually measured. As individual experimental methods evolve, the extrapolations must also be refined. This once again emphasizes the importance of conducting multiple, complementary measurements as a means of evaluating the reliability of the laboratory-derived conversion factors.

Tables I–IV summarize the scope and diversity of the methods available for estimating microbiological parameters in nature. Some methods provide

TABLE I

Summary of Methods for Estimating Microbial Biomass in Environmental Samples

Enumeration
    Culture methods: spread/pour plate, roll tube, dilution MPN, enrichment/selective media
    Microscopic: bright-field/phase-contrast, UV luminescence, electron beam,
        immunofluorescence, autofluorescence, flow cytometry, and fluorescence-activated cell
        sorting/sizing, differential fluorescence.
Chemical/biochemical methods
    "Total" microbial
      Bulk elemental analysis (e.g., C, N, P, S)
      Macromolecular composition (e.g., protein, lipid, lipid-P, DNA)
      Nucleotides (e.g., ATP, total adenylates, FMN)
      Hemin and heme protein
    Prokaryotes: muramic acid, lipopolysaccharide
    Algae: chlorophyll *a*
    Fungi: chitin, lipids
Incubation methods
    Total microbial: fumigation/respiration
    Prokaryotes: electrochemical (e.g., impedance, potentiometric)
    Algae: chlorophyll *a* labeling

TABLE II
*Summary of Methods for Estimating Viability,
Physiological Potential, and Nutritional Status*

Viability
  Vital fluorogenic dyes
  Dye exclusion
  Microautoradiography
  INT labeling
  Yeast extract/nalidixic acid (prokaryotes)
Physiological Potential
  In vitro electron-transport system (ETS) activity
  Specific *in vitro* enzyme activity (e.g., dehydrogenase, alkaline
    phosphatase, glutamine synthetase, nitrogenase)
  Enrichment culture
  Uptake of radiolabeled organics at saturating concentrations
Metabolic Status
  Adenylate energy charge
  PHB accumulation (prokaryotes)
  DCMU-induced fluorescence (algae)
  Adenosine : ATP ratio
  cAMP content
  Patterns of macromolecular labeling (algae) with $^{14}CO_2$
  Triglyceride-glycerol analysis, glycocalyx

an estimate of a total rate or flux, others provide information on biomass yield or growth efficiency, and still others provide a combination thereof. Seldom are the results obtained by the various methods directly comparable. Many experimental methods produce mass-weighted (or mass-activity-weighted) estimates of uptake, growth, or metabolism. The ecological significance of these average community values is reasonably subject to question. Consequently, it is imperative that the assumptions of each method be evaluated

TABLE III
Summary of Methods for Estimating Microbial Metabolic Activity in Environmental Samples

Respirometry: CO_2 and/or O_2 flux
Monitoring fate of isotopically labeled electron donors or acceptors (e.g., specific organic
  substrates, NO_3^-, SO_4^{2-})
Measuring rate of production, intracellular flux or turnover of metabolic intermediates,
  biosynthetic precursors or end products (e.g., turnover of cellular ATP pools, production of
  fermentation metabolites)
Microcalorimetry: heat flux
Measuring spatial (upstream-downstream) or temporal (diel, seasonal) changes in steady state
  concentrations of individual compounds, molecules or elements (e.g., O_2, CO_2, PO_4, total
  organic carbon)

TABLE IV.
*Summary of Methods for Estimating Microbial
Production, Growth, and Cell Division*

Nonincubation methods
  GTP : ATP ratio
  Elemental composition data (e.g., C : N : P ratios)
  Mitotic index or frequency of dividing cells (FDC)
Incubation methods
  Photoautotrophic organic carbon production by light–dark
    $^{14}CO_2$ assimilation, O_2 production (algae)
  Rate of incorporation of precursors into stable macromolecules
    (e.g., lipid, protein, cell walls, nucleic acids)
  Rate of biomass-related changes during timed incubations
    (e.g., chl *a*, ATP, DNA, POC, particle volume)
  Dark CO_2 assimilation (bacteria)
  Assimilatory SO_4^{2-} reduction
  ^{14}C-chl *a* method
  Dialysis/microdiffusion chambers ("caged" cultures)
  Biomass increase after predator removal or inhibition
  Rate of colonization of introduced surfaces (e.g., slides, pedoscope)

carefully and calibrated with laboratory cultures under a variety of environmental conditions and growth states before a particular method is deemed suitable for general field applications. By conducting multiparameter ecological investigations, one can obtain independent estimates of microbiological properties that can be used to set upper and lower limits on the interpretation of the resulting data. For example, total microbial biomass estimated either by direct microscopic enumeration and cell sizing (and subsequent conversion to carbon) or by biochemical analysis (e.g., ATP) should never exceed the total particulate organic carbon (living plus nonliving) as determined by chemical oxidation procedures. If biomass and production are measured simultaneously, specific growth rate (μ) can be calculated to compare field data with the large literature of physiological data based on careful measurement in the laboratory (Eppley, 1980, 1981). Finally, estimates of microbiological nutrient consumption or regeneration (generally from tracer experiments) can be compared with direct measurements of total ecosystem flux (e.g., bell jars in sediments or soils, upstream–downstream flow studies in shallow aquatic ecosystems) or predictions from one dimensional diffusion–advection models (summarized for use in sediments by Reeburgh, 1983). These checks and mass balances are an essential component of microbial ecology as a primary basis for judging the reliability of the field-collected data. Before a more detailed discussion of the individual experimental methods, a few general problems of sampling, subsampling, and incubation of natural microbial communities, as well as some considerations regarding the use of isotopic tracers, are presented.

Sampling, Subsampling, and Incubation Considerations

The first major difficulty in attempting to study microbial communities in nature is that of sampling. Microbes are only rarely present in concentrations ideally suited for analysis. Most samples must be either concentrated (e.g., membrane filtration, reverse filtration, centrifugation) or, on some occasions, diluted. One of the most important considerations is whether sterile equipment and technique are essential. It should be emphasized that the use of sterile samplers and technique may not always provide reliable sampling, as chemical contamination of sterile sampling gear or incubation vessels may often affect ecological measurements. If the objective of a particular study is to determine the presence or absence of a specific organism, an aseptic technique is imperative.

The challenge of obtaining a truly representative sample of the environment is the next sampling consideration. Representative sampling is especially important with regard to spatial and temporal variability of natural microbial populations. The scale of sampling is an important factor if we hope to extrapolate our results to the ecosystem level. Ideally, whole ecosystem measurements are most desirable, since these methods allow indirect evaluation of the role of large organisms and physical processes (e.g., mixing) on microbial activity. This approach is only infrequently attempted. Furthermore, the actual number of total samples obtained in most ecological studies is regrettably small, and it is often virtually impossible to obtain replicate samples of a given environment. If statistical methods are employed, it must be assumed that the microbial populations follow a particular probability distribution (e.g., Poisson, negative binomial, log-normal). However, microorganisms generally exist in localized patches and are rarely, if ever, found in random or uniform distributions over the spatial scales used in most ecological investigations (Karl, 1980). Two recent studies have evaluated the sampling design and enumeration statistics for epifluorescent direct counting of marine bacteria. Using seawater samples, Kirchman et al. (1982b), evaluated the degree of variablility in total bacterial numbers contributed by individual subsamples, by replicate filters from a given sample, and by replicate microscopic fields of water from a given environment. Their study involved the formulation of an optimum sampling scheme which also considered the cost of performing a replicate at each level. The results indicated that proper sample analysis should be based on replication at the highest level, i.e., subsamples of the water body. Unfortunately, in many oceanographic studies, a single sample is often taken to represent a portion of the biosphere covering perhaps several hundred meters vertically and several to tens of kilometers horizontally. In addition, our most commonly used methods of water column sample collection, i.e., water sample bottles, generally select against microorganisms attached to sinking particles and submerged surfaces. These are now considered to be among the most important habitats for microorganisms in the sea (Fellows et al., 1981; Caron et al., 1982).

Many of the techniques currently employed for estimating rates of microbiological processes in nature require incubation or enclosure of a representative subsample of the environment for various periods of time. The underlying assumption of these methods is that the initial sampling procedures and subsequent incubation conditions do not alter the *in situ* rates of metabolism and biosynthesis. This assumption is usually impossible to verify under natural conditions. As a result, most investigators try to duplicate the natural conditions to the best of their ability (e.g., light quantity and quality, temperature, pressure, nutrient concentrations).

When natural populations of aquatic microorganisms are contained in glass bottles for periods of approximately 24 hr, the composition of the population and their rates of metabolism can change drastically and elicit the so-called "bottle effect" (ZoBell and Anderson, 1936; Venrick *et al.*, 1977). Bottle effects are totally unpredictable and vary considerably among taxa, location of sample, and transport and incubation conditions. In this regard, it would seem advantageous to keep incubations to a minimum duration that would still satisfy the prerequisites of the individual method (e.g., sensitivity in uptake, production or release of substance being measured, intracellular radiotracer precursor equilibration). Even during relative short-term incubations (e.g., 1–3 hr), there is the possibility of a nonlinear time course of metabolism as a result of confinement. Goldman *et al.* (1981) recently discussed the relevance of this problem to estimates of the rates of phytoplankton production in the marine environment. These workers stressed the importance of establishing short-term time course assays in order to identify problems associated with nonlinear metabolic responses, even if the problems cannot be eliminated completely.

Studies in deep-sea marine microbiology probably represent the most extreme example of physical changes that might occur during environmental sampling and, as a result, elaborate sampler-incubators have been constructed to conduct meaningful investigations. However, all these so-called *in situ* incubation conditions may still yield potential artifacts resulting from perturbations induced by the initial sampling procedures or subsequent confinement. Consequently, the few methods that do exist for estimating rate processes without an incubation or enclosure period must be considered advantageous, especially for studies of remote habitats. These methods include GTP : ATP ratios, frequency of dividing cells (and related mitotic indices), biochemical fingerprinting of signature compounds, cytofluorometry, analyses of elemental or macromolecular composition of the cells, and *in vitro* analyses of specific enzyme or enzyme system activities (see Tables II–IV). However, each of these methods has its own unique limitations unrelated to sample incubation. These shortcomings are discussed in subsequent sections of this chapter.

Use of Isotopic Tracers

The use of stable and radioisotopic tracers has provided an additional dimension to field studies of microbial growth, metabolism, and rates of biogeochemical cycling of numerous elements and compounds. In general, there are three major categories of isotope research: (1) the use of naturally occurring stable isotope couples (e.g., $^{18}O/^{16}O$, $^{15}N/^{14}N$, $^{13}C/^{12}C$, $^{34}S/^{32}S$, D/H) present in nature in specific molar ratios but that fractionate during microbial metabolism; (2) the use of naturally occurring radioactive nuclides, including long-lived primary radioactive elements and their daughters (e.g., ^{40}K, ^{238}U, ^{230}Th, ^{222}Rn, ^{210}Pb) as well as cosmogenic radionuclides (e.g., ^{14}C, ^{3}H, ^{32}Si); and (3) the use of exogenously supplied stable or radioactive isotopes. The first two categories have been important in whole ecosystem studies, including measurements of oceanic circulation and rates of material transport and exchange within homogeneous fluids and at boundaries, discontinuities, and interfaces (e.g., air-sea, sea-sediment). This chapter does not attempt to provide a detailed review of these methods other than to emphasize their importance in studies of microbial ecology, especially on the ecosystem level. The third category (i.e., use of exogenously added tracers) includes numerous techniques to estimate rates of specific microbial processes on time scales of minutes to days. Tables V and VI summarize the isotopes employed most frequently and successfully in studies of microbial ecology. The use of multiple labels (e.g., $^{15}NDH_3^+$, $^{13}C^{18}O_2$) can enhance the sensitivity and statistical accuracy of rate estimates, especially for naturally occurring stable isotopes, as the probability of encountering these double-labeled molecules in nature is extremely low. Berman (1980) recently reviewed the use of simultaneously added multiple isotopic tracers for the study of microbial growth and metabolism in marine ecosystems.

The details of selected individual methods are discussed separately in subsequent sections of this chapter; however, there are several general considerations regarding the use of stable and radioactive isotope tracers in studies of microbial ecology that merit attention. These include (1) the overall reliability of the added element (or compound) as a tracer, including an evaluation of the site of labeling, its uniqueness, and stability during cellular metabolism and biosynthesis, and isotope discrimination factors; (2) the partitioning of the added tracer with existing exogenous and internal pools of identical atoms, molecules, or compounds and the importance of measuring the specific activity of the incorporated tracer; and (3) the design and implementation of experimental procedures and proper kinetic analysis of the resulting data.

The use of radioisotopic and, to a lesser extent, stable isotopic tracers in ecological studies is often perceived as being straightforward and well documented. Consequently, a detailed working knowledge of the basic chemical, physical, statistical, and analytical principles on which these methods are founded is generally considered unnecessary. However, without such basic background, it is possible to commit inadvertent errors in design or sample analysis that could result in gross misinterpretation of experimental data. In using commercially available isotopes, one relies to a large extent on the manufacturer's claims regarding the radiochemical, radioisotopic, and chemical purity

TABLE V

Radioisotopic Tracers and Applications in Microbial Ecology

Nuclide	Half-life	Emission products	Common chemical forms	Typical specific radioactivities (mCi/mmoles)	Principle and potential ecological applications
^3H	12.35 years	β	H_2, H_2O, specific organic compounds	10^3–10^5	H cycle, indirect tracer for organic carbon
^{13}N	10 min		N_2, NO_2, NO_3, NH_4Cl	10^9–10^{10}	N turnover, nitrification, denitrification, dinitrogen fixation
^{14}C	5730 years	β	$NaHCO_3$, CH_4, CO_2, specific organic compounds	10^0–10^2	C turnover, primary production, tracer for organic carbon
^{32}P	14.3 days	β	H_3PO_4, specific organic compounds	10^1–10^5	P turnover, nucleic acid synthesis
^{33}P	25.5 days	β	H_3PO_4, specific organic compounds	10^1–10^4	P turnover, nucleic acid synthesis, viability studies (autoradiography)
^{35}S	87.4 days	β	$Na_2S_2O_3$, H_2S, H_2SO_4, amino acids, proteins	10^0–10^2	Protein synthesis and degradation, sulfate reduction, sulfide oxidation
^{54}Mn	313 days	γ	$MnCl_2$, $MnCl_3$		Bacterial Mn oxidation and reduction studies
^{55}Fe	2.7 years	X rays, Auger electrons	$FeCl_3$	10^2–10^3	Photosynthetic CO_2 fixation, dinitrogen fixation studies
^{57}Co	271 days	γ	$CoCl_2$	10^3–10^5	Rate of microbial cyanocobalamine (vitamin B_{12}) biosynthesis
^{125}I	60 days	γ	I_2, iodinated proteins	10^3–10^6	Radioimmunoassay, protein degradation studies, viability studies, (^{125}I-INT)
^{31}Si	156 min	β	$Si(OH)_4$		Si metabolism
^{68}Ge	282 days	β	$Ge(OH)_4$, $GeCl_4$	10^2–10^3	Si metabolism

TABLE VI
Stable Isotopic Tracers and Applications in Microbial Ecology

Nuclide	Common chemical forms	Maximum isotope enrichment (at %)	Principle and potential ecological applications
^2H	H_2O, specific organic compounds	95–100	H cycle, indirect tracer for organic carbon
^{13}C	CO_2, CH_4, specific organic compounds	90–99	C turnover, primary production, tracer for organic carbon, and community metabolism
^{15}N	N_2, NH_4Cl, specific organic compounds	95–99.9	N cycle, dinitrogen fixation, denitrification, nitrification
^{18}O	O_2, H_2O, CO_2, specific organic compounds	99.9 (O_2 and H_2O), 20–70% (all others)	Primary production, respiration studies, and metabolic fractionation

of the product, the position and pattern of labeling, and the measured specific radioactivity. Foreign chemicals are sometimes added deliberately to labeled compounds in order to improve the chemical stability (e.g., antioxidants) or radiochemical stability (e.g., radical scavengers) or as bactericidal agents. Impurities may also arise during the chemical, enzymatic, or *in vivo* microbial synthesis of the specific compound or as a result of chemical hydrolysis (especially during long-term storage) and radiolysis. For example, both organic ^{14}C (Williams *et al.*, 1972; D. F. Smith and Horner, 1981) and trace metals, including Mn, Zn, Cu, Ni, Pb, and Fe (Fitzwater *et al.*, 1982), have been detected as contaminants in commercial preparations of [^{14}C]sodium carbonate. The presence of these contaminants may grossly affect the reliability of the tracer for measuring the rate of photosynthesis in aquatic environments. Wand *et al.* (1967) showed that [^3H-methyl]thymidine undergoes decomposition by self-radiolysis and that the resultant breakdown products label cytoplasmic macromolecules other than DNA. Substantial radiolysis was demonstrated to have occurred over a period as short as 3 months. Consequently, even if the radiochemical purity of the isotope is determined at time of purchase, aging may render the isotope unsuitable for certain ecological applications. This potential problem is by no means unique for [^3H]thymidine. It is difficult to comment on an acceptable radiochemical purity level; in theory, even 1% contamination (i. e., an isotope with 99% radiochemical purity) may be totally unacceptable for certain purposes. For example, if the primary radioactive (or stable) isotopic compound is diluted by a very large exogenous pool of nonlabeled material (as is the case for $^{14}CO_2$ or $^{35}SO_4$ in seawater samples), but the contaminant is diluted only slightly (as is the case for ^{14}C-labeled organic contaminants), the actual specific activity of the unexpected tracer (i.e., the contaminant) may be equivalent to or much greater than the anticipated tracer. If less than 1% of the total added radioactivity is assimilated during the incubation period (as is the case for $^{14}CO_2$ or $^{35}SO_4$ in seawater samples), but 100% of the contaminant is consumed, much of the label in the cells may have actually arisen as an experimental artifact.

Isotope discrimination occurs to a finite but variable extent in all tracer experiments. This is because the tracer, by definition, has an atomic mass that is different from that of its naturally most abundant isotope and can be selected for, or against, during cellular metabolism. For many tracers, this effect is negligible, but for others it is substantial, and an appropriate correction factor must be applied. An illustrative, albeit extreme, example is the use of $^3H_2$ as a tracer for H_2. The molecular weight of the tracer in this example is 300% of the molecule nominally being traced, and discrimination would be expected to be substantial.

Another concern regarding the general quantitative reliability of certain tracers (especially radiolabeled organic compounds) is the precise nature and specificity of the label. Radioactive compounds or molecules containing more than one atom of a particular element are either (1) specifically labeled, (2) uniformly labeled, (3) nominally labeled, or (4) generally labeled. The results of a given experiment may depend to a large extent on the specificity and type of labeling. However, even for the so-called specifically labeled compounds (i.e., >95% of the total label incorporated occurs at known positions on the molecule), there is an acknowledged degree of uncertainty regarding the site of labeling if the compound has more than one radioactive atom. For example, a uniform distribution of radioactivity might be assumed to occur in specifically labeled [3,4-^{14}C]glucose, but unfortunately this is neither implied nor guaranteed by the manufacturers' description of the product. Hamilton and Austin (1967) clearly demonstrated the importance of considering the site of labeling when using radiotracers to estimate the efficiency of metabolism in marine bacteria. In experiments using glucose specifically labeled at either the 1, [3,4], or 6 positions, the radiochemical inventory at the end of the experiment (when all glucose had been exhausted) indicated that 74%, 37%, and 8%, respectively, of the label had been converted to $^{14}CO_2$, compared with a value of 40% with uniformly labeled glucose. The differences indicate the degree of participation of the individual carbon atoms in the various pathways of glucose catabolism. These data emphasize the importance of selecting the most appropriate tracer for the particular question of interest. Isotopic tracers are useful ecological tools only if they do, in fact, trace the intended molecules and presumed metabolic pathways.

Another potential problem in tracer experiments results from the common use of ^3H-labeled organic molecules as auxiliary labels for carbon. The primary advantage of the ^3H-labeled compounds is their extremely high specific radioactivities, general availability and relatively low cost (per mCi). The former consideration is especially important for many ecological applications to avoid problems arising from organic nutrient perturbation resulting from the addition of the tracer (Azam and Holm-Hansen, 1973). The major disadvantage is the tendency of certain C-^3H bonds to exhibit exchange reactions with H in the solvent (generally H_2O). This exchange reaction may occur without any chemical change in the organic compound. Such labilization of tritium can occur (1) due to chemical or enzymatic release during the intermediary metabolism of the tracer; (2) during sample storage; or (3) during the extraction, purification, and isolation of intermediate precursors or products. Evans (1972, 1974) discussed isotope exchange reactions at length; anyone involved with the use of ^3H-labeled tracers should consult these sources to become fully cognizant of this potential analytical problem. A second disadvantage in the use of ^3H stems from the nature of the soft β emission ($E_{max} = 0.0816$ MeV), which can create problems of self-absorption when heterogeneous samples are counted (Karl et al., 1981a). This loss of counting efficiency is not compensated for by standard quench corrections and may be substantial for the counting of precipitated macromolecules, such as proteins and nucleic acids.

A very important but often overlooked principle in the use of isotope tracers in ecological studies is the evaluation of the specific activity (or at % enrichment) of the added, incorporated, or metabolized element, molecule, or compound. The ideal tracer is one that can be added without perturbing the steady-state concentration of the ecosystem as a whole. The supplier is generally the source for specific labeling information; however, errors of up to 500% in the quoted specific activities of tritiated nucleosides from one supplier have been reported (Prescott, 1970). A detailed discussion of numerous potential sources of error in the calculation of specific activity has been presented by Monks et al. (1971). In ecological studies, an accurate assessment of the specific activity is further complicated by the dilution of the added tracer with exogenous pools present in the environment, and endogenous pools present in living microbial cells. Without a reliable measurement of the extent of dilution prior to incorporation, tracer uptake data by themselves are of limited use in quantitative microbial ecology. Furthermore, isotope specific activities may change over the course of the labeling period due to the combined effects of depletion (uptake) of the added tracer, and of isotope dilution by a constant regeneration of the exogenous pools (assuming steady-state conditions). In fact, NH_4^+ (Blackburn, 1979; Caperon et al., 1979) and HPO_4^{2-} (Harrison, 1983) regeneration rates have been estimated in environmental samples by measuring the extent of isotope dilution during short-term sample incubation periods.

The final point of concern regards the theoretical bases and mathematical formulations required for the proper interpretation of data arising from the use of isotopes. This topic has been summarized and explicitly discussed by D. F. Smith and Horner (1981), who are of the opinion that ecologists in general and marine biologists in particular are largely ignorant of the vast body of literature available regarding the proper use of stable and radioactive isotopes. Smith and Horner present several multicompartment models and discuss the assumptions, restrictions, and advantages of each approach. This type of rigorous kinetic treatment of tracer data is only now becoming recognized as an essential component of the study of microbial processes in nature.

Attribution of Total Biomass, Metabolic Activity, or Growth to Taxonomic Compartments

For certain assay procedures the parameters measured are specific for one or more groups of microorganisms (e.g., bacteria, algae, fungi, protozoa). However, with many methods there is an unknown and variable contribution from each of the several different groups. Various procedures have evolved in an attempt to distinguish or partition the total microbial community into its component taxonomic categories. The approaches include (1) size fractionation by sieving or membrane filtration; (2) density separation by gradient centrifugation; (3) use of antibiotics, inhibitors, or selective lysis; (4) kinetic separation based on pulse labeling; and (5) separation based on metabolic labeling patterns or pathways. The first and third approaches have been used most extensively in ecological studies.

Fractionation of planktonic communities by size (e.g., Nuclepore filters) has been employed frequently to separate bacteria from unicellular algae (Azam and Hodson, 1977*b*). The success of this approach is dependent on the size spectra of the microorganisms present. Although Azam and Hodson (1977*b*) report that 1-μm filters are 70–90% effective in separating heterotrophic (<1 μm) from autotrophic (>1 μm) activity in open ocean marine plankton, others working with coastal waters indicate that up to 80% of the total heterotrophic activity is associated with a >3-μm-size fraction (Hanson and Wiebe, 1977; Iturriaga and Zsolnay, 1981).

Furthermore, it is now well documented that small (<1–2-μm) phototrophic prokaryotes (Waterbury *et al.*, 1979; Johnson and Sieburth, 1979), and even phototrophic eukaryotes (Johnson and Sieburth, 1982) are widely distributed in the marine environment. Clearly, the reliability of size fractionation procedures is dependent on the habitat in question and may have only limited application in ecological studies. Another uncertainty associated with the size fractionation procedure, especially when subsequent incubation with labeled substrates is required, is whether the sample should be fractionated before or after the incubation procedure. Significant differences have been observed for pre- versus postincubation fractionation procedures (Wheeler *et al.*, 1977), especially with regard to the size distribution of microheterotrophs.

The use of antibiotics has been suggested for the selective inhibition of either prokaryotic, eukaryotic, bacterial, protozoan, or fungal activities. With more than 2000 antibiotics available as biochemical tools (Woodruff and Miller, 1963), the experimental combinations are almost infinite. To be of value in ecological studies, the proposed antibiotic must be broad spectrum but selective for a particular target group, it must react immediately (or nearly so) and must also be active under the given environmental conditions (e.g., in the presence of high particulates, as in soils, or high ionic strength, as in seawater). Furthermore, because most antibiotics are fairly selective in their mode of action (e.g., chloramphenicol and streptomycin inhibit protein synthesis, penicillin inhibits cell wall synthesis, and nalidixic acid inhibits DNA synthesis), it is essential to employ the most suitable inhibitor for the type of study to be undertaken. Ecological data on the use of antibiotics are incomplete and conflicting.

BIOMASS

Observational Methods

Culture Methods

Enumeration of microbial cells by spread or pour plate, most probable number (MPN) techniques, enrichment cultures, or direct microscopy has traditionally been used as a method for estimating total biomass. There are numerous limitations and shortcomings to these techniques, not the least of which is the difficulty of extrapolating cell numbers to biomass. Cell numbers by themselves are of limited ecological value because of the wide range in the dimensions of individual microorganisms. For most field studies, a more relevant expression of viable microorganisms is biomass, or total amount of living organic matter, which is generally expressed in units of dry weight, carbon or nitrogen per unit volume (or area) of habitat. Culture techniques are further complicated by the inability of viable cells to proliferate except under a narrow set of incubation conditions; consequently, the accuracy of these methods is dependent upon the capacity of the selected medium and conditions to satisfy the growth requirements of a significant proportion of the organisms present. Given the physiological diversity of microbial groups and the physical and chemical diversity of the microenvironments, it is not surprising that no single medium or defined set of growth conditions has ever been able to provide the requirements of more than a small percentage of the total bacteria (much less total microorganisms) in any natural habitat. Concerns over the reliability of culture methods are compounded by the well-documented problems arising from the presence of clumps, aggregates, and cells associated with nonliving particulate matter. Consequently, it is concluded that the traditional spread plate technique has little intrinsic value in quantitative synecological studies in nature. This criticism is not new. During the 1972 symposium, "Modern Methods in the Study of Microbial Ecology," a panel was convened to evaluate the current status of the plate-count method. The chairman's summary stated that, "There was ample documentation to verify that the uninitiated, by relying on the plate count method with too little appreciation of its limitations and without supporting techniques may be led to complete waste of time in the accumulation of useless data" (Schmidt, 1973).

Despite its wide acceptance and continued use in ecological studies, this procedure was described by Schmidt as an anachronism in a general discussion of "modern methods." Nevertheless, to the initiated, plate counts are probably not entirely arbitrary or fortuitous (Jensen, 1968), but the quantitative relationship between plate counts and total viable bacteria is likely to vary among different natural habitats. One unique advantage of culture methods is the ability to isolate pure cultures of individual species for subsequent laboratory-based physiological, biochemical, or genetic characterization. One can never be certain, however, that the isolates are truly representative of the microbial assemblages in nature or that they were active *in situ*.

Direct Microscopy

Direct microscopy (e.g., bright-field, dark-field, phase-contrast, reflected, polarized, fluorescence, differential interference) has been used for years as a standard method in microbiology (see also Chapter 1, this volume). Without a doubt, the optical microscope is one of the most important and useful instruments in microbiology. In principle, microscopic observations of cell numbers, in conjunction with estimates of cell size, should provide the absolute standard with which to compare the results derived from more indirect methods. The three major criticisms of direct microscopy are that the method is (1) highly subjective, (2) labor intensive, and (3) relatively insensitive. The first criticism is most important in ecological studies in which the presence of nonliving particulate material often obscures or is mistaken for the microorganisms that are present (especially for soils and sediments) and wherein nutrient limitation often results in a shift of the mean cell size to a value at or below the resolution of the light microscope. This makes enumeration difficult at best and cell sizing impossible. Furthermore, in order to obtain reliable direct microscopic counts of cells from soil, sediments, or highly particulate aquatic environments, separation of the cells from the solid phase and dispersion of clumps and aggregates is absolutely essential. This has been attempted by various techniques, including manual or mechanical shaking and vibration, the use of high shear forces (e.g., Waring blender), ultrasonic treatment, trituration, variations in ionic strength and pH of extracting solution, and treatment with chemical deflocculating agents. Each of these methods has a certain degree of selectivity and variable recovery that depends on the composition of the sample. Furthermore, Brock (1984) recently pointed out that it much easier to see bacteria in aggregates than singly, even if the total density under both conditions is the same. This logically presented argument may have important consequences regarding our ability to evaluate the relative importance of free-living versus attached bacteria in aquatic ecosystems (Brock, 1984). Finally, because direct microscopy is most often performed on preserved samples, it is essential that the microorganisms be stored under proper conditions (including choice of fixatives, their concentrations, and temperature). Although fixation of cells with an aldehyde (e.g., formaldehyde or glutaraldehyde) is most often used as the method of choice, other solutions, such as Lugol's iodine, might offer unique advantages (Pomroy, 1984).

Numerous optical microcopic methods rely on the absorption, refraction, or reflection of light. Although a combination of vital staining and bright-field transmitted illumination is generally the method of choice for visible light microscopy, color infrared photography (Casida, 1968), reflected and reflected-polarized light microscopy (Casida, 1971), and continuously variable amplitude phase-contrast microscopy (Casida, 1976) have all been used to observe bacteria, actinomycetes, fungi, protozoa, and nematodes in unpreserved unstained soil samples.

A fundamental problem in optical microscopy is the limitation imposed by resolution, i.e., 0.25 μm for visible light (Barer, 1974). The twentieth century development and refinement of electron-beam technology, e.g., scanning electron microscope, transmission electron microscope, scanning transmission electron microscope, high-voltage electron microscopy, and analytical electron microscope has provided instruments with resolving power several orders of magnitude greater than conventional optical microscopes. Electron microscopic observation of soil particles and water samples has revealed the presence of bacterial cells of <0.3 μm (dwarf cells, ultrabacterioplankton, minibacteria) at or below the detection limit of the light microscope (Bae *et al.*, 1972; Watson *et al.*, 1977). Unless these cells are extremely abundant, however, their biovolume (hence biomass) must be considered negligible, and there is no assurance that the minicells observed are in fact viable. Nevertheless, the few quantitative comparisons that have been conducted indicate that there is good agreement between light and electron direct microscopic counts of bacteria (Bowden, 1977; Larsson *et al.*, 1978; Watson and Hobbie, 1979).

When microscopic cell counts are compared with plate counts for a given water sample, the former method generally yields values that are two to four orders of magnitude greater than the latter (Jannasch and Jones, 1959; Jensen, 1968). Some controversy has resulted over the interpretation of these data, however. Certain investigators conclude that plating methods underestimate bacterial population densities. Others believe that direct microscopic methods overestimate the true cell densities because of the problems of distinguishing bacterial cells from nonliving particles and the assumption that all recognizable bacteria are viable. This latter point regarding viability is an extremely important one (see Viability).

The various technical problems and limitations inherent in the classic visible light microscopic methods have been alleviated in part by recent developments in fluorescence microscopy, the use of electron-beam instruments, computer-assisted selective microscopic counting, cell sorting and sizing methods, and laser-induced flow cytometry. The use of transmitted and reflected UV fluorescence microscopy greatly facilitates the detection, recognition, and enumeration of certain microbial cells in their natural habitats. Partially obscured organisms are more easily discerned with fluorescence microscopy and, in principle, cells below the theoretical resolution of the optical microscope might also be detected, as cells stained with fluorescent dyes emit light. There are three general applications of UV (and near UV) fluorescence microscopy in microbial ecology: (1) measurement of the primary or autofluorescent characteristics of cells; (2) measurement of secondary fluorescence induced by combination with vital fluorochromes, europium chelate, and fluorescence brighteners (Darken, 1961; Anderson and Westmoreland, 1971) or fluorescent-labeled enzymes, e.g., rhodamine-labeled lysozyme (Millar and Casida, 1970b); and (3) the use of specific fluorescent-labeled antibodies. A limitation of fluorescence techniques, in general, is the background autofluorescence from certain minerals and nonmicrobial organic matter (e.g., higher plants, crustacean exoskeletons).

Autofluorescence has been used successfully to detect and enumerate eukaryotic microalgae, cyanobacteria, photosynthetic bacteria, and methanogenic bacteria. Since autofluorescence decays gradually following cell death, it may be possible to enumerate active cells selectively by this technique. Tsuji and Yanagita (1981) examined the decay of fluorescence in heat-killed samples of several species of marine phytoplankton and reported that within 1 day fluorescence had decreased by 70%. However, the primary fluorescence in microalgae is determined by the cell content of chlorophyll, which is known to vary considerably in response to environmental conditions (especially light). Consequently, the use of fluorescence intensity to distinguish viable algae must yield a lower bound to the true population density. Autofluorescence measurements are best conducted on freshly collected samples. If preservation and storage are required, it is imperative that the activity of the autofluorescent compounds be preserved.

The use of exogenous fluorochromes has evolved from the initial investigations by Strugger (1948) to become the method of choice for direct counting of microorganisms in natural environments. Routine methods now exist for the staining, observation, and enumeration of bacteria, eukaryotic algae and filamentous cyanobacteria (Brock, 1978), and protozoa (Haas, 1982; Sherr and Sherr, 1982). The most frequently employed techniques rely on the use of acridine orange (AO), fluorescein isothiocyanate (FITC), 3,6-diaminoacridine hemisulfate (proflavine), primuline, euchrysine-2GNX, and most recently, the DNA-specific dyes 4,6-diamidino-2-phenylindole (DAPI) and Hoechst dye No. 33258. Most investigators use incidence (i.e., epi-illumination) UV light microscopy, which permits detection of bacterial cells on otherwise optically opaque surfaces. Many refinements, modifications, and methodological improvements have been described over the past few years in order to optimize the overall counting procedures. Two primary concerns with UV epifluorescence counting are (1) nonspecific binding of the fluorochrome to particulate material producing a background autofluorescence that can effectively mask the fluorescence derived from small bacterial cells, and (2) the degree to which nonliving organic matter absorbs added dye. In this respect, the use of DAPI and Hoechst dye No. 33258 have been shown to greatly improve the visualization of cells and overall accuracy of epifluorescence microscopy in seston and particle-rich environments, and for the detection of epiphytic bacteria (Porter and Feig, 1980; Coleman, 1980; Paul, 1982). Although the AO method was originally devised and employed for the purpose of distinguishing living from nonliving cells (distinguished on the basis of fluorescence emission spectrum), subsequent studies indicate that this distinction should not be accepted as a general principle (Korgaonkor and Ranade, 1966; Hobbie *et al.*, 1977).

Recently, Haas (1982) described an epifluorescence staining technique using proflavine that permits the simultaneous enumeration of all groups of planktonic microorganisms. A significant advantage of this method is the ability to distinguish photosynthetic from nonphotosynthetic eukaryotic microflagellates, as the chlorophyll autofluorescence is not masked by the fluorochrome emission, as is the case with AO. This general method for observing both heterotrophic and phototrophic nanoplankton appears to have been further improved by the combined use of the fluorochrome primulin and epifluorescence microscopy (Caron, 1983). This latter method results in a unique separation of the fluorescence of the fluorochrome from that of chlorophyll *a* but requires the use of two separate filter sets, a procedure not necessary for the proflavin procedure. There are a wide variety of individual staining methods and new fluorochromes are constantly being developed and adapted for use in ecological studies. It is not entirely clear whether the individual epifluorescence techniques yield similar quantitative results. Substantial differences in methodology and the subjective nature of direct microscopy in general make standardization of any technique essential before data from different laboratories can be compared routinely with any degree of confidence.

Immunofluorescence (fluorescent antibody) is a specific application of fluorescence microscopy that has been adapted successfully for autecological studies. Methods now exist for the detection of numerous microbial genera (reviewed by Stanley *et al.*, 1979; Bohlool and Schmidt, 1980). The successful application of the fluorescent antibody technique is limited by (1) specificity of the antigenic reaction, (2) interference from background autofluorescence, (3) nonspecific adsorption of the antibody, and (4) the lack of distinction between viable and nonviable cells. The problems of counting a specific bacterial or microalgal population within a mixed assemblage of microorganisms may seem formidable but is actually, in many respects, much easier than estimating total microbial biomass.

Total cell numbers alone cannot be assumed to correlate with biomass because the variation in the size of individual microorganisms is substantial. In principle, electronic particle counters such as the Coulter counter could be used to estimate the biovolume in a particular sample (Cushing and Nicholson, 1966; Kubitscheck, 1969). However, this method can only be applied in the occasional habitat that is devoid of seston and in which the biomass is dominated by relatively large cells. Most often, biomass must be extrapolated from estimates of biovolume, density (weight per unit volume), and percentage dry weight. The first is dependent on microscopic cell sizing, which is tedious and labor intensive. Furthermore, the lower limit of practical particle size measurement with the optical microscope is approximately 0.8 μm; even in the 2–3-μm range, the errors involved are likely to become appreciable (Humphries, 1969). Krambeck *et al.* (1981) described a microcomputer-assisted mensuration method for determining total biovolume by digitizing images of planktonic microorganisms from scanning electron microscopy (SEM) photos. Fuhrman (1981) criticized the use of SEM for the purpose of estimating bacterial cell size distributions. He found that significant (up to 37%) and variable linear shrinkage occurred during sample preparation for SEM as compared with samples prepared for epifluorescence microscopy. The shrinkage problem appears to be especially critical for phototrophic microorganisms due to the extraction of pigments during the solvent-exchange procedures of critical point drying (Montesinos *et al.*, 1983). Recent developments in computer-assisted image microanalysis now provide an alternative to the laborious task of bacterial cell counting and sizing (Sieracki *et al.*, 1985). Nevertheless, the accuracy of cell sizing with optical microscopes is substantially reduced as one approaches the limit of resolution, and the halo effect (Watson *et al.*, 1977) observed with fluorochrome staining may greatly affect the accuracy of all microscopic biovolume extrapolations. Furthermore, it may be difficult to resolve the true dimensions of large cells if only their nucleic acids are stained, e.g., with AO or DAPI.

With the recent development and refinement of laser-based flow cytometry, a new technique for cell counting and sizing has become available to aquatic microbial ecologists. This method has the additional advantages of (1) high statistical precision due to the large number of cells measured, (2) elimination of subjectivity and fluorescence fading inherent in epifluorescence microscopy, and (3) cell sorting capabilities (Muldrow *et al.*, 1982; C. M. Yentsch *et al.*, 1983). The procedure can also be employed in combination with specific fluorochromes for the detection and estimation of cellular macromolecules (e.g., protein, RNA, DNA) or with fluorescent-labeled antigens or antibodies for specific autecological applications.

Once determined, biovolume can subsequently be extrapolated to biomass carbon if the density, percentage dry weight, and dry weight-to-carbon relationships are known. Bakken and Olsen (1983) recently measured the buoyant densities and dry matter contents of selected bacteria and fungi. Their results indicated densities of 1.035 1.093 g cm^{-3} for bacteria and 1.08–1.11 g cm^{-3} for fungal hyphae. These values are close to the frequently stated density of 1.1 g cm^{-3} (Doetsch and Cook, 1973). However, their results for dry matter content were more variable, ranging from 12 to 33% and 18 to 25% (wt/wt) for bacteria and fungi, respectively. A recent report by Bratbak and Dundas (1984) reported that the dry matter content of three different bacterial strains ranged from 31 to 57%, suggesting that the accepted "standard" value of 20% is unacceptable. Furthermore, Romanenko and Dobrynin (1978) report dry weight : wet weight ratios of 0.082–0.291 ($\overline{X} = 0.176$) for a single species of *Pseudomonas*. Newell and Statzell-Tallman (1982) also report a substantial variation for fungal mycelium dry weight ranging from 0.2 to 0.9 g cm^{-3} (expressed as grams dry weight per cm^{-3} fresh volume). Systematic variation was observed between strains and for given strains as a function of mycelium age and environmental conditions. Unfortunately, hyphal age is not readily determined in the field. Furthermore, the common practice of converting total biovolume to biomass assumes that there are no cell-size-related variations in density or percent dry weight. It might be argued, however, that small cells (e.g., minibacteria) would have a much greater surface-to-volume ratio and, because a large percentage of total dry weight in microorganisms is associated with the cell surface, their buoyant densities might be expected to be greater than larger cells of similar composition. Mullin *et al.* (1966) clearly demonstrated that cell carbon per unit volume in unicellular marine algae varies inversely with cell volume. Furthermore, algae armored with SiO_2, $CaCO_2$ or cellulose, bacteria with thick capsules, or microbes containing intracellular storage compounds or metabolic byproducts (e.g., polyphosphate granules, poly-β-hydroxybutyrate, S°) might be expected to have altered buoyant densities and dry matter contents. Bakken and Olsen (1983) reported a range of 7–44% for the dry matter content of isolated capsular materials from three individual bacterial isolates. The density and percent dry matter of individual cells might also be affected by growth rate, although these changes are expected to be small compared with interspecific and interphyletic variations. Finally, if biomass is to be expressed in carbon units, it is also necessary to apply a third correction to the biovolume estimates. Luria (1960) reported a carbon : dry weight ratio of 0.50 ± 0.05 for bacteria, Ferguson and Rublee (1976) used 0.344 at their conversion factor for marine bacteria, and Newell and Statzel-Tallman (1982) gave a range of 0.255–0.470 for marine fungi. One is forced to conclude that the accuracy of a biomass C estimate based solely on direct microscopic analyses cannot be greater than ±100%, considering the extrapolation factors available and the errors in enumeration and cell sizing. Additional applications of optical microscopy are discussed in subsequent sections of this chapter as they relate to determinations of viability and rates of cell division.

Biochemical Methods

Selectivity of culturing methods, subjectivity of microscopical methods, and uncertainties surrounding estimates of carbon biomass from enumeration and cell sizing have hastened the development of alternative approaches for estimating the biomass of either the total microbial community or a specific fraction thereof. The direct measurement of total organic carbon, organic nitrogen, DNA, protein, or other macromolecules is generally unacceptable in ecological studies due to the long residence times of these compounds in the environment following cell death and lysis. These measurements can be used to set an upper bound on total living carbon, however, and in this regard they are extremely useful measurements. In certain aquatic habitats (e.g., eutrophic lakes and coastal seawaters), living microbial carbon can account for 90–100% of the total particulate organic carbon, whereas in other habitats (e.g., deep sea, soils, and sediments), living biomass may be <1% of total carbon.

Ideally, an effective biochemical indicator of living carbon should satisfy the following criteria: (1) the biochemical measured must be present in all living cells or in all organisms of the specific subpopulation of the total microbial assemblage to be assayed; (2) it must be readily metabolized, hydrolyzed, or otherwise eliminated following cell death; (3) it must exist as a uniform and constant percentage of total biomass regardless of environmental conditions; (4) there must exist a convenient method for the extraction and purification (if necessary) of the compound from environmental samples; and (5) a sensitive quantitative assay procedure must be available for routine analysis. Several biomass indicators that have been used or that have potential for use in ecological studies are summarized in Table I. Additional candidates for which there exists an extremely limited (or nonexistent) data base (e.g., diaminopimelic acid for prokaryotes and teichoic acid for gram-positive bacteria) are not discussed here.

Total Microbial Community

ATP. Levin *et al.* (1964) first used ATP measurements as a sensitive method for the detection of living microorganisms in natural samples. Two years later, Holm-Hansen and Booth (1966) proposed that ATP measurements could be used as a measurement of total microbial biomass. Several methodological improvements have since been made during the course of investigations of numerous aquatic and terrestrial environments, and progress in this area has been reviewed several times (Holm-Hansen, 1973; Deming *et al.*, 1979; Karl, 1980). ATP is present in all living cells and occupies a central and obligatory role in metabolism and energy transduction. Many reviewers have criticized the ATP method for its failure to differentiate between individual groups of microorganisms. Although this criticism is justified, one must hasten to add that the real value of the ATP measurement lies in the determination of total microbial community biomass.

ATP functions as an essential link between catabolism and biosynthesis, as a precursor for RNA and DNA synthesis, and, in concert with ADP and AMP, as a regulator of cellular metabolism. Steady-state intracellular concentrations of ATP (i.e., the so-called ATP pool) are regulated by mechanisms that function on various cellular levels and time scales. In general, the pool size is kept at a near basal level which is proportional to biovolume (i.e., at a constant concentration, approximately 1–2 mM). Variable requirements for ATP among different groups of organisms or for a specific organism at different growth rates are seen as changes in the ATP flux (i.e., increases or decreases in ATP pool turnover) rather than by temporal perturbations in the pool size. In theory, ATP measurements should therefore provide an indirect measure of biovolume, although the numerous calibrations of this method have focused exclusively on the stability of the quantitative relationship between ATP and cell carbon.

ATP is biologically labile (turnover time in rapidly growing bacterial cells is less than 1 sec), but it is a very stable molecule on a chemical basis, as evidenced by the variety of methods successfully employed to extract ATP from living cells (Karl, 1980). Consequently, the possibility exists that cell-free ATP may be present in the environment. If unaccounted for, this extracellular ATP would result in an overestimate of the concentration contained exclusively within living cells. Recently, significant concentrations of dissolved ATP have been detected in marine (Azam and Hodson, 1977a) and freshwater (Riemann, 1979; Maki *et al.*, 1983) environments. It is unknown whether this dissolved ATP interferes with the conventional measurement of particulate ATP in aquatic environments (Holm-Hansen and Booth, 1966). Since most samples are concentrated onto membrane filters before extraction, the relative contribution from dissolved ATP should be minimal. However, when working with soils, sediments, or highly particulate aquatic environments, the possibility exists that a variable portion of the total ATP might not have been associated with living organisms at the time of sample extraction. Future studies designed to examine the extent and time course of ATP labeling after the addition $^{32}PO_4$ (i.e., by comparing the ATP γ-^{32}P specific radioactivity with that of the known tracer) should be suitable for evaluating the significance of this potential source of analytical interference (Karl and Bossard, 1985a).

Once extracted from living cells, ATP can be measured at extremely low concentrations (pM, or less) using either the firefly bioluminescence assay (Holm-Hansen and Booth, 1966) or high-performance liquid chromatographic (HPLC) procedures (Davis and White, 1980). The most critical of the five general biomass criteria listed in the preceding section is the validity of the assumption that the measured compound is present as a constant percentage of the total cell mass, or cell carbon. There is no question that the C : ATP ratio in microorganisms varies considerably and, somewhat predictably, between individual taxa and even for a given species as a function of culture conditions (data summarized in Karl, 1980). Among the most conspicuous differences in the C : ATP ratio are those observed between small unicells, i.e., bacteria and microalgae (C : ATP = 220–350) and micrometazoa (C : ATP = 50–150), and the large increases in C : ATP observed during phosphorus limitation and starvation (Karl, 1980). It has been suggested that the total adenine nucleotide concentration (i.e., A_T = [ATP] + [ADP] + [AMP]) might be a better measure of microbial biomass than ATP (Davis and White, 1980); little is known, however, regarding the variability or calibration of C : A_T in microorganisms. The theoretical principles on which the ATP (or A_T) biomass technique is founded suggest that it is most likely a measure of "protoplasm biomass." Therefore, it does not measure the large concentrations of structural carbon (exoskeletons, cell walls), capsular material, slimes, and other extracellular secretions that, under certain environmental conditions, might represent a significant proportion of the total microbial biomass (Hobbie and Lee, 1980; Costerton *et al.*, 1981).

Flavin Nucleotides.   The flavin nucleotides are found in all living cells and function as coenzymes and intermediates in oxidoreduction reactions of electron transport. For the reasons stated for ATP, the total flavin pool might also be expected to be correlated with total microbial biomass. Total flavin can be extracted from cells by the same methods used to extract ATP (and other nucleotides). A sensitive and specific assay system is available that is based on the bacterial luciferase bioluminescence reaction. Chappelle (1975), Okrend *et al.* (1977), and Robrish *et al.* (1979) summarized the existing data on intracellular flavin concentrations in bacteria which, in comparison with the data on ATP, are rather limited. The absence of data on FMN levels from a variety of microbial taxa presents a major limitation on its present application as a total biomass indicator in ecological studies.

Hemin and Hemeprotein.   Protohemin and heme-containing proteins (e.g., peroxidase, catalase, cytochromes) are ubiquitously distributed in living organisms; their concentration has been reported to be proportional to the number of living microorganisms (Oleniacz *et al.*, 1968; Thomas *et al.*, 1977; Miller and Volgelhut, 1978). Relatively few quantitative studies have been made to correlate hemin content with total microbial biomass, but the potential for ecological application exists. Concentrations of hematin compounds as low as 5×10^{-12} M (as protohemin) can be detected using chemiluminescent-linked chemical reactions and commercially available photometers (Ewetz and Thore, 1976). Like the FMN procedure previously described, assay of protohemin may eventually be adapted for routine ecological studies. One common property and unique advantage of the ATP, FMN, and hemin procedures is the fact that they can be automated for continuous-flow analysis of microbiological properties in aquatic environments (Thomas *et al.*, 1977).

Lipid Analysis.   Microbial lipids comprise a complex group of biomolecules that can be used to estimate both total biomass and population structure. Phospholipids, measured as lipid phosphate, are a ubiquitous and major component of all cell membranes. Lipid phosphate has been shown to be an accurate measure of biomass in 13 diverse bacterial monocultures (White *et al.*, 1979c). The phospholipid content ranged from 10 to 100 μmoles and did not change by more than 30–50% for any given species, even in response to extreme environmental stress (White *et al.*, 1979a–c). Phospholipids are isolated from environmental samples by solvent extraction, and the lipid-P is measured using standard spectrophotometric analysis of P. Samples should be extracted immediately after collection because storage can result in significant changes (either increase or decrease, depending on the method of storage) in lipid phosphate concentrations (Federle and White, 1982). Although lipid-P analysis is approximately 100-fold less sensitive than ATP analysis (White *et al.*, 1979*b*), it has been successfully employed to measure total microbial biomass in coastal marine sediments and detrital communities. Phospholipids are degraded in nature with a measured half-life of a few days, hence are not expected to accumulate in nonliving detrital material. The conversion factor used to extrapolate phospholipid to biomass is 50 μmoles phospholipid g^{-1} dry weight (White *et al.*, 1979*a–c*). If biomass-carbon is required, an assumption must be made for the carbon : dry wt ratio. More recently, palmitic acid (16 : 0), a major fatty acid in most organisms, has been suggested as an indicator of total microbial biomass (Federle *et al.*, 1983).

Phospholipid analysis, like ATP, FMN, and hemin, is a nonspecific measure of total microbial biomass. However, relatively simple manipulations of the lipid fraction can yield additional qualitative information regarding community structure and diversity. For example, the proportion of lipid-P not rendered water soluble by mild alkaline methanolysis correlates with the presence of complex prokaryotes (e.g., actinomycetes) and eukaryotes, plasmalogens are restricted to eukaryotes, and anaerobic bacteria and fatty acid methyl ester characterization can be used to detect the presence of gram-positive bacteria, gram-negative bacteria, protozoa, or fungi in mixed microbial assemblages (White *et al.*, 1979*a–c*; King *et al.*, 1977; Bobbie and White, 1980; Federle *et al.*, 1983). This specific organic constituent characterization of natural populations has also been used to estimate the physiological and nutritional status of natural microbial communities, as discussed under Reserve Polymers.

Prokaryotic Biomass

Muramic Acid.   Muramic acid (2-amino-3-0-carboxyethyl-D-glucosamine) is a major component of the peptidoglycan of prokaryotic cell walls (e.g., bacteria, cyanobacteria, prochlorophyta), excluding the archaeobacteria (e.g., halobacteria, methanogenic bacteria). It has been proposed that total prokaryotic biomass can be estimated from measurements of muramic acid (Millar and Casida, 1970*a*; Moriarty, 1975, 1977, 1978; King and White, 1977; Fazio *et al.*, 1979). In its simplest form, this extrapolation assumes a constant muramic acid : dry weight (or carbon) ratio for all prokaryotes and rapid degradation of bacterial cell walls (or at least muramic acid) following cell death. The composition of bacterial cell walls is highly variable among individuals, however, and for a given species as a function of growth rate and environmental conditions (Ellwood and Tempest, 1972). Gram-positive bacteria have far more muramic acid per cell than gram-negative cells, and cyanobacteria, actinomycetes, and bacterial endospores have higher concentrations of muramic acid (on a per cell basis) than do the eubacteria (Millar and Casida, 1970*a*; Moriarty, 1978; White *et al.*, 1979*b*). Consequently, one must know the proportion of gram-positive to -negative bacterial cells and must also have some information regarding the presence or absence of actinomycetes, cyanobacteria, and bacterial spores in each sample. Furthermore, because the muramic acid content of a cell is expected to be proportional to surface area rather than to biomass per se (Ellwood and Tempest, 1972), an estimate of the size distribution of prokaryotic cells is also required to extrapolate the surface area measurement to biovolume and biomass. The use of a single conversion factor is one arbitrary solution to this complex problem, but obviously that approach can lead to unpredictable overestimates or underestimates of the actual prokaryotic biomass.

The measurement of muramic acid is neither simple nor rapid (Moriarty and Hayward, 1982). No standardized protocol exists, and it is possible that the reported variability in muramic acid cell per mass may be due in part to the variation in analytical procedures. For example, King and White (1977) reported that the enzymatic assay of muramic acid (actually of D-lactate derived from alkaline hydrolysis of muramic acid) indicated concentrations that were 10–20 times greater than the colorimetric assay procedure. Because the specificity and sensitivity of the muramic acid method varies considerably among these methods, each investigator must calibrate the method for the procedure used. Direct measurement of muramic acid, either by GLC or HPLC, appears to be superior to the more indirect procedures (e.g., measurement of D-lactate or NADH). Although muramic acid has been shown to be readily degraded by natural microbial communities under certain environmental conditions (King and White, 1977; Moriarty, 1977), direct transmission electron microscopic analysis of sediments revealed that up to 40% of the total muramic acid may be present as empty cell walls ("ghosts"); this obviously affects the accuracy of this method for estimating prokaryotic biomass (Moriarty and Hayward, 1982).

Lipopolysaccharide. The outer membrane of all gram-negative bacteria, including photosynthetic prokaryotes (e.g., sulfur and nonsulfur purple bacteria, green sulfur bacteria, cyanobacteria) contains lipopolysaccharide (LPS). The measurement of LPS has recently been suggested as an indicator of bacterial biomass in the marine environment (Watson *et al.*, 1977; Watson and Hobbie, 1979). Implicit in the application of this technique is the assumption that gram-negative bacteria dominate the total prokaryotic biomass. This assumption may be true for certain seawater samples but is certainly not true for marine sediments (Moriarty and Hayward, 1982), soils, or freshwater habitats. LPS is composed of a lipid A fraction and polysaccharide, both of which exhibit substantial chemical variability among individual bacterial species (Ellwood and Tempest, 1972; Weckesser *et al.*, 1979). Consequently, if total LPS is measured indirectly, as one of the major chemical constitutents, the uncertainty in the biomass extrapolation will be large. Watson *et al.* (1977) adapted the *Limulus* amoebocyte lysate (LAL) test for the *in vitro* detection of bacterial LPS as endotoxin, and a commercially available automatic system for this procedure was recently developed (Jorgensen and Alexander, 1981).

The LAL test reacts with both living and dead cells (Jay, 1977; Jorgensen *et al.*, 1973), and there is at least an order of magnitude variability in the threshold sensitivity of different gram-negative bacteria to this assay (Jay, 1977). Whether this latter observation was the result of variable cell sizes (hence variable absolute concentrations of LPS) cannot be determined from the data presented. In calibrating their method, Watson *et al.* (1977) selected *Escherichia coli* as the laboratory test organism and found that the C : LPS ratio averaged 6.35 (range 3.5–9.0). However, data derived from measurements of *E. coli* may not be applicable to other gram-negative bacteria (Weckesser *et al.*, 1979). Watson and Hobbie (1979) suggest that the LPS content per cell of a "typical" marine bacterium was 2.78 fg compared with 49.6 fg for log phase *E. coli*. Consequently, it may be necessary to investigate the effect of cell size on the C : LPS conversion factor (as for muramic acid), especially for application in oligotrophic environments such as the open ocean.

Although LPS is rapidly metabolized in nature following cell death and lysis, the polysaccharide is more readily attacked than the lipid (Saddler and Wardlaw, 1980); this relative stability of the lipid may have serious consequences for the estimation of LPS by analysis of the lipid A components. Furthermore, cell-free LPS has been detected in aquatic environments and may represent up to 92% of the total LPS in certain habitats (Evans *et al.*, 1978). The production and release of LPS may be related to the physiological state of the cells, and the ratio of free to bound LPS may eventually provide additional information on environmental stress. In addition, detailed quantitative analysis and chemical characterization of the hydroxy fatty acids released from lipid A may also yield information on gram-negative species diversity and overall community structure (Parker *et al.*, 1982), or on taxonomic classification of cyanobacteria populations (Keleti *et al.*, 1979).

Algal Biomass

Measurement of the concentration of photosynthetic pigments (usually as chlorophyll *a* [chl *a*] because it is common to all taxa) is the only biochemical method for estimating microalgal biomass in environmental samples. One basic problem that affects the reliability of this approach is that the chl *a* : C ratio can vary significantly and systematically among algal species and for a given organism as a function of time and day, cell size, temperature, nutrient conditions, growth rate, and especially light history (reviewed by Falkowski, 1981). Variations of 26-fold have been observed for the chl *a* : C ratio of single algal species as a function of environmental conditions and growth rate (Laws *et al.*, 1983). Consequently, reliable extrapolation of chl *a* concentrations to phytoplankton biomass is dependent upon the availability of additional information on species composition, nutrient concentration and limitation, light levels, and growth rate. One approach used in aquatic field studies establishes the relationship between total particulate organic carbon (POC) and chl *a* for a given set of field samples and estimates the mean *in situ* chl *a* : C ratio from the slope of the regression line. Banse (1977) evaluated the assumptions inherent in this method and discusses its problems and limitations. Eppley *et al.* (1977) calculated *in situ* phytoplankton chl *a* : C ratios using a variety of methods for samples collected in the Southern California Bight. The ratios thus calculated ranged from 3.5 to 125 mg chl *a* g^{-1} C with systematic variation in response to seasonal changes in temperature and irradiance.

Chlorophyll is isolated and measured by fluorometry, spectrophotometry, chromatography (thin-layer and HPLC), and, most recently, photoacoustical spectroscopy (Ortner and Rosencwaig, 1977). *In vivo* fluorescence methods have been introduced for the purpose of continuous chl *a* profiling and monitoring in aquatic environments; however, due to the variations observed in the relative fluorescence yield (per unit chl *a*) among different organisms (Strickland, 1968), *in vivo* methods are generally not considered to be acceptable substitutes for quantitative *in vitro* determinations. Although the existence of cell free or detrital chl *a* is possible, chl *a* is generally considered to have a short half-life outside of cells where it is degraded by photo-oxidative, enzymatic, and microbiological (including grazing) activities. However, soluble fluorescence has been reported in certain aquatic environments (Herbland, 1978), and thin-layer chromatographic separations have shown that, on occasion, only 50–70% of the spectrophotometrically determined "chlorophyll" is intact chl *a* (Jeffrey and Hallgraeff, 1980). The use of chromatographic procedures provides additional information on the full pigment spectrum that can yield useful ecological information regarding community structure and perhaps physiological state of the algal populations. Similar information might also be derived from fluorescence spectral signatures as described by Yentsch and Yentsch (1979).

Fungal Biomass

Chitin. Chitin is a hexosamine polymer that is a ubiquitous component of all fungal cell walls except those of Oomycetes; it is also present in the exoskeletons of insects and crustacea. Swift (1973a,b) suggested that, under specified field conditions, chitin measurements might be extrapolated to estimates of total mycelial biomass. Unfortunately, the chitin content of fungi varies from less than 1 up to 25% of the mycelium dry weight (Aaronson, 1966). Furthermore, significant intraspecific variation also occurs depending upon physiological state, nutrient conditions and especially age (Swift, 1973a,b). Consequently, in order to determine reliable estimates of fungal biomass in a particular environmental sample, it is imperative to measure the chitin content (per unit dry weight carbon) of representative organisms separated and extracted from the same habitat.

Chitin can be measured either as *n*-acetylglucosamine following acid hydrolysis or enzymic release, or as chitosan following alkaline hydrolysis. Because the conditions of hydrolysis affect the amount of hexosamine recovered (Ride and Drysdale, 1972), it is essential to establish separate extrapolation factors for each individual detection method. Finally, Frankland (1975) reported that as many as 17–47% of fungal hyphae in woodland soils may be completely devoid of cytoplasm. Consequently, there is no assurance that the total chitin measured in a particular habitat is actually associated with living fungi.

Lipids. White *et al.* (1980) experimentally manipulated a natural detrital microbial population to enrich selectively for fungi in order to define some key biochemical indicators of fungal biomass. Their study included a comprehensive analysis of selected lipid constituents. Of the numerous compounds evaluated, polyenoic fatty acids, wall inositol (rather than glucosamine), and possibly ergosterol-like steroids were considered the most appropriate biomarkers for fungal populations. Ergosterol is the major sterol in most yeasts and fungi and may range from 0.005 to 2.9% of the total dry weight (Weete, 1973). Unfortunately, few data are available to provide reliable extrapolations to dry weight or to total cell carbon, and virtually nothing is known about the decomposition or turnover rates of these chemical tracers in nature. Lee *et al.* (1980) also selected ergosterol as a marker for fungal biomass and provided a factor of 2 mg g^{-1} dry wt as an extrapolation factor, although they suggest that this may be in error by up to a factor of 4.

Miscellaneous Incubation Methods

Conventional culture incubation methods (e.g., spread plate, MPN) for enumerating individual groups of microorganisms in environmental samples were discussed in the section, Culture Methods. This section presents three relatively new approaches for estimating microbial biomass that also require sample incubation. Although each method represents a novel approach, they all suffer from potential artifacts induced by changes in biomass, population structure or metabolic activity during sample incubation. Consequently, it is essential to monitor the kinetics of product formation in order to detect and exclude from consideration any anomalous results. The methods discussed below include the fumigation/respiration method for total soil microbial biomass, $^{14}CO_2$ labeling of chl *a* for algal biomass, and electrochemical (impedance and potentiometric) methods for detection bacterial biomass.

Fumigation/Respiration. In a series of papers, Jenkinson and co-workers introduced a novel method for estimating total microbial biomass in soil samples (Jenkinson, 1966, 1976; Jenkinson and Powlson, 1976; and additional references cited therein). In this method, living microbial carbon is made susceptible to mineralization by fumigation with chloroform vapors. The soil is then inoculated with a small quantity of the original nonfumigated soil ($\leqslant 1\%$ by volume) and incubated at a constant temperature (22°–25°C) for 10 days. The CO_2 evolved during the incubation period, in excess of that evolved in the nonfumigated control, is presumed to have been derived from biomass-

C killed during the fumigation process. By applying an experimentally derived factor (k) equivalent to the proportion of biomass-C converted to CO_2 during the 10-day incubation period, one can estimate the biomass of living organisms present in the original sample. The fumigation method has compared favorably with biomass estimates calculated from microscopic biovolume determinations (Jenkinson et al., 1976).

Three major assumptions are inherent in the use of this technique: (1) k is a constant for all soil types and microbial communities, (2) the fumigation process does not alter nonliving carbon in any manner that will make it more or less susceptible to mineralization, and (3) all living organisms are killed by the fumigation procedure.

Over the years, k has varied from 0.3 (i.e., 30% mineralization during the first 10 days after fumigation) to 0.5 (Jenkinson, 1966, 1976). More recently, Anderson and Domsch (1978) demonstrated that statistically significant differences exist between bacteria and fungi, with remineralization percentages of 33.3% ± 9.9 and 43.7% ± 5.3, respectively. Consequently, the accuracy of the biomass extrapolation is dependent on prior knowledge of the distribution of biomass among the various microbial taxa. Further experimentation should be performed to include other microorganisms such as microalgae and nematodes, as well. Regarding the second assumption, Jenkinson (1976) provided convincing experimental evidence to support his view that chloroform fumigation has no substantial effect on the remineralization of nonliving carbon, a conclusion challenged by Shields et al. (1974) but supported by the experiments of Anderson and Domsch (1978). The susceptibility of naturally occurring microorganisms to the chloroform fumigation treatment has not been systematically evaluated, but the effectiveness might be expected to be habitat dependent. One potential difficulty might arise from the presence of macroorganisms (which might also be expected to contribute to the postfumigation CO_2 flush). However, these organisms could be removed by sieving the soil fumigation. The application of this technique might be limited to environments in which living microbial biomass is expected to represent a significant, if not major, source of utilizable organic matter. Under any other condition, the differences observed between the control and reinoculated samples would be expected to be negligible. Although the use of this method has been restricted to soil samples, there is no a priori reason why the basic approach (with minor modification) could not be used to estimate total microbial biomass in aquatic sediments or intertidal beach sand. A recent study by Ross et al. (1980) of nine diverse topsoils revealed a biomass C (from the fumigation/respiration technique) to ATP ratio of 248 (range 163–423), a value that is remarkably (and perhaps coincidentally) similar to the C : ATP ratio measured in laboratory cultures of microorganisms (reviewed by Karl, 1980).

Chlorophyll a Labeling. Redalje and Laws (1981) described a new method for estimating phytoplankton biomass C in mixed assemblages of microorganisms. The method requires a 6–12-hr incubation with $^{14}CO_2$, followed by

chl *a* extraction, isolation, and determination of the specific radioactivity, i.e.,
Ci (g C)$^{-1}$ in chl *a*, and a separate measurement of total radioactivity incorporated into the total phytoplankton carbon pool. It is assumed (and supported by experimentation) that after a sufficiently long incubation period,
the specific activity of the chl *a* carbon pool is equivalent to the specific activity
of the total phytoplankton carbon pool. The total phytoplankton biomass-C
can thus be calculated by dividing the total ^{14}C activity in the phytoplankton
fraction by the specific activity of the chl *a* pool. The unique distribution of
chl *a* in photoautotrophs and the obligate association of chl *a* with living cells
confer a high degree of selectivity for this method. Redalje (1983) recently
compared the ^{14}C incubation method with biomass estimates derived from
microscopic enumeration and sizing of phytoplankton cells. The direct microscopic values ranged from 89 to 193% ($\overline{X}$ = 146%) of the ^{14}C incubation
method for six samples from Southern California coastal waters.

Electrochemical Methods.   Several electrochemical methods have been devised for estimating the population density of bacteria. Although these methods were originally developed for the potentiometric detection of hydrogen
(Wilkens *et al.,* 1974; Wilkens, 1978), or of changes in electrical impedance
(Silverman and Munoz, 1979) during the growth of microbial populations in
commercial fermentations, the methods have much wider potential application for environmental biomass monitoring. Both methods are simple and
are capable of detecting the growth of single bacterium during an incubation
period of 8–10 hr. For reasons as yet unknown, gram-positive bacteria had
detection end points significantly longer than those observed for gram-negative bacteria. Wilkens *et al.* (1980) recently described a combined membrane
filtration–electrochemical microbial detection method that was successfully
applied to freshwater and estuarine water samples. It is not yet possible to
convert electrochemical potential measurements to estimates of bacterial carbon, but these methods will undoubtedly be explored for future ecological
application.

VIABILITY

One major problem in obtaining accurate estimates of microbial biomass,
whether by direct microscopic or biochemical methods, is the possible presence
of either metabolically inactive or nonviable cells in the environment. Postgate
and Hunter (1963) demonstrated that continuous cultures, at exceedingly low
flow rates, do in fact maintain steady-state populations with a finite proportion
of nonviable microorganisms. It might be argued, however, that nonrepli-
cating cells would be rapidly consumed or otherwise destroyed and consequently should not persist in nature for very long. However, direct attempts
to evaluate the residence time of artificially killed or debilitated microorganisms have produced variable results depending on the type of cell, the en-

vironment studied, and the method used to monitor persistence. Because of the extremely large discrepancies observed among estimates of bacterial cell numbers based on culture methods and direct microscopy and the possibility of microbial dormancy induced by adverse environmental conditions, it is conceivable that only a very small percentage of the recognizable bacterial cells in selected environmental samples are in fact metabolically active, let alone viable.

For unicellular, haploid microorganisms, viability is expressed as cell division, and a viable microorganism is defined as one having the potential for reproduction. Detecting this potential is a major problem in microbial ecology. Classic viable counts by plating dilutions of a sample are not adequate to count all such individuals in a heterogeneous microbial population. The unambiguous test—observation of completion of reproduction—requires at least incubation of the sample and then can prove positive only if appropriate incubation conditions are provided. Maintenance of this potential requires metabolic activity and usually growth at some time before reproduction; it does not require continuous metabolic activity at a measurable level, however, as in the example of a dormant but viable population. By contrast, metabolic activity can continue even after the potential for reproduction has been lost. Consequently, viability cannot be assessed in an unincubated sample, and measurements of metabolic activity yield information that cannot be regarded as rigorously implying either a maximum or a minimum approximation of viability within a naturally occurring microbial community. Furthermore, it is important that we do not equate viability with the community capacity for biogeochemical transformations. The latter is determined by the steady-state metabolic rate rather than merely by the presence of viable cells. However, the potential long-term capacity for biogeochemical transformations is determined by viability.

Several methods for estimating viability have been utilized in field studies, and some of these may eventually prove suitable and reliable indicators. A few of the more promising methods, each involving microscopical procedures to some extent, are reviewed in this section (see also Chapter 1, this volume).

Vital Fluorogenic Dyes

Esterified dyes such as fluorescein diacetate (FDA) and fluorescein dibutyrate (FDB) are nonfluorescent substrates that, upon transport into cells, are hydrolyzed to fluorescein by nonspecific esterases believed to be present in all presumably viable microorganisms. The product is easily visible under incident or transmitted UV light, and fluorescence can be used to distinguish living from nonliving cells including bacteria (Lundgren, 1981), yeasts (Paton and Jones, 1975), and fungi (Soderstrom, 1977). More recently, this method has been adapted to estimate total microbial biomass by extracting and quantitating the fluorescein produced and relating this value to rates determined from known microbial populations (Swisher and Carroll, 1980). Soderstrom

(1977) reported that all 50 species of fungi tested fluoresced in the presence of FDA. However, the intensity of fluorescence was dependent upon growth rate; resting stages, stationary phase cultures, and spores did not stain. Clearly, then, this method cannot rigorously measure viability among fungi.

Bacteria also displayed substantial variability in their response to FDA. Chrzanowski *et al.* (1984) report that FDA is unable to penetrate the outer membrane of the gram-negative cell envelope. Consequently, this measurement may have only limited application in studies of bacterial ecology. Of the 111 soil bacterial isolates tested by Lundgren (1981), only 74 (or 67%) exhibited FDA staining when grown in broth culture. Of the 37 FDA-negative strains, approximately 50% became reactive when grown in sterilized soil. Babiuk and Paul (1970) claimed that bacterial cells more than 4 hr old are impermeable to FDA, an observation that suggests an obvious potential for underestimating the true percentage of viable cells by this method. Nevertheless, a study of 10 natural soil samples showed that FDA-positive (viable) cells made up 27–83% of the AO total count indicating a relatively high but variable level of viability. Since the FDA counts exceeded the plate counts by at least an order of magnitude, one must (again) conclude that culture methods grossly underestimate the number of viable bacteria in soils.

Dye-Exclusion Test

The dye-exclusion test is based on the assumption that certain compounds, such as methylene blue, trypan blue, eosin, Nile blue, and amethyst violet, are able to penetrate cells only if the plasma membrane is destroyed or severely compromised (Bonora and Mares, 1982). The percentage of impermeable cells can be determined by direct microscopic examination or, under ideal conditions of high cell density and absence of nonliving particulate material, by spectrophotometry. Methylene blue, for example, accumulates quickly in dead and injured (nonviable) eukaryotic microorganisms (yeasts and algae). Each viable cell has a threshold tolerance for dye exclusion (Bonora and Mares, 1982); thus it is essential that this limit not be exceeded. Although to my knowledge this method has not been applied to ecological studies, it theoretically has the potential to assist in differentiating between living and nonliving microoranisms and consequently would provide a maximum estimate of viability.

Microautoradiography

In principle, microautoradiography might represent an alternative method for estimating viability, as it can be used to detect *in situ* nutrient uptake and incorporation into macromolecules. If appropriate substrates are employed, viability (i.e., the potential for reproduction) can also be determined. In this procedure, natural populations of microorganisms are exposed to radiolabeled substrates for a brief period of time, preserved with appropriate fixatives, filtered (if necessary), and attached to a microscope slide. The slide is subsequently covered with stripping film or coated with liquid emulsion upon which the radiation produces a latent image that, when developed with standard photographic chemicals, forms metallic silver grains that are readily visible with the light microscope. In the stripping film technique (Hoppe, 1976, 1977), cells (usually on a filter) are exposed to a film plate for a predetermined period of time after which the filters are removed and the plate developed. The film is examined with bright-field optics, and silver grain aggregations consisting of >3–5 spots are registered as active bacterial cells. A separate (usually stained) sample is examined for total cell numbers in order to calculate percent active cells. Kronenberg (1979) described a variation of the single-film technique that can be used to differentiate ^3H from ^{14}C, ^{35}S, ^{33}P, or ^{125}I in double-labeled preparations. An alternate procedure uses liquid emulsion coated on the entire slide. One advantage of this procedure is that the developed autoradiogram can be stained with AO (and presumably other fluorochromes) for simultaneous epifluorescence counting and microautoradiography of a single preparation (Meyer-Reil, 1978a). More recently, Tabor and Neihof (1982b) described a novel procedure that increases the sensitivity and visual clarity of the image by removing the filter before viewing and by arranging the sample so that the cells are not obscured by the opaque silver grains. This preparation improves the recognition of labeled cells and results in a significantly greater number of labeled microorganisms when compared with corresponding samples prepared by the Meyer-Reil (1978a) technique. There are many variables to this general protocol and procedural decisions that must be made, all of which can greatly affect the estimation of percent active cells. The most important decisions regard (1) choice of substrate and isotope label, (2) isotope concentration and specific radioactivity, (3) incubation time, (4) exposure time, and (5) sensitivity and resolving power of the emulsion.

Obviously the choice of substrate is dictated by the specific ecological question. For example, if one desires to know the percentage of viable autotrophic cells in a particular sample, a logical isotope would be $^{14}CO_2$. For total heterotrophs, the choice is somewhat arbitrary and the results selective. For example, Hoppe (1976) found that 29% of the cells in Kiel Bay were active when assayed with [^{14}C]glucose, whereas 100% were active when a [^3H]amino acid mixture was used. Furthermore, Ramsay (1974) reported that the percentage of active bacteria with respect to [^3H]thymidine was always less than that measured with [^3H]glucose. However, the ratio of thymidine : glucose positive cells approached unity as the active percentage of the total population, as measured by [^3H]glucose, increased. Using three different substrates (glucose, glutamate, and thymidine), Novitsky (1983a) showed variable seawater depth-dependent labeling patterns. At all depths sampled, however, the percentage of labeled cells was greatest with glutamate (e.g., at

10 m, 35% glutamate versus 14% thymidine versus 4% glucose). He also commented on the fact that the nonlabeled cells appeared "healthy" in the sense of having a bright uniform fluorescence, regular shape, and apparently intact cell envelopes. In a companion study of microbial activity at the sea-water–sediment interface, Novitsky (1983*b*) examined changes in cell numbers and percentage labeling over incubation periods as long as 10 days. By observing increases in total cell number, he was able to set a lower bound on the number of viable cells. The percentge of these known viable cells that were responsive to radiolabeled glutamate and thymidine was only 5.5% and 2.0%, respectively. Clearly, dependence on microautoradiography alone can result in a gross underestimate of viable cells. Ideally, one would like to add a complex mixture of organic substrates in order to eliminate the selectivity based on metabolic differences of the endogenous populations. The use of high specific activity ^{3}H-labeled substrates permits the addition of the required radioactivity without adding substantial amounts of organic carbon. In this regard, the use of essential inorganic substrates (e.g., $^{33}PO_4$ or $^{35}SO_4$) might offer some unique advantages in total microbial viability-assessment studies.

By analogy with photographic development, the quantitative results of autoradiography are influenced to a large extent by exposure time because the appearance of a "positive" cell is a threshold-dependent phenomenon. The optimum exposure period is a function of the external specific activity of the isotope, the cell-specific uptake rate (which is dependent on the degree of metabolic activity), and the incubation period. Latent image fading, chemography (both negative and positive), and the natural variability of radioactive emissions from the source materials are also important factors that influence the quality of the autoradiogram. Although various attempts have been made to standardize the autoradiographic procedure with respect to these variables, in practice there can be no single set of optimum conditions. Even in pure culture studies, the apparent percentage of viable cells is clearly related to isotope concentration, incubation, and exposure (Ramsay, 1974). Consequently, mixed populations of microorganisms require various incubation and exposure times that must be determined empirically for each set of environmental samples.

The criterion for deciding whether a cell is labeled varies among individual investigators. The common practice of requiring the presence of a minimum number of silver grains (e.g., a cluster of three to five grains) may underestimate the actual percentage of labeled cells. England *et al.* (1973) proposed a simple analytical method (based on the assumption that both background and sample grain densities follow Poisson distributions) that can be used to detect radioactive cells without any grains over them at all. Scintillation autoradiography (fluorography) has also been suggested as a means of increasing the sensitivity (i.e., lower detection limit), although this procedure has not been used extensively in ecological studies.

Microautoradiography can also be used to estimate the actual quantity of isotope taken up by individual cells in the natural microbial assemblage. This approach has been used extensively for measuring rates of primary production in selected algal species (Watt, 1971; Maguire and Neill, 1971; Knoechel and Kalff, 1978). In general, two different approaches have been described: grain density and track autoradiography. In the former, the number of silver grains is used as a measure of the quantity of radioactivity incorporated, and in the latter a characteristic series of silver grains (i.e., a track) records the path of each radioactive disintegration through a thick layer emulsion (Knoechel and Kalff, 1976*b*). Knoechel and Kalff (1976*a*) criticized the use of grain-density autoradiography for quantitative investigations; however, Paerl and Stull (1979) argued that the criticisms presented were based primarily on theoretical considerations and were not entirely valid for normal environmental applications. By far the largest analytical problem with either type of quantitative autoradiography appears to be the uncontrollable loss or translocation of radioactivity during sample fixation and processing (Silver and Davoll, 1978; R. E. Smith and Kalff, 1983). Paerl (1984) recently suggested that rapid freezing of filtered samples in liquid N_2 followed by freeze drying prior to autoradiographic preparation may reduce, or eliminate, potential artifacts. An alternative solution might be to look specifically at the amount of radioactivity incorporated into selected macromolecular constituents (e.g., lipids or nucleic acids); this could be ascertained by combining either track or grain density autoradiography with prior selective chemical treatments (M. L. Brock and Brock, 1968), thereby eliminating the need for preservation per se. In this way, one might also be able to distinguish metabolizing cells from those that are also growing or dividing. The procedure may ultimately form the basis for a rigorous test of viability.

INT Labeling

The active electron-transport system of respiring microorganisms reduces 2-(*p*-iodophenyl)-3-(*p*-nitrophenyl)-5 phenyltetrazolium chloride (INT) to INT-formazan, which accumulates in the cell as an optically dense deposit that can be detected using bright-field illumination. When used in combination with AO epifluorescence microscopy (Zimmermann *et al.*, 1978) or malachite green staining (Dutton *et al.*, 1983), an INT staining method can be used to estimate the proportion of actively metabolizing cells. This technique is fairly simple and straightforward and is clearly less tedious and time consuming than microautoradiography, which might be expected to yield comparable ecological data. Only a brief incubation period is required (usually 20 min), so that perturbations in the *in situ* metabolic state are considered minimal. Cells pretreated with preservative (formaldehyde, $HgCl_2$ or Lugol's solution) or dry heat (40 min at 70°C) lose their ability to reduce INT. Tabor and Neihof (1982*a*) recently described a modification of the INT-formazan method that represents a substantial improvement by eliminating several potential artifacts involved in the microscopic examination of formazan-labeled bacterial cells.

Several limitations of this technique compromise the present ecological interpretations of *in vivo* INT labeling. First, the INT-formazan must be deposited in a quantity sufficient to be visualized with the optical microscope. This sets a lower limit to the size of the deposit that can be detected and restricts the use of this method to environments dominated by large (0.5–1-μm) bacteria. Unfortunately, much of the current debate regarding viability and respiratory activity is directed toward the so-called minibacteria, or "dwarfs" that predominate in oligotrophic environments. Newell (1984) recently reported the presence of numerous nonformazan (but formazanlike) dark spots in association with organic flocs that were formaldehyde-fixed before INT incubation. This potential for false-positive bacterial cells must be carefully evaluated for all environmental samples. It may be possible to use radiolabeled INT and microautoradiography to increase the sensitivity and specificity of this assay procedure, but this may have no real advantage over the use of more conventional isotopes to assess viability (e.g., $^{33}PO_4$, $^{35}SO_4$). INT is also known to act as a metabolic inhibitor; unpublished studies in the author's laboratory indicate that the addition of INT (at concentrations recommended by Zimmermann *et al.*, 1978) to marine and freshwater microbial communities results in an immediate inhibition of cellular metabolism, and hence loss of viability, as detected by several independent criteria, e.g., rates of RNA and DNA synthesis and amino acid assimilation (D. Karl, unpublished data). Consequently, the ability to produce a detectable (visible) intracellular INT-formazan deposit prior to inhibition of cellular metabolism is clearly related to the intensity of *in vivo* respiratory activity per unit biomass (or instantaneous rate of INT-formazan production) rather than to the mere presence or absence of *in vivo* respiration. This interpretation is consistent with the time course of labeling, which generally indicates a maximum value within a few minutes of incubation with no additional accumulation of labeled cells (Zimmermann *et al.*, 1978; Maki and Remsen, 1981). Consequently, the proportion of INT positive cells must represent the number of cells at or above some specific, but unknown, threshold metabolic activity. Dormant or slowly growing but viable microorganisms would obviously be overlooked by this technique. The extremely low percentage (5–35%) of actively respiring cells observed in environments presumed to be dominated by metabolically active microbial populations (e.g., eutrophic lakes, coastal seawaters) is in sharp contrast to the results obtained from microautoradiographic analyses; this finding suggests that the two methods may be enumerating different populations of microorganisms. INT-positive cells must be considered to represent a lower bound on the actual proportion of viable cells.

The INT-formazan technique has recently been combined with the use of specific fluorescent antibodies to study the autecology of *Legionella pneumophila* under *in situ* conditions (Fliermans *et al.*, 1981). Another adaptation is that reported by Paerl and Bland (1982), who proposed a microscopic method that uses a suite of tetrazolium salts, each having its own characteristic redox potential, in order to localize specific sites of biological reduction (e.g., N_2 fixation, CO_2 fixation, H_2 uptake) in filamentous cyanobacteria.

Yeast Extract/Nalidixic Acid

Kogure *et al.* (1979) recently proposed a novel technique for enumerating viable bacterial cells in seawater. The method involves the addition of yeast extract (0.025%) as a growth substrate and nalidixic acid (0.002%) as an inhibitor of DNA synthesis. The implicit assumption is that viable cells will elongate in response to the combined effects of nutrient addition and inhibition of cell division. After a suitable incubation period (generally 6 hr), the samples are stained and observed with epifluorescence microscopy. Elongated cells are enumerated as viable cells. This method would be expected to underestimate the percentage of viable bacterial cells in environments in which gram-positive (or nalidixic acid-insensitve) bacteria are relatively abundant and may consequently have a somewhat limited application in general ecological studies. Although the elongated cells from coastal marine environments may reach lengths of 3–4 μm (and sometimes up to 10 μm) during the 6-hr incubation period, the differences between viable and nonviable cells are expected to be less dramatic in more oligotrophic environments. Consequently, a decision based strictly on cell dimensions becomes increasingly subjective, since the distinction between a large nonviable cell and an elongate viable cell becomes impossible. Maki and Remsen (1981) compared the yeast extract/nalidixic acid technique with the INT procedure described previously. No significant differences were observed for counts of viable bacteria. Total bacterial counts in the yeast extract method were generally higher, however, an observation that these investigators interpreted as evidence for the presence of cell division by nalidixic acid insensitive bacteria in response to nutrient addition. When the two techniques were employed simultaneously, most cells that appeared to be enlarged also contained formazan deposits and most cells with formazan appeared enlarged (Maki and Remsen, 1981). However, the presence of enlarged cells without formazan and small (nonelongate) formazan-positive cells indicated that the two methods were not directly comparable, an observation made in cultures as well as natural samples (Maki and Remsen, 1981). The direct viable count ranged from 1.5 to 39.8% ($\overline{X}$ = 11.2%) of the total direct count in eutrophic Tokyo Bay, and 0.7–7.9% ($\overline{X}$ = 2.8%) in offshore waters (Kogure *et al.,* 1980). In all environments examined, the direct viable count exceeded plate count, occasionally by a factor of up to three orders of magnitude.

PHYSIOLOGICAL POTENTIAL/NUTRITIONAL STATE

Several methods are available with which to determine the physiological potential or metabolic status of natural microbial communities. Certain approaches are well documented and frequently employed in ecological studies (e.g., measurement of electron-transport system activity). Other methods are more specialized or lack an extensive data base to support them. When used in conjunction with other *in situ* measurements of metabolism or growth, however, the information derived from the methods described below may prove invaluable for defining the nutritional or metabolic state of the microbial communities being investigated.

Enzymic Activity or Potential

Measurement of the *in vivo* or *in vitro* activity of specific enzymes in environmental samples can be used to provide useful information regarding the physiological state of microbial populations in nature and to assess the potential for various biochemical transformations. Certain enzymes (e.g., lipase, amylase, protease, dehydrogenase) are either known or believed to be present in all microorganisms and might be expected to be correlated with biomass or metabolic activity. Others are inducible/derepressible (e.g., alkaline phosphatase, nitrate reductase), or unique to certain microbial groups (e.g., luciferase, nitrogenase, chitinase) and might be used to define the nutritional status or metabolic/physiologic diversity of the population. For example, the ratio of the two major carboxylating enzymes, ribulose-1, 5-bisphosphate carboxylase (RuBPCase), and phosphoenol pyruvate carboxylase (PEPCase), may be used as an approximate index of the relative contribution of C-3 versus C-4 plants (Morris, 1980) or of the occurrence of chemolithotrophic microbial activity (Glover, 1983; Tuttle *et al.*, 1983). Furthermore, the ratio of photosynthesis to total carboxylase activity in marine algae varies with physiological state (Glover and Morris, 1979); thus, the use of this ratio might help define the growth state of natural populations under certain environmental conditions. The presence of the enzyme alkaline phosphatase has also been used to indicate environmental conditions of inorganic phosphate (P_i) limitation. This approach should be viewed with caution because the ability to synthesize alkaline phosphatase is not a universal characteristic and because alkaline phosphatase may also be present as a constitutive enzyme in many microbial taxa (Karl and Craven, 1980). Measurement of chitinase, cellulase, DNase, and polyphenol oxidase can provide information on the degradative potential of the communities present or changes in community structure.

It is important to keep in mind that the activity measured in most environmental samples actually represents enzymes derived from plant, animal, and microbial sources. Particulate enzymic activity may be derived from growing cells, metabolically active but nonproliferating cells, intact dead cells, cell fragments, or as free enzymes adsorbed onto particles. It is also well documented that extracellular enzymes are also present in both terrestrial (R. G. Burns, 1978) and aquatic (Kim and ZoBell, 1974; Somville and Billen, 1983; Hoppe, 1983) habitats. Because the ultimate source of the enzyme activity is not always known, it may be misleading to assume an exclusively microbial origin. Two specific areas in which enzyme assays have proved extremely useful are in the study of microbial nitrogen assimilation and in the measurement of total electron-transport system activity.

Microbial Nitrogen Cycle

A variety of nitrogen-assimilating enzymes are found in microorganisms, including nitrate reductase (NaR), nitrite reductase (NiR), glutamate dehydrogenase (GDH), glutamine synthetase/glutamine 2-oxyglutarate aminotransferase (GS/GOGAT), and nitrogenase. One approach to the study of the N cycle in nature is to determine the *in vitro* rates of these individual enzymes. For example, NaR is an inducible enzyme; growth with NO_3^- is required for the detection of cellular NaR activity. Consequently, the presence of NaR in natural phytoplankton assemblages indicates utilization of NO_3^-, while a negative result might imply growth on NH_4^+, or N limitation (Eppley *et al.*, 1969). Eppley *et al.* (1970) showed that the specific activity of NaR in the euphotic zone of the Peru current (NaR activity/unit chl *a*) increased with increasing NO_3^- concentration. NaR activity also exhibited a fourfold diel variation, with maximum rates during the period of maximum photosynthesis. The activity of NaR extracted from surface seawater samples is quickly repressed by the addition of NH_4^+ (Packard and Blasco, 1974), which adds support to its use as an indicator of recent environmental conditions. In principle, *in vitro* activity might be expected to correlate with *in vivo* activity, since intracellular NO_3^- concentrations may be present at saturating levels. However, the rate of nitrogen assimilation estimated from the *in vitro* activity of NaR, in certain cases, is as low as 15% of the rate measured by $^{15}NO_3^-$ tracer experiments (Eppley *et al.*, 1970).

Dortch *et al.* (1979) likewise reported that NH_4^+ assimilation rates, as estimated by GDH activity, represent approximately 5% of the actual NH_4^+ uptake and assimilation rates. Consequently, it may be premature to use *in vitro* activity values to estimate N fluxes in nature. It is very likely that the GS/GOGAT system of NH_4^+ assimilation is favored over the GDH system under most environmental conditions (Miflin and Lea, 1976; Dortch *et al.*, 1979; D. J. Burns, 1983). In fact, the GS/GOGAT system is now believed to be responsible for the assimilation of NO_3^-, NO_2^-, and N_2, in addition to NH_4^+. Unfortunately, no field measurements of GS activity are available at the present time, although suitably sensitive techniques now exist (D. J. Burns, 1983; Thomas *et al.*, 1984; Bressler and Ahmed, 1984). Finally, nitrogenase activity (as measured by acetylene reduction) has also been used to estimate the rate of dinitrogen fixation in selected natural habitats. Unfortunately, the molar ratio of acetylene reduced to N_2 assimilated varies by at least one order of magnitude from the theoretical value of 3 (Hardy *et al.*, 1973; Mague *et al.*, 1974; Graham *et al.*, 1980). Therefore, reliable extrapolations are not possible without additional ecological information. Investigators are strongly urged to employ an experimentally determined conversion factor obtained from simultaneous measurements of acetylene reduction and $^{15}N_2$ reduction rather than relying on a theoretical value (Peterson and Burris, 1976).

Respiratory Electron-Transport System Activity

Electron-transport system (ETS) activity is present in most living cells and functions to channel electrons produced during cellular metabolism to an appropriate terminal acceptor, which most frequently is molecular oxygen. During this process, a portion of the chemical energy is coupled to ATP production (oxidative phosphorylation). Since the pioneering work of Packard (1971), measurements of *in vitro* ETS activity have been used to estimate the potential *in situ* respiration of bacteria, eucaryotic algae and zooplankton in soil, sediment and aquatic environments (Trevors, 1984). In principle, the method is fairly straightforward, requires no sample incubation, and has the sensitivity required to provide estimates of oxygen consumption in even the most oligotrophic environments (Packard *et al.*, 1971). Unfortunately, the actual value obtained for a particular sample is dependent to a large extent upon the extraction and assay procedures employed, and differences of greater than one order of magnitude have been revealed by a recent comparison of the individual methods (Christensen and Packard, 1979). This variability obviously affects the accuracy of much of the published data as well as the reliability of empirically determined extrapolation factors.

ETS enzyme activity is estimated in sample preparations by rates of *in vitro* INT reduction under saturating substrate levels and an incubation temperature of 35°C. Under these conditions, the system is probably operating at or near V_{max}; for this reason, several scaling factors must be applied in order to extrapolate such data to *in situ* oxygen consumption estimates. These include a correction for the combined effects of changes in the physical and chemical conditions (e.g., temperature, pressure, light, nutrient concentrations, time of day) and a factor to relate *in vitro* activity to *in vivo* rates. In order to calibrate the ETS assay procedure, the measured rates of INT-formazan production must be stoichiometrically converted to "oxygen equivalents" (Kenner and Ahmed, 1975a,b) and then compared to direct measurements of oxygen consumption (R). R : ETS data, derived from laboratory experiments, range from 0.11 to 0.32 for exponentially growing log-phase marine diatoms (Kenner and Ahmed, 1975a,b), 0.79–2.79 for a variety of marine zooplankton (King and Packard, 1975), and 2.91–8.25 for several species of heterotrophic marine bacteria (Christensen *et al.*, 1980). Significantly lower ratios occur in physiologically depressed cells, with typical values of <0.05 for carbon-starved heterotrophic bacteria and senescent algae (Kenner and Ahmed, 1975a,b; Christensen *et al.*, 1980). Direct comparison of these values may be inappropriate due to the previously mentioned analytical differences used by the various investigators. Nevertheless, it should be obvious that prior knowledge of the growth rate, nutritional status and taxonomic structure of the community is necessary in order to derive meaningful respiration rates from ETS data.

Christensen (1983) compared published bell-jar oxygen-consumption rate data to ETS values for a variety of marine sediments. For 10 shallow-water stations (0–200 m), the estimated R : ETS was 0.17 ± 0.030, comparable to the data previously reported for growing cells. In deeper waters, however, the R : ETS decreased dramatically to a value of 0.021 at 1850 m and 0.00036 at 5000 m, a value three orders of magnitude lower than the ratios found in shallow-water sediments. Recently, Packard and Williams (1981) have compared measurements of ETS activity with direct measurements of oxygen consumption for surface seawater samples from the northwest Atlantic Ocean. The two sets of data were fitted by the regression equation

$$\text{ETS} = 2.92\,R + 99 \qquad r = 0.89; \quad n = 21$$

It was suggested that an "on location" direct calibration be made rather than relying on the laboratory-derived extrapolation factors. If this is actually necessary for all ETS applications, then measurement of the rate of O_2 consumption by direct respirometry should be sufficient. In theory, one might expect that changes in respiration would be more immediate than changes in the synthesis and degradation of various ETS enzymes. This is consistent with the data indicating a significant positive intercept of ETS activity at zero respiration rate (Packard and Williams, 1981). Consequently, the use of *in vitro* ETS activity data may be restricted to habitats where the physiologic state of the population is well defined. The high ETS values measured in the deep sea might also indicate the persistence of cell-free ETS activity. High extractable ETS activity has been detected in zooplankton after death (Bamstedt, 1980), which indicates that detrital materials may contain ETS activity that could bias the estimation of *in vivo* microbial community respiration.

Nucleotide Ratios, Nucleosides, and Related Intracellular Compounds

More than 100 different nucleosides and nucleotides have been isolated from living cells (Jonsen and Laland, 1960), ranging from the abundant and familiar 5'-ribonucleotides of the bases adenine, guanine, uracil, and cytosine (and their derivatives) to more exotic and less well-studied structurally related compounds. As the maximum potential rate of energy production in microorganisms is generally much greater than their energy requirements, living systems have evolved efficient mechanisms to ensure integration, coordination, and control of cellular processes. Several of these metabolic control systems are believed to involve cellular nucleotides (reviewed by Karl, 1980). In theory, the measurement of nucleotide ratios, expressed as follows, might serve as important metabolic variables to help define the physiological status of the population:

Adenylate energy charge: $EC_A = [\text{ATP}] + \tfrac{1}{2}[\text{ADP}]/[\text{ATP}] + [\text{ADP}] + [\text{AMP}]$
Phosphorylation state: $PS = [\text{ATP}]/[\text{ADP}]\,[\text{HPO}_4^{2-}]$
Catabolic reduction charge: $CRC = [\text{NADH}]/[\text{NADH} + \text{oxidized NAD}^+]$
Anabolic reduction charge: $ARC = [\text{NADPH}]/[\text{NADPH} + \text{oxidized NADP}]$

Of these nucleotide ratios, only EC_A has been used in ecological studies. Even under optimum conditions of extraction and measurement, however, the interpretation of environmental EC_A data is rather limited. Since the EC_A ratio is unitless, no information is provided on concentrations or turnover rates, both of which can vary significantly at a fixed EC_A ratio (Knowles, 1977). Despite occasional claims to the contrary, EC_A measurements cannot be used to estimate *in situ* growth rates or rates of cellular metabolism, as these rates are independent of EC_A ratio (Laws *et al.*, 1983). On the other hand, EC_A measurements can provide information on the relative metabolic status of naturally occurring microbial communities and may facilitate the detection of populations exposed to sublethal environmental stress. Specific applications have been reviewed elsewhere (Karl, 1980). Recently, Davis and White (1980) suggested that measurements of cellular adenosine : ATP ratio might be more useful in ecological studies than the measurement of EC_A, but additional laboratory and field observations are required to substantiate this claim. Nevertheless, their HPLC-fluorometric method for the simultaneous detection of the full spectrum of adenosine nucleotide derivatives (adenine, adenosine, AMP, ADP, ATP, NAD, cAMP) clearly represents an improvement over the previously employed enzymatic procedures.

Many important cellular reactions, especially those coupled to biosynthesis and growth, are dependent on energy derived from nucleotide triphosphates other than ATP. GTP, for example, is required for activation and interconversion of precursors for bacterial cell wall biosynthesis, for DNA replication (as dGTP), for RNA transcription as well as for initiation, binding and translocation processes of protein synthesis. This indirect coupling of cellular energy during biosynthesis predicts that the total flux of energy through the GTP pool may be directly correlated with cellular growth rates. Unlike the ATP pool, which is maintained at a relatively constant intracellular concentration regardless of growth rate, the concentration of nonadenine nucleotide triphosphates fluctuates in proportion to their biosynthetic requirements (Franzen and Binkley, 1961; Smith and Maaloe, 1964; Karl, 1978). Karl (1978, 1979*a*) demonstrated that the GTP : ATP ratio in microorganisms is positively correlated with growth rate and suggested that this ratio may be useful for estimating and comparing the relative community growth rates in naturally occurring microbial assemblages.

Synthesis and Catabolism of Storage or Reserve Polymers

Many microorganisms have evolved the abilities to synthesize and accumulate a variety of storage products during periods of nutritional feast and then catabolize these materials during times of famine, thereby maintaining viability (Dawes and Senior, 1973). Three main classes of compounds have been defined: polysaccharides (glycogen, starch), lipids (triglycerides, wax esters, β-hydroxy fatty acids), and polyphosphates. Poly-β-hydroxybutyrate (PHB) or, more precisely, poly-β-hydroxyalkanoate (PHA) (Findlay and White,

1983) is an endogenous reserve polymer found in most prokaryotes, including phototrophs and chemolithotrophs. In laboratory monocultures, PHA accumulates during sporulation or cyst formation and under conditions of unbalanced growth induced by inorganic nutrient limitation with the supply of carbon and energy from exogenous sources exceeding that required for growth (see Matin *et al.*, 1979, for a recent exception to this rule). It has been suggested that direct measurements of PHA (or PHB) concentrations and rates of synthesis might provide useful information regarding the nutritional status of microbial populations in nature (Herron *et al.*, 1978; Nickels *et al.*, 1979; Findlay and White, 1983). When unbalanced growth was experimentally induced in natural communities, an accumulation of PHB and a decrease in PHB catabolism were observed (Nickels *et al.*, 1979; Findlay and White, 1983), implying that the PHB : lipid-P (as a measure of biomass) ratio might be indicative of balanced versus unbalanced growth. Unfortunately, PHA (or PHB) accumulation is limited to environments in which inorganic nutrients (most notably N) are limiting and utilizable reduced carbon is in excess. Many heterotrophic microbial communities are presumed to be carbon limited and, under such conditions, carbonaceous storage products would not be expected to accumulate. More recently, Gehron and White (1982) suggested that triglyceride analysis might be used to define the nutritional status of micro- and macroeukaryotes in environmental samples; however, neither the rationale nor the existing data base is well defined.

Photosynthetic Potential

The herbicide 3-(3,4-dichlorophenyl)-1,1-dimethyl urea (DCMU) is known to inhibit specifically the reoxidation of reduced coenzyme Q, the primary electron acceptor of photosynthetic reaction center II. The addition of DCMU to algal cultures causes an immediate (sec to min) increase in the chl *a* fluorescence due to cessation of normal fluorescence quenching. Samuelsson and Oquist (1977) proposed that the relative increase in chl *a* fluorescence upon the addition of DCMU might be used as an indirect measure of the photosynthetic capacity of the cells. In several independent studies of unialgal batch cultures, strong correlations were observed between DCMU-enhanced fluorescence and rates of photosynthesis (Roy and Legendre, 1979; Fukazawa *et al.*, 1980; Samuelsson and Oquist, 1977; Samuelsson *et al.*, 1978). Significant quantitative differences were observed among species, however, and for a given species as a function of environmental conditions (Roy and Legendre, 1979). Consequently, the data derived from natural algal populations cannot be interpreted without further laboratory experimentation (Cullen and Renger, 1979; Harris, 1980; Roy and Legendre, 1980). Nevertheless, Cullen and Renger (1979) observed systematic variations in the fluorescence response index (FRI = [(DCMU-induced)-control]/(DCMU-induced)) with water depth and concluded that "we may not know exactly what we are measuring but the patterns are too strong to ignore." It is not entirely clear whether DCMU-enhanced fluorescence is related to photosynthetic rate, capacity, or efficiency or simply a measure of the gross physiological state of the algal populations. In any case, it seems evident that a high (>0.6) FRI is characteristic of growing cells and low (<0.25) values are indicative of debilitated or physiologically stressed cells (Cullen and Renger, 1979).

Chemical Composition and Growth State

The genotype of any cell determines not a unique phenotype but a range in phenotypic capacity controlled to a large extent by environmental factors (Neidhart, 1963). This inherent variability manifests itself as systematic and, therefore, predictable changes in the chemical composition of the cell. Consequently, the physiologic state, growth potential, and even relative growth rates of microbial populations in nature have been estimated by measuring specific compositional variables.

RNA : DNA Ratios

The results of early studies of macromolecular synthesis in bacteria indicated that the cell quotas of RNA, DNA, protein, and total mass all increase with increasing growth rate (reviewed in Maaloe and Kjeldgaard, 1966) and are independent of the chemical composition of the medium or nature of the growth rate-limiting nutrient. Furthermore, it was observed that the ratio of RNA to DNA was positively correlated with specific growth rate over a finite range of μ (Rosset *et al.*, 1966; Dennis and Bremer, 1974; Brunschede *et al.*, 1977; Dortch *et al.*, 1983). It has also been observed, however, that under conditions of extremely slow growth rate, *E. coli* maintains "extra" RNA (Koch, 1971), such that the cellular RNA : DNA ratio deviates from the expected growth rate dependent relationship. Whether this is a common feature of starved or slow-growing cells in nature remains unknown.

It has been proposed that the RNA : DNA production rate ratio of microorganisms can be used as a physiological index for estimating and comparing the productivities of microbial assemblages in nature (Karl, 1981). The synthesis rate ratio is conveniently measured using [^{3}H]adenine, which has been shown to label both RNA (via ATP) and DNA (via dATP) at equivalent precursor specific radioactivities (see the section under Rates of Nucleic Acid Synthesis, [2,^{3}H]Adenine). Consequently, the radioactivity incorporation ratio is identical to the synthesis rate ratio (Karl, 1981). One major advantage of the [^{3}H]adenine tracer technique is that it measures the nucleic acids of growing cells only. Thus, the RNA : DNA ratio is affected neither by the presence of debilitated, starved, and senescent cells nor by the ubiquitous occurrence of detrital (nonliving) nucleic acids. Alternatively, the ratio of RNA : DNA can be determined in single cells using dual laser flow cytometry (C. M. Yentsch *et al.*, 1983), although this latter method has not yet been exploited in ecological studies.

C : N : P Ratios

Goldman and colleagues recently reviewed the existing data base and presented new data on the effects of growth rate on the bulk elemental composition of phytoplankton cells (Goldman *et al.*, 1979; Goldman, 1980). These workers concluded that the characteristic Redfield ratio of 106 C : 16 N : 1 P (by atoms) is achieved only under conditions of high relative growth rate (>90% μ_{max}), i.e., under conditions of nutrient saturation. Predictable and significant deviations from the Redfield ratio are observed when growth rates are altered by nutrient limitation. It was emphasized that the Redfield ratio is attained at or above a particular relative rather than absolute growth rate, usually $\mu/\mu_{max} > 0.9$. Organisms with widely different maximum growth rates show similar C : N : P ratios when growing at or near their own nutrient-saturated rates. Consequently, without additional independent measurements of absolute μ_{max}, it is impossible to estimate rates of carbon production or nutrient flux strictly on the basis of elemental composition. One disadvantage of relying on bulk particulate C : N : P compositional data is the difficulty in differentiating phytoplankton cells from other living and nonliving organic materials. It is conceivable that fluorescence-activated cell sorters might prove useful for future studies in this area.

Patterns of Autotrophic ^{14}C Assimilation

A careful inventory and quantitative evaluation of the major end products of $^{14}CO_2$ assimilation by natural populations of photoautotrophic microorganisms can be used: (1) to describe the physiological state of the cells, (2) to identify the environmental factors that control phytoplankton growth, and (3) to define the relative growth rates under *in situ* conditions (Morris, 1980, 1981). Convenient differential solvent extraction techniques yield major subcellular fractions containing either low molecular weight metabolites, lipids, polysaccharides, nucleic acids, or proteins (Morris *et al.*, 1974). In this type of analysis, it is essential to differentiate rates of synthesis from rates of incorporation, because the metabolic stability (i.e., intracellular turnover) of different classes of compounds may be quite variable. Furthermore, the condition of isotopic equilibration (i.e., equal specific radioactivities for all carbon pools) is implicit, although not rigorously evaluated or confirmed, in this type of comparison. DiTullio and Laws (1983) recently devised a method based on $^{14}CO_2$ incorporation into protein that can be used to estimate both phytoplankton N assimilation and the relative growth rate (i.e, μ/μ_{max}) of N-limited autotrophic populations. The method is based on the fact that the C : N ratio in phytoplankton protein is remarkably constant (3.3 ± 0.1 by weight) and that under N limitation, phytoplankton allocate 85 ± 5% of their cellular N to protein. Thus, phytoplankton N assimilation can be estimated from the rate of ^{14}C incorporation into protein divided by [3.3 × 0.85] = 2.8. The total cellular C : N ratio (hence μ/μ_{max}) can also be determined by dividing

2.8 by the fraction of ^{14}C activity in the protein fraction after a specified incubation period (DiTullio and Laws, 1983).

METABOLIC ACTIVITY

Attempts to measure total ecosystem metabolic activity are generally concerned with evaluating the flux of carbon, energy, or both rather than with quantifying production or growth (i.e., efficiency of metabolism) per se. However, accurate measurements of total metabolic activity in natural environments are virtually impossible to perform due to the diversity of physiologic groups and metabolic processes. In principle, we need a common bioenergetic metric that can be used to integrate simultaneously the metabolic activities of all microorganisms present and at the same time to exclude the activities of macroorganisms and cell-free enzymes. In an attempt to approximate total microbial metabolic activity, several methods have been developed to estimate the rates of biological decomposition of organic matter by measuring fluxes of O_2 and CO_2, the turnover rate of specific organic substrates, or total ecosystem heat production.

O_2 and CO_2 Fluxes

Direct measurement of changes in the concentration of either O_2 or CO_2 or both, has been used to estimate rates of photosynthetic and heterotrophic processes as will be seen. Implicit in these studies is a quantitative relationship among O_2 consumption, CO_2 evolution, and microbial activity. On a molar basis, the amount of CO_2 liberated should be approximately equal to the amount of O_2 consumed; however, measured respiratory quotients $(RQ = CO_2/O_2)$ range from 0.27 to 1.35 for a variety of benthic marine ecosystems (Pamatmat, 1968). Production of CO_2 from anaerobic processes (e.g., fermentation, SO_4 reduction, denitrification), precipitation or dissolution of calcium carbonate, differential atmospheric exchange of CO_2 versus O_2, abiotic consumption of oxygen, and the presence of active chemolithotrophic microorganisms all may affect the expected stoichiometric relationships. Simultaneous measurement of both O_2 and CO_2 flux is an obvious advantage over measuring only one of these gases. K. M. Johnson *et al.* (1981) summarized much of the literature and presented their own novel data on dirct diel CO_2 and O_2 fluxes in seawater. Evolution (or flux) of CO_2 can be measured using static chambers, flow systems, conventional Warburg respirometers, all used for soils (deJong *et al.*, 1979), by measuring CO_2 soil profiles and *in situ* gas-diffusion coefficients in order to estimate total CO_2 flux (Lai *et al.*, 1976), or by monitoring temporal changes in CO_2 by removing and analyzing discrete samples (K. M. Johnson *et al.*, 1981). The last approach avoids the problems of confinement artifacts. Van Cleve *et al.* (1979) recently compared four different methods for measuring CO_2 evolution in forest litter and discussed the advantages, limitations, and systematic errors inherent in each approach.

Oxygen concentrations in aquatic environments have been measured manometrically, by direct chemical techniques (e.g., Winkler titration), and by polarographic electrodes. Electrodes have the advantage of providing a continuous record of O_2 changes and are less susceptible to chemical interferences, but they are less precise than the Winkler method. In the past, marine plankton samples had to be concentrated up to several thousandfold and incubated for long periods of time in order to obtain reliable respiration rate data (Pomeroy and Johannes, 1966). The effects on cell-specific respiration rates of concentrating and incubating samples are not entirely known. However, with recent advances in photometric endpoint determinations and automated microprocessor-controlled titration devices, the Winkler procedure can now yield data with typical coefficients of variation of 0.03–0.1% (Williams and Jenkinson, 1982). This system permits detection of changes in O_2 concentration on the order of $<1\%$ day^{-1}, which is suitable for estimating the total respiration rate in unconcentrated samples of microorganisms from all but the most oligotrophic aquatic ecosystems.

In many aquatic environments, the bulk of microbial biomass, metabolic activity, and hence O_2 flux is associated with particles and occur at the water–sediment interface. Bell jars, domes, and benthic chambers of various dimensions have been used both in shallow-water and deep-sea (to 5200 m) environments. One major problem with this enclosure approach is the unknown effects that local hydrodynamic conditions may exert on the actual *in situ* O_2 flux or the impact of stratification beneath the dome. Several investigators have attempted to mimic the water motion by stirring the trapped water at a rate approaching the ambient current velocity. As a compromise, it may be best to perform paired duplicate treatments whenever possible (one stirred, one static) with an understanding that the real value lies somewhere between the two extremes. Another limitation is the problem of assessing chemical (abiotic) oxygen demand (COD), which may account for a variable percentage of the total measured O_2 flux. The use of formalin or antibiotics in replicate domes has been attempted but, because much of the COD is ultimately derived from end products of anaerobic decomposition of organic matter (i.e., through the oxidation of HS^- and Fe^{2+}), these methods most likely yield an underestimate of the COD present in the unpoisoned treatment.

Revsbech and co-workers recently devised a fundamentally different method for estimating O_2 flux in sediments and soils (Revsbech *et al.*, 1980; Revsbech and Jorgensen, 1983; Revsbech and Ward, 1983). The method is based on direct measurements of the O_2 gradient with depth during light to dark transitions rather than by monitoring changes in the O_2 content of overlying waters. These measurements are made possible through the use of extremely small membrane-covered platinum microelectrodes (tip diameter 2–8 μm). Their small size, extremely low rates of O_2 consumption (10^{-13} moles O_2 min $^{-1}$), and quick response times (90% response in 0.2 sec) make them virtually insensitive to stirring and uniquely suited for resolution of small-scale spatial and temporal variations.

Uptake and Assimilation of [14]C- and ^{3}H-Labeled Organic Compounds

In 1962, Parsons and Strickland suggested that the uptake of ^{14}C-radiolabeled organic compounds by natural microbial populations could be analyzed kinetically using saturation adsorption isotherm models similar to those previously applied to the study of enzyme catalyzed reactions (e.g., Michaelis–Menten equation). Wright and Hobbie (1965, 1966) further discussed the use of this method as applied to the analysis of aquatic microbial communities and introduced the use of a modified Lineweaver–Burk equation, that is,

$$t/f = (K + S_n)/V_{\max} + A/V_{\max}$$

where t is the incubation time in h, f is the fraction of available substrate utilized during the incubation, K is a constant related to uptake, $V_{\max}$ is the maximum uptake rate at substrate saturation, S_n is the natural substrate concentration, and A is the concentration of added substrate. This formula permits estimation of three important metabolic variables of the community: (1) $V_{\max}$, the maximum uptake velocity at saturating substrate concentrations, i.e., the "heterotrophic potential" of the ecosystem; (2) T_t or t/f, the turnover time, i.e., a value theoretically equivalent to the time required for complete utilization of the existing substrate pool assuming no production or regeneration; and (3) $[K_t + S_n]$, i.e., a value equivalent to the transport constant plus the natural substrate concentration. The latter is of limited value other than to set an upper limit on either of the two independent and unknown variables. If $K_t < S_n$, an assumption that is sometimes made, the value derived for $[K_t + S_n]$ is an approximation of S_n.

This method was immediately called into question by Hamilton and Austin (1967), who criticized this kinetic approach because no attempt was made to account for the substantial and variable release of respired [14]CO$_2$. This analytical problem was subsequently solved (Williams and Askew, 1968; Hobbie and Crawford, 1969), but even at the present time, little effort has been made to estimate other extracellular ^{14}C-labeled metabolites (e.g., fermentation end products) that may accumulate under certain enviromental conditions (Novitsky and Kepkay, 1981; King and Klug, 1982).

The simultaneous measurement of the amount of ^{14}C-labeled organic matter incorporated into cellular materials and that portion respired (or remineralized) has been used extensively to calculate the assimilation (growth) efficiency of natural microbial populations, i.e., [dpm incorporated]/[dpm incorporated + dpm mineralized]. This analysis is only valid for uniformly labeled organic substrates. Williams (1973) reported that the apparent growth yields for marine plankton populations growing on ^{14}C-labeled glucose and a mixture of amino acids were 0.71 ± 0.10 ($N = 40$) and 0.77 ± 0.08 ($N = 92$), respectively. This finding led to the conclusion that marine bacteria are highly efficient in converting dissolved organic matter into biomass. These high apparent growth yields may be a bit misleading, however, as we do not know the full menu of marine microheterotrophs. Many compounds may be used preferentially to generate energy (i.e., <20% incorporation), whereas others are salvaged intact for biosynthesis (>70% incorporation). Furthermore, Hanson and Wiebe (1977) pointed out that it is not clear whether the $^{14}CO_2$ respired during a 1- to 2-hr incubation period is directly proportional to $^{12}CO_2$ respiration. If it is not, $^{14}CO_2$ evolution is not a valid estimate of assimilation efficiency. In this regard, King and Klug (1982) recently showed that the rate constant for [^{14}C]glucose uptake is an order of magnitude greater than the rate constant for ^{14}C-end product formation, a result they interpret as evidence for dilution of the tracer-specific radioactivity by existing intracellular pools of glucose and glucose metabolites. This level of sophistication in the analysis of kinetic tracer data in ecological studies is long overdue.

The kinetic approach to the study of heterotrophic activity in nature has been widely used, primarily in aquatic environments, and certainly represents one of the most significant methodological developments in modern microbial ecology. The use of the kinetic model is not without difficulty however, and limitation (reviewed by Wright, 1973; Thompson and Hamilton, 1974; Wright and Burnison, 1979; D. F. Smith *et al.*, 1984). One immediate and practical problem is the selection of an appropriate substrate. Nearly all the uptake experiments conducted thus far have used simple carbohydrates (monomers, expecially hexoses), dissolved free amino acids (either singly or as commercially available amino acid mixtures), or short-chain fatty acids as the introduced tracer. With the bulk of the total dissolved organic matter in aquatic environments uncharacterized, it is impossible to know how representative any one substrate might be for the analysis of microbial activity in a particular ecosystem. It is likely that a large number of substrates are used simultaneously. If microorganisms in nature prefer to use dissolved polymeric compounds (e.g., protein, nucleic acids, cellulose, chitin) or particulate organic matter, we may be grossly underestimating total metabolic activity in many environments by our present techniques.

In applying the kinetic approach, a mixed population is treated conceptually as a finite number of equivalent responsive metabolic sites that conform to a uniform set of kinetic variables (Azam and Holm-Hansen, 1973). Unfortunately, deviations from the expected Michaelis–Menten kinetics are frequently observed, especially in oligotrophic waters, suggesting that each individual cell, or perhaps subpopulation, has its own characteristic set of kinetic parameters. The theoretical implications of simultaneous uptake by diverse groups of microorganisms have been discussed by Williams (1973); however, the validity of his conclusions was challenged by Wright and Burnison (1979). Nevertheless, Azam and Hodson (1981) clearly demonstrated the existence of eurykinetic patterns in several aquatic environments and suggested that nonlinear plots of t/f versus [A] are the rule rather than the exception in naturally occurring microbial assemblages. Even if microbial heterotrophic activity data appear to conform to Michaelis–Menten kinetics, a careful consideration of theoretical error analysis of the linearly transformed, but unweighted data indicate that the experimental results and subsequent ecological interpretations may be statistically unsatisfactory (Li, 1983).

Azam and Holm-Hansen (1973) proposed using high specific activity ^{3}H-labeled organics (typically Ci mmoles^{-1}) rather than ^{14}C substrates (typically mCi mmoles^{-1}). The use of tritium makes it possible to measure directly the turnover time (t/f) of selected compounds at ambient concentrations, even in the most oligotrophic environments. This eliminates the need for simplifying assumptions and extrapolations (Azam and Holm-Hansen, 1973). As S_n is generally unknown, however, one can never be certain that the amount of tracer substrate added was in fact negligible. If the availability of organic substrates is actually the growth rate-limiting factor for heterotrophic microorganisms, by definition, their concentrations will be expected to be at undetectable (or at least at values below the K_m) levels. Another disadvantage with this method is the inability to assess respiration and, as recently pointed out by Kuparinen and Tamminen (1982), the ^{3}H-tracer method as described by Azam and Holm-Hansen (1973) yields a systematic underestimate of the total uptake, hence of turnover time. Although ^{3}H$_2$O might be collected and measured, the production of ^{3}H$_2$O is not quantitatively coupled to CO_2 production due to the possibility of ^{3}H-isotope exchange reactions during cellular metabolism. All tracer experiments, whether they employ ^{14}C- or ^{3}H-labeled substrates, should provide evidence for a full radiochemical inventory and mass balance during the uptake, assimilation, excretion, and mineralization of the exogenous organic compounds.

Dietz *et al.* (1977) described an "alternative" approach to studying microheterotrophic activities in nature. Their new experimental method essentially amounts to a combination of the use of ^{3}H-labeled organic compounds (as previously described by Azam and Holm-Hansen, 1973) with the kinetic approach (as previously described by Wright and Hobbie, 1965, 1966) and in principle is identical to the isotope-dilution method described previously by Forsdyke (1968). Wright and Burnison (1979) concluded that the fundamental assumption of the Dietz *et al.* approach, *viz.* "velocity of utilization is constant since background amounts of substrate are added," is untenable. Laws (1983) recently presented a careful theoretical analysis of the isotope-dilution methods conducted by Dietz *et al.* (1977) and Forsdyke (1968); he concluded that these methods are based on faulty mathematics and can provide only upper bounds to *in situ* substrate concentrations. The parameter they define as S_n (natural substrate concentration) is actually equivalent to $(S_n + K_t)$, and their V_n (velocity of uptake at natural substrate levels) is equivalent to V_{max} as defined in the original papers on the kinetic approach (Wright and Hobbie, 1965, 1966). Furthermore, D. F. Smith *et al.* (1984) recently discussed additional obstacles to the unequivocal interpretation of data arising from the heterotrophic potential method. They cited several artifacts of the method, as currently employed, eloquently concluding that the applicability of Michaelis–Menten kinetics to data obtained from natural populations of microorganisms is actually irrelevant as well as incorrect.

Finally, if turnover time (t/f) estimates are combined with direct measurements of S_n, the *in situ* mass flux (e.g., g liter^{-1} hr^{-1}) of carbon (i.e., flux $= S_n/t_f$) can be determined. This approach has been used by several investigators (Andrews and Williams, 1971; Burnison and Morita, 1974; Williams *et al.*, 1976; Dawson and Gocke, 1978; Hanson and Gardner, 1978; Meyer-Reil, 1978*b*) and in principle is the most useful absolute metric to be derived from ^{14}C- or ^{3}H-organic tracer experiments. Unfortunately, several assumptions implicit in this procedure are difficult, if not impossible, to verify. First, it is imperative that the added radiotracer(s) equilibrate with the existing substrate pool(s). If any microorganisms exist in diffusion-controlled microenvironments, as we know they do, this assumption may not be valid, because the microenvironments would essentially represent compartmentalized substrate pools. Furthermore, microorganisms associated with certain particulate organic matter might never even "see" the isotope tracer at all. That is to say, the concentration gradient at the particle–water boundary would favor diffusion of ^{12}C organics from the microhabit rather than diffusion of ^{14}C organics toward the active metabolic sites.

Another critical assumption of this approach is that the concentration of substrate obtained by chemical procedures (S_n) is identical to the amount of substrate available for heterotrophic uptake (i.e., the so-called free or available pool). This is impossible to evaluate because methods do not exist for differentiating between total and available pools of dissolved organics. It is conceivable that the chemical extraction and purification methods themselves affect the measured pools by altering the pH or ionic strength or by affecting the equilibrium between complexed, adsorbed, particulate, colloidal, and dissolved states. If S_n is overestimated, the flux will likewise be affected. Burnison and Morita (1974) concluded that the chemically determined concentrations of amino acids were actually the sum of free and adsorbed substrates, because the values obtained for S_n were always greater than the estimates of $[K_t + S_n]$ extrapolated from kinetic analyses. Christensen and Blackburn (1980) also concluded that much of the dissolved alanine measured in nearshore marine sediments (total concentration, 800 nM) is probably unavailable to bacteria and, on the basis of independently measured turnover rate constants, conclude that a more realistic biologically available concentration is closer to 10 nM. By contrast, Jorgensen and Sondergaard (1984) recently reported that 47–116% of the total dissolved free amino acids in a variety of untreated water samples was recovered in the low-molecular-weight fraction (<700 M_r), suggesting that the bulk of the amino acids are truly dissolved free molecules. Gocke *et al.* (1981) also reported that $[K_t + S_n]$ for glucose was consistently lower than S_n measured directly (6.4–30% of total). Since the uptake kinetics and linear transformations looked acceptable and internal chemical standards gave excellent results with their analytical procedures, they ruled out most potential methodological problems. It was concluded that only a fraction of the glucose was actually available and consequently that the use of chemically determined S_n to estimate flux may be misleading. In a recent study of glucose metabolism of lake sediments, King and Klug (1982) observed a significant disparity between the specific radioactivity (nCi pmoles^{-1}) of the extracellular glucose concentration and the specific radioactivity of the intracellular pool. King and Klug concluded that if any metabolic reaction contains one or more intermediates, mass flux through the whole system cannot be estimated simply by determining the specific radioactivity of the initial reactants. A full inventory of the labeling patterns of the immediate precursor and products must be made in order to obtain reliable and accurate flux data.

Microcalorimetry

During the process of microbial metabolism, there is a measurable evolution of heat that represents the summation of both the exothermic processes of catabolism and endothermic anabolic reactions. Direct measurements of heat production and flux, by microcalorimetry, is a general method for studying enthalpy changes (ΔH) during microbial metabolism. This is probably the least specific measure of metabolic activity because heat is a common by-product of all physiological groups. As such, total heat flux measurements are extremely useful for investigations of the complex and diverse communities typically encountered in soils and sediments. Although the use of calorimetry to study metabolism dates back to the eighteenth century, recent developments in solid-state electronics and the availability of sensitive commercial microcalorimeters have provided the instrumentation required to conduct ecologically relevant experiments with natural microbial communities. Several different designs are available (e.g., titration calorimeter, batch calorimeter, flow calorimeter); the choice is dependent on the specific application. The batch (also static or heat-flow) calorimeter is generally the method of choice for soils and sediments (Mortensen et al., 1973; Pamatmat et al., 1981; Pamatmat, 1982, Sparling, 1981; Pamatmat and Findlay, 1983).

The factors affecting heat production in bacterial cultures are poorly understood or at least not well documented (Gordon and Millero, 1980). The recorded thermogram (plot of heat production versus time) is an integration of the thermodynamics of the system and is affected by culture history, degree of aeration, growth rate, and medium composition. The area enclosed by a thermogram is proportional to the heat produced by (1) catabolism of substrates supplying energy, (2) heats of dilution and ionization of reaction products, and (3) enthalpy changes associated with growth (Monk, 1979).

There are several limitations to the application of direct microcalorimetry to ecological studies. First, each sample requires a period of 4–8 hr for thermal equilibration prior to the period of incubation and measurement (6–12 hr). During this time, the heat flux generally decreases (Pamatmat et al., 1981), making it difficult to extrapolate the measured values back to in situ estimates.

Furthermore, unless an investigator has several colorimeters available (commercial units cost approximately $20,000–40,000), it is virtually impossible to perform replicate measurements or to provide meaningful synoptic data (e.g., depth profiles or spatial patterns). Storage of samples for periods of days can lead to substantial changes in total metabolic activity. Unavoidable exposure of anoxic sediments to air can result in substantial perturbations in the rate of heat production. Finally, it is impossible to distinguish metabolic heat flux from that produced by chemical reactions which may occur simultaneously, especially during the production of organic acids. Nevertheless, direct microcalorimetry has already provided us with useful ecological information regarding the intensity of degradation of chemical potential energy in complex communities where other techniques and methodologies cannot be easily applied (Pamatmat, 1982).

GROWTH, PRODUCTION, AND CELL DIVISION

Most models of microbial biosynthesis, growth, and cell division are based on extensive laboratory investigations of the Enterobacteriaceae, especially *E. coli.* The extent to which bacteria in natural terrestrial and aquatic habitats conform to these conceptualized growth models is largely unknown.

As natural habitats and their associated microbial communities are diverse and complex, we must anticipate equally complex and diverse growth responses from the individuals that comprise a natural microbial assemblage. Natural ecosystems may be described as either open or closed. In open ecosystems, the time available for growth is unlimited with respect to environmental conditions. By definition, open systems have a continuous flow of energy and bioelements, and even though nutrient availability may ultimately limit the growth rate of microorganisms in open ecosystems, neither severe nutrient depletion, starvation, discontinuous growth, nor a high proportion of nonviable cells is expected in ideal open ecosystems. These ideal environments are considered to have achieved steady state, a dynamic time-independent condition in which production and consumption are balanced, and in which the concentrations of all elements within the system remain constant despite a continuous material turnover and energy flux (Brock, 1967*b*).

Two important considerations in steady-state theory regard the spatial and temporal constants for the particular system or habitat. Sampling and incubation experiments must be performed on spatial scales believed to be appropriate to the steady-state assumption. Many microbial populations exhibit discontinuous or non-steady-state growth over periods of hours or days but may display steady-state growth over much longer time scales. Chemostat cultures (a special case of an open, or continuous culture system) have been employed to study the steady-state growth kinetics of a variety of microorganisms under well-defined nutrient conditions and have occasionally been proposed as laboratory analogs to natural microbial populations. However, Jannasch (1974) emphatically pointed out that the chemostat is neither meant to reproduce, nor capable of reproducing a natural open ecosystem.

By contrast, closed systems provide only a limited period of time for the growth of microorgansims because of restricted rates of matter and energy exchange with their surroundings. Microbial growth and decomposition associated with a decaying animal carcass or within the diffusion-limited microenvironment of an organic particle suspended in an oligotrophic lake or ocean are common examples of closed natural environments. Depending on the successional stage, closed ecosystems may be dominated by dormant, debilitated, nonviable, or otherwise inactive microbial populations with metabolic signatures indicative of extreme nutrient stress or starvation. It is conceivable, and quite likely, that any random sample of a selected natural habitat will contain both open and closed microecosystems. This simple fact underlies the magnitude and complexity of the problem of measuring microbial community growth processes in nature.

Primary Production of Organic Matter

The bulk of the total primary production occurs through the process of oxygenic photosynthesis (i.e., light-dependent reduction of CO_2 to $[CH_2O]_n$ with concomitant evolution of O_2); it should be noted, however, that either anoxygenic photosynthesis (i.e., light-dependent reduction of CO_2 to $[CH_2O]_n$ without the evolution of O_2) or chemolithotrophic production (i.e., "dark" reduction of CO_2 to $[CH_2O]_n$ at the expense of reduced inorganic substrates) may be the dominant pathway in certain environments. From a thermodynamic viewpoint, it had been assumed that chemolithotrophy represented secondary production, since the substrates required to drive these reactions were derived from the anaerobic decomposition of organic matter that was originally produced via photosynthesis. However, the recent discovery of deep-sea primary production occurring at the expense of geothermal energy (Jannasch and Wirsen 1979; Karl *et al.*, 1980) has modified our views of the flow of carbon and energy in the biosphere. The global significance of this aphotic primary production is not known at the present time. The more conventional chemolithotrophy (e.g., S^{2-}, NH_4^+, H_2, or Fe^{2+} oxidation), which is ubiquitous in nature, may be an extremely important source of newly reduced carbon for the local food web, regardless of whether it represents primary or secondary production (Karl *et al.*, 1984). In either case, it should be noted that most chemolithotrophy is O_2 dependent and in that respect is dependent on oxygenic photosynthesis. By comparison to the large data base on rates of photosynthetic carbon production, there exist only a few reliable estimates of the rates of total chemolithotrophic production in environmental samples.

The rate of oxygenic photosynthetic carbon fixation can be estimated by the following approaches:

1. Direct measurement of O_2 evolution
2. $^{14}CO_2$ tracer techniques
3. CO_2 changes (e.g., in seawater CO_2 = [free CO_2] + [H_2CO_3] + [HCO_3^-] + [CO_3^{2-}], either calculated from measurements of pH and carbonate alkalinity or measured by direct methods (such as non-dispersive infrared photometry)
4. ^{13}C-Isotope fractionation
5. Production of $^{18}O_2$ from $H_2^{18}O$

Of these various approaches, the first two techniques, and especially the radiocarbon procedure (Steemann-Nielsen, 1952), have been used most extensively in ecological studies.

The ^{14}C technique has until recently been considered the absolute standard for measuring primary production in aquatic habitats. Several excellent reviews of this method, including technical details, assay limitations, and a summary of the historical debate over data interpretation have recently appeared (Yentsch, 1974; Peterson, 1980; Dring and Jewson, 1982). The most important concerns are (1) the well-documented complexity of the current $^{14}CO_2/^{14}C$-organic carbon flow model; (2) the production, excretion and subsequent uptake of ^{14}C-labeled dissolved organic carbon; (3) the occurrence, magnitude, and significance of dark $^{14}CO_2$ uptake; and (4) the uncertainty regarding whether the ^{14}C method represents gross or net carbon productivity. Despite its widespread use and general acceptance, there remains a great deal of uncertainty regarding the accuracy of photosynthetic rate data derived from the ^{14}C method. A number of observations over the past decade, most notably from oligotrophic oceanic environments, indicate that the ^{14}C method is underestimating gross production by factors in excess of an order of magnitude (Tijssen, 1979; Jenkins, 1982; Shulenberger and Reid, 1981).

A technique that relies on recent improvements in the photometric end-point detection and on computer-assisted chemical titrations now provides an accurate measurement of photosynthetically derived O_2 changes, even in oligotrophic environments (Williams and Jenkinson, 1982). One advantage of this approach is the opportunity to assess net, as well as gross community production. A recent comparison of the ^{14}C and O_2 methods for samples collected in oligotrophic seawaters near the Hawaiian Islands indicated that the ^{14}C-determined rates were more closely related to gross than to net production (Williams *et al.*, 1983). Furthermore, the calculated photosynthetic quotients (i.e., the molar flux ratio of gross O_2 production to $^{14}CO_2$ estimated carbon fixation over the light period) were not significantly different from the expected ratio 1–1.8. In order to reconcile our historical data base, however, it is imperative to determine whether these methods will also yield comparable rate data in waters from the oceanic subtropical latitudes at which *in situ* oxygen dynamics presently suggest a gross underestimate of primary production by the ^{14}C methods (Shulenberger and Reid, 1981; Jenkins, 1982).

Measurements of photosynthetic carbon production are not so useful as simultaneous assessments of production and biomass that together permit estimation of phytoplankton growth rate, which can then be compared with the extensive literature based on careful laboratory experiments (reviewed by Eppley, 1981). (The difficulties in estimating phytoplankton standing stock have been discussed in the section, Algal Biomass.) One of the most promising new techniques for estimating phytoplankton growth rates is the chl *a* labeling method of Redalje and Laws (1981).

The underlying assumption in the application of this radiochemical ($H^{14}CO_3$) method is that the specific radioactivity of carbon (i.e., $^{14}C/^{12}C$) in the isolated chl *a* (a uniquely phototrophic product) pool is identical to the specific radioactivity of total phytoplankton carbon. If this assumption is correct, one can easily determine the biomass of phytoplankton (measured as mol C) even in the presence of large and variable concentrations of nonalgal particulate matter (measured in Ci ^{14}C) by the specific activity (measured in Ci ^{14}C/moles C) of the chl *a*. Recent results from unialgal culture (*Mantoniella* sp.) laboratory experiments indicate that carbon isotopic equilibration is achieved after relatively short incubation periods (Redalje, 1983). It should be pointed out, however, that this method is based on $^{14}CO_2$ labeling and is consequently susceptible to most, if not all, of the general criticisms leveled at that analytical approach (Peterson, 1980; Dring and Jewson, 1982). Laws (1984) recently provided a thoughtful discussion of the chl *a* labeling technique, paying particular attention to the accuracy of the method under field conditions. He concluded that an important advantage of the chl *a* technique is that the derived phytoplankton growth rate estimate is unaffected by zooplankton grazing and bacterial assimilation of excreted organic carbon, as neither of these processes alters the specific radioactivity of the chl *a* pool. On the contrary, the estimate of phytoplankton carbon is affected by the activities of zooplankton and bacteria but, with appropriate corrections, accurate algal biomass estimates are still possible.

The measurement of the rate of total chemolithotrophic production in natural habitats is complicated by the diversity of possible metabolic pathways and the impossibility of separating the microorganisms from their energy source (i.e., there is nothing analogous to the "dark bottle" of a photosynthesis experiment). The dark uptake of $^{14}CO_2$ cannot be used as a measure of chemolithotrophic carbon production because all heterotrophs also assimilate CO_2 as a normal part of their metabolism. Careful analysis of the metabolic patterns of $^{14}CO_2$ labeling (intermediates and products), however, may permit differentiation of chemolithotrophic from chemoheterotrophic CO_2 assimilation (Karl, unpublished results). One approach that has been employed in ecological studies is to monitor the increase in $^{14}CO_2$ assimilation upon the addition of substrates known to support the growth of chemolithotrophs (e.g., $S_2O_3^{2-}$, NH_4^+, H_2). However, it is often difficult to interpret these enrichment culture data, especially with regard to extrapolating back to *in situ* rate of total chemolithotrophic production.

Dark CO_2 Assimilation

During microbial growth, there is a significant removal of Krebs (tricarboxylic acid) cycle intermediates for the synthesis of macromolecular precursors such as porphyrins and amino acids of the glutamate and aspartate families. Light-independent CO_2 fixation serves to regenerate the C_4 dicarboxylic acids essential for the continuation of cellular metabolism. These enzymatic carboxylations are referred to collectively as the anaplerotic reactions (Kornberg, 1966). Romanenko (1964) reasoned that because the rate of CO_2 fixation via anaplerotic reactions should be directly coupled to the rate of bacterial growth, measurement of dark CO_2 fixation might provide useful quantitative data on the total rate of bacterial production. The detailed methodology was presented by Romanenko *et al.* (1972) and is basically identical to the "dark bottle" in ^{14}C photosynthetic production protocol. In the Romanenko technique, bacterial production is extrapolated from $^{14}CO_2$ incorporation, assuming that CO_2 supplies 6% of the total cell carbon during chemoheterotrophic growth.

As early as 1932, Walker demonstrated that *E. coli* can grow on a complex medium in the absence of CO_2 but that CO_2 was an absolute growth requirement when the cells were grown in a simple lactose–tartrate–mineral salts medium. Subsequent $^{14}CO_2$-labeling experiments confirmed that the amount of CO_2 fixed relative to total cell carbon produced decreased in more complex media (Prescot *et al.*, 1965). This suggests that when the dark CO_2 assimilation technique is applied to natural habitats, the amount of $^{14}CO_2$ assimilated will be dependent not only on the total rate of carbon production but on the chemical composition of the surrounding environment and the intensity of metabolic salvage pathways as well. Consequently, the various attempts to calibrate the dark CO_2 assimilation method using bacterial cultures in defined media (Overbeck, 1979; Overbeck and Daley, 1973; Li, 1982) may yield misleading extrapolation factors. Even if all heterotrophs were shown to have identical assimilation patterns under a given set of environmental conditions, which is unlikely, it would still be impossible to extrapolate the $^{14}CO_2$ data directly to heterotrophic bacterial production because algae are also known to assimilate CO_2 in the dark (Craigie, 1963; Galloway *et al.*, 1974). (Difficulties encountered in the separation of algae from bacteria, whether by size fractionation or through the use of antibiotics, are discussed in the section, Attribution of Total Biomass, Metabolic Activity, or Growth. . . .) Another potential problem when attempting to estimate the actual rates of $^{14}CO_2$ uptake (whether in the dark or in the light) is the variable, but often potentially substantial, isotope dilution caused by respiratory CO_2 production (Galloway *et al.*, 1974). Finally, the presence of chemolithotrophic bacteria in a given environmental sample will obviously result in an overestimate of bacterial productivity, especially considering the fact that certain chemolithotrophs are capable of deriving their entire cell quota of carbon from CO_2. Although the distribution of certain types of chemolithotrophic bacteria is thought to be restricted to oxic–anoxic interfaces, the possible presence of anaerobic microenvironments within suspended particulate materials and in otherwise "oxidized" environments might provide an acceptable habitat for the growth of chemolithotrophic and mixotrophic bacteria (Karl, 1982a). There is no convenient method for estimating the contribution of chemolithotrophy to total dark CO_2 uptake, hence no rigorous or precise interpretation of data derived from the Romanenko procedure.

Mitotic Index/Frequency of Dividing Cells

In principle, the specific growth rate (μ) of any growing population of microorganisms can be estimated from the fraction of cells undergoing mitosis, if the time required for this process is known. Mitosis can be assessed directly via nuclear stains or indirectly by any other recognizable developmental stage or morphological trait occupying a constant fraction of the total generation time. It has long been recognized that certain species of marine phytoplankton restrict their cell division activities to a relatively short and predictable period of time each day (i.e., synchronous or phased cell division). When synchronous division occurs, species-specific growth rates can be estimated from integrated values of cell division frequency/unit time, although it is still essential to know the duration of nuclear division in order to obtain a precise estimate of μ (Weiler and Chisholm, 1976; Weiler and Eppley, 1979). Weiler (1980) recently applied this technique to the north Pacific Ocean in order to examine the population structure and individual species specific growth rates for the dinoflagellate genus *Ceratium*. Water samples were collected over closely spaced time intervals (0.5–1 hr), stained with the nuclear stain acetocarmine, and examined microscopically for paired nuclei. The temporal pattern of division stages, the increase in recently divided cells, and the maximum frequency of cell division (F_{max}) were all used to compute the *in situ* growth rates. An obvious advantage of this technique is the fact that no sample incubation period is required. Attempts to calibrate the technique with laboratory cultures of *Ceratium* were relatively successful—μ estimated from F_{max} averaged 79.2% (range: 52.9–111.7%) of μ calculated from the increase in cell numbers. However, it was noted that the laboratory-derived values for the time required for nuclear division was, for some unknown reasons(s), shorter than in natural samples (Weiler and Eppley, 1979). Clearly, more data are needed on the effects of environmental conditions on nuclear division rate. McDuff and Chisholm (1982) recently reviewed the assumptions, limitations, and uncertainties in the use of the mitotic index methods for calculating phytoplankton growth rates; Coats and Heinbokel (1982) discussed the extension of this method to study the reproduction dynamics of planktonic tintinnid ciliates.

Hagstrom *et al.* (1979) proposed an analogous method that they claim can be used to estimate the mean growth rate of bacterial communities in nature. This method requires an estimate of the frequency of dividing cells

(FDC) determined by epifluorescence microscopic examination of AO-stained samples. If a cell showed a visible invagination, it was considered to be in a stage of division. In addition to being tedious and highly subjective, there are several more serious limitations and untested assumptions that limit the general application of this method. First, the theoretical and experimental bases for relating FCD to μ are derived from data on exponentially growing asynchronous cultures (e.g., Woldringh, 1976), which indicated that the time needed for cell constriction and separation (T) is constant over a range of growth rates. Woldringh (1976) reported that for *E. coli*, T is only constant for $\mu > 1$ hr^{-1} but increased with decreasing μ. Chung *et al.* (1973) described a novel method for estimating T in chloramphenicol inhibited cultures of *E. coli* and also reported that the time required for division varied substantially under different environmental conditions. These workers concluded that the ecologist using this procedure must have a sufficiently large number of independent determinations of T in order to estimate the distribution of this characteristic. If T is a function of μ, this method has limited application except to set upper bounds on the population μ.

In applying the Hagstrom *et al.* (1979) method, one must assume that all cells are growing and that T is identical for all individuals in that community. This might be valid for a given species but certainly cannot be expected to hold true for an entire natural assemblage. Hagstrom *et al.* (1979) attempted to calibrate the FDC method with laboratory cultures of bacteria. Although they observed considerable scatter in their plots of μ versus FDC, they were able to fit linear regressions to the data in order to generate extrapolation factors for estimating μ from FDC in natural populations. Although never explicitly stated in their paper, most of their environmental FDC values (see their Fig. 3) were $\leq 2-3\%$, a value that is lower than the extrapolated FDC at zero μ (see their Fig. 4). They mention, but do not dwell on, the importance of two substantial limitations to the ecological application of these laboratory-based methods: (1) if any significant percentage of the total bacterial population is nonviable, and numerous studies claim it is, this would substantially reduce the effective FDC of the community; and (2) starved cells have been shown to divide (several times) without increasing their mass (Novitsky and Morita, 1977, 1978). If the latter process is occurring in a particular field sample, this process severely compromises the interpretation of mean community μ from FDC. The existence of diel or diurnal growth processes and the finite resolution of the light microscope for observing morphological features of very small fluorescent cells further limit the general ecological application of this approach.

More recently, Newell and Christian (1981) reexamined the relationship between FDC and μ and concluded that the linear fit proposed by Hagstrom *et al.* (1979) was simply fortuitous and would not have been found if higher growth rates were examined. Newell and Christian recommended that a regression of the natural logarithm of μ versus FDC be used; however, a fair amount of scatter was still observed, which obviously affects the precision of the estimate of μ derived from FDC. A recent comparison of the FDC technique with a method designed to measure bacterial production based on DNA synthesis (see the section, Methyl-[^{3}H]Thymidine) indicated that the FDC method resulted in production estimates that were 7.5–59 times greater than those based on DNA synthesis (Fallon *et al.*, 1983). Despite the lack of an absolute standard for estimating the accuracy of either of the methods, Fallon and co-workers concluded that the FDC estimates were unreasonably high on the basis of independent estimates of primary production and O_2 flux. Clearly, the FDC approach for estimating bacterial production in environmental samples cannot be used without evaluating the validity of the assay assumptions and the relative significance of the major potential sources of analytical interference. The FDC method might best be restricted to autecological studies (as in the application of this technique in phytoplankton ecology) in which individual species can be identified and enumerated (by immunofluorescence) and in which the laboratory extrapolations might be justified. By combining microautoradiography (with a known radiolabeled growth substrate) and immunofluorescence microscopy, one might also determine, and correct for, the presence of nonviable or starved cells of that particular species. This might represent a fairly straightforward and powerful autecological tool for use in future studies.

Assimilatory Sulfur Metabolism

Sulfur is an essential bioelement, and most microorganisms (bacteria, algae, yeasts, and other fungi) have the ability to transport, reduce (if necessary), and assimilate sulfur as required for cellular biosynthesis. The major quantitative contribution of assimilatory sulfur metabolism is the production of the S-containing amino acids, cysteine and methionine, and consequently the bulk of the S in a cell is contained in protein. Unlike most bacteria, algae also have the ability to synthesize sulfolipids and polysaccharides containing sulfate-esters. Roberts *et al.* (1963) demonstrated, at least for *E. coli*, that SO_4^{2-} uptake was proportional to growth provided no partially reduced inorganic or organic sulfur sources were present in the medium. For example, $S_2O_3^{2-}$ reduced the uptake of SO_4^{2-} by 50% when its concentration in the medium was approximately 1000-fold less than that of SO_4^{2-} (Roberts *et al.*, 1963). Competition was also observed for SO_3^{2-}, S^{2-}, and certain organic sulfur compounds as well.

In 1974, Monheimer proposed that $^{35}SO_4^{2-}$ uptake of natural populations of aquatic microorganisms could be used as an indirect measurement of heterotrophic and autotrophic microbial growth and production. Monheimer (1974, 1975) originally reasoned that by performing simultaneous measurements of $^{35}SO_4^{2-}$ uptake (total autotrophic plus heterotrophic organic sulfur production) and $^{14}CO_2$ uptake (total autotrophic carbon production) and by assuming a C : S ratio by weight of 500 : 1 for the algae, the percentage of total $^{35}SO_4^{2-}$ uptake by heterotrophic bacteria could be determined, i.e., [measured SO_4^{2-} uptake] − [measured CO_2 uptake/500] = [estimated hetero-trophic SO_4^{2-} uptake]. By assuming a C : S ratio of 500 : 1 for heterotrophs as well, total heterotrophic carbon production could be estimated. Unfortu-nately, this approach has a serious limitation later acknowledged by Mon-heimer (1978). The C : S ratio, previously assumed to be constant at a value of 500 : 1, is now recognized as a physiological variable with bounds ranging from 100 : 1 to 10,000 : 1.

A modified approach using $^{35}SO_4^{2-}$ was suggested by Jassby (1975). He contended that algae were incapable of SO_4^{2-} uptake in the dark; consequently, dark $^{35}SO_4^{2-}$ assimilation must be exclusively heterotrophic. Assuming a C : S ratio of 50 : 1, Jassby extrapolated dark $^{35}SO_4^{2-}$ uptake data to bacterial carbon-production estimates. Microbial C : S ratios exhibit considerable variability, though. In support of his approach, Jassby reported that dark uptake of $^{35}SO_4^{2-}$ was negligible in two cultures of unicellular algae, *Chlorella* and *Sce-nedesmus*. However, subsequent laboratory investigations have revealed sub-stantial dark uptake of $^{35}SO_4^{2-}$ by *Chlorella*, *Nanochloris*, and *Scenedesmus*, rang-ing from 0.8 to 53% of the maximum light uptake values (Monheimer, 1978). Bates (1981) confirmed the existence of algal SO_4^{2-} uptake in the dark and found that it varied considerably among species and as a function of physi-ological state. Obviously, dark uptake cannot be equated exclusively with bac-terial sulfur metabolism. Bates (1981) suggested size fractionation and the use of selective inhibitors to separate algal from bacterial protein sulfur assimi-lation. Although these approaches may be useful under certain environmental conditions, their casual use should be discouraged.

The most recent development in the study of microbial S metabolism has been a comprehensive investigation of the feasibility of measuring rates of total microbial protein synthesis by incorporation of $^{35}SO_4^{2-}$ (Cuhel *et al.*, 1981a–d; 1982a,b). Despite the acknowledged and unpredictable variability in whole-cell C : S ratios, Cuhel and co-workers emphasized the relative constancy of the protein-S : total protein ratio. Their data indicate an average sulfur con-tent of $1.09 \pm 0.14\%$ (by weight) for 134 pure cultures of bacteria (equivalent to a protein C : protein S ratio of 50 : 1), a value that is in excellent agreement with the 1.1% content calculated from the amino acid composition of the "average" protein (Cuhel *et al.*, 1982b). More recently, Cuhel *et al.* (1984) verified that the protein C : protein S ratio holds for *Dunaliella tertiolecta* (a unicellular marine alga) and for field samples collected from marine and freshwater ecosystems. Because the C : S ratio in protein appears to be rel-atively invariant among species and as a function of physiological state and growth rate, the incorporation of $^{35}SO_4^{2-}$ into protein might be used as an accurate measure of the rate of *de novo* protein synthesis in environmental samples. Furthermore, because sulfolipid biosynthesis in algae appears to be a relatively constant percentage of total S assimilation, measurement of its rate of synthesis might ultimately be used to help attribute the total protein synthesis to autotrophic and heterotrophic components (Cuhel, personal com-munication). Alternatively, an estimate of the rate of heterotrophic protein synthesis in complex natural microbial communities may be derived utilizing the $^{14}CO_2$ technique of DiTullio and Laws (1983) as an independent measure of strictly autotrophic protein synthesis.

There are, nevertheless, several limitations and potential analytical difficulties in the use of $^{35}SO_4^{2-}$ to measure microbial production in environmental samples. First, $^{35}SO_4^{2-}$ is not the immediate precursor to protein sulfur. There is a possibility that the precursor-specific radioactivity is different from that of the external SO_4^{2-} pool. This could result from preferential uptake of partially reduced sulfur compounds known to compete with SO_4^{2-} or from intracellular turnover of proteins or other sulfur-containing compounds. Cuhel *et al.* (1982*b*) indicated that dilution of $^{35}SO_4^{2-}$ by endogenous pools of low-molecular-weight organic S compounds in laboratory-grown bacteria is minimal. Another restriction is the need to use extremely high levels of radioactivity (typically >500 μCi $^{35}SO_4^{2-}$ liter^{-1} for freshwater to >20 mCi $^{35}SO_4^{2-}$ liter^{-1} for marine environments) in order to obtain statistically reliable incorporation data. This creates an unacceptably high noise : signal ratio which has been, in part, alleviated by improved filtration methods and experimental protocol (Jordan *et al.*, 1978). However, for oligotrophic or even mesotrophic environments, this is a major limitation of the present technique. R. L. Cuhel (personal communication) suggested that the use of $^{35}S_2O_3^{2-}$ rather than $^{35}SO_4^{2-}$ might help lower the specific activity barrier that presently exists, especially for seawater samples, where $[SO_4^{2-}] = 28$ mM, and would enable investigators to use much lower levels of radioactivity. Alternatively, it has been suggested that HPLC of the individual amino acids resulting from hydrolysis of the isolated protein fraction may provide a unique separation of $^{35}SO_4^{2-}$-labeled contaminants from the $^{35}SO_4^{2-}$ actually incorporated into proteins (Laws, personal communication).

Rates of Nucleic Acid Synthesis

RNA and DNA are vital cellular constituents; in microorganisms, their rates of synthesis are directly proportional to rates of growth and cell division, respectively (Maaloe and Kjeldgaard, 1966; Ingraham *et al.*, 1983). Consequently, measurement of rates of nucleic acid synthesis should provide a meaningful experimental approach for the study of the metabolism and *in situ* growth kinetics of natural microbial assemblages. Brock (1967*a*) described a novel method for determining the *in situ* growth rates of selected bacteria in aquatic ecosystems. His technique was based on direct analysis of natural water samples by [^{3}H]thymidine autoradiography and required a prior laboratory calibration of the relationship between growth rate and rate of accumulation of radioactive cells determined using pure cultures. As originally outlined, this autecological approach was an eloquent, albeit fairly specialized, time-consuming, and tedious procedure; the latter features may have contributed to its lack of widespread use by microbial ecologists.

More recently, there has been renewed interest in the measurement of nucleic acid synthesis, especially in aquatic microbial communities. In principle, rates of microbial RNA or DNA synthesis can be measured by monitoring the rate of incorporation (expressed as g or moles hr^{-1}) of radiolabeled nucleosides (e.g., thymidine, uridine), nucleic acid bases (e.g., adenine), or phosphate into the respective purified nucleic acid fractions of the total microbial community. However, the mere incorporation of these selected radiotracers into total acid-insoluble cellular materials cannot be accepted as even a "relative rate" of nucleic acid synthesis in natural microbial assemblages due to (1) an unknown and variable dilution of the added radiotracer with existing extracellular and intracellular pools of structurally related compounds, (2) an unknown balance between *de novo* synthesis of purine and pyrimidine bases and salvage pathways (i.e., the utilization of exogenous pools), and (3) an unknown specificity of macromolecular labeling. In order to derive meaningful estimates of nucleic acid synthesis, it is imperative to measure both the specific radioactivity (as Ci g^{-1} or Ci $moles^{-1}$) of the immediate precursor pool(s) (i.e., the ribonucleotide and deoxyribonucleotide triphosphates for RNA and DNA, respectively) and the rate of accumulation of radioactivity (as Ci hr^{-1}) into the individual nucleic acids. The importance of measuring the specific radioactivity of the immediate precursor pool cannot be overemphasized for application of radiolabeled experiments in microbial ecology. Two additional concerns with the use of nucleic acid precursors include (1) absence of evidence that the ability to assimilate the exogenous tracer is universal among microorganisms, if total microbial production is to be estimated, or—in the case of thymidine—absence of evidence that all bacteria and only bacteria will incorporate this tracer into DNA; and (2) the uncertainty regarding how to extrapolate rates of RNA or DNA synthesis to reflect the total carbon production or growth rate.

There are presently three separate and distinct approaches for measuring rates of microbial nucleic acid synthesis in environmental samples. The first involves the use of [methyl-^{3}H]thymidine to assess the rate of heterotrophic bacterial DNA synthesis. The specific radioactivity of the immediate intracellular precursor pool (TTP in the case of [^{3}H]thymidine incorporation) is measured indirectly by performing a separate isotope dilution experiment with each environmental sample (Moriarty and Pollard, 1981; Moriarty, 1984). The second approach employs [2,^{3}H]adenine as a measure of total microbial nucleic acid synthesis. A unique advantage of this method is the ability to measure both RNA and DNA synthesis, as adenine serves as a precursor to both macromolecules. The specific radioactivity of the immediate intracellular precursor pools is measured directly by purifying and measuring the total concentration (pmoles) and radioactivity (nCi) of cellular ATP (Karl, 1979*b*, 1980; Karl and Winn, 1984). The third approach involves the use of $^{32}PO_4$ to assess rates of total microbial nucleic acid synthesis. Like the [^{3}H]adenine technique, $^{32}PO_4$ can be used to measure rates of RNA and DNA synthesis provided the specific radioactivity of the α-P position of ATP is directly measured (Karl and Bossard, 1985*a,b*). Karl and Bossard (1985*b*) recently compared the results of the [^{3}H]adenine and ^{32}PO$_4$ procedures and report excellent agreement between the two independent methods. Although such comparisons do not directly evaluate the accuracy of the methods involved (there is no absolute standard for comparison), the agreement of field-derived estimates tends to add a certain amount of credibility to both methods. (A more elaborate discussion of the ^{32}PO$_4$ method is presented in the section, ATP and Adenine Nucleotide Pool Turnover.) Recently, the relative merits of the [^{3}H]thymidine and [^{3}H]adenine techniques have been evaluated, compared, and contrasted (Fuhrman and Azam, 1980, 1982; Karl, 1982b; Karl and Winn, 1984; Moriarty, 1984). It should be emphasized, however, that the individual methods yield different and unique information regarding the growth of microorganisms in nature and should be considered as complementary methods in microbial ecology.

[Methyl-^{3}H]Thymidine

The uptake and incorporation of [methyl-^{3}H]thymidine have been proposed as a measure of (heterotrophic) bacterial DNA synthesis. This method assumes that all (or in practice, most) bacteria assimilate exogenous thymidine, and conversely, that eukaryotes do not. The ability of chemautotrophic bacteria, oxygenic and anoxygenic photosynthetic prokaryotes, and mixotrophically growing bacterial cells to assimilate thymidine has not been discussed. The principal assumption seems inconsistent with certain data on the assimilation of [^{3}H]thymidine by eukaryotic algae, protozoa, yeasts, other fungi, and slime molds (summarized by Karl, 1982b). The mere absence of thymidine kinase in eukaryotes is insufficient to argue against [^{3}H]thymidine uptake and, in fact, when thymidine kinase is absent there is a greater probability of nonspecific labeling with [^{3}H]thymidine (Grivell and Jackson, 1968). It has been argued, however, that at the concentrations employed (generally nM) and for the relatively short incubation periods used, the uptake of [^{3}H]thymidine by eukaryotes should be negligible (Fuhrman and Azam, 1980; Moriarty, 1984). Recently, uptake of ^{3}H-thymidine by eukaryotic algae in natural mixed populations of microorganisms has been clearly demonstrated (Rivkin, 1986).

The quantitative results of most published [^{3}H]thymidine uptake experiments (Thomas et al., 1974; Straat et al., 1977; Tobin and Anthony, 1978; Fuhrman and Azam, 1980, 1982; Kirchman et al., 1982a; Newell and Fallon, 1982; Fallon et al., 1983; Newell et al., 1983; Bell et al., 1983) must be interpreted with caution because there was no attempt to evaluate and correct for either or both of (1) the effects of either extracellular or intracellular dilution of the introduced precursor, and (2) the extent of nonspecific macromolecular labeling. Fuhrman and Azam (1980) suggested that isotope dilution by external pools must be small, since addition of thymidine at levels above 5 nM did not significantly increase incorporation of [^{3}H]thymidine into acid-insoluble materials. However, this protocol does not evaluate the extent of intracellular dilution (by internal recycling and de novo synthesis), which could account for the greatest reduction in the specific activity of the immediate nucleotide precursor pools (Rosenbaum-Oliver and Zamenhof, 1972; Karl *et al.*, 1981*a,b*). Furthermore, it now appears well documented that radioactivity in DNA represents a variable percentage (from <20% to >95%) of the total [^{3}H]thymidine-derived, acid-insoluble radioactivity in environmental samples (Karl, 1982*b;* Fuhrman and Azam, 1982; Riemann *et al.*, 1982; Moriarty, 1984; Witzel and Graf, 1984). It must be accepted that there is no shortcut: Purification of DNA is essential. This point has been stressed by Moriarty and Pollard (1982) from the start but is still ignored in many field studies. An even more rigorous approach might be adopted to confirm that the radioactivity incorporated into DNA is, in fact, contained within the TMP base moiety of DNA. This is especially important in environments in which non-specific labeling of protein and RNA, from [methyl-^{3}H]thymidine, is evident. Under these conditions, there is no *a priori* assurance that the presumed "DNA" labeling is any more reliable than the general non-specific labeling of all cellular macromolecules. If the radioactivity incorporated into DNA resides in bases other than TMP, the expected labeling pathways and presumed precursor–product relationships may be inappropriate and the DNA synthesis estimates, derived from these incorporation data, inaccurate.

Since it is difficult (if not impossible) to measure the specific radioactivity of the TTP pool (immediate precursor to DNA) directly, Moriarty and Pollard (1981) proposed a novel indirect approach. By adding a series of different quantities of unlabeled thymidine along with a constant amount of radioactivity and by measuring the magnitude of the change (i.e., the dilution) of the radiotracer ultimately incorporated into DNA, one can estimate the specific radioactivity of the immediate precursor, [^{3}H]TTP, pool (Moriarty, 1984). Using this approach, Moriarty and Pollard (1981, 1982) detected a substantial but highly variable dilution of the added [^{3}H]thymidine by extracellular and intracellular pools of thymidine or precursor molecules. In addition to spatial variability, there was also significant diurnal variation in isotope dilution (Moriarty and Pollard, 1982). Since the measured degree of participation is actually a community average, a large population variance will result in an equally large variance in the estimated rates of DNA synthesis. This would be true even if the average specific radioactivity of the community TTP pool could be measured by direct techniques. Pollard and Moriarty (1984) reported an isotope dilution of 76% for a seawater sample incubated with 5 nM [^{3}H]thymidine. Since this concentration has been recommended for most environmental studies (Fuhrman and Azam, 1980), as the concentration at which isotope dilution is negligible, the accuracy of previous estimates of bacterial DNA synthesis based on this assumption remains uncertain. Clearly, isotope dilution effects are probably time and space dependent such that *a priori* predictions are of little use in ecological studies.

Although more rigorous than previous techniques utilizing [^{3}H]thymidine, the isotope dilution procedure of Moriarty and Pollard (1981) has a number of additional problems. First, the method frequently produces bi- or multiphasic plots of 1/DPM incorporated into DNA versus thymidine added (rather than the expected linear plots); this obviously complicates the interpretation of the isotope dilution experiments (Moriarty and Pollard, 1981; Riemann *et al.*, 1982; Riemann, 1984). Furthermore, the theory of isotope dilution suffers from general criticisms, described previously (Wright and Burnison, 1979; Laws, 1983; D. F. Smith *et al.*, 1984), regarding the validity of the assumption that the velocity of substrate utilization is completely independent of the concentration of substrate added. As currently employed, this technique can only set an upper limit on the extent of isotope dilution (Laws, 1983). Finally, this method represents a substantial investment in terms of time and effort (e.g., it typically requires a minimum of seven separate analyses per sample per time point; see Moriarty and Pollard, 1981); however, this latter criticism (i.e., time investment) is appropriate only if accurate estimates of precursor-specific radioactivity can be determined by simpler procedures. Pollard and Moriarty (1984) recently presented the results of a set of laboratory chemostat experiments designed to evaluate the accurary of the isotope dilution method for estimating bacterial DNA synthesis. For most of their test bacteria, there was close agreement among growth rates: imposed by dilution rate, determined by direct microscopic counts, and estimated from [^{3}H]thymidine incorporation into DNA.

Most of the investigators using [^{3}H]thymidine have elected to extrapolate their incorporation data (DNA synthesis) to bacterial production in units of cells liter^{-1} hr^{-1}, rather than g C liter^{-1} hr^{-1}. This extrapolation assumes that TMP comprises 25 mol% of bacterial DNA and that the amount of DNA per cell ranges from 7.47×10^{-16} to 4.82×10^{-15} g; this assumption yields a conversion factor ranging from 2.0×10^{17} to 1.3×10^{18} cells produced per mol of thymidine incorporated (Fuhrman and Azam, 1980). This extrapolation is highly dependent on the characteristic size of the bacterial cells; consequently, the factor is expected to vary among environments. Kirchman *et al.* (1982a) recently devised a method for routinely calibrating the conversion factor for a given environment. Their approach requires that growth and DNA synthesis be coupled throughout the observation period (up to 24 hr). The extrapolation factors derived from a variety of freshwater and marine environments ranged from 1.9 to 68×10^{18} cells produced per mole thymidine incorporated, which emphasizes the need for frequent calibration of this method when used in ecological studies.

Although [^{3}H]thymidine uptake measurements sometimes correlate well with other independent estimates of microbial production, the quantitative results indicate a general lack of agreement (Newell and Fallon, 1982; Riemann and Sondergaard, 1984; Bell *et al.*, 1983; Riemann *et al.*, 1984). These results indicate that the thymidine method underestimates bacterial production when compared with increases in cell numbers in prefiltered (<3 μm) samples and frequency of dividing cell measurements by up to an order of magnitude; however, neither of these latter methods can be considered an absolute standard for comparison of methods. A recent report by Riemann

(1984) indicated that direct microscopic counts of increases in bacterial cells during timed incubations yielded estimates that ranged from 30 to 231% of those derived from thymidine uptake measurements. The lack of agreement between the two methods was even greater when [^{3}H]thymidine incorporation into purified DNA was considered (Riemann, 1984).

[^{3}H]Adenine

The use of radiolabeled adenine in ecological studies was first proposed by Karl (1979*b*) as a relatively simple and sensitive technique for measuring rates of microbial stable RNA synthesis. Subsequently, this method was expanded in scope in order to provide simultaneous measurements of total microbial RNA and DNA synthesis in water samples (Karl, 1981) as well as sediments (Craven and Karl, 1984). A review of the basic protocol, assay assumptions, experimental verification, laboratory and field calibration and a summary of selected ecological applications of the [^{3}H]adenine method has recently appeared (Karl and Winn, 1984), and thus will be discussed only briefly here.

Most microorganisms (i.e., bacteria, microalgae, yeasts, and other fungi) possess the ability to utilize exogenous supplies of adenine, and certain other nucleic acid bases, as a supplement to *de novo* synthesis. The mechanisms and implications of these so-called salvage pathways have been discussed elsewhere (Hochstadt, 1974). When adenine is transported into a cell, it is rapidly incorporated into a number of derivatives including ATP, dATP (deoxyadenosine triphosphate), ADP, and AMP and ultimately into the metabolically stable nucleic acids. It has been demonstrated that adenine uptake is directly coupled to nucleic acid synthesis such that the rate of entry of adenine into the cell does not exceed the biosynthetic requirements for ATP and dATP.

When radioactive adenine is added to an environmental sample, it is diluted to an unknown extent by existing extracellular pools of adenine and adenine-containing compounds such as ATP, ADP, AMP, and adenosine. This process decreases the specific radioactivity of the introduced radiotracer prior to transport into the cells. Since the salvage pathways generally do not supply the total amount of adenine required by the cells, this process must be supplemented by *de novo* synthesis of adenine. This further decreases the specific activity of the introduced radiotracer after transport into the cells but before incorporation into nucleic acids. Fortunately, the specific radioactivity of total microbial ATP can be measured directly, and the combined effects of extracellular and intracellular dilution of the radiotracer can be determined (Karl, 1979*b*). Furthermore, since the specific radioactivity of dATP is at all times identical to that of ATP (Karl, 1980), this value need not be measured independently.

There are several testable assumptions implicit in the application of the [^{3}H]adenine method to the analysis of natural microbial assemblages, including (1) all (or in practice, most) microorganisms assimilate exogenous adenine by a common pathway and exhibit a similar salvage response under *in situ* conditions; (2) addition of radioactive adenine does not affect the ATP cell quota, ATP turnover rate, or the rate of microbial RNA and DNA synthesis; and (3) there is no intracellular compartmentalization of microbial ATP pools or, if there is, such compartmentalization does not affect the accuracy of the measured rates. A thorough discussion of the validity of each of these assumptions has been presented elsewhere (Karl and Winn, 1984). More recently, the [^{3}H]adenine method has been calibrated using chemostat and batch cultures of representative marine microorganisms (Winn and Karl, 1984). The results show excellent agreement between the [^{3}H]adenine-derived rates of nucleic acid synthesis and the rates calculated from direct measurements of growth rate and concentrations of cellular RNA and DNA.

As currently employed, the [^{3}H]adenine method assumes that all individual cellular ATP pools are in approximate isotopic equilibrium, since the procedure requires extraction of the total microbial ATP pool in order to determine the precursor specific radioactivity. If, for whatever reason, there were a large proportion of the living (i.e., ATP-containing) microbial population that was not responsive to the exogenous [^{3}H]adenine, or if there were a subtantial variation among or intense competition between bacteria and algae for the uptake of nanomolar concentrations of adenine, the measured community ATP pool-specific radioactivity would be lower than that of the individuals actually incorporating the radiotracer. This would result in a systematic overestimate in the rates of RNA and DNA synthesis (i.e., rate = nCi incorporation into RNA or DNA/specific radioactivity of the ATP pool [nCi pmoles^{-1}]). It is difficult to evaluate this assumption directly, but it is encouraging to note that laboratory cultures of algae yield ATP pool-specific radioactivities comparable to those measured in bacteria grown under identical laboratory conditions. This implies that the salvage pathways in both bacteria and unicellular algae supply similar percentages of the required cell quota of adenine. In fact, this appears to be true under field conditions as well, since there is no substantial difference in the magnitude of the ATP pool specific activities between different size fractions of microbial populations found in seawater samples analyzed to date (Karl *et al.*, 1981*b;* Karl and Winn, 1986). Magnitude discrepancies would be expected if only a portion (e.g., bacteria) of the community was actively assimilating [^{3}H]adenine. Furthermore, the results of nutrient addition experiments, the results of double-labeled ^{32}PO$_4$/[^{3}H]adenine incubations and isotope-dilution experiments (Karl, 1982*b;* Karl and Winn, 1984; Karl and Bossard, 1985*b*), also support the field validity of the isotope equilibration assumption. Finally, it is informative to point out that the magnitude of the field-derived ATP pool-specific radioactivity at saturating concentrations of [^{3}H]adenine is indistinguishable from that measured in pure cultures of bacteria (Karl, 1982*b;* Karl *et al.*, 1981*a*); this argues against the presence of a substantial population of viable (i.e., ATP-containing) microorganisms which are incapable of [^{3}H]adenine assimilation. All the above-mentioned observations and data support the assumption that marine microorganisms assimilate adenine by a common and uniform pathway under *in situ* conditions.

Most investigations concerned with the production, flux and turnover of organic matter in natural ecosystems have relied on carbon as the basic unit of cell mass and growth. If the assumptions inherent in the application of nucleic acid precursors to ecological studies were determined to be valid under *in situ* conditions, the [³H]adenine method as described would yield reliable estimates of the total rates of microbial RNA and DNA synthesis under natural environmental conditions. Whereas one might argue that nucleic acid synthesis rate data, by themselves, provide useful information regarding the growth of microorganisms in nature, it is often necessary and/or desirable to extrapolate the DNA synthesis data to estimates of total microbial production (g C produced per unit volume per unit time) for this information itself, or for comparison with other experimental data. The precision and accuracy of the estimates are adversely affected because the extrapolation factors are, for the most part, highly variable. Calculations to date have assumed (1) that dAMP represents 25 mol% of total microbial DNA, and (2) a mean DNA : C ratio of 50, which is based upon laboratory experiments (Holm-Hansen, 1969; Mandelstam and McQuillen, 1976) and direct field measurements of total DNA and organic carbon. The latter approach (i.e., direct-field calibration) is applicable only where the amount of nonliving particulate DNA and carbon are expected to be negligible, a fact that may severely limit the use of an empirically determined extrapolation factor. Finally, measurements of RNA and DNA synthesis, when used in conjunction with steady-state RNA and DNA concentration data, may be used to estimate the mean community turnover time and specific growth rate (μ). This must be considered a minimum value for the growth rate because the presence of detrital (i.e., nonliving) RNA or DNA tends to overestimate the value for microbial standing stock and underestimate the growth rate (Karl and Winn, 1984).

ATP and Adenine Nucleotide Pool Turnover

Previous measurements of ATP (i.e., steady-state concentrations) in microorganisms and in environmental samples (data reviewed in Karl, 1980) have emphasized the futility of extrapolating these static pool measurements to estimates of metabolic activity or growth. Although adenine nucleotide turnover rate might be expected to be positively correlated with growth rate, it is well documented that the intracellular pool turnover rates are independent of steady-state ATP concentrations. Consequently, the direct measurement of ATP and adenine nucleotide pool turnover should provide useful information regarding growth rates in natural microbial communities. Karl and Bossard (1985a) outlined a novel method for the direct measurement of the turnover rate of ATP. The method relies on the use of $^{32}PO_4$ and is designed to monitor changes in the specific radioactivities of the γ-P and α-P positions of ATP. From time course information on the change in specific activity of the γ-P, the turnover rate of ATP is derived. This information can then be used in conjunction with ATP pool measurements to calculate the total biological energy flux (e.g., kcal liter^{-1} hr^{-1}) for comparison with other indirect measurements such as O_2 respirometry (for aerobically growing heterotrophs), O_2 production or CO_2 reduction (for photoautotrophs) or microcalorimetry (for total microbial communities). Rates of change of the α-P position of ATP (Karl and Bossard, 1985*b;* Bossard and Karl, 1986) or rates of accumulation of [^{3}H]adenine into [^{3}H]ATP (Karl, 1985) can be used to estimate the turnover time of the adenine nucleotide pool. Data compiled from laboratory experiments using bacterial cultures indicate that the adenine nucleotide pool turnover rate is equivalent to 2–3% of the generation time. Consequently, mean community doubling times, and hence specific growth rates, can be extrapolated directly from the turnover time estimates. This method can also be used to measure the rates of total microbial community RNA and DNA synthesis (Karl and Bossard, 1985*b;* see also Section 8.5). It is too early to predict the overall importance or significance of these new methods; however, preliminary results and interpretations are promising (Bossard and Karl, 1986; Karl, 1985).

Increase in Cell Numbers/Biomass during Timed Incubations

Growth (or death) rates of microorganisms in non-steady-state conditions can be determined by monitoring changes in cell numbers or biomass with time in isolated samples. However, under strictly steady-state conditions, it is impossible to measure growth rate or production from changes in cell numbers or mass, since, by definition, d biomass/$dt = 0$. Under these ideal steady states, growth is balanced by death or removal (i.e., advection, sinking, predation). Of these mechanisms, predation is generally considered the most important factor in natural habitats. By selectively removing predation pressure, the *in situ* growth rate of the prey can be determined by measuring increases in cell numbers or biomass during timed incubations. With appropriate modification, this basic method has been used to estimate the growth rate and production of bacterial cells (Gambaryan, 1965; Gak *et al.*, 1972; Fuhrman and Azam, 1980; Kirchman *et al.*, 1982*a;* Newell *et al.*, 1983; Wright and Coffin, 1984), unicellular algae (Eppley, 1968; Falkowski and Owens, 1982; Bienfang and Takahashi, 1983), and total microbial community based on total particle volume (Cushing and Nicholson, 1966; Sheldon *et al.*, 1973) and total microbial ATP (Sieburth *et al.*, 1977; Sheldon and Sutcliff, 1978). More recently, this approach has also been used to estimate the impact of predation by protozoa and micrometazoa on microbial populations in seawater (Landry and Hassett, 1982). The various methods that have been used to eliminate or reduce the influence of predators include (1) direct removal by screening or membrane filtration, (2) extinction dilution with microbe-free water from the same habitat, (3) chemical inhibition of protozoan grazing activity, and (4) advertent or inadvertent statistical elimination of predators (especially large predators) by taking an extremely small sample size.

Several limitations to this approach compromise the accuracy of the ecological interpretations theoretically derived. First, the application of this method presupposes that growth in the absence of predation is continuous and exponential, although Fuhrman and Azam (1980) assumed linear growth. The presence of a lag phase, diurnal periodicity or phased growth, or progressive changes in growth rate due to containment, nutrient depletion, adaptation, succession or selective species-specific growth or death tend to invalidate the results (Eppley, 1968). Consequently, incubation periods must be kept short in order to minimize secondary growth effects but long enough to allow for statistically significant increases in the microbial population. A desirable incubation period is approximately equal to 10% of the mean community doubling time (T. D. Brock, 1971). In this regard, incubation in dialysis or diffusion chambers, rather than in enclosed containers, may prolong the expected period of exponential growth by reducing the impact of nutrient depletion. Another area of concern is the selection of the most appropriate metric to assess microbial biomass changes. Fuhrman and Azam (1980) suggested that direct epifluorescent cell counts can be used to estimate bacterial growth rates in prescreened (<3 μm) seawater samples. Unfortunately, this method of analysis (i.e., enumeration of cells) is complicated by changes in mean cell volume during incubation. Christian *et al.* (1982) documented a significant increase in bacterial cell volume from 0.06 to 0.17 μm^3 and a concomitant increase in microbial ATP during the first 6 hr of incubation with <3 μm seawater filtrates before there was any detectable change in cell numbers. The presence of nondividing cells will produce a concavity in the plot of log bacterial abundance versus time, a result that is indistinguishable from a situation in which cells experience a *bona fide* lag period (Kirchman *et al.*, 1982a). Last, but certainly not least, is the practical problem of nonlinear curve fitting, which may result in substantial data interpretation error as recently demonstrated by Sheldon (1979).

Ferguson *et al.* (1984) evaluated the response of marine bacterioplankton to differential filtration and confinement. In their analysis, they differentiate between cultivable bacteria, i.e., colony-forming units (CFU) on complex nutrient agar and total cell number as determined by epifluorescence microscopy. Their results indicate that the cultivable bacteria in the <3.0-μm filtrate increase from an insignificant proportion of the total bacterial assemblage (<0.1%) immediately after collection, to a major (>40%) proportion at 32-hr confinement. This shift of numerical proportion was the result of differential net growth, and not death, with estimated doubling times of 53 hr versus 2.2 hr for the culturable and noncultural groups, respectively. It is concluded that the results obtained are based on an analytical artifact which is, in part, caused by the release of nutrients during the size fractionation procedures. In any case, the results derived from growth rate or production experiments in the absence of predation must be considered an upper bound on the microbial community production. Under normal environmental conditions, the potential exponential growth rate of microorganisms is rarely, if ever, fully realized. During ideal steady-state conditions, all new microbial production is immediately consumed by higher trophic levels or lost by mechanical removal.

SUMMARY AND PROSPECTS FOR THE FUTURE

In reading this chapter, an apprentice microbial ecologist might be discouraged by the numerous difficulties, uncertainties and limitations of the methods currently employed in field studies. This is certainly the wrong attitude to adopt. Although the complexities of natural ecosystems and their resident microbial populations still overwhelm our technological and analytical capabilities, prospects for the future are bright and should be viewed with optimism. However, the discipline must maintain a high level of self-criticism and protect its present rigorous standards for careful and thorough evaluation of new methodologies prior to general acceptance.

Many of the experimental methods presently used in microbial ecology are operationally straightforward and easy to perform, although not necessarily very accurate. Other approaches are much more complex, both in theory and in practice, although not necessarily any more accurate. In preparing this chapter, I have tried to remain objective in my criticisms, a goal that was not easy, as I have a personal investment in several of the methods described herein. Nevertheless, we should not consider the individual methods as competing with one another or as mutually exclusive techniques, since each approach yields valuable information when interpreted in light of the recognized methodological limitations. The complementary nature of the multiple methods approach to microbial ecology cannot be overemphasized. As the ecological questions and hypotheses become more refined, even the present-day state-of-the-art techniques will become obsolete. New methods must be devised continuously and evaluated for use.

It is dangerous to predict the future progress of any scientific discipline; in the case of microbial ecology, however, there is already an indication that the field will experience a rapid increase in technology and instrumentation during the next decade. Research in microbial ecology should anticipate, and gratefully accept, an increasing level of automation, data processing capabilities and computerization. Remote sensing capabilities (both satellite and tetherless unmanned submersibles) and applications of continuous monitoring of microbial biomass, metabolic activity, and growth will probably become commonplace during the next decade. Furthermore, advanced statistical treatment of ecological data and the development of comprehensive ecological models should provide us with a better understanding of the complex interactions among microorgansims and between microbes and their environment.

I leave the reader with the challenge to refine the existing experimental methods, or to devise better ones.

ACKNOWLEDGMENTS. I wish to thank P. Bossard, D. Burns, G. Knauer, J. Novitsky, K. Orrett, and G. Taylor for their helpful criticisms and valuable suggestions for the improvement of this review, L. Takeuchi, L. Wong, and P. Sexton for their cheerful and flawless preparation of several versions of the manuscript, and T. Tagami for her assistance with the reference list. I would especially like to express my sincere gratitude and appreciation to the numerous investigators, both past and present, who challenged the existing methods and introduced innovative alternatives for the study of microorganisms in nature. A portion of the original research summarized in this chapter was supported by grants OCE 78-20721, OCE 78-25446, OCE 80-05180, OCE 81-09256, OCE 82-16673, and PYI83-51751 from the National Science Foundation. The primary literature search for this manuscript was completed in July 1983, although selected portions of this review have been further updated.

REFERENCES

Aaronson, J. M., 1966, The cell wall, in: *The Fungi*, Vol. 1 (G. C. Ainsworth and A. S. Sussonan, eds.), pp. 49–76, Academic Press, London.

Anderson, J. R., and Domsch, K. H., 1978, Mineralization of bacteria and fungi in chloroform-fumigated soils, *Soil Biol. Biochem.* **10**:207–213.

Anderson, J. R., and Westmoreland, D., 1971, Direct counts of soil organisms using a fluorescent brightener and a europium chelate, *Soil Biol. Biochem.* **3**:85–87.

Andrews, P., and Williams, P. J. leB., 1971, Heterotrophic utilization of dissolved organic compounds in the sea. III. Measurement of the oxidation rates and concentrations of glucose and amino acids in sea water, *J. Mar. Biol. Assoc. U.K.* **51**:111–125.

Azam, F., and Hodson, R. E., 1977a, Dissolved ATP in the sea and its utilization by marine bacteria, *Nature (Lond.)* **267**:696–698.

Azam, F., and Hodson, R. E., 1977b, Size distribution and activity of marine microheterotrophs, *Limnol. Oceanogr.* **22**:492–501.

Azam, F., and Hodson, R. E., 1981, Multiphasic kinetics for D-glucose uptake by assemblages of natural marine bacteria, *Mar. Ecol. Prog. Ser.* **6**:213–222.

Azam, F., and Holm-Hansen, O., 1973, Use of tritiated substrates in the study of heterotrophy in seawater, *Mar. Biol.* **23**:191–196.

Babiuk, L. A., and Paul, E. A., 1970, The use of fluorescein isothiocyanate in the determination of the bacterial biomass of glassland soil, *Can. J. Microbiol.* **16**:57–62.

Bae, H. C., Cota-Robles, E. H., and Casida, L. E., 1972, Microflora of soil as viewed by transmission electron microscopy, *Appl. Microbiol.* **23**:637–648.

Bakken, L. R., and Olsen, R. A., 1983, Buoyant densities and dry-matter contents of microorganisms: Conversion of a measured biovolume into biomass, *Appl. Environ. Microbiol.* **45**:1188–1195.

Bamstedt, U., 1980, ETS activity as an estimator of respiratory rate of zooplankton populations. The significance of variations in environmental factors, *J. Exp. Mar. Biol. Ecol.* **42**:267–283.

Banse, K., 1977, Determining the carbon-to-chlorophyll ratio of natural phytoplankton, *Mar. Biol.* **41**:199–212.

Barer, R., 1974, Microscopes, microscopy, and microbiology, *Annu. Rev. Microbiol.* **28**:371–389.

Bates, S. S., 1981, Determination of the physiological state of marine phytoplankton by use of radiosulfate incorporation, *J. Exp. Mar. Biol. Ecol.* **51:**219–239.

Bell, R. T., Ahlgren, G. M., and Ahlgren, I., 1983, Estimating bacterioplankton production by measuring [³H]thymidine incorporation in a eutrophic Swedish lake, *Appl. Environ. Microbiol.* **45:**1709–1721.

Berman, T., 1980, Multiple isotopic tracer methods in the sstudy of growth, dynamics, and metabolic processes in marine ecosystems, in: *Primary Productivity in the Sea* (P. G. Falkowski, ed.), pp. 213–230, Plenum Press, New York.

Bienfang, P., and Takahashi, M., 1983, Ultraplankton growth rates in a subtropical ecosystem, *Mar. Biol.* **76:**213–218.

Blackburn, T. H., 1979, Method for measuring rates of NH_4^+ turnover in anoxic marine sediments, using a^{15}N-NH_4^+dilution technique, *Appl. Environ. Microbiol.* **37:**760–765.

Bobbie, R. J., and White, D. C., 19809, Characterization of benthic microbial community structure by high-resolution gas chromatography of fatty acid methyl esters, *Appl. Environ. Microbiol.* **39:**1212–1222.

Bohlool, B. B., and Schmidt, E. L., 1980, The immunofluorescence approach in microbial ecology, in: *Advances in Microbial Ecology,* Vol. 4 (M. Alexander, ed.), pp. 203–241, Plenum Press, New York.

Bonora, A., and Mares, D., 1982, A simple colorimetric method for detecting cell viability in cultures of eukaryotic microorganisms, *Curr. Microbiol.* **7:**217–222.

Bossard, P., and Karl, D. M., 1986, The direct measurement of ATP and adenine nucleotide turnover in microorganisms: A new method for environmental assessment of metabolism, energy flux and phosphorus dynamics, *J. Plankton Res.* **8:**1–13.

Bowden, W. B., 1977, Comparison of two direct-count techniques for enumerating aquatic bacteria, *Appl. Environ. Microbiol.* **33:**1229–1232.

Bratbak, G., and Dundas, I., 1984, Bacterial dry matter content and biomass estimations, *Appl. Environ. Microbiol.* **48:**755–757.

Bressler, S. L., and Ahmed, S. I., 1984, Detection of glutamine synthetase activity in marine phytoplankton: Optimization of the biosynthetic assay. *Mar. Ecol. Prog. Ser.* **14:**207–217.

Brock, M. L., and Brock, T. D., 1968, The application of microautoradiographic techniques to ecological studies, *Mitt. Int. Verein. Limnol.* **15:**1–27.

Brock, T. D., 1967*a*, Bacterial growth rate in the sea: Direct analysis by thymidine autoradiography, *Science* **155:**81–83.

Brock, T. D., 1967*b*, The ecosystem and the steady state, *BioScience* **17:**166–169.

Brock, T. D., 1971, Microbial growth rates in nature, *Bacteriol. Rev.* **35:**39–58.

Brock, T. D., 1978, Use of fluorescence microscopy for quantifying phytoplankton, especially filamentous blue-green algae, *Limnol. Oceanogr.* **23:**158–160.

Brock, T. D., 1984, How sensitive is the light microscope for observation on microorganisms in natural habitats?, *Microb. Ecol.* **10:**297–300.

Brunschede, H., Dove, T. L., and Bremer, H., 1977, Establishment of exponential growth after a nutritional shift-up in *Escherichia coli* B/r: Accumulation of deoxyribonucleic acid, ribonucleic acid and protein, *J. Bacteriol.* **129:**1020–1033.

Burnison, B. K., and Morita, R. Y., 1974, Hetetrophic potential for amino acid uptake in a naturally eutrophic lake, *Appl. Microbiol.* **27:**488–495.

Burns, D. J., 1983, Glutamine sythetase activity in marine phytoplankton, M.S. thesis, University of Hawaii at Manoa.

Burns, R. G. (ed.), 1978, *Soil Enzymes,* U.S. edition, Academic Press, New York.

Caperon, J., Schell, D., Hirota, J., and Laws, E., 1979, Ammonium excretion rates in Kaneohe Bay, Hawaii, measured by a ^{15}N isotope dilution technique, *Mar. Biol.* **54:**33–40.

Caron, D. A., 1983, Technique for enumeration of heterotrophic and phototrophic nanoplankton, using epifluorescence microscopy, and comparison with other procedures, *Appl. Environ. Microbiol.* **46:**491–498.

Caron, D. A., Davis, P. G., Madin, L. P., and Sieburth, J. McN., 1982, Heterotrophic bacteria and bacterivorous protozoa in oceanic macroaggregates, *Science* **218:**795–797.

Casida, L. E., 1968, Infrared color photograph: Selective demonstration of bacteria, *Science* **159**:199–200.

Casida, L. E., 1971, Microorganisms in unamended soil as observed by various forms of microscopy and staining, *Appl. Microbiol.* **21**:1040–1045.

Casida, L. E., 1976, Continuously variable amplitude contrast microscopy for the detection and study of microorganisms in soil, *Appl. Environ. Microbiol.* **31**:605–608.

Chappelle, E., 1975, Determination of bacterial flavins by bacterial bioluminescence, in: *ATP Methodology Seminar* (G. A. Borun, ed.), pp. 62–103, Science Applications Incorporated Technology Co., San Diego, California.

Christensen, D., and Blackburn, T. H., 1980, Turnover of tracer (^{14}C, ^{3}H labelled) alanine in inshore marine sediments, *Mar. Biol.* **58**:97–103.

Christensen, J. P., 1983, Electron transport system activity and oxygen consumption in marine sediments, *Deep-Sea Res.* **30**:183–194.

Christensen, J. P., and Packard, T. T., 1979, Respiratory electron transport activity in phytoplankton and bacteria: Comparison of methods, *Limnol. Oceanogr.* **24**:576–583.

Christensen, J. P., Owens, T. G., Devol, A. H., and Packard, T. T., 1980, Respiration and physiological state in marine bacteria, *Mar. Biol.* **55**:267–276.

Christian, R. R., Hanson, R. B., and Newell, S. Y., 1982, Comparison of methods for measurement of bacterial growth rates in mixed batch cultures, *Appl. Environ. Microbiol.* **43**:1160–1165.

Chrzanowski, T. H., Crotty, R. D., Hubbard, J. G., and Welch, R. P., 1984, Applicability of the fluorescein diacetate method of detecting active bacteria in freshwater, *Microb. Ecol.* **10**:179–185.

Chung, K.-T., Nilson, E. H., Case, M. J., Marr, A. G., and Huntgate, R. E., 1973, Estimation of growth rate from the mitotic index. *Appl. Microbiol.* **25**:778–780.

Coats, D. W., and Heinbokel, J. F., 1982, A study of reproduction and other life cycle phenomena in planktonic protists using an acridine orange fluorescence technique, *Mar. Biol.* **67**:71–79.

Coleman, A. W., 1980, Enhanced detection of bacteria in natural environments by fluorochrome staining of DNA, *Limnol. Oceanogr.* **25**:948–951.

Costerton, J. W., and Colwell, R. R. (eds.), 1979, *Native Aquatic Bacteria: Enumeration, Activity, and Ecology,* American Society for Testing Materials, Baltimore.

Costerton, J. W., Irvin, R. T., and Cheng, K. J., 1981, The bacterial glycocalyx in nature and disease, *Annu. Rev. Microbiol.* **35**:299–324.

Craigie, J. S., 1963, Dark fixation of C^{14}-bicarbonate by marine algae, *Can. J. Bot.* **41**:317–325.

Craven, D. B., and Karl, D. M., 1984, Microbial RNA and DNA synthesis in marine sediments, *Mar. Biol.* **83**:129–139.

Cuhel, R. L., Taylor, C. D., and Jannasch, H. W., 1981a, Assimilatory sulfur metabolism in marine microorganisms: Sulfur metabolism, growth, and protein synthesis of *Pseudomonas halodurans* and *Alteromonas luteo-violaceus* during sulfate limitation, *Arch. Microbiol.* **130**:1–7.

Cuhel, R. L., Taylor, C. D., and Jannasch, H. W., 1981b, Assimilatory sulfur metabolism in marine microorganisms: Sulfur metabolism, protein synthesis, and growth of *Pseudomonas halodurans* and *Alteromonas luteoviolaceus* during unperturbed batch growth, *Arch. Microbiol.* **130**:8–13.

Cuhel, R. L., Taylor, C. D., and Jannasch, H. W., 1981c, Assimilatory sulfur metabolism in marine microorganisms: Characteristics and regulation of sulfate transport in *Pseudomonas halodurans* and *Alteromonas luteoviolaceus, J. Microbiol.* **147**:340–349.

Cuhel, R. L., Taylor, C. D., and Jannasch, H. W., 1981d, Assimilatory sulfur metabolism in marine microorganisms: A novel sulfate transport system in *Alteromonas luteo-violaceus, J. Bacteriol.* **147**:350–353.

Cuhel, R. L., Taylor, C. D., and Jannasch, H. W., 1982a, Assimilatory sulfur metabolism in marine microorganisms: Sulfur metabolism, protein synthesis, and growth of *Alteromonas luteo-violaceus,* and *Pseudomonas halodurans* during perturbed batch growth, *Appl. Environ. Microbiol.* **43**:151–159.

Cuhel, R. L., Taylor, C. D., and Jannasch, H. W., 1982b, Assimilatory sulfur metabolism in marine microorganisms: Considerations for the application of sulfate incorporation into protein as a measurement of natural population protein synthesis, *Appl. Environ. Microbiol.* **43**:160–168.

Cuhel, R. L., Ortner, P. B., and Lean, D. R. S., 1984, Night synthesis of protein by algae, *Limnol. Oceanogr.* **29:**731–744.

Cullen, J. J., and Renger, E. H., 1979, Continuous measurement of the DCMU-induced fluorescence response of natural phytoplankton populations, *Mar. Biol.* **53:**13–20.

Cushing, D. H., and Nicholson, H. F., 1966, Method of estimating algal production rates at sea, *Nature (Lond.)* **212:**310–311.

Darken, M. A., 1961, Applications of fluorescent brighteners in biological techniques, *Science* **133:**1704–1705.

Davis, W. M., and White, D. C., 1980, Fluorometric determination of the adenosine nucleotide derivatives as measures of the microfouling, detrital, and sedimentary microbiol biomass and physiological status, *Appl. Environ. Microbiol.* **40:**539–548.

Dawes, E. A., and Senior, P. J., 1973, The role and regulation of energy reserve polymers in micro-organisms, *Adv. Microb. Physiol.* **10:**135–266.

Dawson, R., and Gocke, K., 1978, Heterotrophic activity in comparison to the free amino acid concentrations in Baltic sea water samples, *Ocean. Acta* **1:**45–54.

deJong, E., Redmann, R. E., and Ripley, E. A., 1979, A comparison of methods to measure soil respiration, *Soil Sci.* **127:**300–306.

Deming, J. W., Picciolo, G. L., and Chappelle, E. W., 1979, Important factors in adenosine triphosphate determinations using firefly luciferase: Applicability of the assay to studies of native aquatic bacteria, in: *Native Aquatic Bacteria: Enumeration, Activity and Ecology* (J. W. Costerton and R. R. Colwell, eds.), pp. 89–98, American Society for Testing Materials, Baltimore.

Dennis, P. P., and Bremer, H., 1974, Macromolecular composition during steady-state growth of *Escherichia coli* B/r, *J. Bacteriol.* **119:**270–281.

Dietz, A. S., Albright, L. J., and Tuominen, T., 1977, Alternative model and approach for determining microbial heterotrophic activities in aquatic systems, *Appl. Environ. Microbiol.* **33:**817–823.

DiTullio, G., and Laws, E. A., 1983, Estimates of photoplankton N uptake based on $^{14}CO_2$ incorporation into protein, *Limnol. Oceanogr.* **28:**177–185.

Doetsch, R. N., and Cook, T. M., 1973, *Introduction to Bacteria and Their Ecobiology,* University Park Press, Baltimore.

Dortch, Q., Ahmed, S. I., and Packard, T. T., 1979, Nitrate reductase and glutamate dehydrogenase activities in *Skeletonema costatum* as measures of nitrogen assimilation rates, *J. Plankton Res.* **1:**169–186.

Dortch, Q., Roberts, T. L., Clayton, J. R., and Ahmed, S. I., 1983, RNA/DNA ratios and DNA concentrations as indicators of growth rate and biomass in planktonic marine organisms, *Mar. Ecol. Prog. Ser.* **13:**61–71.

Dring, M. J., and Jewson, D. H., 1982, What does ^{14}C uptake by phytoplankton really measure? A theoretical modelling approach, *Proc. R. Soc. Lond.* **214:**351–368.

Dutton, R. J., Bitton, G., and Koopman, B., 1983, Malachite green-INT (MINT) method for determining active bacteria in sewage, *Appl. Environ. Microbiol.* **46:**1263–1267.

Ellwood, D. C., and Tempest, D. W., 1972, Effects of environment on bacterial wall content and composition, *Adv. Microb. Physiol.* **7:**83–117.

England, J. M., Rogers, A. W., and Miller, R. G., 1973, The identification of labelled structures of autoradiographs, *Nature (Lond.)* **242:**612–613.

Eppley, R. W., 1968, An incubation method for estimating the carbon content of phytoplankton in natural samples, *Limnol. Oceanogr.* **13:**574–582.

Eppley, R. W., 1980, Estimating phytoplankton growth rates in the central oligotrophic oceans, in: *Primary Productivity in the Sea* (P. G. Falkowski, ed.), pp. 231–242, Plenum Press, New York.

Eppley, R. W., 1981, Relations between nutrient assimilation and growth in phytoplankton with a brief review of estimates of growth rate in the ocean, in: *Physiological Bases of Phytoplankton Ecology* (T. Platt, ed.), pp. 251–263, Canadian Bulletin of Fish. Aquatic Science 210, Ottawa, Canada.

Eppley, R. W., Coatsworth, J. L., and Solorzano, L., 1969, Studies of nitrate reductase in marine phytoplankton, *Limnol. Oceanogr.* **14**:194–205.

Eppley, R. W., Packard, T. T., and MacIsaac, J. J., 1970, Nitrate reductase in Peru current phytoplankton, *Mar. Biol.* **6**:195–199.

Eppley, R. W., Harrison, W. G., Chisholm, S. W., and Stewart, E., 1977, Particulate organic matter in surface waters off Southern California and its relationship to phytoplankton, *J. Mar. Res.* **35**:671–696.

Evans, E. A., 1972, Purity and stability of radiochemical tracers in autoradiography, *J. Microsc.* **96**:165–180.

Evans, E. A., 1974, *Tritium and Its Compounds,* 2nd ed., Butterworth, London.

Evans, T. M., Schillinger, J. E., and Stuart, D. G., 1978, Rapid determination of bacteriological water quality by using *Limulus* lysate, *Appl. Environ. Microbiol.* **35**:376–382.

Ewetz, L., and Thore, A., 1976, Factors affecting the specificity of the luminol reaction with hermatin compounds, *Anal. Biochem.* **71**:564–570.

Falkowski, P. G., 1981, Light-shade adaptation and assimilation numbers, *J. Plankton Res.* **3**:203–216.

Falkowski, P. G., and Owens, T. G., 1982, A technique for estimating phytoplankton division rates by using a DNA-binding fluorescent dye, *Limnol. Oceanogr.* **27**:776–782.

Fallon, R. D., Newell, S. Y., and Hopkinson, C. S., 1983, Bacterial production in marine sediments: Will cell-specific measures agree with whole-system metabolism?, *Mar. Ecol. Prog. Ser.* **11**:119–127.

Fazio, S. D., Mayberry, W. R., and White, D. C., 1979, Muramic acid assay in sediments, *Appl. Environ. Microbiol.* **38**:349–350.

Federle, T. W., and White, D. C., 1982, Preservation of estuarine sediments for lipid analysis of biomass and community structure of microbiota, *Appl. Environ. Microbiol.* **44**:1166–1169.

Federle, T. W., Hullar, M. A., Livingston, R. J., Meeter, D. A., and White, D. C., 1983, Spatial distribution of biochemical parameters indicating biomass and community composition of microbial assemblies in estuarine mud flat sediments, *Appl. Environ. Microbiol.* **45**:58–63.

Fellows, D. A., Karl, D. M., and Knauer, G. A., 1981, Large particle fluxes and the vertical transport of living carbon in the upper 1500 m of the northeast Pacific Ocean, *Deep-Sea Res.* **28A**:921–936.

Ferguson, R. L., and Rublee, P., 1976, Contribution of bacteria to standing crop of coastal plankton, *Limnol. Oceanogr.* **21**:141–145.

Ferguson, R. L., Buckley, E. N., and Palumbo, A. V., 1984, Response of marine bacterioplankton to differential filtration and confinement, *Appl. Environ. Microbiol.* **47**:49–55.

Findlay, R. H., and White, D. C., 1983, Polymeric beta-hydroxyalkanoates from environmental samples and *Bacillus megaterium, Appl. Environ. Microbiol.* **45**:71–78.

Fitzwater, S. E., Knauer, G. A., and Martin, J. H., 1982, Metal contamination and its effect on primary production measurements, *Limnol. Oceanogr.* **27**:544–551.

Fliermans, C. B., Soracco, R. J., and Pope, D. H., 1981, Measure of *Legionella pneumophila* activity in situ, *Curr. Microbiol.* **6**:89–94.

Forsdyke, D. R., 1968, Studies of the incorporation of [5-^{3}H]uridine during activation and transformation of lymphocytes induced by phytohaemagglutin, *Biochem. J.* **107**:197–205.

Fox, G. E., Stackebrandt, E., Hespell, R. B., Gibson, J., Manihoff, J., Dyer, T. A., Wolfe, R. S., Balch, W. E., Tanner, R. S., Magrum, L. J., Zablen, L. B., Blackemore, R., Gupta, R., Bonen, L., Lewis, B. J., Stahl, D. A., Luehrsen, K. R., Chen, K. N., and Woese, C. R., 1980, The phylogency of the prokaryotes, *Science* **209**:457–463.

Frankland, J. C., 1975, Estimation of live fungal biomass, *Soil Biol. Biochem.* **7**:339–340.

Franzen, J. S., and Binkley, S. B., 1961, Comparison of the acid-soluble nucleotides in *Escherichia coli* at different growth rates, *J. Biol. Chem.* **236**:515–519.

Fuhrman, J. A., 1981, Influence of method on the apparent size distribution of bacterioplankton cells: Epifluorescence microscopy compared to scanning electron microscopy, *Mar. Ecol. Prog. Ser.* **5**:103–106.

Fuhrman, J. A., and Azam, F., 1980, Bacterioplankton secondary production estimates for coastal waters of British Columbia, Antarctica, and California, *Appl. Environ. Microbiol.* **39:**1085–1095.

Fuhrman, J. A., and Azam, F., 1982, Thymidine incorporation as a measure of heterotrophic bacterioplankton production in marine surface waters: Evaluation and field results, *Mar. Biol.* **66:**109–120.

Fukuzawa, N., Ishimaru, T., Takahashi, M., and Fujita, Y., 1980, A mechanism of "red tide" formation. I. Growth rate estimate by DCMU-induced fluorescence increase, *Mar. Ecol. Prog. Ser.* **3:**217–222.

Gak, D. S., Romanova, E. P., Romanenko, V. I., and Sorokin, Y. I., 1972, Estimation of changes in number of bacteria in the isolated water samples, in: *Techniques for the Assessment of Microbial Production and Decomposition in Fresh Waters* (Y. I. Sorokin and H. Kadota, eds.), pp. 78–82, Blackwell Scientific, London.

Galloway, R. A., Rolle, I., and Soeder, C. J., 1974, CO_2 fixation and biosynthetic activity of darkened synchronous *Chlorella fusca, Arch. Hydrobiol.* **73:**1–13.

Gambaryan, M. E., 1965, Method of determining the generation time of microorganisms in benthic sediments, *Microbiology* **34:**939–943.

Gehron, M. J., and White, D. C., 1982, Quantitative determination of the nutritional status of detrital microbiota and the grazing fauna by triglyceride glycerol analysis, *J. Exp. Mar. Biol.* **64:**145–158.

Glover, H. E., 1983, Measurement of chemautotrophic CO_2 assimilation in marine nitrifying bacteria: An enzymatic approach, *Mar. Biol.* **74:**295–300.

Glover, H. E., and Morris I., 1979, Photosynthetic carboxylating enzymes in marine phytoplankton, *Limnol. Oceanogr.* **24:**510–519.

Gocke, K., Dawson, R., and Liebezeit, G., 1981, Availability of dissolved free glucose to heterotrophic microorganisms, *Mar. Biol.* **62:**209–216.

Goldman, J. C., 1980, Physiological processes, nutrient availability, and the concept of relative growth rate in marine phytoplankton ecology, in: *Primary Productivity in the Sea* (P. G. Falkowski, ed.), pp. 179–194, Plenum Press, New York.

Goldman, J. C., McCarthy, J. J., and Peavey, D. G., 1979, Growth rate influence on the chemical composition of phytoplankton in oceanic waters, *Nature (Lond.)* **279:**210–215.

Goldman, J. C., Taylor, C. D., and Gilbert, P. M., 1981, Nonlinear time-course uptake of carbon and ammonium by marine phytoplankton, *Mar. Ecol. Prog. Ser.* **6:**137–148.

Gordon, A., and Millero, F. J., 1980, Use of microcalorimetry to study the growth and metabolism of marine bacteria, *Thalassia Jugosl.* **16:**405–414.

Graham, B. M., Hamilton, R. D., and Campbell, N. E. R., 1980, Comparison of the Nitrogen-15 uptake and acetylene reduction methods for estimating the rates of nitrogen fixation by freshwater blue-green algae, *Can. J. Fish. Aquat. Sci.* **37:**488–493.

Grivell, A. R., and Jackson, J. F., 1968, Thymidine kinase: Evidence from *Neurospora crassa* and some other microorganisms and relevance of this to the specific labelling of deoxyribonucleic acid, *J. Gen. Microbiol.* **54:**307–317.

Haas, L. W., 1982, Improved epifluorescence microscopy for observing planktonic micro-organisms, *Ann. Inst. Oceanogr. Paris (Suppl.)* **58:**261–266.

Hagstrom, A., Larsson, U., Horstedt, P., and Normark, S., 1979, Frequency of dividing cells, a new approach to the determination of bacterial growth rates in aquatic environment, *Appl. Environ. Microbiol.* **37:**805–812.

Hamilton, R. D., and Austin, K. E., 1967, Assay of relative heterotrophic potential in the sea: The use of specifically labelled glucose, *Can. J. Microbiol.* **13:**1165–1173.

Hanson, R. B., and Gardner, W. S., 1978, Uptake and metabolism of two amino acids by anaerobic microorganisms in four diverse salt-marsh soils, *Mar. Biol.* **46:**101–107.

Hanson, R. B., and Wiebe, W. J., 1977, Heterotrophic activity associated with particulate size fractions in a *Spartina alterniflora* salt-marsh estuary, Sapelo Island, Georgia, USA, and the continental shelf waters, *Mar. Biol.* **42:**321–330.

Hardy, R. W. F., Burns, R. C., and Holsten, R. D., 1973, Applications of the acetylene-ethylene assay for measurement of nitrogen fixation, *Soil Biol. Biochem.* **5:**47–81.

Harris, G. P., 1980, The relationship between chlorophyll *a* fluorescence, diffuse attenuation changes and photosynthesis in natural phytoplankton populations, *J. Plankton Res.* **2:**109–127.

Harrison, W. B., 1983, Uptake and recycling of soluble reactive phosphorus by marine microplankton, *Mar. Ecol. Prog. Ser.* **10:**127–135.

Herbland, A., 1978, The soluble fluorescence in the open sea: Distribution and ecological significance in the equatorial Atlantic Ocean, *J. Exp. Mar. Biol. Ecol.* **32:**275–284.

Herron, J. S., King, J. D., and White, D. C., 1978, Recovery of poly-β-hydroxybutyrate from estuarine microflora, *Appl. Environ. Microbiol.* **35:**251–257.

Hobbie, J. E., and Crawford, C. C., 1969, Respiration corrections for bacterial uptake of dissolved organic compounds in natural waters, *Limnol. Oceanogr.* **14:**528–532.

Hobbie, J. E., and Lee, C., 1980, Microbial production of extracellular material: Importance in benthic ecology, in: *Marine Benthic Dynamics*, (K. R. Tenore and B. C. Coull, eds.), pp. 341–346, University of South Carolina Press, Georgetown, South Carolina.

Hobbie, J. E., Daley, R. J., and Jaspar, S., 1977, Use of Nuclepore filters for counting bacteria by fluorescence microscopy, *Appl. Environ. Microbiol.* **33:**1225–1228.

Hochstadt, J., 1974, The role of the membrane in the utilization of nucleic acid precursors, *CRC Crit. Rev. Biochem.* **2:**259–310.

Holm-Hansen, O., 1969, Algae: Amounts of DNA and organic carbon in single cells, *Science* **163:**87–88.

Holm-Hansen, O., 1973, Determination of total microbial biomass by measurement of adenosine triphosphate, in: *Estuarine Microbial Ecology* (L. H. Stevenson and R. R. Colwell, eds.), pp. 73–89, University of South Carolina Press, Columbia, South Carolina.

Holm-Hansen, O., and Booth, C. R., 1966, The measurement of adenosine triphosphate in the ocean and its ecological significance, *Limnol. Oceanogr.* **11:**510–519.

Hoppe, H.-G., 1976, Determination and properties of actively metabolizing heterotrophic bacteria in the sea, investigated by means of microautoradiography, *Mar. Biol.* **36:**291–302.

Hoppe, H.-G., 1977, Analysis of actively metabolizing bacterial populations with the autoradiographic method, in: *Microbial Ecology of a Brackish Water Environment* (G. Rheinheimer, ed.), pp. 179–197, Springer-Verlag, Heidelberg.

Hoppe, H.-G., 1983, Significance of exoenzymatic activities in the ecology of brackish water: Measurements by means of methylumbelliferyl-substrates, *Mar. Ecol. Prog. Ser.* **11:**299–308.

Humphries, D. W., 1969, Mensuration methods in optical microscopy, *Adv. Optical Electron Microsc.* **3:**33–98.

Ingraham, J. L., Maaloe, O., and Neidhardt, F. C., 1983, *Growth of the Bacterial Cell*, Sinauer Associates, Sunderland, Massachusetts.

Iturriaga, R., and Zsolnay, A., 1981, Differentiation between auto- and heterotrophic activity: Problems in the use of size fractionation and antibiotics, *Bot. Mar.* **24:**399–404.

Jannasch, H. W., 1974, Steady state and the chemostat in ecology, *Limnol. Oceanogr.* **19:**716–720.

Jannasch, H. W., and Jones, G. E., 1959, Bacterial populations in sea water as determined by different methods of enumeration, *Limnol. Oceanogr.* **4:**128–139.

Jannasch, H. W., and Wirsen, C. O., 1979, Chemosynthetic primary production at East Pacific sea floor spreading centers, *BioScience* **29:**592–598.

Jassby, A. D., 1975, Dark sulfate uptake and bacterial productivity in a subalpine lake, *Ecology* **56:**627–636.

Jay, J. M., 1977, The Limulus lysate endotoxin assay as a test of microbial quality of ground beef, *J. Appl. Bacteriol.* **43:**99–109.

Jeffrey, S. W., and Hallegraeff, G. M., 1980, Studies of phytoplankton species and photosynthetic pigments in a warm core eddy of the east Australian current. II. A note on pigment methodology, *Mar. Ecol. Prog. Ser.* **3:**295–301.

Jenkins, W. J., 1982, Oxygen utilization rates in North Atlantic subtropical gyres and primary production in oligotrophic systems, *Nature (Lond.)* **300:**246–248.

Jenkinson, D. S., 1966, Studies on the decomposition of plant material in soil. II. Partial sterilization of soil and the soil biomass, *J. Soil Sci.* **17:**280–302.

Jenkinson, D.S., 1976, The effects of biocidal treatments on metabolism in soil. IV. The decomposition of fumigated organisms in soil, *Soil Biol. Biochem.* **8:**203–208.

Jenkinson, D. S., and Powlson,D. S. 1976, The effects of biocidal treatments on metabolism in soil. V. A method for measuring soil biomass, *Soil Biol. Biochem.* **8:**209–213.

Jenkinson, D. S., Powlson, D. S., and Wedderburn, R. W. M., 1976, The effects of biocidal treatments on metabolism in soil. III. The relationship between soil biovolume, measured by optical microscopy, and the flush of decomposition cause by fumigation, *Soil Biol. Biochem.* **8:**189–202.

Jensen, V., 1968, The plate count technique, in: *Ecology of Soil Bacteria* (T. R. G. Gray and D. Parkinson, eds.), pp. 158–170, University of Toronto Press, Toronto, Canada.

Johnson, K. M., Burney, C. M., and Sieburth, J. McN., 1981, Engmatic marine ecosystem metabolism measured by direct diel CO_2 and O_2 flux in conjunction with DOC release and uptake, *Mar. Biol.***65:**49–60.

Johnson, P.W., and Sieburth, J. McN., 1979, Chroococcoid cyanobacteria in the sea: A ubiquitous and diverse phototrophic biomass, *Limnol. Oceanogr.* **24:**928–935.

Johnson, P. W., and Sieburth, J. McN., 1982, In-situ morphology and occurrence of eucaryotic phototrophs of bacterial size in the picoplankton of estuarine and oceanic waters, *J. Phycol.* **18:**318–327.

Jonsen, J., and Laland, S., 1960, Bacterial nucleosides and nucleotides. *Adv. Carbohydr. Chem.* **15:**210–234.

Jordan, M. J., Daley, R. J., and Lee, K., 1978, Improved filtration procedures for freshwater [^{35}S]SO_4 uptake studies, *Limnol. Oceanogr.* **23:**154–157.

Jorgensen, J. H., and Alexander, G. A., 1981, Automation of the *Limulus* amoebocyte lysate test by using the Abbott MS-2 microbiology system, *Appl. Environ. Microbiol.* **41:**1316–1320.

Jorgensen, J. H., Carvajal, H. F., Chipps, B. E., and Smith, R. F., 1973, Rapid detection of gram-negative bacteriuria by use of the *Limulus* endotoxin assay, *Appl. Microbiol.* **26:**38–42.

Jorgensen, N. O. G., and Sondergaard, M., 1984, Are dissolved free amino acids free?, *Microb. Ecol.* **3:**301–316.

Karl, D. M., 1978, Occurrence and ecological significance of GTP in the ocean and in microbial cells, *Appl. Environ. Microbiol.* **36:**349–355.

Karl, D. M., 1979a, Adenosine triphosphate and guanosine triphosphate determinations in intertidal sediments, in: *Methodology for Biomass Determinations and Microbial Activities in Sediments* (C. D. Litchfield and P. L. Seyfried, eds.), pp. 5–20, American Society for Testing and Materials, Baltimore.

Karl, D. M., 1979b, Measurement of microbial activity and growth in the ocean by rates of stable ribonucleic acid synthesis, *Appl. Environ. Microbiol.* **38:**850–860.

Karl, D. M., 1980, Cellular nucleotide measurements and application in microbial ecology, *Microbiol. Rev.* **44:**739–796.

Karl, D. M., 1981, Simultaneous rates of ribonucleic acid and deoxyribonucleic acid syntheses for estimating growth and cell division of aquatic microbial communities, *Appl. Environ. Microbiol.* **42:**802–810.

Karl, D. M., 1982a, Microbial transformations of organic matter at oceanic interfaces: A review and prospectus, *EOS Trans. Am. Geophys. Un.* **63:**138–140.

Karl, D. M., 1982b, Selected nucleic acid precursors in studies of aquatic microbial ecology, *Appl. Environ. Microbiol.* **44:**891–902.

Karl, D. M., 1985, Specific growth rates of natural microbial populations, *EOS, Trans. Amer. Geophys. UN.* **66:**1322.

Karl, D. M., and Bossard,P. 1985a, Measurement and significance of ATP and adenine nucleotide pool turnover in microbial cells and environmental samples, *J. Microbiol. Meth.* **3:**125–139.

Karl, D. M., and Bossard, P., 1985b, A field comparison of two independent methods for estimating microbial nucleic acid synthesis and specific growth rates, *Appl. Environ. Microbiol.* **50:**706–709.

Karl, D. M., and Craven, D. B., 1980, Effects of alkaline phosphatase activity on nucleotide measurements in aquatic microbial communities, *Appl. Environ. Microbiol.* **40:**549–561.

Karl, D. M., and Winn, C. D., 1984, Adenine metabolism and nucleic acid synthesis: Applications to microbiological oceanography, in: *Heterotrophic Activity in the Sea* (J. E. Hobbie and P. J. leB. Williams, eds.), pp. 197–215, Plenum Press, New York.

Karl, D. M., and Winn, C. D., 1986, Does adenine incorporation into nucleic acids measure total microbial production? A response to comments by Fuhrman *et al. Limnol. Oceanogr.* (in press).

Karl, D. M., Wirsen, C. O., and Jannasch, H. W., 1980, Deep-sea primary production at the Galapagos hydrothermal vents, *Science* **207**:1345–1347.

Karl, D. M., Winn, C. D., and Wong, D. C. L., 1981a, RNA synthesis as a measure of microbial growth in aquatic environments. I. Evaluation, verification and optimization of methods, *Mar. Biol.* **64**:1–12.

Karl, D. M., Winn, C. D., and Wong, D. C. L., 1981b, RNA synthesis as a measure of microbial growth in aquatic environments. II. Field applications, *Mar. Biol.* **64**:13–21.

Karl, D. M., Knauer, G. A., Martin, J. H., and Ward, B. B., 1984, Bacterial chemolithotrophy in the ocean is associated with sinking particles, *Nature (Lond.)* **309**:54–56.

Keleti, G., Sykora, J. L., Lippy, E. C., and Shapiro, M. A., 1979, Composition and biological properties of lipopolysaccharides isolated from *Schizothrix calcicola* (Ag.) gomont (cyanobacteria), *Appl. Environ. Microbiol.* **38**:471–477.

Kenner, R. A., and Ahmed, S. I., 1975a, Measurements of electron transport activities in marine phytoplankton, *Mar. Biol.* **33**:119–127.

Kenner, R. A., and Ahmed, S. I., 1975b, Correlation between oxygen utilization and electron transport activity in marine phytoplankton, *Mar. Biol.* **33**:129–133.

Kim, J., and ZoBell, C. E., 1974, Occurrence and activities of cell-free enzymes in oceanic environments, in: *Effect of the Ocean Environment on Microbial Activities* (R. R. Colwell and R. Y. Morita), pp. 368–385, University Park Press, Baltimore.

King, G. M., and Klug, M. J., 1982, Glucose metabolism in sediments of a eutrophic lake: Tracer analysis of uptake and product formation, *Appl. Environ. Microbiol.* **44**:1308–1317.

King, F. D., and Packard, T. T., 1975, Respiration and the activity of the respiratory electron transport system in marine zooplankton, *Limnol. Oceanogr.* **20**:849–854.

King, J. D., and White, D. C., 1977, Muramic acid as a measure of microbial biomass in estuarine and marine samples, *Appl. Environ. Microbiol.* **33**:777–783.

King, J. D., White, D. C., and Taylor, C. W., 1977, Use of lipid composition and metabolism to examine structure and activity of estuarine detrital microflora, *Appl. Environ. Microbiol.* **33**:1177–1183.

Kirchman, D., Ducklow, H., and Mitchell, R., 1982a, Estimates of bacterial growth from changes in uptake rates and biomass, *Appl. Environ. Microbiol.* **44**:1296–1307.

Kirchman, D., Sigda, J., Kapuscinski, R., and Mitchell, R., 1982b, Statistical analysis of the direct count method for enumerating bacteria, *Appl. Environ. Microbiol.* **44**:376–382.

Knoechel, R., and Kalff, J., 1976a, The applicability of grain density autoradiography to the quantitative determination of algal species production: A critique, *Limnol. Oceanogr.* **21**:583–590.

Knoechel, R., and Kalff, J., 1976b, Track autoradiography: A method for the determination of phytoplankton species productivity, *Limnol. Oceanogr.* **21**:590–596.

Knoechel, R., and Kalff, J., 1978, An in situ study of the productivity and population dynamics of five freshwater planktonic diatom species, *Limnol. Oceanogr.* **23**:195–218.

Knowles, C. J., 1977, Microbial metabolic regulation by adenine nucleotide pools. *Symp. Soc. Gen. Microbiol.* **27**:241–283.

Koch, A. L., 1971, The adaptive responses of *Escherichia coli* to a feast and famine existence, *Adv. Microb. Physiol.* **6**:147–217.

Kogure, K., Simidu, U., and Taga, N., 1979, A tentative direct microscopic method for counting living marine bacteria, *Can. J. Microbiol.* **25**:415–420.

Kogure, K., Simidu, U., and Taga, N., 1980, Distribution of viable marine bacteria in neritic seawater around Japan, *Can. J. Microbiol.* **26**:318–323.

Korgaonkar, K. S., and Ranade, S. S., 1966, Evaluation of acridine orange fluorescence test in viability studies on *Escherichia coli*, *Can. J. Microbiol.* **12**:185–190.

Kornberg, H. L., 1966, Anaplerotic sequences and their role in metabolism, *Essays Biochem.* **2**:1–31.

Krambeck, C., Krambeck, H.-J., and Overbeck, J., 1981, Microcomputer-assisted biomass determination of plankton bacteria on scanning electron micrographs, *Appl. Environ. Microbiol.* **42:**142–149.

Kronenberg, L. H., 1979, Radioautography of multiple isotopes using color negative films, *Anal. Biochem.* **93:**189–195.

Kubitschek, H. E., 1969, Counting and sizing microorganisms with the Coulter counter, in: *Methods in Microbiology,* Vol. 1 (R. Norris and D. W. Ribbons, eds.), pp. 593–610, Academic Press, New York.

Kuparinen, J., and Tamminen, T., 1982, Respiration of tritiated substrates in heterotrophic activity assays, *Appl. Environ. Microbiol.* **43:**806–809.

Lai, S.-H., Tiedje, J. M., and Erickson, A. E., 1976, In situ measurement of gas diffusion coefficient in soils, *Soil Sci. Soc. Am. J.* **40:**3–6.

Landry, M. R., and Hassett, R. P., 1982, Estimating the grazing impact of marine microzooplankton, *Mar. Biol.* **67:**283–288.

Larsson, K., Weibull, C., and Cronberg, G., 1978, Comparison of light and electron microscopic determinations of the number of bacteria and algae in lake water, *Appl. Environ. Microbiol.* **35:**397–404.

Laws, E. A., 1983, Plots of turnover times versus added substrate concentrations provide only upper bounds to *in situ* substrate concentration, *J. Theor. Biol.* **101:**147–150.

Laws, E. A., 1984, Improved estimates of phytoplankton carbon based on ^{14}C incorporation into chlorophyll *a, J. Theor. Biol.* **110:**425–434.

Laws, E. A., Karl, D. M., Redalje, D. G., Jurick, R. S., and Winn, C. D., 1983, Variability in ratios of phytoplankton carbon and RNA to ATP and Chl *a* in batch and continuous cultures, *J. Phycol.* **19:**439–445.

Lee, C., Howarth, R. W., and Howes, B. L., 1980, Sterols in decomposing *Spartina alterniflora* and the use of ergosterol in estimating the contribution of fungi to detrital nitrogen, *Limnol. Oceanogr.* **25:**290–303.

Leftley, J. W., Bonin, D. J., and Maestrini, S. Y., 1983, Problems in estimating marine phytoplankton growth, productivity and metabolic activity in nature: An overview of methodology, *Oceanogr. Mar. Biol. Annu. Rev.* **21:**23–66.

Levin, G. V., Clendenning, J. R., Chappelle, E. W., Heim, A. H., and Rocek, E., 1964, A rapid method for detection of microorganisms of ATP assay, *BioScience* **14:**37–38.

Li, W. K. W., 1982, Estimating heterotrophic bacterial productivity by inorganic radiocarbon uptake: Importance of establishing time courses of uptake, *Mar. Ecol. Prog. Ser.* **8:**167–172.

Li, W. K. W., 1983, Consideration of errors in estimating kinetic parameters based on Michaelis-Menten formalism in microbial ecology, *Limnol. Oceanogr.* **28:**185–190.

Litchfield, C. D., and Seyfried, P. L. (eds.), 1979, *Methodology for Biomass Determinations and Microbial Activities in Sediments,* American Society for Testing and Materials, Baltimore.

Lundgren, B., 1981, Fluorescein diacetate as a stain of metabolically active bacteria in soil, *Oikos* **36:**17–22.

Luria, S. E., 1960, The bacterial protoplasm: Composition and organization, in: *The Bacteria,* Vol. 1 (I. C. Gunsalus and R. Y. Stanier, eds.), pp. 1–34, Academic Press, New York.

Maaloe, O., and Kjeldgaard, N. O., 1966, *Control of Macromolecular Synthesis,* W. A. Benjamin, New York.

Mague, T. H., Weare, N. M., and Holm-Hansen, O., 1974, Nitrogen fixation in the North Pacific Ocean, *Mar. Biol.* **24:**109–119.

Maguire, B., and Neill, W. E., 1971, Species and individual productivity in phytoplankton communities, *Ecology* **52:**903–907.

Maki, J. S., and Remsen, C. C., 1981, Comparison of two direct-count methods for determining metabolizing bacteria in freshwater, *Appl. Environ. Microbiol.* **41:**1132–1138.

Maki, J. S., Sierszen, M. E., and Remsen, C. C., 1983, Measurements of dissolved adenosine triphosphate in Lake Michigan, *Can. J. Fish. Aquat. Sci.* **40:**542–547.

Mandelstam, J., and McQuillen, K., 1976, *Biochemistry of Bacterial Growth,* Wiley, New York.

Matin, A., Veldhuis, C., Stegeman, V., and Veenhuis, M., 1979, Selective advantage of a *Spirillum* sp. in a carbon-limited environment. Accumulation of poly-hydroxybutyric acid and its role in starvation, *J. Gen. Microbiol.* **112:**349–355.

McDuff, R. E., and Chisholm, S. W., 1982, The calculation of in situ growth rates of phytoplankton populations from fractions of cells undergoing mitosis: A clarification, *Limnol. Oceanogr.* **27:**783–788.

Meyer-Reil, L.-A., 1978a, Autoradiography and epifluorescence microscopy combined for the determination of number and spectrum of actively metabolizing bacteria in natural waters, *Appl. Environ. Microbiol.* **36:**506–512.

Meyer-Reil, L.-A., 1978b, Uptake of glucose by bacteria in the sediment, *Mar. Biol.* **44:**293–298.

Miflin, B. J., and Lea, P. J., 1976, The pathway of nitrogen assimilation in plants, *Phytochemistry* **15:**873–885.

Millar, W. N., and Casida, L. E., 1970a, Evidence for muramic acid in soil, *Can. J. Microbiol.* **16:**299–304.

Millar, W. N., and Casida, L. E., 1970b, Microorganisms in soil as observed by staining with rhodamine-labeled lysozyme, *Can. J. Microbiol.* **16:**305–307.

Miller, C. A., and Volgelhut, P. O., 1978, Chemiluminescent detection of bacteria: Experimental and theoretical limits, *Appl. Environ. Microbiol.* **35:**813–816.

Monheimer, R. H., 1974, Sulfate uptake as a measure of planktonic microbial production in freshwater ecosystems, *Can. J. Microbiol.* **20:**825–831.

Monheimer, R. H., 1975, Planktonic microbial heterotrophy: Its significance to community biomass production, *Verh. Int. Verein. Limnol.* **19:**2658–2663.

Monheimer, R. H., 1978, Difficulties in interpretation of microbial heterotrophy from sulfate uptake data: Laboratory studies, *Limnol. Oceanogr.* **23:**150–154.

Monk, P. R., 1979, Thermograms of *Streptococcus thermophilus* and *Lactobacillus bulgaricus* in single and mixed culture in milk medium, *J. Dairy Res.* **46:**485–496.

Monks, R., Oldham, K. G., and Tovey, K. C., 1971, Labelled nucleotides in biochemistry, Review No. 12, The Radiochemical Centre Ltd., Amersham, Arlington Heights, Illinois.

Monod, J., 1949, The growth of bacterial cultures, *Annu. Rev. Microbiol.* **3:**371–394.

Montesinos, E., Esteve, I., and Guerrero, R., 1983, Comparison between direct methods for determination of microbial cell volume: Electron microscopy and electronic particle sizing, *Appl. Environ. Microbiol.* **45:**1651–1658.

Moriarty, D. J. W., 1975, A method for estimating the biomass of bacteria in aquatic sediments and its application to trophic studies, *Oecologia* **20:**219–229.

Moriarty, D. J. W., 1977, Improved method using muramic acid to estimate biomass of bacteria in sediments, *Oecologia* **26:**317–323.

Moriarty, D. J. W., 1978, Estimation of bacterial biomass in water and sediments using muramic acid, in: *Microbial Ecology* (M. W. Loutit and J. A. R. Miles, eds.), pp. 31–33, Springer-Verlag, Heidelberg.

Moriarty, D. J. W., 1984, Measurement of bacterial growth rates in marine systems using nucleic acid precursors, in: *Heterotrophic Activity in the Sea* (J. E. Hobbie and P. J. leB. Williams, eds.), pp. 217–231, Plenum Press, New York.

Moriarty, D. J. W., and Hayward, A. C., 1982, Ultrastructure of bacteria and the proportion of gram-negative bacteria in marine sediments, *Microb. Ecol.* **8:**1–14.

Moriarty, D. J. W., and Pollard, P. C., 1981, DNA synthesis as a measure of bacterial productivity in seagrass sediments, *Mar. Ecol. Prog. Ser.* **5:**151–156.

Moriarty, D. J. W., and Pollard, P. C., 1982, Diel variation of bacterial productivity in seagrass (*Zostera capricorni*) beds measured by rate of thymidine incorporation into DNA, *Mar. Biol.* **72:**165–173.

Morris, I., 1980, Paths of carbon assimilation in marine phytoplankton, in: *Primary Production in the Sea* (P. G. Falkowski, ed.), pp. 139–159, Plenum Press, New York.

Morris, I., 1981, Photosynthetic products, physiological state, and phytoplankton growth, in: *Physiological Bases of phytoplankton Ecology* (T. Platt, ed.), pp. 83–102, Canadian Bulletin of Fish. Aquatic Sciences 210, Ottawa, Canada.

Morris, I., Glover, H. E., and Yentsch, C. S., 1974, Products of photosynthesis by marine phytoplankton: The effect of environmental factors on the relative rates of protein synthesis, *Mar. Biol.* **27:**1–9.

Mortensen, U., Noren, B., and Wadso, I., 1973, Microcalorimetry in the study of the activity of microorganisms, *Bull. Ecol. Res. Comm. (Stockh.)* **17:**189–197.

Muldrow, L. L., Tyndall, R. L., and Fliermans, C. B., 1982, Application of flow cytometry to studies of pathogenic free-living amoebae, *Appl. Environ. Microbiol.* **44:**1258–1269.

Mullin, M. M., Sloan, P. R., and Eppley, R. W., 1966, Relationship between carbon content, cell volume, and area in phytoplankton, *Limnol. Oceanogr.* **11:**307–311.

Neidhardt, F. C., 1963, Effects of environment on the composition of bacterial cells, *Annu. Rev. Microbiol.* **17:**61–86.

Newell, S. Y., 1984, Modification of the gelatin–matrix method for enumeration of respiring bacterial cells for use with salt-marsh water samples, *Appl. Environ. Microbiol.* **47:**873–875.

Newell, S. Y., and Christian, R. R., 1981, Frequency of dividing cells as an estimator of bacterial productivity, *Appl. Environ. Microbiol.* **42:**23–31.

Newell, S. Y., and Fallon, R. D., 1982, Bacterial productivity in the water column and sediments of the Georgia (USA) coastal zone: Estimates via direct counting and parallel measurement of thymidine incorporation, *Microb. Ecol.* **8:**33–46.

Newell, S. Y., and Statzell-Tallman, A., 1982, Factors for conversion of fungal biovolume values to biomass, carbon and nitrogen: Variation with mycelial ages, growth conditions, and strains of fungi from a salt marsh, *Oikos* **39:**261–268.

Newell, S. Y., Sherr, B. F., Sherr, E. B., and Fallon, R. D., 1983, Bacterial response to presence of eukaryote inhibitors in water from a coastal marine environment, *Mar. Environ. Res.* **10:**1–11.

Nickels, J. S., King, J. D., and White, D. C., 1979, Poly-hydroxybutyrate accumulation as a measure of unbalanced growth of the estuarine detrital microbiota, *Appl. Environ. Microbiol.* **37:**459–465.

Novitsky, J. A., 1983a, Heterotrophic activity throughout a vertical profile of seawater and sediment in Halifax Harbor, Canada, *Appl. Environ. Microbiol.* **45:**1753–1760.

Novitsky, J. A., 1983b, Microbial activity at the sediment–water interface in Halifax Harbor, Canada, *Appl. Environ. Microbiol.* **45:**1761–1766.

Novitsky, J. A., and Kepkay, P. E., 1981, Patterns of microbial heterotrophy through changing environments in a marine sediment, *Mar. Ecol. Prog. Ser.* **4:**1–7.

Novitsky, J. A., and Morita, R. Y., 1977, Survival of a psychophilic marine vibrio under long-term nutrient starvation, *Appl. Environ. Microbiol.* **33:**635–641.

Novitsky, J. A., and Morita, R. Y., 1978, Possible strategy for the survival of marine bacteria under starvation conditions, *Mar. Biol.* **48:**289–295.

Okrend, H., Thomas, R. R., Deming, J. W., Chappelle, E. W., and Picciolo, G. L., 1977, Methodology for photobacteria luciferase FMN assay of bacterial leves, in: *Second Bi-Annual ATP Methodology Symposium* (G. A. Borun, ed.), pp. 525–546, Science Applications Incorporated Technology Co., San Diego, California.

Oleniacz, W. S., Pisano, M. A., Rosenfield, M. H., and Elgart, R. L., 1968, Chemiluminescent method for detecting microorganisms in water, *Environ. Sci. Technol.* **2:**1030–1033.

Ortner, P. B., and Rosencwaig, A., 1977, Photoacoustic spectroscopic analysis of marine phytoplankton, *Hydrobiologia* **56:**3–6.

Overbeck, J., 1979, Dark CO_2 uptake—biochemical background and its relevance to in situ bacterial production, *Arch. Hydrobiol. Beih. Ergebn. Limnol.* **12:**38–47.

Overbeck, J., and Daley, R. J., 1973, Some precautionary comments on the Romanenko technique for estimating heterotrophic bacterial production, *Bull. Ecol. Res. Comm. (Stockh.)* **17:**342–344.

Packard, T. T., 1971, The measurement of respiratory electron transport activity in marine phytoplankton, *J. Mar. Res.* **29:**235–244.

Packard, T. T., and Blasco, D., 1974, Nitrate reductase activity in upwelling regions. 2. Ammonia and light dependence, *Tethys* **6**:269–280.

Packard, T. T., and Williams, P. J. leB., 1981, Rates of respiratory oxygen consumption and electron transport in surface seawater from the northwest Atlantic, *Ocean. Acta* **4**:351–358.

Packard, T. T., Healy, M. L., and Richards, F. A., 1971, Vertical distribution of the activity of the respiratory electron transport system in marine plankton, *Limnol. Oceanogr.* **16**:60–70.

Paerl, H. W., 1984, An evaluation of freeze fixation as a phytoplankton preservation method for microautoradiography, *Limnol. Oceanogr.* **29**:417–426.

Paerl, H. W., and Bland, P. T., 1982, Localized tetrazolium reduction in relation to N_2 fixation, CO_2 fixation, and H_2 uptake in aquatic filamentous cyanobacteria, *Appl. Environ. Microbiol.* **43**:218–226.

Paerl, H. W., and Stull, E. A., 1979, In defense of grain density autoradiography, *Limnol. Oceanogr.* **24**:1166–1169.

Pamatmat, M. M., 1968, Ecology and metabolism of a benthic community on an intertidal sandflat, *Int. Rev. Ges. Hydrobiol.* **53**:211–298.

Pamatmat, M. M., 1982, Heat production by sediment: Ecological significance, *Science* **215**:395–397.

Pamatmat, M. M., and Findlay, S., 1983, Metabolism of microbes, nematodes, polychaetes, and their interactions in sediment, as detected by heat flow measurements, *Mar. Ecol. Prog. Ser.* **11**:31–38.

Pamatmat, M. M., Graf, G., Bengtsson, W., and Novak, C. S., 1981, Heat production, ATP concentration and electron transport activity of marine sediments, *Mar. Ecol. Prog. Ser.* **4**:135–143.

Parker, J. H., Smith, G. A., Fredrickson, H. L., Vestal, J. R., and White, D. C., 1982, Sensitive assay, based on hydroxyfatty acids from lipopolysaccharide lipid A, for gram-negative bacteria in sediments, *Appl. Environ. Microbiol.* **44**:1170–1177.

Parsons, T. R., and Strickland, J. D. H., 1962, On the production of particulate organic carbon by heterotrophic processes in sea water, *Deep-Sea Res.* **8**:211–222.

Paton, A. M., and Jones, S. M., 1975, The observation and enumeration of microorganisms in fluids using membrane filtration and incident fluorescence microscopy, *J. Appl. Bacteriol.* **38**:199–200.

Paul, J. H., 1982, Use of Hoechst dyes 33258 and 33342 for enumeration of attached and planktonic bacteria, *Appl. Environ. Microbiol.* **43**:939–944.

Peterson, B. J., 1980, Aquatic primary productivity and the ^{14}C-CO_2 method: A history of the productivity problem, *Annu. Rev. Ecol. Syst.* **11**:359–385.

Peterson, R. B., and Burris, R. H., 1976, Conversion of acetylene reduction rates to nitrogen fixation rates in natural populations of blue-green algae, *Anal. Biochem.* **73**:404–410.

Pollard, P. C., and Moriarty, D. J. W., 1984, Validity of the tritiated thymidine method for estimating bacterial growth rates: Measurement of isotope dilution during DNA synthesis, *Appl. Environ. Microbiol.* **48**:1076–1083.

Pomeroy, L. R., and Johannes, R. E., 1966, Total plankton respiration, *Deep-Sea Res.* **13**:971–973.

Pomroy, A. J., 1984, Direct counting of bacteria preserved with lugol iodine solution, *Appl. Environ. Microbiol.* **47**:1191–1192.

Porter, K. G., and Feig, Y. S., 1980, The use of DAPI for identifying and counting aquatic microflora, *Limnol. Oceanogr.* **25**:943–948.

Postgate, J. R., and Hunter, J. R., 1963, The survival of starved bacteria, *J. Appl. Bacteriol.* **26**:295–306.

Prescott, D. M., 1970, Frustrations of mislabeling, *Science,* 168:1285.

Prescott, J. M., Ragland, R. S., and Hurley, R. J., 1965, Utilization of CO_2 and acetate in amino acid synthesis by *Streptococcus bovis*, *Proc. Soc. Exp. Biol. Med.* **119**:1097–1102.

Ramsay, A. J., 1974, The use of autoradiography to determine the proportion of bacteria metabolizing in an aquatic habitat, *J. Gen. Microbiol.* **80**:363–373.

Redalje, D. G., 1983, Phytoplankton carbon biomass and specific growth rates determined with the labeled chlorophyll *a* technique, *Mar. Ecol. Prog. Ser.* **11**:217–225.

Redalje, D. G., and Laws, E. A., 1981, A new method for estimating phytoplankton growth rates and carbon biomass, *Mar. Biol.* **62**:73–79.

Reeburgh, W. S., 1983, Rates of biogeochemical processes in anoxic sediments, *Annu. Rev. Earth Planet. Sci.* **11**:269–298.

Revsbech, N. P., and Jorgensen, B. B., 1983, Photosynthesis of benthic microflora measured with high spatial resolution by the oxygen microprofile method: Capabilities and limitations of the method, *Limnol. Oceanogr.* **28**:749–756.

Revsbech, N. P., Sorensen, J., Blackburn, T. H., and Lomholt, J. P., 1980, Distribution of oxygen in marine sediments measured with microelectrodes, *Limnol. Oceanogr.* **25**:403–411.

Revsbech, N. P., and Ward, D. M., 1983, Oxygen microelectrode that is insensitive to medium composition: Use in an acid microbial mat dominated by *Cyanidium caldarium, Appl. Environ. Microbiol.* **45**:755–759.

Ride, J. P., and Drysdale, R. B., 1972, A rapid method for the chemical estimation of filamentous fungi in plant tissue, *Physiol. Plant Pathol.* **2**:7–15.

Riemann, B., 1979, The occurrence and ecological importance of dissolved ATP in fresh water, *Freshwater Biol.* **9**:481–490.

Riemann, B., 1984, Determining growth rates of natural assemblages of freshwater bacteria by means of ^{3}H-thymidine incorporation into DNA: Comments on methodology, *Arch. Hydrobiol. Beih. Ergebn. Limnol.* **19**:67–80.

Riemann, B., and Sondergaard, M., 1984, Measurements of diel rates of bacterial secondary production in aquatic environments, *Appl. Environ. Microbiol.* **47**:632–638.

Riemann, B., Fuhrman, J., and Azam, F., 1982, Bacterial secondary production in freshwater measured by ^{3}H-thymidine incorporation method, *Microb. Ecol.* **8**:101–114.

Riemann, B., Nielsen, P., Jeppesen, M., Marcussen, B., and Fuhrman, J. A., 1984, Diel changes in bacterial biomass and growth rates in coastal environments, determined by means of thymidine incorporation into DNA, frequency of dividing cells (FDC), and microautoradiography, *Mar. Ecol. Prog. Ser.* **17**:227–235.

Rivkin, R. B., 1986, Incorporation of tritiated thymidine by eucaryotic microalgae, *J. Phycol.* (in press).

Roberts, R. B., Cowie, D. B., Abelson, P. H., Bolton, E. T., and Britten, R. J., 1963, *Studies of Biosynthesis in Escherichia coli*, Publication 607, Carnegie Institution, Washington, D.C.

Robrish, S. A., Kemp, C. W., Adderly, D. C., and Bowen, W. H., 1979, The flavin mononucleotide content of oral bacteria related to the dry weight of dental plaque obtained from monkeys, *Curr. Microbiol.* **2**:131–134.

Romanenko, V. I., 1964, Heterotrophic assimilation of CO_2 by bacterial flora of water, *Microbiology* **33**:679–683.

Romanenko, V. I., and Dobrynin, E. G., 1978, Specific weight of the dry biomass of pure bacteria cultures, *Microbiology* **47**:220–221.

Romanenko, V. I., Overbeck, J., and Sorokin, Y. I., 1972, Estimation of production of heterotrophic bacteria using ^{14}C, in: *Techniques for the Assessment of Microbial Production and Decomposition in Fresh Waters*, Handbook No. 23 (Y. I. Sorokin and H. Kadota, eds.), pp. 82–85, Blackwell Scientific, London.

Rosenbaum-Oliver, D., and Zamenhof, S., 1972, Degree of participation of exogenous thymidine in the overall deoxyribonucleic acid synthesis in *Escherichia coli, J. Bacteriol.* **110**:585–591.

Ross, D. J., Tate, K. R., Cairns, A., and Pansier, E. A., 1980, Microbial biomass estimations in soils from tussock grasslands by three biochemical procedures, *Soil Biol. Biochem.* **12**:375–383.

Rosset, R., Julien, J., and Monier, R., 1966, Ribonucleic acid composition of bacteria as a function of growth rate, *J. Mol. Biol.* **18**:308–320.

Rosswall, T. (ed.), 1973, *Modern Methods in the Study of Microbial Ecology*, Ecological Research Committee, Vol. 17, Rotobeckman, Stockholm.

Roy, S., and Legendre, L., 1979, DCMU-enhanced fluorescence as an index of photosynthetic activity in phytoplankton, *Mar. Biol.* **55**:93–101.

Roy, S., and Legendre, L., 1980, Field studies of DCMU-enhanced fluorescence as an index of in situ phytoplankton photosynthetic activity, *Can. J. Fish. Aquat. Sci.* **37**:1028–1031.

Saddler, J. N., and Wardlaw, A. C., 1980, Extraction, distribution and biodegradation of bacterial lipopolysaccharides in estuarine sediments, *Antonie van Leeuwenhoek J. Microbiol. Serol.* **46:**27–39.

Samuelsson, G., and Oquist, G., 1977, A method for studying photosynthetic capabilities of unicellular algae based on *in vivo* chlorophyll fluorescence, *Physiol. Plant.* **40:**315–319.

Samuelsson, G., Oquist, G., and Halldal, P., 1978, The variable chlorophyll *a* fluorescence as a measure of photosynthetic capacity in algae, *Mitt. Int. Verein. Limnol.* **21:**207–215.

Schmidt, E. L., 1973, The traditional plate count technique among modern methods: Chairman's summary, *Bull. Ecol. Res. Comm. (Stockh.)* **17:**453–454.

Sheldon, R. W., 1979, Measurement of phytoplankton growth by particle counting, *Limnol. Oceanogr.* **24:**760–767.

Sheldon, R. W., and Sutcliff, W. H., 1978, Generation times of 3 h for Sargasso Sea microplankton determined by ATP analysis, *Limnol. Oceanogr.* **23:**1051–1055.

Sheldon, R. W., Sutcliff, W. H., and Prakash, A., 1973, The production of particles in the surface waters of the ocean with particular reference to the Sargasso Sea, *Limnol. Oceanogr.* **18:**719–733.

Sherr, B. F., and Sherr, E. B., 1982, Enumeration of heterotrophic microprotozoa by epifluorence microscopy, *Est. Coastal Shelf Sci.* **16:**1–7.

Shields, J. A., Paul, E. A., Lowe, W. E., 1974, Factors influencing the stability of labelled microbial materials in soils, *Soil Biol. Biochem.* **6:**31–37.

Shulenberger, E., and Reid, J. L., 1981, The Pacific shallow oxygen maximum, deep chlorophyll maximum, and primary productivity, reconsidered, *Deep Sea Res.* **28A:**901–919.

Sieburth, J. McN., Johnson, K. M., Burney, C. M., and Lavoie, D. M., 1977, Estimation of *in situ* rates of heterotrophy using diurnal changes in dissolved organic matter and growth rates of picoplankton in diffusion culture, *Helgol. Wiss. Meeresunters.* **30:**565–574.

Sieracki, M. E., Johnson, P. W., and Sieburth, J. McN., 1985, The detection, enumeration and sizing of aquatic bacteria by image analysis epifluorescence microscopy, *Appl. Environ. Microbiol.* **49:**799–810.

Silver, M.W., and Davoll, P.J., 1978, Loss of ^{14}C activity after chemical fixation of phytoplankton: Error source for autoradiography and other productivity measurements, *Limnol. Oceanogr.* **23:**362–368.

Silverman, M. P., and Munoz, E. F., 1979, Automated electrical impendance technique for rapid enumeration of fecal coliforms in effluents from sewage treatment plants, *Appl. Environ. Microbiol.* **37:**521–526.

Smith, D. F., and Horner, S. M. J., 1981, Tracer kinetic analysis applied to problems in marine biology, in: *Physiological Bases of Phytoplankton Ecology* (T. Platt, ed.), pp. 113–129, Canadian Bulletin of Fish. Aquatic Science 210, Ottawa, Canada.

Smith, D. F., Wiebe, W. J., and Higgins, H. W., 1984, Heterotrophic potential estimates: an inherent paradox in assuming Michaelis-Menten kinetics, *Mar. Ecol. Prog. Ser.* **17:**49–56.

Smith, R. C., and Maaloe, O., 1964, Effect of growth rate on the acid-soluble nucleotide composition of *Salmonella typhimurium, Biochim. Biophys. Acta* **86:**229–234.

Smith, R. E., and Kalff, J., 1983, Sample preparation for quantitative autoradiography of phytoplankton, *Limnol. Oceanogr.* **28:**383–389.

Soderstrom, B. E., 1977, Vital staining of fungi in pure cultures and in soil with fluorescein diacetate, *Soil Biol. Biochem.* **9:**59–63.

Somville, M., and Billen, G., 1983, A method for determining exoproteolytic activity in natural waters, *Limnol. Oceanogr.* **28:**190–193.

Sparling, G. P., 1981, Microcalorimetry and other methods to assess biomass and activity in soil, *Soil Biol. Biochem.* **13:**93–98.

Stanley, P. M., Gage, M. A., and Schmidt, E. L., 1979, Enumeration of specific populations by immunofluorescence, in: *Native Aquatic Bacteria: Enumeration Activity and Ecology* (J. W. Costerton and R. R. Colwell, eds.), pp. 46–55, American Society for Testing and Materials, Baltimore.

Steemann-Nielsen, E., 1952, The use of radioactive carbon (^{14}C) for measuring organic production in the sea, *J. Cons. Cons. Int. Explor. Mer.* **18:**117–140.

Straat, P. A., Wolochow, H., Dimmick, R. L., and Chatigny, M. A., 1977, Evidence for incorporation of thymidine into deoxyribonucleic acid in airborne bacterial cells, *Appl. Environ. Microbiol.* **34:**292–296.

Strickland, J. D. H., 1968, Continuous measurement of *in vivo* chlorophyll *a:* Precautionary note, *Deep Sea Res.* **15:**225–227.

Strugger, S., 1948, Fluorescence microscope examination of bacteria in soil, *Can. J. Res. Sect. C* **26:**188–193.

Swift, M. J., 1973*a*, The estimation of mycelial biomass by determination of the hexosamine content of wood tissue decayed by fungi, *Soil Biol. Biochem.* **5:**321–332.

Swift, M. J., 1973*b*, Estimation of mycelial growth during decomposition of plant litter, *Bull. Ecol. Res. Comm. (Stockh.)* **17:**323–328.

Swisher, R., and Carroll, G. C., 1980, Fluorescein diacetate hydrolysis as an estimator of microbial biomass on coniferous needle surfaces, *Microb. Ecol.* **6:**217–226.

Tabor, P. S., and Neihof, R. A., 1982*a*, Improved method for determination of respiring individual microorganisms in natural waters, *Appl. Environ. Microbiol.* **43:**1249–1255.

Tabor, P. S., and Neihof, R. A., 1982*b*, Improved microautoradiographic method to determine individual microorganisms active in substrate uptake in natural waters, *Appl. Environ. Microbiol.* **44:**945–953.

Taylor, C. D., Molongoski, J. J., and Lohrenz, S. E., 1983, Instrumentation for the measurement of phytoplankton production, *Limnol. Oceanogr.* **28:**781–787.

Thomas, D. R., Richardson, J. A., and Dicker, T. J., 1974, The incorporation of tritiated thymidine into DNA as a measure of the activity of soil microorganisms, *Soil. Biol. Biochem.* **6:**293–296.

Thomas, R. R., Picciolo, G. L., Chappelle, E. W., Jeffers, E. L., and Taylor, R. E., 1977, Use of luminol assay for the determination of bacteria iron porphyrins: Flow techniques for wastewater effluent, in: *Second Bi-Annual ATP Methodology Symposium* (G. A. Borun, ed.), pp. 569–599, Science Applications Incorporated Technology Co., San Diego, California.

Thomas, T. E., Turpin, D. H., and Harrison, P. J., 1984, γ-glutamyl transferase activity in marine phytoplankton, *Mar. Ecol. Prog. Ser.* **14:**219–222.

Thompson, B., and Hamilton, R. D., 1974, Some problems with heterotrophic-uptake methodology, in: *Effect of the Ocean Environment on Microbial Activities* (R. R. Colwell and R. Y. Morita, eds.), pp. 566–575, University Park Press, Baltimore.

Tijssen, S. B., 1979, Diurnal oxygen rhythm and primary production in the mixed layer of the Atlantic Ocean at 20°N, *Neth. J. Sea Res.* **13:**79–84.

Tobin, R. S., and Anthony, D. H. J., 1978, Tritiated thymidine incorporation as a measure of microbial activity in lake sediments, *Limnol. Oceanogr.* **23:**161–165.

Trevors, J. T., 1984, Electron transport system activity in soil, sediment, and pure cultures, *Crit. Rev. Microbiol.* **11:**83–99.

Tsuji, T., and Yanagitra, T., 1981, Improved fluorescent microscopy for measuring the standing stock of phytoplankton including fragile components, *Mar. Biol.* **64:**207–211.

Tuttle, J. H., Wirsen, C. O., and Jannasch, H. W., 1983, Microbial activities in the emitted hydrothermal waters of the Galapagos Rift Vents, *Mar. Biol.* **73:**293–299.

van Cleve, K., Coyne, P. I., Goodwin, E., Johnson, C., and Kelley, M., 1979, A comparison of four methods for measuring respiration in organic material, *Soil Biol. Biochem.* **11:**237–246.

van Es, F. B., and Meyer-Reil, L.-A., 1982, Biomass and metabolic activity of heterotrophic marine bacteria, in: *Advances in Microbial Ecology*, Vol. 6 (K. C. Marshall, ed.), pp. 111–170, Plenum Press, New York.

van Leeuwenhoek, A., 1677, Observations concerning little animals observed in rain, well, sea and snow water, *Phil. Trans. R. Soc. Lond.* **11:**821–831.

van Niel, C. B., 1949, *The Chemistry and Physiology of Growth*, Princeton University Press, Princeton, New Jersey.

Venrick, E. L., Beers, J. R., and Heinbokel, J. F., 1977, Possible consequences of containing microplankton for physiological rate measurements, *J. Exp. Mar. Biol. Ecol.* **26:**55–76.

Walker, H. H., 1932, Carbon dioxide as a factor affecting lag in bacterial growth, *Science* **76:**602—604.

Wand, M., Zeuthen, E., and Evans, E. A., 1967, Tritiated thymidine: Effect of decomposition by self-radiolysis on specificity as a tracer for DNA synthesis, *Science* **157:**436–438.

Waterbury, J. B., Watson, S. W., Guillard, R. R., and Brand, L. E., 1979, Widespread occurrence of a unicellular marine planktonic cyanobacterium, *Nature (Lond.)* **277:**293–294.

Watson, S. W., and Hobbie, J. E., 1979, Measurement of bacterial biomass as lipopolysaccharide, in: *Native Aquatic Bacteria: Enumeration, Activity and Ecology* (J. W. Costerton and R. R. Colwell, eds.), pp. 82–88, American Society for Testing and Materials, Baltimore.

Watson, S. W., Novitsky, T. J., Quinby, H. L., and Valois, F. W., 1977, Determination of bacterial number and biomass in the marine environment, *Appl. Environ. Microbiol.* **33:**940–946.

Watt, W. D., 1971, Measuring the primary production rates of individual phytoplankton species in natural mixed poopulations, *Deep Sea Res.* **18:**329–339.

Weckesser, J., Drews, G., and Mayer, H., 1979, Lipopolysaccharides of photosynthetic prokaryotes, *Annu. Rev. Microbiol.* **33:**215–239.

Weete, J. D., 1973, Sterols of the fungi: Distribution and biosynthesis, *Phytochemistry* **12:**1843–1864.

Weiler, C. S., 1980, Population structure and in situ division rates of *Ceratium* in oligotrophic waters of the north Pacific central gyre, *Limnol. Oceanogr.* **25:**610–619.

Weiler, C. S., and Chisholm, S. W., 1976, Phased cell division in natural populations of marine dinoflagellates from shipboard cultures, *J. Exp. Mar. Biol. Ecol.* **25:**239–247.

Weiler, C. S., and Eppley, R. W., 1979, Temporal pattern of division in the dinoflagellate genus *Ceratium* and its application to the determination of growth rate, *J. Exp. Mar. Biol. Ecol.* **39:**1–24.

Wheeler, P., North, B., Littler, M., Stephens, G., 1977, Uptake of glycine by natural phytoplankton communities, *Limnol. Oceanogr.* **22:**900–910.

White, D. C., 1983, Analysis of microorganisms in terms of quantity and activity in natural environments, in: *Microbes in Their Natural Environments*, Symposium 34, (J. H. Slater, R. Whittenbury, and J. W. T. Wimpenny, eds.), pp. 37–66, Society for General Microbiology Ltd., Cambridge University Press, Cambridge.

White, D. C., Bobbie, R. J., King, J. D., Nickels, J., and Amoe, P., 1979a, Lipid analysis of sediments for microbial biomass and community structure, in: *Methodology for Biomass Determinations and Microbial Activities in Sediments* (C. D. Litchfield and P. L. Seyfried, eds.), pp. 87–103, American Society for Testing and Materials, Baltimore.

White, D. C., Bobbie, R. J., Herron, J. S., King, J. D., and Morrison, S. J., 1979b, Biochemical measurements of microbial mass and activity from environmental samples, in: *Native Aquatic Bacteria: Enumeration, Activity and Ecology* (J. W. Costerton and R. R. Colwell, eds.), pp. 69–81, American Society for Testing and Materials, Baltimore.

White, D. C., Davis, W. M., Nickels, J. S., King, J. D., Bobbie, R. J., 1979c, Determination of the sedimentary microbial biomass by extractible lipid phosphate, *Oecologia* **40:**51–62.

White, D. C., Bobbie, R. J., Nickels, J. S., Fazio, S. D., and Davis, W. M., 1980, Nonselective biochemical methods for the determination of fungal mass and community structure in estuarine detrital microflora, *Bot. Mar.* **23:**239–250.

Wilkins, J. R., 1978, Use of platinum electrodes for the electrochemical detection of bacteria, *Appl. Environ. Microbiol.* **36:**683–687.

Wilkins, J. R., Stoner, G. E., and Boykin, E. H. 1974, Microbial detection method based on sensing molecular hydrogen, *Appl. Microbiol.* **27:**949 952.

Wilkins, J. R., Grana, D. C., and Fox, S. S., 1980, Combined membrane filtration-electrochemical microbial detection method, *Appl. Environ. Microbiol.* **40:**852–853.

Williams, P. J. leB., 1973, On the question of growth yields of natural heterotrophic populations, *Bull. Ecol. Res. Comm. (Stockh.)* **17:**400–401.

Williams, P. J. leB., and Askew, C., 1968, A method of measuring the mineralization by microorganisms of organic compounds in sea-water, *Deep-Sea Res.* **15:**365–375.

Williams, P. J. leB., and Jenkinson, N. W., 1982, A transportable microprocessor-controlled precise Winkler titration suitable for field station and shipboard use, *Limnol. Oceanogr.* **27:**576–584.

Williams, P. J. leB., Berman, T., and Holm-Hansen, O., 1972, Potential sources of error in the measurement of low rates of planktonic photosynthesis and excretion, *Nature (New Biol.)* **236:**91–92.

Williams, P. J. leB., Berman, T., and Holm-Hansen, O., 1976, Amino acid uptake and respiration by marine microheterotrophs, *Mar. Biol.* **35**:41–47.

Williams, P. J. leB., Heinemann, K. R., Marra, J., and Purdie, D. A., 1983, Phytoplankton production in oligotrophic waters: Measurements by the ^{14}C and oxygen techniques, *Nature (Lond.)* **305**:49–50.

Winn, C. D., and Karl, D. M., 1984, Laboratory calibrations of the [^{3}H]adenine technique for measuring rates of RNA and DNA synthesis in marine microorganisms, *Appl. Environ. Microbiol.* **47**:835–842.

Witzel, K.-P., and Graf, G., 1984, On the use of different nucleic acid precursors for the measurement of microbial nucleic acid turnover, *Arch. Hydrobiol. Beih. Ergebn. Limnol.* **19**:59–65.

Woldringh, C. L., 1976, Morphological analysis of nuclear separation and cell division during the life cycle of *Escherichia coli*, *J. Bacteriol.* **125**:248–257.

Woodruff, H. B., and Miller, I. M., 1963, Antibiotics, in: *Metabolic Inhibitors. A Comprehension Treatise* (R. M. Hochster and J. H. Quastel, eds.), pp. 23–51, Academic Press, New York.

Wright, R. T., 1973, Some difficulties in using ^{14}C-organic solutes to measure heterotrophic bacterial activity, in: *Estuarine Microbial Ecology* (L. H. Stevenson and R. R. Colwell, eds.), pp. 199–217, University of South Carolina Press, Columbia, South Carolina.

Wright, R. T., and Burnison, B. K., 1979, Heterotrophic activity measured with radiolabelled organic substrates, in: *Native Aquatic Bacteria: Enumeration, Activity and Ecology* (J. W. Costerton and R. R. Colwell, eds.), pp. 140–155, American Society for Testing Materials, Baltimore, Maryland.

Wright, R. T., and Coffin, R. B., 1984, Measuring microzooplankton grazing on planktonic marine bacteria by its impact on bacterial production, *Micro. Ecol.* **10**:137–149.

Wright, R. T., and Hobbie, J. E., 1965, The uptake of organic solutes in lake water, *Limnol. Oceanogr.* **10**:22–28.

Wright, R. T., and Hobbie, J. E., 1966, Use of glucose and acetate by bacteria and algae in aquatic ecosystems, *Ecology* **47**:447–464.

Yentsch, C. M., Horan, P. K., Muirhead, K. Chisholm, S. W., Dortch, Q., Haugen, E., Legendre, L., Murphy, L. S., Olson, R. J., Perry, M. H., Phinney, M. D., Spinrad, R. W., Wood, M., Yentsch, C. S., and Zahuranec, B. J., 1983, Flow cytometry and cell sorting: A powerful technique with potential applications in aquatic sciences, *Limnol. Oceanogr.* **28**:1275–1280.

Yentsch, C. S., 1974, Some aspects of the environmental physiology of marine phytoplankton: A second look, *Oceanogr. Mar. Biol. Annu. Rev.* **12**:41–75.

Yentsch, C. S., and Yentsch, C. M., 1979, Fluorescence spectral signatures: The characterization of phytoplankton populations by the use of excitation and emission spectra, *J. Mar. Res.* **37**:471–483.

Zimmermann, R., Iturriaga, R., and Becker-Birck, J., 1978, Simultaneous determination of the total number of aquatic bacteria and the number thereof involved in respiration, *Appl. Environ. Microbiol.* **36**:926–935.

ZoBell, C. E., and Anderson, D. Q., 1936, Observations on the multiplication of bacteria in different volumes of stored sea water and the influence of oxygen tension and solid surfaces, *Biol. Bull.* **70–71**:324–342.

4

QUANTITATIVE PHYSICOCHEMICAL CHARACTERIZATION OF BACTERIAL HABITATS

David C. White

The work summarized in this chapter shows that detailed chemical characterization of the biomass of the microbiota and its extracellular products, the community structure of the microbiota, the nutritional status, as well as the metabolic activity of the microbiota reflect, and thus can be used to infer, the physicochemical condition of the particular habitat. This chapter demonstrates how direct and indirect measurements can be used to overcome problems introduced by habitat disturbance during the sampling procedure (the disturbance artifact) and the fact that microbes by their metabolic activities create microhabitats with properties different from those of the surrounding milieu. The creation of anaerobic microhabitats in highly aerated environments is of particular importance, as this can greatly expand the metabolic potential of the consortia of microbes in a given habitat.

INTRODUCTION

The detailed physicochemical characterization of the habitat of a bacterial cell is particularly difficult because it is of very small size. Small size requires the application of new concepts for understanding. The continuous behavior we intuitively associate with the chemical properties of matter, with its theoretical basis in statistical mechanics, requires a large number of molecules

David C. White • Center for Biomedical and Toxicological Research, Florida State University, Tallahassee, Florida 32306.

because the calculations are essentially statistical. Many of the concepts by which microbial environments are defined, such as temperature, pH, redox potential (Eh), and pressure, are statistical concepts that have no meaning on the scale of the individual bacterium. For example, the hydrogen ion concentration in a single bacterium with a volume of 1–2 μm^3 is a fraction of one hydrogen ion at neutral pH values. Nevertheless, proton gradients across membranes can furnish the metabolic energy of the cells.

In addition, bacterial habitats are remarkably diverse. These habitats involve the widest extremes of temperature, pressure, pH, Eh, ionic strength, and irradiation density of any portion of the biosphere (Kushner, 1978). Bacterialike organisms probably were the first self-reproducing units to evolve on earth, and their metabolic activities may well have provided the basis of essential elementary biogeochemical cycling necessary to maintain the relatively stable and mild climate of this planet. Despite the diversity and longevity of bacterial niches, there are constraints inherent in the bacterial solution to the problems of structural organization that limit these organisms.

The first of these constraints is that imposed by size. Bacteria exist for the most part in three simple shapes: spheres, rods, and curved rods (Murray, 1978). The cocci with spherical symmetry are almost all between 1 and 2.5 μm in diameter; starvation forms of marine bacteria may be considerably smaller (Morita, 1982). Small rods generally measure 0.8 μm in diameter and 1.2 μm long (the size of the touchstone microbe *Escherichia coli*). Larger bacilli can be 1.6 μm in diameter and 5 μm long. The average volume of these organisms is 10^{-12} cm^3, numbering more than 10^{12} cells per g dry wt (Luria, 1960).

Consequences of Small Size

Being small provides a huge surface area/unit volume. This is essential for organisms that rely on nutrients of molecular dimensions, but the small size means that these organisms exist in a milieu very different from that of larger animals. To move, a large animal pushes fluid backward against the inertia of the fluid. Motion results as the momentum is conserved. The viscous resistance is only important at the boundary layer. This can be expressed as the Reynolds number. The Reynolds number R is related to the length L and speed v of the moving object, the density of the fluid d and the viscosity μ by the relationship: $R = Lvd/\mu$. For a swimming human, $R = 10^4$. For a microbe 1 μm long with a velocity of 30 $\mu m/sec$ v swimming in water ($d = 1$, $\mu = 0.01$ poise), $R = 3 \times 10^{-5}$. For bacteria, the problem is viscosity, because the inertia is nearly zero. The coasting distance after cessation of propulsion of a bacterium is 0.01 mm (Purcell, 1977).

Imagine swimming in a pool of molasses so that no part of your body can move faster than 1 cm/sec and one can picture life at low Reynolds numbers (Purcell, 1977). Life in these conditions requires different methods for swimming. All that is required for swimming at low Reynolds numbers is that an asymmetrical body be rotated. The best studied propulsion system is the rotating stiff flagellum used by *E. coli.* At these scales, the efficiency of this propulsion system was calculated to be about 1%. However, moving requires only a small portion of the energy required by the cell. A speed of 30 μm/sec is 6×10^{-6} mph or 25 cell lengths/sec. A cheetah at its fastest moves at four body lengths/sec; thus, the relative speed of bacterial swimming is impressive. A second problem resulting from life at small scales is that stirring the surrounding medium is no help in increasing the rate of delivery of nutrient molecules to the bacterium. The time for transporting anything by a distance 1 is approximated by the ratio of 1 to the stirring speed v. Transport by diffusion is equal to $1^2/D$, where D is the diffusion constant (for bacterial substrates $D =$ about 10^{-5} cm^2/sec). The effectiveness of stirring is $1v/D$. At micrometer dimensions, the effectiveness of stirring is much smaller than 1. Thus, the microbe is the prisoner of its local environment. Calculations by Purcell (1977) established that bacteria cannot swim fast enough to increase their food supply by more than 10%. It is possible to find new areas of higher nutrient concentrations. To do this, the organism must swim at least far enough to outrun diffusion. To do this, it must swim D/v, where D is the diffusion constant and v is the velocity of swimming, or 30 μm/sec, or about 30 μm. This is the typical swimming distance for an enteric bacterium before there is a change in direction (Berg, 1975, *a,b*).

Microbial Modification of Microenvironments

A description of the local environment of a microbe is complicated by the ability of the organism to modify its own environment. Microbes can alter the pH level of the growth medium by altering the products of their metabolic fermentations. Alkalinity is countered by the production of acids, and acid production is balanced under some conditions by the production of bases. Probably the most dramatic change in pH value is that produced by the thiobacilli, by which reduced sulfur is transformed to sulfuric acid or nitrifiers that produce nitric acid. In these cases, the microbes alter the environment to favor their own proliferation and possibly to inhibit the growth of competing microbes. The acidification of the environment facilitates solution of the reduced sulfur and, in some cases, other elements. The organisms are quite resistant to the acidification and can create an environment that precludes the growth of competitive organisms. Because the majority of S-oxidizing thiobacilli are aerobes, this problem is particularly intense where the reduced sulfur and iron interfaces associated with surface coal mining are exposed in oxygen.

Microbes can create an environment that is poisonous not only by altering the pH but by secreting toxins. Bacteriocins may inhibit closely related species. Clostridia and bacilli are well known to damage tissues with toxins and thus to create a new growth substrate; these toxins can even lead to the death of the host. The presence of the toxins often preserves the carrion for the microbe. Similarly, the production of aflatoxins by molds may prevent other organisms from utilizing the substrate.

Bacteria have the ability to exploit anaerobic niches much more effectively than do eukaryotes. Modification of the Eh can be an effective deterrent to eukaryotes. The sulfate-reducing bacteria can create the ghostly white fringe that surrounds large animal carcasses deposited under the ice in Antarctica. The sulfate-reducing bacteria, acting as the terminal reactants in an anaerobic degradation process, are able to create such a toxic surface that the microinvertebrates that usually comminute animal carcasses in the sea are rendered ineffective. In the cold water, the resultant slow anaerobic processes permit these carcasses to exist for very long periods.

Bacteria have the capacity to adhere to surfaces. This property allows them to escape from a major consequence of being small. The microbes in the water column are trapped in the 30-μm diffusion sphere surrounding each organism, which limits the transport of nutrients to the cell. It would appear possible for a bacterium to swim to areas of higher nutrient concentration, but these are difficult to find. By adhering to a surface, however, a bacterium can escape from the imprisoning 30-μm diffusion sphere and let the microcurrents provide continuously renewed sources of nutrients at very little energy cost. The mechanism of adhesion is thought to involve extracellular polymers; these polymers may ultimately prove to be a primary mechanism by which the microbes modify the microhabitat.

Extracellular polymers, particularly in marine systems, often involve the secretion of acidic polysaccharides (Corpe, 1973; Sutherland, 1977); these polymers have been shown to protect organisms from desiccation. The addition of a hydrophilic polymer is a traditional mechanism whereby the viability of lyophilized cultures can be increased. Such polymers may provide hysteresis, preventing too rapid drying or reentry (Wilkinson, 1958) and provide the aqueous environment necessary for the activity of extracellular enzymes (Darbyshire, 1974).

Certain nitrogen-fixing bacteria can utilize their extracellular polysaccharides as a storage polymer (Patel and Gerson, 1974). However, use of extracellular polymers as reserve by the secreting organism is unusual in the organisms thus far studied (Dudman, 1977).

Extracellular polymers are closely related to virulence in bacterial infections. Higher organisms resist bacterial infections by both cellular (phagocytic) and humoral components (complement, antibodies) of the blood and extracellular tissues. Virulent bacteria resist destruction by macrophages by surviving and resisting digestion within the phagocyte or by secreting extracellular polymers to resist phagocytosis. The evidence for the effectiveness of these capsules in such resistance is overwhelming (Dudman, 1977). Newly isolated virulent organisms are often capsulated, and virulence is often associated with the presence of capsular antigens; loss of the capsule generally decreases the virulence. Capsular materials were called "impedins" by Glynn (1973). They generally lower the contact angle between saline and the dried smooth layers of bacterial cells. Smaller contact angles are correlated with decreased susceptibility to phagocytosis (Van Oss and Gillman, 1972). Opsonization of the bacteria results in an increase in the contact angle and in susceptibility to phagocytosis. Polysaccharide polymers provide protection from both humoral antibody and complement attack; the Vi antigens of the enteric pathogenic bacteria provide the classic demonstration of resistance to agglutinability and resistance to the bactericidal effects of complement. Extracellular polysaccharides related to those that contribute to virulence by providing protection from cellular and humoral attack also appear to protect bacteria from predation by soil protozoa and possibly from the monoflagellated bacterial predators, the bdellovibrios.

Many microbes in nature secrete acidic extracellular and intramural polyanions with ion-exchange capacity. This is particularly noticeable in recent isolates from nature (Costerton et al., 1981). Most gram-positive bacteria contain teichoic acids, which are glycerol or ribitol polyphosphate polymers to which carbohydrates and amino acids are attached (Duckworth, 1977). Even under conditions of phosphate limitation, gram-positive bacteria such as *Bacillus subtilis* replace the teichoic acids with other anionic polymers formed of teichuronic acids (Tempest et al., 1968; Wright and Heckels, 1975). Some gram-positive organisms secrete lipomannans with covalently-bound succinate anionic groups. Gram-negative organisms possess surface lipopolysaccharide–protein complexes in which phosphate groups are present (Leive, 1973). All three types of anionic polymers can act as ion-exchange polymers (Damadian, 1971; Harold and Altendorf, 1974), a property that can protect some bacteria from toxic metallic salts (Tornabene and Edwards, 1972; Corpe, 1974), basic peptides such as lysozyme (Sutherland, 1972), or inhibitors such as bile salts (Poland and Odell, 1974). When isolated, these slimes and capsules show selective ion exchange properties. The alginic acids with glucuronic acid residues are selective for Sr^{2+} over Mg^{2+} and Ca^{2+} over Mg^{2+} and show a marked selectivity for Cu^{2+} over other ions (Smidsrop and Haug, 1968). Teichoic acids are selective for Mg^{2+} over K^+ and Na^+ or Ca^{2+} (Lambert et al., 1975).

Capsules and slime layers can modify the specific microbial habitat surrounding the microcolony. Ionic interactions between bacterial polyanionic polymers and clays are particularly important in soils (Daniels, 1972). Binding of montmorillonite clay particles to bacteria by the capsules of *Rhizobium* protected the bacteria from high temperatures, desiccation and X-irradiation (Marshall, 1964; Muller and Schmidt, 1966). Capsules can also serve as diffusion barriers, molecular sieves, and adsorbants.

The area within 3 μm of the cell surface can be considered as relatively unstirred in consideration of diffusion at 25°C (Bull, 1964). Many capsules (which are, physically, gels) can exceed this size. In these gels, diffusion may be considered to be decreased by the volume occupied by the gel times the coefficient relating to the order of the linear molecules. This coefficient is 5/3 for randomly oriented rods. The volume occupied by the gel must include the gel material itself and the shells of hydration in which water motion is restricted. The structure of these gels can create "pores." In 1–2% agar gels, these "pores" appear to be 44–95 nm in diameter (Ackers and Steers, 1962). This could impose an upper size limit on diffusion and cause the gel to act as a molecular sieve.

Not only can extracellular polysaccharides act as barriers to macromolecules and ions, but they can increase adsorption. Zoogleal flocs concentrate proteins and amino acids (Dugan *et al.*, 1970) and some chlorinated hydrocarbon insecticides (Lechinowsky *et al.*, 1970). These extracellular polymers can stimulate and protect extracellular enzymes such as the fatty acid synthesis complex which is activated by components of the exopolymers of *Mycobacterium phlei* (Vance *et al.*, 1973) and the exoenzymes secreted by *Micrococcus sodonesis* (Braatz and Heath, 1974).

Microbial Modification of the Biosphere

Microbes can modify their local environments by the secretion of enzymes, metabolites or extracellular polymers as described above. Microbes have also profoundly affected the earth's biosphere, responding in adaptive ways to both major and minor geologic and geochemical changes. One of the major differences between the planet Earth and the nearby planets of Mars and Venus is the presence of the biosphere, since the basic composition and early history of these planets appear to be very similar. It is believed that once microbes appeared on Earth, the subsequent history of the planetary surface was irreversibly changed. Microbes were able to capture hydrogen as water and organic molecules and thus prevent its rapid loss as it outgassed from the cooling earth. With the development of continental shelves came photosynthesis and the photolytic cleavage of water, the major source of atmospheric molecular oxygen, O_2. With the accumulation of dioxygen the atmosphere became oxidizing and profoundly altered the metabolic potentialities of the biosphere (Walker, 1977). Life on Earth appears to be a buffer that prevents drastic climatic changes that appeared in the waterless deserts of Mars or the superheated surface of Venus. The biogeochemical cycles so essential to the maintenance of the modern biosphere are absolutely dependent on microbial activity, such as volatilization of nitrogen, carbon, and sulfur; mineralization of phosphate and iron; and the creation of soil; for most of Earth's history, these cycles proceeded in the absence of any higher organisms.

DIRECT MEASUREMENT OF THE MICROBIAL WORLD

Microelectrodes

An exciting new development in microbial ecology is the direct measurement of parameters by microelectrodes. Clark-type platinum electrodes with tips of 2–20 μm in diameter have been constructed by Revsbech (Revsbech and Ward, 1983) for the polarographic measurement of oxygen tension. These microelectrodes require a small current (<50 pA) and have a short response time of 90% response in 0.2 sec. Although these electrodes are current-utilizing devices, stirring increased the currents only by <5%. Using these electrodes, the highest oxygen concentration in acid (pH 2.5) algal mats of *Cyanidium* in a thermal spring was 4 mm into the 10-mm photic zone. Maximum photosynthetic activity was 0.5 mm below the surface (Revsbech and Ward, 1983). A similar profile of oxygen content was detected in microbial slimes on rocks taken from swiftly flowing streams. These films showed a rapid loss of oxygen in the dark (Bungay and Chen, 1981). In muddy fine clay and sand estuarine sediments, the oxygen penetration depth was 1–5.5 mm (Revsbech *et al.*, 1980). Oxygen penetrates deeper into the sediments near animal burrows (Sorensen *et al.*, 1979). In general, the oxygenated zones detected by the microelectrodes are shallower than the superficial brown zone in organic-rich sediments. The depth of the brown zone is often taken as the aerobic zone.

Microelectrodes that measure pH, H_2S and Eh, may also be included in these measuring devices. Combined electrodes have been used to show how the giant bacterium *Thiovulum* creates a veil of exopolymer 100 μm thick as an unstirred boundary between oxygen and hydrogen sulfide in which it moves to create steep gradients of oxygen and hydrogen sulfide that maximize metabolic efficiency (Jorgensen and Revsbech, 1983). Rates of sulfate reduction in the bacterial habitat can be estimated with an ingenious technique developed by Cohen (1983) in which a silver wire is coated with $Na_2{}^{35}SO_4$ and inserted into the microbial mass. The sulfate reducers form $H_2{}^{35}S$, which reacts to form the silver salt. Analysis of the wire gives both the rate of reduction and the localization with the microbial mass. The use of microelectrodes and short time exposures minimizes the poisoning of platinum electrodes that causes distortion in the measurement of Eh (Whitfield, 1969).

Epifluorescence Microscopy

A significant advance in the use of light microscopy in the understanding of the microbial habitat was made by Zimmerman and Meyer-Reil (1974). They combined staining with acridine orange (3,6-tetramethyl diaminoacridine) (AO) with epifluorescence illumination and black membrane filters to minimize background fluorescence. With this technique (see also Chapter 1, this volume), microcolonies on particulate detritus could be detected, since bacterial cells, with their nucleic acids stained, were readily recognized. The AO stains DNA green and RNA red (the color is concentration dependent). It also binds nonspecifically to detritus, clays, colloids, and extracellular polymers producing an orange background. This technique has subsequently been improved by the substitution of 4′6-diamidino-2-phenylindole (DAPI) for the acridine orange. DAPI produces a much decreased nonspecific staining of components other than the microbial DNA (Porter and Feig, 1980). The epifluorescent technique can be combined with autoradiographic analysis. In this technique the material is incubated with ^{3}H-labeled precursor molecules, washed, fixed, stained, and covered with a photographic nuclear emulsion. The emulsion is developed, and the grains reduced by the radiation are localized to the active organisms by epifluorescence microscopy (Meyer-Reil, 1978*a,b;* see also Chapter 1, this volume). This can be used to give an estimate of the proportion of the bacterial community that is metabolically active (see also Chapter 3, this volume). The uptake of [^{3}H]thymidine by individual organisms has been a particularly useful technique in showing growth of microbes in environmental samples (Brock and Brock, 1966, 1968).

Epifluorescent counting has proved an effective method for examining free living bacteria in the water column. The technique has been extended to sediments in which the cells are observed *in situ.* However, attempts to estimate the microbial biomass quantitatively by so-called direct measurement have either assumed that a proportion of the community is on the underside of the sediment particles and thus not visible (Rublee and Dornseif, 1978) or the sediment was placed in a high-speed blender, with the supernatant of the blending filtered and the number of microbes counted with epifluorescence illumination. This assumes that microbes can be quantitatively removed from the sediments. In the one reported case in the literature, in which the sediment was analyzed before and after blending by an independent chemical method for the presence of bacteria (in this case, muramic acid), microbe detachment by blending of silty sands and coral sands was neither quantitative nor reproducible (Moriarty, 1980; Moriarty and Hayward, 1982). Recovery ranged from 10 to 100%. Using the technique, it was impossible to detect differences between sediment cores because the subsample variation in replicate tests from the individual cores was so large (Montagna, 1982). In examining one sample blended by Meyer-Reil (1978*a,b*) of fine marine sand, we showed that the blending decreased the muramic acid detected by 93% (R. H. Findlay, unpublished data). The possibility of rupture of the bacterial cells should be assessed by examination for the release of cytoplasmic enzymes. The use of properly designed dilution fluids with components such as pyrophosphate buffers that partially compete with the bacteria for attachment sites apparently can effect removal, but the effectiveness and possible selectivity must be established by an independent assay for bacterial components in each case. Unfortunately, the blending method destroys any details of associations of distribution in the sediment.

In the so-called direct measurement, the second assumption—the conversion from bacterial numbers to bacterial "biomass"—was even more tenuous. In the past, some magic average size for the bacteria was estimated with the microscope and used to calculate some average value for the biomass. This would be quite a subjective estimation, particularly when there is a 500-fold variation in size in the microbes on a single piece of coral reef detritus. It may now be possible, with image analyzing, to obtain accurate estimates of the biomass of the organisms visible in the microscopic field (Caldwell and Germida, 1984).

Electron Microscopy

The substitution of electrons for light energy greatly increases the resolving power of microscopes. With electron microscopy, it has become possible to examine the detailed structure of the microbes themselves as well as their immediate habitats. The specimens can be examined in thin sections at very high vacuum and must be stained with heavy-metal atoms to increase contrast (see Chapter 2, this volume). With the electron microscope it has become possible to examine directly some bacterial habitats fixed *in situ*. Two features of microbial habitats become immediately apparent: (1) microbes often exist in microcolonies of mixed types, and (2) these microcolonies are bound to surfaces by extensive extracellular polymers. One such colony is illustrated in Fig. 1A, taken from coral sands (Moriarty and Hayward, 1982), which shows microbes, with both gram-negative and gram-positive wall morphologies.

The scanning electron microscope provides a deeper field of view when examining surfaces. The material must be fixed, dehydrated, coated with a heavy-metal film, and examined in high vacuum. With the great depth of field and a magnification of more than five orders of magnitude, the scanning electron microscope has provided a view of surface morphology that has revolutionized ideas of microbial interactions. Sieburth's (1975) atlas of marine seascapes is a work of art, as well as revealing the morphological complexities of surfaces exposed to seawater. Organisms of remarkable diversity can be detected on the surfaces. Figure 1B represents the microorganisms collected on a titanium surface exposed for 4 weeks to rapidly flowing seawater.

The problems of defining microbial habitats by scanning electron microscopy (SEM) are severalfold: First, not all microbes are quantitatively preserved in the fixation procedures. Protozoa, for example, are often destroyed during the dehydration and critical point drying process. A second problem lies in the irreproducible shrinkage that accompanies the dehydration process. The third problem is the lamentable lack of relationship between form and function in most of the prokaryotic world. For example, methane-forming bacteria come in all shapes (Zeikus, 1977). Not only can the type of microbial activity not be defined accurately by the morphology, but the level of metabolic activity of the organisms is often impossible to detect. Occasionally, dividing cells can be seen, but usually no other clue to metabolic activity is obvious. The proportion of dividing bacteria in a population has been used as a measure of bacterial activity (Hagstrom *et al.*, 1979). A combination of autoradiography and electron or epifluorescence microscopy (Stanley and Staley, 1977) is more informative than either method alone.

Despite the lack of variation in form that might provide clues to function or ecological niche in small bacteria, microscopic studies can yield insights

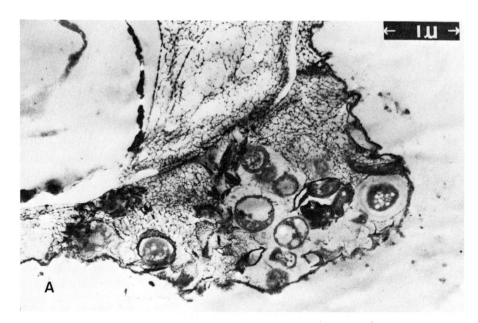

FIGURE 1. (A) Transmission electron micrograph (TEM) of a microcolony thin section attached to a diatom frustule from the Great Barrier Reef stained with ruthenium red and osmium tetroxide. (From Moriarty and Hayward, 1982.) (B) Scanning electron micrograph (SEM) of microbes attached to a titanium surface exposed to rapidly flowing seawater (1.5 m/sec) for 29 days at Panama City, Florida. (White, D. C., 1982.)

regarding the habits and habitats of some of the larger microbes, as was summarized by Brock (1971). Pelagic microbes have morphological adaptations to control the sinking rates. Some dinoflagellates modify the specific gravity of their cells with the phase of the cell cycle: Thick-walled forms sink and thin-walled forms live planktonically. Gas vacuoles in which content is coupled to metabolic activity are found in halophilic and photosynthetic bacteria. Secretion of a gelatinous sheath or the development of elongated shapes with spicules or protrusions can greatly decrease the sinking rates. Large structurally complex protozoa are virtually absent from soils. The drier the soil, the smaller the protozoans that are found inhabiting it. In the interstitial environment of beach sands, large protozoa with tentacles inhabit areas with large sand grains, while small flexible forms with adhesive areas are found in finer-grained sands (Borror, 1968). Diatoms that move are pennate with flat or concave shapes if they are found on flat surfaces (rocks or leaves), whereas diatoms with central bulges and raphe at each end are much better adapted to a particulate environment. With *Leucothrix mucor* and other morphologically complex microbes, the presence of migratory gonidia indicates that conditions are relatively adverse, i.e., too cold, too low a nutrient supply, or partially anaerobic.

Surface Analysis

The scanning electron microscope can be coupled to an X-ray diffraction apparatus in the technique called energy dispersive spectrometry (EDS). With this technique, it is possible to localize the concentration of heavy atoms in scanning electron micrographs of uncoated specimens. The concentration of iron in the extracellular polymers of *Gallionella* has been demonstrated (Ridgeway *et al.*, 1981), as have the changes in metal concentrations in microbial biofouling (Gerchakov *et al.*, 1978).

The detailed examination of surfaces has benefited from the introduction of new techniques. With these techniques, beams of electrons, ions, or photons interact with surfaces and by their effects on the surface atoms can provide the details of structure. The most commonly used techniques can be performed on the same instrument. In Auger electron spectroscopy (AES), atoms at the surface are excited by finely focused 2–3 keV electron beams. The electrons in the surface atoms are raised to excited states that emit characteristic energies permitting their identification. This technique is often combined with SEM to provide a map of the distribution of the elements existing at the surface. This technique provides information on many more elements than EDS, and the map is given at much greater resolution. In X-ray photoelectron spectroscopy (XPS), soft X rays are used to activate the surface atoms, thereby yielding information relative to the chemical state of the surface atoms. In secondary ion mass spectroscopy (SIMS), a beam of low-energy inert gas ions sputter (knock off) clusters of ions, which are then analyzed by mass spectrometry. These techniques can provide detailed analysis of the top

10 atomic layers of a specimen. AES yields detailed information about the geometry of the surface with high spatial resolution, XPS provides information about the chemical state of the surface atoms and is particularly useful in the examination of surface corrosion products or adsorbed surfactants, and the ion sputtering of SIMS can be used to remove contaminants collected during sample handling and to expose successive atomic layers to depths of several microns. The problems with these techniques lie in the necessity for a very high vacuum and the limitations of the analysis to the uppermost surface of the sample. Often the sample surfaces must be exposed in the vacuum chamber of the instruments to eliminate artifacts of sample handling. Surface-analyzing instruments represent the fastest-growing area in scientific instruments and, although they are far beyond the means of most microbiological laboratories, continued development may eventually enable bacterial habitats to be probed with these sophisticated instruments.

Fourier Transforming Infrared Spectrometry

Recent developments in instrumentation and computer technology have made available a new technique for the study of bacterial habitats that does not require a high vacuum and that can monitor biofilms up to 500 nm thick. The technique is based on infrared (IR) spectroscopy, which essentially defines the vibration, rotational, and, to some extent, the electronic environment of chemical bonds in molecules. In general the interpretation can be relatively straightforward; the analysis is nondestructive and does not require the specimen be in a high vacuum. Especially important is the fact that living organisms can be examined. Previously, the low energy of the IR region of the spectrum resulted in a relative insensitivity that made the usefulness of IR spectrometry in situations as complex as nature very limited; these limitations have been overcome by new instruments and their computer components.

The new instruments use interfering beams of IR radiation in order to create a spectrum. This is achieved by moving one mirror very accurately. The beams are recombined, and the interference effects are observed as a phase difference between the two beams. Since the whole spectrum is observed continuously, the signal-to-noise (S/N) ratio is greatly increased over that of grating instruments in which a narrow beam of radiation was moved across the dispersion generator; although the noise was continuously generated, the signal for a given spectral element was measured for only a short time. These advantages generate an S/N ratio 2500 times greater than dispersion instruments (Griffiths, 1975, 1978). The combination of these optical advantages plus the computer manipulations in translating data from Fourier space to frequency space make this instrument exceedingly powerful. The application of FT/IR spectroscopy to microbial biofilms was recently summarized by Nichols *et al.* (1985*c*).

The remarkable advances in instrumentation achieved by the use of the FT/IR can provide direct information regarding the chemical environments of bacteria. Each of these techniques has specific advantages and, because each is nondestructive, combinations of several techniques can be used to analyze each sample.

INDIRECT INFERENCES FROM MICROBES

It is possible to infer properties of a bacterial habitat if the microbial composition of that site in nature can be determined. The inference of the habitat characteristics is based on the knowledge of the physiologic potential of the component members of the microbial consortia found in that habitat and thus is an indirect measure of the habitat properties.

Over the past few years, this laboratory has concentrated on the development of a suite of biochemical methods by which microbial consortia from environmental samples can be measured. These measures are without the errors induced by cultural methods that depend on growth of component members of the microbial community under artificial conditions, or the estimates of microbial volumes or separation from substrates as needed in microscopical methods (Findlay et al., 1983). These methods are nonselective in that they use measurements of biochemical components of cells. Biomass of microbes from a particular sample can be estimated by measuring cellular components universally found in cells. Components restricted to particular subsets of the microbial community can be used as "signatures" of those specific groups in the determination of the community structure. Actually a continuum exists between components common to all cells and to the more or less specific "signature" compounds.

Since these measures of biomass and community structure involve the isolation and assay of specific components, it is possible to determine the metabolic activity of the community and its various subsets by following the rates of incorporation or the turnover after pulse-chase exposures to labeled precursors. In this regard, the use of mass-labeled precursors containing ^{15}N, ^{13}C, ^{18}O, and in some cases ^{2}H are particularly valuable because the precursors are not radioactive and the specific activity can be nearly 100%. The high specific activity makes it possible to utilize the precursor molecules at substrate levels just above the background and thus provoke less distortion of the microbial consortia than the higher concentrations needed with radioactive isotopes. Using ^{15}N-ammonium ion, it proved possible to show bacterial growth by the reproducible detection of enrichments of as little as 1 at% excess ^{15}N in as little as 90 fmoles of D-alanine isolated from the bacterial cell wall (Tunlid et al., 1985). The sensitive detection of these labels in "signature" molecules can be achieved by capillary gas–liquid chromatography/mass spectrometry with the formation of electron withdrawing pentafluorobenzoate ester at room temperature in glass capillaries, followed by capillary gas chromatography/mass spectrometry (GC-MS). With soft (chemical) ionization and detection of negative ions, it proved possible to detect fatty acid profiles at the 2-fmole level, i.e., at a level equivalent to 100 bacteria the size of *E. coli* (Odham *et al.*, 1985). The nature of these assays as well as their validation and application to various environmental situations was recently reviewed (White, 1983).

Biomass and Community Structure

One of the most useful techniques for the analysis of microbial community structure is based on chloroform-methanol extraction from which the lipid fraction can be concentrated and examined for its components. Among the most useful components in detecting the presence of microbial cells are the phospholipids. Phospholipids constitute a part of practically every cellular membrane and form a relatively constant proportion of the membranes of various microbes (White *et al.*, 1978*a,b*). Phospholipids are not used as endogenous storage products and have a relatively rapid turnover in both living and killed cells added to the environment (King *et al.*, 1977; White *et al.*, 1979*a,c*). They make up about 98% of the extractable lipids in eubacteria and about 50% of membranes from other organisms. Phospholipids can be readily detected as lipid phosphate (White *et al.*, 1979*c*) or glycerol phosphate (Gehron and White, 1983). Either ^{14}C or $^{32}PO_4$ can be used to detect growth of microbes by increases in phospholipids; this is useful, for example, in determining whether certain processes such as the corrosion of mild steel in water involve the growth of microbes.

The presence of microeukaryotes within the bacterial community can be used to help define habitat properties, since the life histories and physiological requirements of some species are known in considerable detail. The presence of microeukaryotes (algae, fungi, protozoa, or micrometazoa) can be detected by a. ratio of glucosamine to muramic acid greater than 1 : 1 in the gas-chromatographic analysis of muramic acid preparations (Findlay *et al.*, 1983). The bacterial cell wall is a copolymer of muramic acid and glucosamine; the walls of many microeukaryotes are chitin, a polymer containing glucosamine. Microeukaryotes are also more likely than prokaryotes to contain polyenoic fatty acids longer than 20 carbons that are ester-linked to the phospholipids, steroids, phosphonolipids, and specific sulfolipids (White *et al.*, 1980). Certain gliding bacteria, such as cytophaga, contain sulfonolipids (Godchaux and Leadbetter, 1983). Polyenoic fatty acids with more than 20 carbon atoms and with α-linolenic unsaturation patterns (ω 3 series) are characteristic of the eukaryotic microflora. The polyenoic fatty acids with the γ-linolenic unsaturation patterns (ω 6 series) are characteristic of microfauna (Erwin, 1973; Bobbie and White, 1980). The absence of C_{18} polyenoic fatty acids in marine diatoms (Bobbie *et al.*, 1981) and the composition of the phytopigments (Gillan and Johns, 1983) can also indicate the types of microflora present. Triglycerides are found exclusively in the microeukaryotes and can be readily measured as glycerol in the neutral lipids (Gehron and White, 1983). The monohydroxy steroids phosphatidyl inositol, wall inositol, and sulfolipid synthesis are signatures of microeukaryotes (White *et al.*, 1980).

The biomass and types of microeukaryotes can often greatly affect the composition of the bacterial habitat. Grazing can markedly change the composition and metabolic activity of the microbiota (Morrison and White, 1980). There seems to be a grazing pressure that provides an optimal metabolic activity. Grazing pressures that are too low result in communities that contain slower-growing microbes of lower activity. Grazing pressures that are high result in a much depressed microbial biomass. With the types of quantitative assays just described, it proved possible to demonstrate quantitatively that sympatric amphipods with different feeding appendages partition the detrital microbiota (G. A. Smith *et al.*, 1982a). Manipulations of the top epibenthic predators of an estuarine mud flat can be shown to affect the bacteria in the sediments (Federle *et al.*, 1983). In this study, removal of the epibenthic predators permitted proliferation of specific polychaete worms, stimulating a group of microbes with a high proportion of branched-chain C_{15} atom fatty acids and inhibited those containing normal C_{15} atom fatty acids in ester linkage in the phospholipids.

The succession of microbiota colonizing a newly created bacterial habitat can also be followed using biochemical methods. In marine systems, the initial colonizers appear to be gram-negative bacteria that are followed by bacteria and microeukaryotes with more complex structures (Morrison *et al.*, 1977; Nickels *et al.*, 1981a; Odham *et al.*, 1985). Just as gram-negative bacteria are the initial colonizers of newly created surfaces in the oceans, instabilities in the physical and chemical properties of bacterial niches seem to favor gram-negative bacteria. The proportion of gram-positive bacteria increases as one proceeds deeper into sediments or soils (Gehron *et al.*, 1984). The proportion of gram-negative bacteria can be quantitatively determined by measuring the hydroxy fatty acids of the lipid A of the lipopolysaccharide (Parker *et al.*, 1982). Different types of gram-negative bacteria can often be distinguished from each other by the patterns of their LPS-lipid A fatty acid profiles (Parker *et al.*, 1982; Wilkinson and Caldwell, 1980). The presence of gram-positive bacteria in the environment can be measured by taking advantage of the specificity of cold concentrated hydrofluoric acid in the hydrolysis of phosphodiester polymers. The glycerol and ribitol liberated after this specific hydrolysis provides an estimate of the gram-positive teichoic acid containing bacteria present (Gehron *et al.*, 1984). Contamination of the subsurface aquifer sediment with polyaromatics stimulated growth in the vadose zone and drastically decreased the proportion of gram-positive bacteria (G. A. Smith *et al.*, 1985). Effects of oil and gas well drilling fluids on the colonization of marine sands can be followed by charting phospholipid fatty acid patterns (Smith *et al.*, 1982b).

The type and concentration of terminal electron acceptors present in the bacterial niche can determine the organisms that are present (Guckert *et al.*, 1985). The competition for hydrogen and substrates between sulfate reducers, methane formers, and denitrifiers in anaerobic sediments has received wide study (Abram and Nedwell, 1978a,b; Cappenberg and Jongejan, 1978; Winfrey and Zeikus, 1977; Oremland and Taylor, 1978; R. L. Smith and Klug, 1981). The presence of methane-forming bacteria in environmental samples can be assayed by detecting their di- and tetraphytanyl glycerol ether phospholipids after extraction and purification of the phospholipids, saponification, chromatography, acid methanolysis, derivatization, and separation by high-pressure chromatography (Martz *et al.*, 1983). Preliminary evidence indicates that it may even be possible to differentiate among different groups of methane-forming bacteria by examination of these ethers. Sulfate-reducing bacteria contain unusual fatty acids in their phospholipids. *Desulfovibrio desulfuricans* contains high levels of a branched-chain C_{17} monounsaturated fatty acid (Boon *et al.*, 1977; Parkes and Taylor, 1983; Edlund *et al.*, 1985). Sediments apparently do not contain this specific lipid but contain other branched-chain fatty acids more characteristic of the newly isolated acetate-utilizing strains (Parkes and Taylor, 1983). These sulfate-reducing bacteria contain high levels of ester-linked hydroxy fatty acids in the phospholipids (Fredrickson, 1981). The anaerobic denitrifiers have proved a difficult group in which to find signature lipids because they are a diverse group of organisms. Many of the most active denitrifiers in soils are pseudomonads, and many of them have unusual patterns of hydroxy fatty acids linked to the lipid A of their lipopolysaccharides (Parker *et al.*, 1984; Wilkinson and Caldwell, 1980).

By determining the position and configuration of the unsaturation of monoenoic fatty acids ester-linked in the extractable phospholipid, the signature profiles of specific groups of bacteria can be greatly increased (Nichols *et al.*, 1985a). For example, the pathogen *Francisella tularensis* contains long monoenoic fatty acids with the unsaturation in unusual positions (Nichols *et al.*, 1985b). Myron Sasser of the University of Delaware in combination with Hewlett Packard has been able to distinguish accurately among some 8000 strains of bacteria. This has been achieved with the aid of an innovative multicomponent-analysis computer program that compares profiles of fatty acids with the standard organisms from the Data Bank. Anaerobes are important organisms in fermentations such as those in the rumen, where the transformation of biopolymers into hydrogen, carbon dioxide, short-chain alcohols, and acids involves often the rate-limiting steps (Wolin, 1979). Plasmalogen phospholipids (phospholipids with a glycerol backbone and a vinyl ether in the 1 position) are among the signature lipids of anaerobic bacteria (Goldfine and Hagen, 1972; Kamio *et al.*, 1969). They can be analyzed as the phosphate that is rendered water soluble by mild acid hydrolysis after resisting mild alkaline hydrolysis or by isolating the fatty aldehydes derived from the vinyl ethers after acid hydrolysis (White *et al.*, 1979b). Plasmalogens are largely confined to gram-positive anaerobic fermentative bacteria and to some ciliate protozoa found in the rumen. They can account for 2–10% of the phospholipids of soils and sediments. As one goes deeper into the soil, there is a general increase in the proportion of plasmalogen phospholipid.

The detection of abundant anaerobic bacteria in aerobic zones of sediments and films by the presence of their signature (White, 1983) is not fully in agreement with the interpretations of findings with microelectrodes in defining the total activity of sedimentary layers (see the section, Microelectrodes). These biochemical findings of multiple microbial types in microcolonies are supported, however, by the transmission electron micrographs of different wall morphologies bound together with extracellular materials of sediments (Moriarty and Hayward, 1982).

Phosphosphingolipids are very rare in bacteria (Karlsson, 1970), but unique phosphosphingolipids have been found among the non-spore-forming gram-negative anaerobic bacteria (LaBach and White, 1969; White and Tucker, 1969; Rizza et al., 1970). These components can be isolated from bacterial habitats as nonsaponifiable phospholipids, and the sphinganines released by acid methanolysis can, after derivatization, be assayed by GC-MS. These lipids are enriched in habitats such as dental plaque or human feces.

The redox potentials of the respiratory quinones suggest that the terminal electron acceptors of those bacteria containing ubiquinones (benzoquinones) should be of high potential as compared with those of bacteria containing naphthoquinones. Bacteria capable of forming both types of respiratory quinones such as *Proteus rettqeri* form ubiquinones when grown aerobically and menaquinone (naphthoquinones) when grown anaerobically (Kroger et al., 1971). Aerobes contain benzoquinones; some, but not all, anaerobes contain naphthoquinones (Whistance et al., 1969; Hollander et al., 1977). Using HPLC with electrochemical detection, D. Hedrick of this laboratory showed that manipulation of sediments between aerobic and anaerobic conditions shifts the ratios of naphthoquinones to benzoquinones from 0.03 for the aerated consortia to 3.0 for the fermenters precisely as expected from the studies on monocultures. The sensitivity and specificity of the electrochemical detection of these respiratory quinones now makes possible the predication of the terminal electron acceptors for the bacterial community (Hedrick and White, 1986).

Microbial Metabolic Activity

Measurements of metabolic activity of the microbes in a habitat can help in the understanding of the physical and chemical environment of that site. Decreasing the availability of phosphate can induce increases in the activity of phosphatase and other enzymes in the detrital microbiota (Morrison and White, 1980). If levels of phosphate are limiting, at least in monocultures, gram-positive bacteria respond by forming teichuronic acids in place of the wall teichoic acids (Ellwood and Tempest, 1972; Ward, 1981), and some gram-negative pseudomonads form glucosuronosyl diglyceride and glucosyl-glucuronosyl diglyceride instead of phosphatidyl glycerol (Minnikin et al., 1974). The availability of a biodegradable surface induces distinctly different enzymatic activities in the microorganisms than a surface with similar microtopography that is not biodegradable (Bobbie et al., 1978).

Relative activities of most of the prokaryotes, compared with those of the microeukaryotes present, can be determined by following the incorporation

of [³H]thymidine into DNA relative to the incorporation of $^{32}PO_4$ or [¹⁴C]acetate
or other precursor into phospholipids. With the notable exceptions of some
anaerobes and autotrophs, prokaryotes generally incorporate thymidine into
DNA via salvage pathways. Prokaryotic microbial activity can be assayed accurately if care is taken to determine isotope dilution factors and if the [³H]-
DNA is purified before assay (Moriarty and Pollard, 1981) (see Chapter 1,
this volume for a more extensive discussion of the use of [³H]thymidine).
Phospholipid synthesis occurs in the total community. A convenient method
for measuring ^{32}P incorporation into phospholipid-utilizing disposable sy-
ringes has been described (Moriarty et al., 1985). Sulfolipids apparently are
rare in sediment bacteria, and their rates of formation give an indication of
the activities of the microeukaryotes present (White et al., 1980; Bobbie et al.,
1981).

Measurements of the rates of incorporation of radioactive or mass-labeled
isotopes into the specific signature lipids of the various subgroups of microbes
in the habitat can indicate the activity of that group (Moriarty et al., 1985).
When applied to sediments, there is a disturbance associated with the application of the labeled precursors. In work done in this laboratory, Findlay et
al., (1985) have shown that a disturbance such as raking the surface of a mud
flat could only be detected by injecting radioactive acetate into carefully manipulated cores. The effect of the surface disturbance was lost in the artifacts
produced by perfusing isolated cores with labeled pore water or after mixing
the sediment with labeled buffer, as is usually done in classical heterotrophic
potential measurements (Findlay et al., 1985).

Metabolism of cellular components can be followed if pulse-chase exposures to labeled precursors are allowed. The unique cell-wall component
of prokaryotes, muramic acid, shows complex turnover patterns with both
slow (3-day half-times) and fast (3-hr half-times) components (King and White,
1977). Turnover patterns after short pulse exposures of lipids such as phosphatidyl glycerol show relationships to the rates of growth in both monocultures (White and Tucker, 1969) and the detrital microbiota (King et al., 1977).
These estimates of growth agree with rates of phospholipid synthesis, muramic
acid turnover, and respiratory activity (Morrison et al., 1977). Lipids showing
little or no turnover, such as the glycerol phosphoryl choline derived from
phosphatidyl choline, can be used to estimate the rates of grazing on the
detrital microbiota (Morrison and White, 1980).

Nutritional Status

The nutritional status of the microbes in a habitat can often be used to
define the recent history of the organisms at that site. The ATP levels have
been used both as a measure of biomass and metabolic activity (Holm-Hansen,
1973). This assumes that ATP has a short half-life both inside and outside
the cells. The adenylate energy charge, a ratio among the adenosine nucleotides in the cells, is a more sensitive indicator of the nutritional status (Knowles,

1971). The development of an even more sensitive indicator is based on the homeostatic mechanisms by which the microbes maintain the energy charge under conditions of metabolic stress. This mechanism involves compensation for the lowering rate of formation of ATP by lowering the intercellular concentration of AMP and ADP. One mechanism by which organisms lower the concentration of intercellular AMP and ADP is by its hydrolysis to adenosine. With the development of a HPLC method for the separation and detection of the adenosine containing components of cells by forming fluorescent 1-N^6-ethenoadenosine derivatives (Davis and White, 1980), it was possible to show that stress on exponentially growing *E. coli* by gentle filtration markedly increased the ratio of adenosine to ATP but had no effect on the adenylate energy charge.

Longer-term measurements of the nutritional status involve the formation and turnover of the endogenous storage lipids. Triglyceride in the microeukaryotes can be readily measured as the glycerol in the neutral lipid (Gehron and White, 1982). Measuring the ratio of triglyceride glycerol to the phospholipid provides an indication of the nutritional status. Using this ratio, it was possible to show that amphipods grazing the detrital microbiota in nature are essentially equivalent to starving organisms confined in the laboratory. In marine organisms, wax esters form a more slowly metabolized endogenous storage pool (Sargent *et al.*, 1976) that can be used to show metabolic stress on reef-building corals (Parker *et al.*, 1984).

Prokaryotes form a unique endogenous storage lipid, poly β-hydroxy-butyrate during conditions of unbalanced growth (Nickels *et al.*, 1979). Findlay and White (1983) showed this polymer also contains several other short-chain hydroxy fatty acids and recommended that it be called poly-β-hydroxy alkanoate (PHA). Different growth conditions alter the ratios of the various component hydroxy fatty acids in PHA. In the sedimentary environment, PHA is rapidly formed under conditions of ample carbon source and oxygen and some other nutrient limiting growth. The rates of formation can be greatly increased by the incorporation of a chelator such as humic acid or EDTA into the medium. With a proper balance of carbon, nitrogen, phosphate, and terminal electron acceptor, PHA is rapidly degraded, and the organisms grow as indicated by the synthesis of DNA or phospholipid. High ratios of phospholipid to PHA can be generated by aerating anaerobic sediments (White *et al.*, 1979b). Findlay used the lowering of the ratio of phospholipid to PHA to define the artifacts introduced into measurements of metabolic activity by manipulation of sediments (Findlay *et al.*, 1983). The findings of high ratios of PHA to muramic acid in the microbiota on the surface of plant roots or seagrass leaves indicates that the plants are witholding some nutrient other than carbon. The highest ratios of PHA to phospholipids have been found in the groundwater aquifer microbiota (White *et al.*, 1983, 1984).

The roots of the rape plant (*Brassica napus* L.) can be shown to stimulate growth of attached bacteria by the formation of a specific complement of ester-linked fatty acids isolated from the phospholipids (Tunlid *et al.*, 1986).

The attached bacteria appear to sustain balanced growth, since no PHA is formed. This is in marked contrast to the bacteria in the rhizosphere not attached to the roots; these bacteria form large amounts of PHA, indicating a carbon adequate habitat that fosters unbalanced growth.

Another indicator of nutritional stress in microbes is the formation of extracellular polysaccharide glycocalyx (Sutherland, 1972; Costerton *et al.*, 1981). Exopolymer formation in *Pseudomonas atlantica* has been shown to increase greatly with nutritional stress as detected by the adenylate energy charge (Uhlinger and White, 1983). Very high ratios of exopolymer polysaccharide to phospholipid ratios have been found in the groundwater aquifer microbiota (White *et al.*, 1983, 1984).

The formation of microbial biofilms with large amounts of extracellular polymers on the surface of stainless steel exposed to seawater has been shown to stimulate corrosion reversibly (White *et al.*, 1985) and to inhibit markedly the transfer of heat across metal condenser tubes (White and Benson, 1984).

EFFECTS OF SURFACES

It is possible to show that the nature of the surface to which microbes are attached can affect the community structure. Aluminum, titanium, stainless steel, and copper surfaces exposed to seawater developed morphologically different microbial films (Berk *et al.*, 1981). The succession of the microbial film on titanium also differed markedly from that occurring on aluminum (Nickels *et al.*, 1981a). The addition of bentonite clay or barite to azoic marine sands showed markedly different microbiota after exposure to unfiltered pelagic seawater for 8 weeks (G. A. Smith *et al.*, 1982a). The microtopography of the sediments of the same grain size can affect the composition of the microbial surface film. The loss of microdiscontinuities decreased the bacterial and microfloral biomass but increased the microfaunal biomass, as estimated by the γ-linolenic series of polyenoic fatty acids (Nickels *et al.*, 1981b).

CONCLUSIONS

With the development of ever more sophisticated analytical methods to directly measure the physical and chemical characteristics of microbial niches, indirect methods based on measuring cellular components and activities will become more definitive. These chemical methods can then be applied to new habitats to gain insight into the properties of the microbes that proliferate there. One area that will benefit greatly from the application of these methods is in the definition of the role of microbes in the facilitation of corrosion. Some obvious roles have been clearly identified, such as the role of sulfuric acid generated by the aerobic reduced sulfur oxidizers. More subtle effects involve the generation of unstable reactants that greatly increase the rate of corrosion of steel in the presence of the anaerobic sulfate reducers (Iverson and Olson, 1983). The role of organisms that generate extracellular polymers that can both chelate metals very strongly as well as induce concentration cells of different cathodic activity awaits further study.

ACKNOWLEDGMENTS. The research reported in this chapter was supported by contracts NOOO14-82-0404 and NOOO14-83-K-0056 from the Department of the Navy, Office of Naval Research; grants CR-810292 and CR-80994-02 from the Gulf Breeze Environmental Research Laboratory and Robert S. Kerr Environmental Research Laboratory of the U. S. Environmental Protection Agency; grant NAG-2-149 from the Advanced Life Support Office, National Aeronautics and Space Administration; and grant OCE-80-19757 from the Biological Oceanography section of the National Science Foundation.

REFERENCES

Abram, J. W., and Nedwell, D. B., 1978a, Inhibition of methanogenesis by sulphate reducing bacteria competing for transferred hydrogen, *Arch. Microbiol.* **117:**89–92.

Abram, J. W., and Nedwell, D. B., 1978b, Hydrogen as a substrate for methanogenesis and sulphate reduction in anaerobic saltmarsh sediment, *Arch. Microbiol.* **117:**93–97.

Ackers, G. K., and Steers, R. L., 1962, Restricted diffusion of macromolecules through agar-gel membranes, *Biochim. Biophys. Acta* **59:**137–149.

Berg, H. C., 1975a, Bacterial behaviour, *Nature (Lond.)* **254:**389–392.

Berg, H. C., 1975b, How bacteria swim, *Sci. Am.* **233:**36–44.

Berk, S. G., Mitchell, R., Bobbie, R. J., Nickels, J. S., and White, D. C., 1981, Microfouling on metal surfaces exposed to seawater, *Int. Biodeterior. Bull.* **17:**29–37.

Bobbie, R. J., and White, D. C., 1980, Characterization of benthic microbial community structure by high resolution gas chromatography of fatty acid methyl esters, *Appl. Environ. Microbiol.* **39:**1212–1222.

Bobbie, R. J., Morrison, S. J., and White, D. C., 1978, Effects of substrate biodegradability on the mass and activity of the associated estaurine microbiota, *Appl. Environ. Microbiol.* **35:**179–184.

Bobbie, R. J., Nickels, J. S., Smith, G. A., Fazio, S. D., Findlay, R. H., Davis, W. M., and White, D. C., 1981, Effect of light on the biomass and community structure of the estaurine detrital microbiota, *Appl. Environ. Microbiol.* **42:**150–158.

Boon, J. J., de Leeuw, J. W., v. d. Hoek, C. J., and Vosjan, J. H., 1977, Significance and taxonomic value of iso and anteiso monoenoic fatty acids and branched-hydroxy acids in *Desulfovibrio desulfuricans, J. Bacteriol.* **129:**1183–1191.

Borror, A. C., 1968, Ecology of interstitial ciliates. *Trans. Am. Microsc. Soc.* **87:**233–243.

Braatz, J. A., and Heath, E. C., 1974, The role of polysaccharide in the secretion of protein by *Micrococcus sodonesis, J. Biol. Chem.* **249:**2636–2647.

Brock, T. D., 1971, Microbial growth rates in nature, *Bacteriol. Rev.* **35:**39–58.

Brock, T. D., and Brock, M. L., 1966, Autoradiography as a tool in microbial ecology, *Nature (Lond.)* **209:**734–736.

Brock, T. D., and Brock, M. L., 1968, Measurement of steady-state growth rates of a thermophilic alga directly in nature, *J. Bacteriol.* **95:**811–815.

Bull, H. B., 1964, *An Introduction to Physical Biochemistry,* F. A. Davis, Philadelphia.

Bungay, H. R., and Chen, Y. S., 1981, Dissolved oxygen profiles in photosynthetic microbial slimes, *Biotechnol. Bioeng.* **23:**1883–1895.

Caldwell, D. E., and Germida, J. J., 1984, Evaluation of difference imagery for visualizing and quantitating microbial growth, *Can. J. Microbiol.* **31:**35–44.

Cappenberg, T. E., and Jongejan, E., 1978, Microenvironments for sulfate reduction and methane production in fresh water sediments, in: *Environmental Biogeochemistry and Geobiology,* Vol. 1 (W. E. Krumbein, ed.), pp. 129–138, Ann Arbor Science Publishers, Ann Arbor, Michigan.

Cohen, Y., 1983, Oxygenic photosynthesis, anoxygenic photosynthesis and sulfate reduction in cyanobacterial mats, *Abstracts of the Third International Symposium on Microbial Ecology.*

Corpe, W. A., 1973, Microfouling: Role of primary film-forming marine bacteria, in: *Proceedings of the Third International Congress on Marine Corrosion and Fouling,* pp. 598–609, Northwestern University Press, Evanston, Illinois.

Corpe, W. A., 1974, Periphytic marine bacteria and the formation of microbial films on solid surfaces, in: *Effect of the Ocean Environment on Microbial Activities* (R. R. Colwell and R. Y. Morita, eds.), pp. 397–417, University Park Press, Baltimore.

Costerton, J. W., Irvin, R. T., and Cheng, K. J., 1981, The bacterial glycocalyx in nature and disease, *Annu. Rev. Microbiology* **35:**299–324.

Damadian, R., 1971, Biological ion exchange resins. I. Quantitative electrostatic correspondence of fixed charge and mobile counter ion, *Biophys. J.* **11:**739–760.

Daniels, S. L., 1972, The adsorption of microorganisms onto solid surfaces: A review, *Dev. Indust. Microbiol.* **13:**211–253.

Darbyshire, B., 1974, The function of the carbohydrate units of three fungal enzymes in their resistance to dehydration, *Plant Physiol.* **54:**717–721.

Davis, W. M., and White, D. C., 1980, Fluorometric determination of adenosine nucleotide derivatives as measures of the microfouling, detrital and sedimentary microbial biomass and physiological status, *Appl. Environ. Microbiol.* **40:**539–548.

Duckworth, M., 1977, Teichoic acids, in: *Surface Carbohydrates of the Prokaryotic Cell* (I. W. Sutherland, ed.), pp. 177–208, Academic Press, New York.

Dudman, W. F., 1977, The role of surface polysaccharides in natural environments, in: *Surface Carbohydrates of the Prokaryotic Cell* (I. W. Sutherland, ed.), pp. 357–414, Academic Press, New York.

Dugan, P. R., Macmillan, C. B., and Pfister, R. M., 1970, Aerobic heterotrophic bacteria indigenous to pH 2.8 acid mine water: Microscopic examination of acid streamers, *J. Bacteriol.* **101:**973–981.

Edlund, A., Nichols, P. D., Roffey, R., and White, D. C., 1985, Extractable and lipopolysaccharide fatty acid and hydroxy acid profiles from *Desulfovibrio* species, *J. Lipid Res.* **26**(8):982–988.

Ellwood, D. C., and Tempest, D. W., 1972, Effects of environment on bacterial wall content and composition, *Adv. Microb. Physiol.* **7:**83–117.

Erwin, J. A., 1973, Fatty acids in eukaryotic microorganisms, in: *Lipids and Biomembranes of Eukaryotic Microorganisms* (J. A. Erwin, ed.), pp. 41–143, Academic Press, New York.

Federle, T. W., Livingston, R. J., Meeter, D. A., and White, D. C., 1983, Modification of estuarine sedimentary microbiota by exclusion of top predators, *J. Exp. Marine Biol. Ecol.* **73:**81–94.

Findlay, R. H., and White, D. C., 1983, Polymeric beta-hydroxyalkanoates from environmental samples and *Bacillus megaterium, Appl. Environ. Microbiol.* **45:**71–78.

Findlay, R. H., Moriarty, D. J. W., and White, D. C., 1983, Improved method of determining muramic acid from environmental samples, *Geomicrobiology* **3:**135–150.

Findlay, R. H., Pollard, P. C., Moriarty, D. J. W., and White, D. C., 1985, Quantitative determination of microbial activity and community nutritional status in estuarine sediments: Evidence for a disturbance artifact, *Can. J. Microbiol.* **31:**493–498.

Fredrickson, H. L., 1981, Lipid characterization of sedimentary sulfate-reducing communities, *Abst. Am. Soc. Microbiol.* 205.

Gehron, M. J., and White, D. C., 1982, Quantitative determination of the nutritional status of detrital microbiota and the grazing fauna by triglyceride glycerol analysis, *J. Exp. Mar. Biol. Ecol.* **64:**145–158.

Gehron, M. J., and White, D. C., 1983, Sensitive measurements of phospholipid glycerol in environmental samples, *J. Microbiol. Methods* **1:**23–32.

Gehron, M. J., Davis, J. D., Smith, G. A., and White, D. C., 1984, Determination of the gram-positive bacterial content of soils and sediments by analysis of teichoic acid components, *J. Microbiol. Methods* **2:**165–176.

Gerchakov, S. M., Roth, F. J., Sallman, B., and Udey, L. R., 1978, Observations on microfouling applicable to OTEC systems, in *Proceedings of the Ocean Thermal Energy Conversion Biofouling and Corrosion Symposium, 1977*, pp. 63–75.

Gillan, F. T., and Johns, R. B., 1983, Normal-phase HPLC analysis of microbial carotenoids and neutral lipids, *J. Chromatogr. Sci.* **21**:34–38.

Glynn, A. A., 1973, in: *Non-specific Factors Influencing Host Resistance* (W. Braun and J. Unger, eds.), pp. 350–353, S. Karger, Basel.

Godchaux, III, W., and Leadbetter, E. R., 1983, Unusual sulfonolipids are characteristic of the *Cytophaga-Flexibacter* group, *J. Bacteriol.* **153**:1238–1246.

Goldfine, H., and Hagen, P. O., 1972, Bacterial plasmalogens, in: *Ether Lipids, Chemistry and Biology* (F. Snyder, ed.), pp. 329–350, Academic Press, New York.

Griffiths, P. R., 1975, *Chemical Infrared Fourier Transform Spectroscopy*, Wiley, New York.

Griffiths, P. R., 1978, Fourier transform infrared spectrometry: Theory and instrumentation, in: *Transform Techniques in Chemistry* (P. R. Griffiths, ed.), pp. 109–139, Plenum Press, New York.

Guckert, J. B., Antworth, C. B., Nichols, P. D., and White, D. C., 1985, Phospholipid, ester-linked fatty acid profiles as reproducible assays for changes in prokaryotic community structure of estuarine sediments, *F.E.M.S. Microb. Ecol.* **31**:147–158.

Hagstrom, A., Larsson, U., Horstedt, P., and Normark, S., 1979, Frequency of dividing cells, a new approach to the determination of bacterial growth rates in aquatic environments, *Appl. Environ. Microbiol.* **37**:805–812.

Harold, F. M., and Altendorf, K., 1974, Cation transport in bacteria: K^+ ion, Na^+ ion and H^+ ion, in: *Current Topics in Membranes and Transport*, Vol. 5, pp. 1–50, Academic Press, New York.

Hedrick, D. B., and White, D. C., 1986, Microbial respiratory quinones in the environment. I. A sensitive liquid chromatographic method, *J. Microbiol. Methods* **5** (in press).

Hollander, R., Wolf, G., and Mannheim, W., 1977, Lipoquinones of some bacteria and mycoplasmas, with considerations on their functional significance, *Antonie van Leeuwenhoek* **43**:177–185.

Holm-Hansen, O., 1973, Determination of total microbial biomass by measurement of adenosine triphosphate, in: *Estuarine Microbial Ecology* (L. H. Stevenson and R. R. Colwell, eds.), pp. 73–89, University of South Carolina Press, Columbia.

Iverson, W. P., and Olson, G. J., 1983, The mechanism of anaerobic (microbial) corrosion, *NBSIR 83-2738*, U. S. Department of Commerce.

Jorgensen, B. B., and Revsbech, N. P., 1983, Colorless sulfur bacteria, *Beggiatoa* spp., in O_2 and H_2S microgradients, *Appl. Environ. Microbiol.* **45**:1261–1270.

Kamio, Y., Kanegasaki, S., and Takahashi, H., 1969, Occurrence of plasmalogens in anaerobic bacteria *J. Gen. Appl. Microbiol.* **15**:439–451.

Karlsson, K. A., 1970, Sphingolipid long chain bases, *Lipids* **5**:878–891.

King, J. D., and White, D. C., 1977, Muramic acid as a measure of microbial biomass in estuarine and marine samples, *Appl. Environ. Microbiol.* **33**:777–783.

King, J. D., White, D. C., and Taylor, C. W , 1977, Use of lipid composition and metabolism to examine structure and activity of estuarine detrital microflora, *Appl. Environ. Microbiol.* **33**:1177–1183.

Knowles, C. J., 1971, Microbial metabolic regulation by adenine nucleotide pools, *Symp. Soc. Gen. Microbiol.* **27**:241–283.

Kroger, A., Dadak, V., Klingenberg, M., and Diemer, F., 1971, On the role of quinones in bacterial electron transport: Differential roles of uniquinone and menaquinone in *Proteus rettgeri, Eur. J. Biochem.* **21**:322–333.

Kushner, D. J., 1978, *Microbial Life in Extreme Environments*, Academic Press, New York.

LaBach, J. P., and White, D. C., 1969, Identification of ceramide phosphorylethanolamine and ceramide phosphorylglycerol in the lipids of an anaerobic bacterium, *J. Lipid Res.* **10**:528–534.

Lambert, P. A., Hancock, I. C., and Baddiley, J., 1975, The interaction of magnesium ions with teichoic acid, *Biochem. J.* **149**:519–524.

Lechinowsky, W. O., Dugan, P. R., Pfister, R. M., Frea, J. I., and Randles, C. I., 1970, Aldrin: Removal from lake water by flocculent bacteria, *Science* **169**:993–995.

Leive, L., 1973, *Bacterial Membranes and Walls*, Marcel Decker, New York.

Luria, S. E., 1960, The bacterial protoplasm: Composition and organization, in: *The Bacteria*, Vol. 1 (I. C. Gunsalus and R. Y. Stanier, eds.), pp. 1–34, Academic Press, New York.

Marshall, K. C., 1964, Survival of root-nodule bacteria in dry soils exposed to high temperatures, *Aust. Agric. Res.* **15**:273–281.

Martz, R. F., Sebacher, D. K., and White, D. C., 1983, Biomass measurements of methane-forming bacteria in environmental samples, *J. Microbiol. Methods* **1**:53–61.

Meyer-Reil, L. A., 1978*a*, Autoradiography and epifluorescence microscopy combined for the determination of number and spectrum of actively metabolizing bacteria in natural waters, *Appl. Environ. Microbiol.* **36**:506–512.

Meyer-Reil, L. A., 1978*b*, Uptake of glucose by bacteria in the sediments, *Mar. Biol.* **44**:293–298.

Minnikin, D. E., Abdolrahimzadeh, H., and Baddiley, J., 1974, Replacement of acidic phospholipids by acidic glycolipids in *Pseudomonas diminuta, Nature (Lond.)* **249**:268–269.

Montagna, P. A., 1982, Sampling design and enumeration statistics for bacteria extracted from marine sediments, *Appl. Environ. Microbiol.* **43**:1366–1372.

Moriarty, D. J. W., 1980, Problems in the measurement of bacterial biomass in sandy sediments, in: *Biogeochemistry of Ancient and Modern Sediments* (P. A. Trudinger, M. R. Walter, and B. J. Ralph, eds.), pp. 131–139, Australian Academy of Science, Canberra.

Moriarty, D. J. W., and Hayward, A. C., 1982, Ultrastructure of bacteria and the proportion of gram-negative bacteria in marine sediments, *Microb. Ecol.* **8**:1–14.

Moriarty, D. J. W., and Pollard, P. C., 1981, DNA synthesis as a measure of bacterial productivity in seagrass sediments, *Mar. Ecol.* **5**:151–156.

Moriarty, D. J. W., White, D. C., and Wassenberg, T. J., 1985, A convenient method for measuring rates of phospholipid synthesis in seawater and sediments: Its relevance to the determination of bacterial productivity and the disturbance artifacts introduced by measurements, *J. Microbiol. Methods* **3**:321–330.

Morita, R. Y., 1982, Starvation-survival of heterotrophs in the marine environment, *Adv. Microbial Ecol.* **6**:171–198.

Morrison, S. J., and White, D. C., 1980, Effects of grazing by estuarine gammaridean amphipods on the microbiota of allochthonous detritus, *Appl. Environ. Microbiol.* **40**:659–671.

Morrison, S. J., King, J. D., Bobbie, R. J., Bechtold, R. E., and White, D. C., 1977, Evidence for microfloral succession on allochthonous plant litter in Apalachicola Bay, Florida, USA, *Marine Biol.* **41**:229–240.

Muller, H. P., and Schmidt, L., 1966, Kontinuierliche Atmungsmessungen an *Azobacter chroococcum* Beij in Montmorillonit unter chizonischer Rontgenbesjrahlung, *Arch. Mikrobiol.* **54**:70–79.

Murray, R. G. E., 1978, Form and function: I: Bacteria, in: *Essays in Microbiology* (J. R. Norris and M. H. Richmond, eds.), pp. 2/1–2/31, Wiley, New York.

Nichols, P. D., Shaw, P. M., and Johns, R. B., 1985*a*, Determination of monoenoic double bond position and geometry in complex microbial and environmental samples by capillary GC/MS, *J. Microbiol. Methods* **3**(5–6):311–319.

Nichols, P. D., Mayberry, W. R., Antworth, C. P., and White, D. C., 1985*b*, Determination of monounsaturated double bond position and geometry in the cellular fatty acids of the pathogenic bacterium *Francisella tularensis, J. Clin. Microbiol.* **21**:738–740.

Nichols, P. D., Henson, J. M., Guckert, J. B., Nivens, D. E., and White, D. C., 1985*c*. Fourier transform-infrared spectroscopic methods for microbial ecology: Analysis of bacteria, bacteria-polymer mixtures and biofilms. *J. Microbiol. Methods* **4**:79–94.

Nickels, J. S., King, J. D., and White, D. C., 1979, Poly-beta-hydroxybutyrate metabolism as a measure of unbalanced growth of the estuarine detrital microbiota, *Appl. Environ. Microbiol.* **37**:459–465.

Nickels, J. S., Bobbie, R. J., Lott, D. F., Martz, R. F., Benson, P. H., and White, D. C., 1981*a*, Effect of manual brush cleaning on the biomass and community structure of the microfouling film formed on aluminun and titanium surfaces exposed to rapidly flowing seawater, *Appl. Environ. Microbiol.* **41**:1442–1453.

Nickels, J. S., Bobbie, R. J., Martz, R. F., Smith, G. A., White, D. C., and Richards, N. L., 1981*b*, Effect of silicate grain shape, structure and location on the biomass and community structure of colonizing marine microbiota, *Appl. Environ. Microbiol.* **41**:1262–1268.

Odham, G., Tunlid, A., Westerdahl, G., Larsson, L., Guckert, J. B., and White, D. C., 1985, Determination of microbial fatty acid profiles at femtomolar levels in human urine and the initial marine microfouling community by capillary gas chromatography-chemical ionization mass spectrometry with negative ion detection, *J. Microbiol. Methods* **3**:331–344.

Oremland, R. S., and Taylor, B. F., 1978, Sulfate reduction and methanogenesis in marine sediments, *Geochim. Cosmochim. Acta* **42**:209–219.

Parker, J. H., Smith, G. A., Fredrickson, H. L., Vestal, J. R., and White, D. C., 1982, Sensitive assay, based on hydroxy-fatty acids from lipopolysaccharide lipid A for gram negative bacteria in sediments, *Appl. Environ. Microbiol.* **44**:1170–1177.

Parker, J. H., Nickels, J. S., Martz, R. F., Gehron, M. J., Richards, N. L., and White, D. C., 1984, Effect of oil and well-drilling fluids on the physiological status and microbial infection of the reef building coral *Montastrea annularis.*, *Arch. Environ. Contam. Toxicol.* **13**:113–118.

Parkes, R. J., and Taylor, J., 1983, The relationship between fatty acid distributions and bacterial respiratory types in comtemporary marine sediments, *Estuarine Coastal Shelf Sci.* **16**:173–189.

Patel, J. J., and Gerson, T., 1974, Formation and utilization of carbon reserves by *Rhizobium,* *Arch. Microbiol.* **101**:211–220.

Poland, R. L., and Odell, G. B., 1974, The binding of bilirubin to agar *Proc. Soc. Exp. Biol. Med.* **146**:1114–1118.

Porter, K. G., and Feig, Y. S., 1980, The use of DAPI for identifying and counting aquatic microflora, *Limnol. Oceanogr.* **25**:943–948.

Purcell, E. M., 1977, Life at low Reynolds number, *Am. J. Phys.* **45**:3–11.

Revsbech, N. P., and Ward, D. M., 1983, Oxygen microelectrode that is sensitive to medium chemical composition: Use in an acid microbial mat dominated by *Cyanidium caldarium,* *Appl. Environ. Microbiol.* **45**:755–759.

Revsbech, N. P., Jorgensen, B. B., and Balckburn, T. H., 1980, Oxygen in the sea bottom measures with a microelectrode, *Science* **207**:1355–1356.

Ridgeway, H. F., Means, E. G., and Olson, B. H., 1981, Iron bacteria in drinking-water distribution systems: Elemental analysis of *Gallionella* stalks, using X-ray energy-dispersive microanalysis, *Appl. Environ. Microbiol.* **41**:288–297.

Rizza, V., Tucker, A. N., and White, D. C., 1970, Lipids of *Bacteroides melaninogenicus, J. Bacteriol.* **101**:84–91.

Rublee, P., and Dornseif, B. E., 1978, Direct counts of bacteria in the sediments of a North Carolina salt marsh, *Estuaries* **1**:188–191.

Sargent, J. R., Lee, R. F., and Nevenzel, J. C., 1976, Marine waxes, in: *Chemistry and Biochemistry of Natural Waxes* (P. E. Kolattukudy, ed.), pp. 49–91, Elsevier, New York.

Sieburth, J. McN., 1975, *Microbial Seascapes—a Pictorial Essay on Marine Microorganisms and Their Environments,* University Park Press, Baltimore.

Smidsrop, O., and Haug, A., 1968, Dependence upon uronic acid composition of some ion exchange properties of alginates, *Acta Chem. Scand.* **22**:1989–1997.

Smith, G. A., Nickels, J. S., Davis, W. M., Martz, R. F., Findlay, R. H., and White, D. C., 1982*a*, Perturbations of the biomass, metabolic activity, and community structure of the estuarine detrital microbiota: Resource partitioning by amphipod grazing, *J. Exp. Mar. Biol. Ecol.* **64**:125–143.

Smith, G. A., Nickels, J. S., Bobbie, R. J., Richards, N. L., and White, D. C., 1982*b*, Effects of oil and gas well drilling fluids on the biomass and community structure of the microbiota that colonize sands in running seawater, *Arch. Environ. Contam. Toxicol.* **11**:17–23.

Smith, G. A., Nickels, J. S., Davis, J. D., Findlay, R. H., Vashio, P. S., Wilson, J. T., and White, D. C., 1985, Indices identifying subsurface microbial communities that are adapted to organic pollution, in: *Second International Conference On Ground Water Quality Research* (N. N. Durham and A. E. Redelfs, eds.), pp. 210–213, Oklahoma State University Printing Services, Stillwater.

Smith, R. L., and Klug, M. J., 1981, Reduction of sulfur compounds in the sediments of a eutrophic lake basin, *Appl. Environ. Microbiol.* **41:**1230–1237.

Sorensen, J., Jorgensen, B. B., and Revsbech, N. P., 1979, A comparison of oxygen, nitrate and sulfate respiration in coastal marine sediments, *Microb. Ecol.* **5:**105–115.

Stanley, P. M., and Staley, J. J., 1977, Acetate uptake by aquatic bacterial communities measured by autoradiography and filterable radioactivity, *Limnol. Oceanogr.* **22:**26–37.

Sutherland, I. W., 1972, Bacterial exopolysaccharides, *Adv. Microbiol. Physiol.* **8:**143–214.

Sutherland, I. W., 1977, Bacterial exopolysaccharides—Their nature and production, in: *Surface Carbohydrates of the Prokaryotic Cell* (I. W. Sutherland, ed.), pp. 27–96, Academic Press, New York.

Tempest, D. W., Dicks, J. W., and Ellwood, D. C., 1968, Influence of growth condition on the concentration of potassium in *Bacillus subtilis* var. *Niger* and its possible relationship to cellular ribonucleic acid, teichoic acid and teichuronic acid, *Biochem. J.* **106:**237–243.

Tornabene, T. G., and Edwards, H. W., 1972, Microbial uptake of lead, *Science* **176:**1334.

Tunlid, A., Baird, B. H., Trexler, M. B., Olsson, S., Findlay, R. H., Odham, G., and White, D. C., 1986, Determination of phospholipid ester-linked fatty acids and polybetahydroxybutyrate for the estimation of bacterial biomass and activity in the rhizosphere of the rape plant *Brassica napus* (L.). *Can. J. Microbiol.* **31:**1113–1119.

Tunlid, A., Odham, G., Findlay, R. H., and White, D. C., 1985, Precision and sensitivity in the measurement of ^{15}N enrichment in D-alanine from bacterial cell walls using positive/negative ion mass spectrometry, *J. Microbiol. Methods* **3:**237–246.

Ulhinger, D. J., and White, D. C., 1983, Relationship between the physiological status and the formation of extracellular polysaccharide glycocalyx in *Pseudomonas atlantica*, *Appl. Environ. Microbiol.* **45:**64–70.

Vance, D. E., Mitsuhashi, O., and Bloch, K., 1973, Purification and properties of the fatty acid synthetase from *Mycobacterium phlei*, *J. Bacteriol.* **248:**2303–2309.

Van Oss, C. J., and Gillman, C. F., 1972, Phagocytosis as a surface phenomenon: I. Contact angles and phagocytosis of non-opsonized bacteria, *Res. J. Reticuloendothel. Soc.* **12:**283–292.

Walker, J. C. G., 1977, *Evolution of the Atmosphere*, Macmillan, New York.

Ward, J. B., 1981, Teichoic and teichuronic acids, biosynthesis, assembly and location, *Microbiol. Rev.* **45:**211–243.

Whistance, G. R., Dillon, J. F., and Threlfall, D. R., 1969, The nature, intergeneric distribution and biosynthesis of iosprenoid quinones and phenols in gram-negative bacteria, *Biochem. J.* **111:**461–472.

White, D. C., 1982, *Microbial faciliation of corrosion, Corrosion 82*, pp. 1–16, National Association of Corrosion Engineers, Houston, Texas.

White, D. C., 1983, Analysis of microorganisms in terms of quantity and activity in natural environments, in: *Microbes in Their Natural Environments* (J. H. Slater, R. Whittenbury, and J. W. T. Wimpenny, eds.), *Soc. Gen. Microbiol. Symp.* **34:**37–66.

White, D. C., and Benson, P. H., 1984, Determination of the biomass, physiological status, community structure and extracellular plaque of the microfouling film, in: *Marine Biodeterioration: An interdisciplinary study* (J. D. Costlow and R. C. Tipper, eds.), pp. 68–74, U.S. Naval Institute Press, Annapolis, Maryland.

White, D. C., and Tucker, A. N., 1969, Phospholipid metabolism during bacterial growth, *J. Lipid Res.* **10:**220–233.

White, D. C., Bobbie, R. J., Herron, J. S., King, J. D., and Morrison, S. J., 1979a, Biochemical measurements of microbial mass and activity from environmental samples, in: *Native Aquatic Bacteria: Enumeration, Activity and Ecology* (J. W. Costerton and R. R. Colwell, eds.), ASTM STP 695, pp. 69–81, American Society for Testing and Materials, Philadelphia.

White, D. C., Bobbie, R. J., King, J. D., Nickels, J. S., and Amoe, P., 1979b, Lipid analysis of sediments for microbial biomass and community structure, in: *Methodology for Biomass Determinations and Microbial Activities in Sediments* (C. D. Litchfield and P. L. Seyfried, eds.), ASTM STP 673, pp. 87–103, American Society for Testing and Materials, Philadelphia.

White, D. C., Davis, W. M., Nickels, J. S., King, J. D., and Bobbie, R. J., 1979c, Determination of the sedimentary microbial biomass by extractible lipid phosphate, *Oecologia* **40:**51–62.

White, D. C., Bobbie, R. J., Nickels, J. S., Fazio, S. D., and Davis, W. M., 1980, Nonselective biochemical methods for the determination of fungal mass and community structure in estuarine detrital microflora, *Bot. Mar.* **23:**239–250.

White, D. C., Smith, G. A., Gehron, M. J., Parker, J. H., Findlay, R. H., Martz, R. F., and Fredrickson, H. L., 1983, The ground water aquifer microbiota: biomass, community structure and nutritional status, *Dev. Indust. Microbiol.* **24:**201–211.

White, D. C., Wilson, J. T., McNabb, J. F., Ghiorse, W. G., and Balkwill, D. L., 1984, Ground water sediment microbiota—an unusual consortia, *Proceedings of the Sixth International Environmental Biogeochemistry, Sante Fe, New Mexico*, pp. 1–12.

White, D. C., Nivens, D. E., Nichols, P. D., Kerger, B. D., Henson, J. M., Geesey, G. G., and Clarke, C. K., 1985, Corrosion of steels induced by aerobic bacteria and their exocellular polymers, *Proceedings of the International Conference on Biology-Induced Corrosion, NACE, Washington, D. C., June 10–12.*

Whitfield, M., 1969, Eh as an operational parameter in estuarine studies, *Limnol. Oceanogr.* **14:**547–558.

Wilkinson, J. F., 1958, The extracellular polysaccharides of bacteria, *Bacteriol. Rev.* **22:**46–73.

Wilkinson, S. G., and Caldwell, P. F., 1980, Lipid composition and chemotaxonomy of *Pseudomonas putrefaciens (Alteromonas putrefaciens)*, *J. Gen. Microbiol.* **118:**329–341.

Winfrey, M. R., and Zeikus, J. G., 1977, Effect of sulfate on carbon and electron flow during microbial methanogenesis in freshwater sediments, *Appl. Environ. Microbiol.* **33:**275–281.

Wolin, M. J., 1979, The rumen fermentation: A model for microbial interactions in anaerobic ecosystems, *Adv. Microbial. Ecol.* **3:**49–77.

Wright, J., and Heckels, J. E., 1975, The teichuronic acid of cell walls of *Bacillus subtilis* WZ3 grown in a chemostat under phosphate limitation, *Biochem. J.* **147:**187–189.

Zeikus, J. G., 1977, The biology of methanogenic bacteria, *Bacteriol. Rev.* **41:**514–541.

Zimmerman, R., and Meyer-Reil, L. A., 1974, A new method for the fluorescent staining of bacterial populations on membrane filters, *Kiel. Meersforsch* **30:**24–27.

5

GNOTOBIOTIC AND GERMFREE ANIMAL SYSTEMS

Rolf Freter

INTRODUCTION

After the early studies of various pioneers during the first half of this century (reviewed by Reyniers, 1959), it was only in the late 1950s that research with germfree animals was made accessible to large numbers of investigators with the development by Trexler (1959) of a reliable and relatively inexpensive flexible vinyl isolator. During the next two decades, this resulted in a phenomenal outpouring of new and exciting data based on the observations of differences between germfree and conventional animals. It soon became obvious that the germfree (GF) animal does not at all resemble its conventional (CV) counterpart minus the microbes. Indeed, probably few physiological and immunological systems are not affected in some way by the GF status. It seems reasonable to assume that much of this conceptually straightforward comparative type of work has already been accomplished; one may also expect that it will be continued profitably in a number of areas such as carcinogenesis (Pollard, 1981; Reddy, 1981) and nutrition (Wostmann, 1981), i.e., in areas in which progress has to await further advances in those specialties rather than in gnotobiotics.

In contrast to this initial type of approach, investigators soon began to ask more precise questions. For example, it appeared important to know which of the several hundred bacterial species that populate the mammalian body is responsible for a given effect of the indigenous microflora, i.e., for a feature, such as the small cecum or resistance to colonization by other bacteria, that is found in the CV but lacking in the GF animal. This problem was and still is approached most often by one of the following two experimental designs: (1) bacterial strains are isolated from the indigenous flora that have certain properties of interest (e.g., the ability to inhibit the growth of pathogens or to synthesize carcinogens *in vitro;* or, alternatively, (2) germfree animals are

Rolf Freter • Department of Microbiology and Immunology, The University of Michigan Medical School, Ann Arbor, Michigan 48109.

associated with one or a limited number of known bacterial strains, and the resulting effects on the gnotobiotic animal and/or on the individual bacterial populations that colonize the formerly germfree host are observed. While data based on the comparison of CV with GF animals can usually be interpreted in a relatively uncomplicated manner, extrapolations from the latter experimental approaches can be made only with considerable difficulty. For this reason, germfree and gnotobiotic animals are discussed in different sections of this chapter.

DEFINITIONS

Germfree is the term most often used in the English literature to describe animals not associated with microbes (within the technical limitations to be discussed later) and maintained in a sterile environment. This is contrasted with the free-living conventional animal, which harbors a complete undisturbed indigenous microflora. GF animals may be associated with known strains of microorganisms introduced into the isolator and subsequently colonizing the animals. The term *contaminated,* sometimes used synonymously with *associated,* usually refers to unintentional association brought about by faulty technique. Animals may be described as *monoassociated, diassociated,* or *polyassociated,* depending on the number of microbial strains with which they have been brought into contact. Animals so associated are usually referred to as gnotobiotic animals, or gnotobiotes, from the Greek word *gnotos,* meaning known. The bacteria colonizing a gnotobiotic animal may be described accordingly as monoassociates, diassociates, and so forth. Most authorities consider the germfree state a special case of the gnotobiotic. GF animals may be associated with a complete (and undefined) indigenous microflora, e.g., by caging with CV animals. Such animals, useful as experimental controls, are often described as *conventionalized,* a term that is more precise than *ex-germfree.*

A parallel terminology was introduced by Baker and Ferguson (1942), who coined the term *axenic* (from the Greek privative prefix *a,* without, and *xenos,* denoting a stranger) because they considered it more correct than the adjective germfree, with which it is now used synonymously. Raibaud *et al.* (1966) introduced this Greek root into the French literature and coined related terms such as *gnotoxenique* and *holoxenique,* which are synonymous with gnotobiotic and conventional, respectively. Anglicized versions of these French terms are also found in the English literature. Additional discussions of terminology may be found in Luckey (1963).

GERMFREE ANIMALS

Techniques and Tests

Much has been written about the problem that arises as a consequence of the fact that the definition of germfree is a negative one, which therefore makes it impossible to demonstrate experimentally with absolute certainty that a GF condition has indeed been achieved in any given animal colony (e.g., Wagner, 1959a; Luckey, 1963; Fuller, 1968; Gordon and Pesti, 1972; Coates and Fuller, 1977). These investigators also reviewed and recommended specific procedures for sterility testing.

The problem is less serious when only the absence of bacteria and higher forms is to be established. More extensive tests are necessary in animal colonies that are newly established with offspring recovered by cesarean section from conventional gravid females. In such instances, fastidious anaerobic bacteria and even worm infestations have been reported, which were detected only with difficulty and after considerable periods of time (Wagner, 1959a); it is likely that such contaminations resulted from infection of the fetus *in utero*. As pointed out by Luckey (1963), among others, exogenous contamination from the environment through faulty technique is less likely to involve fastidious microorganisms. Consequently, simplified tests for sterility are usually employed after more extensive initial studies have revealed no contamination of a given newly established colony. Microscopic observations of feces or intestinal contents are a simple and especially useful adjunct to bacteriological cultures. Autoclaved standard diets contain numerous (dead) bacterial cells that can be seen microscopically in fecal smears, but these can be distinguished without difficulty from the large numbers of bacteria that appear in smears from animals that harbor one of the common exogenous bacterial contaminants.

The problem of testing for possible viral contamination of germfree animals is considerably more difficult. Early attempts at demonstrating viruses in germfree animals were negative. However, the report of Pollard and Matsuzawa (1964) of induction of leukemia in X-irradiated GF mice and that of Ashe *et al.* (1965) of viruses in salivary glands of GF rats suggested that vertically transmitted occult viruses may indeed be present in some GF colonies. Many strains of GF mice harboring congenitally transmitted viruses are known, but no clearly defined virus has been isolated from GF rats (Pollard, 1981). The difficulties inherent in detecting the possible presence of unknown viruses have been discussed in some detail by Wagner (1959a). It is apparent that there is no method that can rule out conclusively the presence of occult viruses in GF animals that show no clinical or serological abnormalities.

One must conclude that it remains the responsibility of the individual investigator to decide what type of possible contaminant would interfere with the interpretation of a given experiment and to carry out appropriate sterility tests. For example, the presence of occult virus in the GF colony is not likely to interfere with experiments involving the interaction of bacteria in the gut. In this instance, simple testing for the absence of bacteria in the gut of the GF animals may well suffice. By contrast, the possible presence of congenitally transmitted viruses is a constant source of concern to experimental oncologists (Pollard, 1981).

The most commonly used isolation methods—plastic isolators, air filtered through glass wool, and sterilization by peracetic acid—have proved reliable in preventing bacterial contamination from exogenous sources; isolators will remain sterile for several years when maintained properly. The question whether viruses can penetrate glass wool filters has not been studied as extensively as one might desire, but the studies that have been done showed retention of viruses by glass wool (reviewed by Sacquet, 1968). This writer is unaware of any reports that viruses from the external environment have penetrated through glass wool filters into GF isolators. This excellent record must be viewed with some reservation, however, because relatively few laboratories would have detected such contamination. This problem can be avoided using incineration as the method for sterilization of incoming air (Gustafsson, 1959; Miyakawa, 1959). Methods for isolation, production, and maintenance of GF animals have frequently been reviewed (e.g., Luckey, 1963; Coates *et al.*, 1968; National Academy of Sciences, 1970; Pleasants, 1974; Trexler, 1982; Yale, 1982; Lattuada *et al.*, 1982; Vieira *et al.*, 1982). Among these, the article by Pleasants gives the most explicit technical details.

The antibiotic-treated animal has been used in the past as a simple substitute for the strictly gnotobiotic animal, especially in short-term experiments concerned with bacterial competition in the gut. Streptomycin was used originally for this purpose (Freter, 1954; Bohnhoff *et al.*, 1954, 1964) because it is not absorbed from the intestine. This drug causes the animal to acquire GF-like characteristics within 24 hr or less: an enlarged, liquid-filled cecum, elevated pH and a striking increase in the oxidation-reduction potential of the cecum. Gram-stained smears of cecal contents from mice that had been treated with streptomycin 24 hr earlier show none of the normally numerous fusiform bacteria; only scanty gram-positive bacteria and occasional yeasts remain (Meynell, 1963; Freter, unpublished observations). This ability of streptomycin to eliminate most of the indigenous microflora is somewhat surprising because this drug is known to be relatively ineffective *in vitro* against anaerobic bacteria (e.g., Wilkins and Appleman, 1976). One may speculate that streptomycin inhibits a sufficient number of facultatively anaerobic bacteria that function to keep the contents of the large intestine in a reduced state and that the resulting rise in oxidation-reduction potential causes the elimination of the strict anaerobes. Many other antibiotics also bring about GF-like characteristics when administered to CV animals (Gustafsson and Norin, 1977).

When used as a substitute for gnotobiotes, antibiotic-treated animals are usually infected with known antibiotic-resistant bacteria, and administration of the drug is continued throughout the experiment, usually via the drinking water (e.g., Freter, 1962). Alternatively, the antibiotic is administered once; the animals are challenged 24 hr later with bacteria that are not necessarily antibiotic resistant. In these circumstances, challenge must be done at a time when most of the antibiotic has already been excreted from the gut, but the indigenous flora is still repressed (Bohnhoff *et al.*, 1964). If, as is usually the case, the challenge inoculum vastly outnumbers the remnants of the indigenous flora, antibiotic-treated animals may be regarded as equivalent to true gnotobiotes for the purposes of many short-term experiments (i.e., those lasting for a few days). This technique cannot be recommended for experiments of longer duration because the possible overgrowth of drug-resistant indigenous or environmental bacteria and yeasts in the antibiotic-treated animals cannot be controlled.

The Function of a Complete Indigenous Microflora: Comparison of GF and CV Animals

Most of our current knowledge of the effects of the indigenous microflora on the host come from comparative studies of GF and CV animals, and most of the literature on gnotobiotics is therefore concerned with data obtained in this manner. The various aspects of the subject have frequently been reviewed (Coates, 1968; Coates and Fuller, 1977; Coates *et al.*, 1968; Fliedner *et al.*, 1979; Gordon and Pesti, 1972; Heneghan, 1973; Luckey, 1963, 1965; Mirand and Back, 1968; Miyakawa and Luckey, 1968; Pleasants, 1974; Reyniers, 1959; Sasaki *et al.*, 1982; Wostmann, 1981). For this reason, only a brief summary of the more important differences between CV and GF animals is given here, in order to allow the reader to appreciate the fact that GF animals must be regarded as a class apart from their CV counterparts.

Cecal enlargement and increased hydration of the contents of cecum and colon are probably the most striking abnormalities in GF rodents. Other animals (e.g., dog, pig, chicken, and probably man) do not show similarly enlarged intestinal structures but still show increased hydration of large intestinal contents (Gordon, 1981). The major cause of this defect in water absorption from the GF large intestine appears to be the accumulation of endogenous "mucus," a mixture consisting of various glycoproteins that include hexuronic acid, hexosamine, and sialic acid (Wostmann *et al.*, 1973; Gordon and Wostmann, 1973; Gordon, 1981). In the absence of a large intestinal flora that normally degrades this material (Miller and Hoskins, 1981), the colloid pressure rises, resulting in water attraction. Moreover, the negatively charged nonabsorbable macromolecules also reduce the concentration of diffusible anions (Cl^-, HCO_3^-) in the large intestine. These diffusible anions are required for the functioning of Na^+-dependent water transport out of the cecum. Consequently, in spite of the consistent finding that the Na^+ concentration in the GF cecum is only slightly depressed, net water transport from this organ is inhibited or even negative, resulting in net tissue to lumen flux (Wostmann *et al.*, 1973; Gordon and Wostmann, 1973; Gordon, 1981).

Concomitant with these changes in intestinal contents is an increase in the concentration of a number of enzymes (reviewed in Kawai and Morotomi, 1978). The cause of this increase is not fully understood. It has been attributed to (1) the slower peristaltic movement of the GF gut and the consequent slower elimination of intestinal contents, and to (2) the slower turnover of intestinal epithelial cells, resulting in the presence in GF animals of more mature cells with a more active synthesis of certain enzymes (Wostmann and Bruckner-

Kardoss, 1981; Savage *et al.*, 1981). The higher concentration of proteolytic enzymes in the GF animal leads to more rapid destruction of intestinal immunoglobulins (Fubara and Freter, 1972).

In addition to the above-mentioned features, cholesterol and bile acid metabolism is drastically affected by the indigenous microflora. In the GF state, the primary conjugated bile acids remain intact, and more bile acids are resorbed from the gut. By contrast, in CV animals bile acids are deconjugated, dehydroxylated, and converted in various other ways to secondary bile acids. Since the modified bile acids are usually less efficiently reutilized, relatively more bile acids are excreted in the CV. This, in turn, results in a higher rate of catabolism of cholesterol from which bile acids are derived (reviewed in Wostmann, 1981; Hill, 1980). Modified bile acids may have inhibitory effects on certain intestinal bacteria (reviewed by Savage, 1977). Similarly, various metabolites of cholesterol and of amino acids, some of which are carcinogenic, are found in the gut of the CV animal (Hill, 1980). These, too, are absent in the GF animal.

As reviewed by Kim (1981), the immunological state of GF animals is characterized by various deficiencies that can be found, for example, in the .sparse development of the lymphoid systems, in reduced levels of most immunoglobulins (Nielsen and Friis, 1980), and in the relatively low levels or absence of normal antibodies (Wagner, 1959*b*). GF mice fed a chemically defined diet of minimal antigenicity have barely detectable circulating IgG levels (Pleasants *et al.*, 1982). GF animals show no wasting disease after neonatal thymectomy, and some delayed-type hypersensitivity reactions are absent (MacDonald *et al.*, 1979).

These immunological deficiencies certainly contribute to the exquisite sensitivity of GF animals to colonization and infection by bacteria with which they come in contact. Most important is the fact that microorganisms that come into contact with a CV animal are prevented from colonizing it by the antagonistic action of the indigenous microflora. This phenomenon has been discovered and rediscovered several times, and terms such as "bacterial antagonism" (Freter, 1956), "bacterial interference" (Dubos, 1963; Aly and Shinefield, 1982), and "colonization resistance" (van der Waaij *et al.*, 1971) have been applied to it by various investigators. This effect was demonstrated initially by the finding that the administration of streptomycin permitted the implantation of antibiotic-resistant bacteria into the gut of mice (Freter, 1954; Bohnhoff *et al.*, 1954). Several earlier workers (Meynell, 1963; Bohnhoff *et al.*, 1964; Hentges, 1970) had presented evidence that the strict anaerobes were responsible for the control of "invading" bacteria in the CV animal. Subsequent work (Freter and Abrams, 1972) showed that the mechanisms of control are more complex, and that enterobacteria such as *Escherichia coli* can participate in such control to some extent, in spite of the fact that these antagonistic *E. coli* are themselves suppressed by other members of the indigenous microbiota. The possible mechanisms by which populations of the indigenous flora control each other's multiplication and inhibit the implantation of "invaders" have been reviewed recently (Freter, 1983).

The review by Wostmann (1981) is the most recent account of the effects of a complete indigenous microflora on animal nutrition. The microflora synthesizes certain vitamins but has a growth-depressing effect on young animals, which commercial breeders counteract by the administration of subtherapeutic doses of antibiotics in the feed. The differences between GF and CV animals with respect to bile acid and cholesterol metabolism and the effect of an antigen-free diet on immune mechanisms have been noted. In addition, GF rats and mice have smaller hearts and smaller cardiac output and use less oxygen per unit of body weight (Wostmann, 1981).

It is obvious that comparisons between GF and CV animals have been a major source of our understanding of the interactions between an animal host and its indigenous microflora. Work along these lines is likely to continue in the future, as the various aspects of the subject will be investigated in increasing detail. The interpretation of data obtained by this method is usually straightforward, but a few points should be mentioned that require attention.

Although the CV animal has been glibly held up as the standard to which GF and gnotobiotic animals are compared, such comparative studies are only as valid as this standard of comparison. Since diet and environmental factors affect many physiological and microbiological parameters, it has been common practice to house the CV control animals in the same type of isolator as the GF controls and to feed both groups the same sterilized food and water. Animals in both groups should be of the same genetic stock. This may pose problems of genetic drift if a long time has passed since the GF colony was derived from the CV stock. This problem may be avoided by using conventionalized rather than CV animals as controls. GF animals may be conventionalized by simply caging them with CV animals of the same species. In these circumstances, CV characteristics and flora develop in the ex-germfree animals within a period of time ranging from a few hours to several days (Luckey, 1963; Khoury et al., 1969). The donor animal of the CV flora does not have to be of the same genetic stock as the GF recipients, but the donors obviously must carry a CV indigenous microflora. Obtaining suitable donors for this purpose poses serious difficulties, at least in the United States because CV animals of certain species (e.g, mice and rats) are not available commercially. To the knowledge of this writer, all commercially bred rats and mice are of the specific pathogen-free (SPF) variety. Such animals are derived from GF stock but are conventionalized in a way that results in their intestinal flora being incomplete and especially lacking the strictly anaerobic bacteria that predominate in the intestine of CV animals (Freter et al., 1983d). Consequently, SPF animals are unsuitable for most types of experiments likely to be undertaken by microbial ecologists. The only sources of authentic CV rat or mouse indigenous microflora in the United States appear to be small private animal colonies that did not originate from GF stock, and animals caught in the wild.

Observance of the above precautions is probably sufficient for many comparative studies of GF and CV animals. Those workers interested in microbial interactions among the indigenous microflora face an additional problem, namely, the variability of the flora among individual hosts (Finegold *et al.*, 1974; Holdeman *et al.*, 1976), which can obscure possible differences among host populations. In attempts to define "normal" population sizes of enterobacteria, the present author found that population levels of *E. coli* in the CV mouse colony varied from month to month (Freter and Abrams, 1972). These changes were uniform among all animals tested; i.e., at one period in time all animals tested had no detectable *E. coli*, whereas at another sampling period the *E. coli* population varied from 3×10^5 to 2.1×10^7 per cecum in individual animals. These changes in *E. coli* population presumably reflected fluctuations in the composition of the predominant anaerobic flora, which is the prime factor controlling *E. coli* populations in the large intestine. This variability in the indigenous flora of the CV controls made it impossible to determine whether the objective of our study had been obtained, namely, to identify a collection of pure cultures isolated from the gut of CV mice, which, when introduced into GF mice, were able to reverse all GF abnormalities studied. The best one can say is that the indigenous microflora assumes various "states" with respect to its composition and/or physiological effects and that the collection of strains under study was perhaps able to mimic one of the states that "is sometimes found in CV individuals" (Freter and Abrams, 1972). It is not known whether other (e.g., physiological, immunological) parameters are also affected by such fluctuations in the composition of the indigenous microflora, but it is obvious that the concept of a definable CV microflora becomes increasingly difficult to realize as one's inquiry delves into increasingly finer details. This problem is one reason why a defined microflora that can reverse all GF abnormalities when introduced into GF mice has never been assembled from pure cultures of indigenous bacteria.

Comparisons between CV and GF animals of metabolic activities on the basis of total body weight must be corrected for the large amounts of fluid that accumulate in the large intestine. For example, Levenson *et al.* (1969) found considerably decreased oxygen consumption in GF rats on a total weight basis. After correction for gut weight, these differences became much less impressive (although remaining statistically significant). Likewise, when cecectomy of GF rats is found to reverse the abnormally low rate of oxygen consumption (e.g., Wostmann *et al.*, 1968), one must make sure that this reversal is not simply a consequence of a more realistic determination of body weight in the cecectomized GF animals. Similar considerations apply when comparing rates of weight gain in young GF and CV animals.

The above are only the most obvious examples of the interrelationships among the various primary and secondary effects that occur in GF animals. For example, the inability of the GF animal to degrade endogenous mucopolysaccharides leads to disturbances in water and electrolyte balances. The resulting enlarged cecum apparently is in some way responsible for certain physiologic effects, such as lower cardiac output and reduced oxygen consumption. The absence of bacteria or their products is probably not directly responsible for the latter effects because these abnormalities disappear upon cecectomy of GF animals (reviewed by Gordon, 1981). As pointed out by Wostmann (1981), "the animal without its conventional microflora has become a changed animal, in which different homeostatic conditions prevail. Only by taking these differences into account can we make intelligent use of an animal model that makes it possible to control virtually every conceivable external parameter." Such "intelligent use" has not been overly difficult in studies comparing GF with CV animals. As discussed in the next section, the complexities introduced by such primary and secondary changes are a major problem in the design and interpretation of experiments involving gnotobiotic animals.

GNOTOBIOTIC ANIMALS

In a sense, it may be appropriate to regard the development of GF technologies as one of the most fruitful technical advances in microbial ecology, on a par perhaps with computer modeling. Our understanding of the interactions between an animal host and its indigenous flora has progressed in quantum leaps since GF techniques became generally available. It must be emphasized, however, that all studies reviewed above were concerned with the entire complex indigenous microflora, i.e., the flora was treated like the proverbial "black box," and little attention was paid to the contents of the box, i.e., to the various individual microbial populations that are the components of the indigenous microflora. Microbial ecologists have asked more precise questions, of which three general types are discussed in this chapter:

1. What are the interactions between individual bacterial species (especially pathogens) and the host?
2. Which components (genera, species, strains) of the indigenous microflora are responsible for its various functions (e.g., reducing cecal size)?
3. What are the mechanisms which determine the composition of the indigenous microflora in various areas of the body, and which determine its protective effect against colonization by invading bacteria?

Gnotobiotic animals would appear to be natural tools for these types of studies. This is indeed true in many instances; however, the interpretation of data obtained with gnotobiotic animals is often quite difficult.

Etiology and Pathogenic Mechanisms (Models of Disease)

Robert Koch was the first to attempt a rigorous definition of an animal model of a human infectious disease. In his famous postulates, he asked for the reproduction of a human disease in experimental animals as proof of the pathogenicity of a given bacterial species. Unfortunately, such a requirement can rarely be fulfilled, and if one chooses to be rigorous in the definition of what constitutes the reproduction of a human disease, one may well state that no such model has ever been discovered. Most human pathogens simply do not produce a condition in animals that resembles the human disease in all major aspects. This problem does not exist for veterinary microbiologists, who can usually work with the natural host species. The latter workers are therefore in an excellent position to compare the interaction of bacterial pathogens with gnotobiotic animals and with the CV counterparts. Such studies confirm what one would suspect intuitively: The gnotobiotic animal is a very different creature. For example, the K88 adhesin is an essential virulence factor for enteropathogenic *E. coli* in CV piglets, the natural host. By contrast, the K88 adhesin is not required by enteropathogenic *E. coli* to cause diarrhea in monoassociated gnotobiotic piglets (Jones and Rutter, 1972). One must assume that the slower intestinal motility and more rapid growth of the monoassociate (in the absence of competition from an indigenous microflora) obviate the need for this pathogen to adhere to the intestinal wall, whereas colonization of the CV host would be impossible without such adhesive ability.

It is well known that most bacteria, even those that are not constituents of the CV indigenous microflora, can rapidly colonize gnotobiotic animals. Because of the large bacterial population sizes that develop in monoassociated animals, and because of the undeveloped immune status of these hosts, microorganisms that are entirely innocuous or even unable to colonize CV animals may cause systemic infection and death as monoassociates (reviewed in Luckey, 1965). The converse may also occur. Some strict anaerobes of the intestinal flora are unable to colonize as monoassociates but will do so as di- or triassociates in animals previously associated with other bacterial species (Gibbons *et al.*, 1964; Syed *et al.*, 1970). Likewise, some pathogens such as *Entamoeba histolytica* or *Treponema hyodysenteriae* will not cause disease as monoassociates, whereas they are pathogenic in CV hosts of the same species (Phillips and Wolfe, 1959; Whipp *et al.*, 1979).

In view of the difficulties in relating the interaction of a given bacterium with a monoassociated host animal to that in a CV host of the same species, it is even more problematic to draw conclusions concerning the possible pathogenicity for man of a given bacterium from the finding that such a bacterium causes disease in a monoassociated gnotobiotic animal. Gorbach and Onderdonk (1979), for example, discussed in an editorial an experimental paper in which the authors had shown that certain strains of *E. coli* that had been isolated from patients with tropical sprue caused a diarrheal disease in monoassociated gnotobiotic rats. Gorbach and Onderdonk noted that there was no evidence that the gnotobiotic rat was a valid model for human tropical sprue. It seemed just as likely that the condition of the monoassociated rats had no counterpart in nature. They concluded correctly that such a finding does not permit any judgments as to the pathogenicity of this bacterium for man, and that the most one could say was that the *E. coli* strain studied was apparently capable of producing an enterotoxin *in vivo*, which, however, may or may not be synthesized by this bacterium in man, and which may or may not be toxic for man.

Strict insistence on the validity of animal models, although correct within the proper context, is sometimes carried to extremes and may then lead to an unnecessarily defeatist attitude concerning the usefulness of animals in microbiological research. The extreme position along this line of reasoning has led some to the profound conclusion that "man is not a mouse," implying that research based on animal models of human infections is for this reason of questionable value. Although such opinions can still be heard today, one must nevertheless realize that little progress in the understanding of human infections could be made without the use of laboratory animals, both of the CV and of gnotobiotic kinds.

The obvious means of resolving the above-stated extreme positions lies in the proper critical interpretations and perspectives that must be applied to the data obtained with models of infection. For good reason, those interested in public health, etiology or vaccine development have tended to adopt the legacy of Robert Koch, and insisted on more or less complete reproduction of a human disease in an experimental animal in order for this model to be considered valid. On the other hand, we have come to realize during the past decades that the pathogenesis of any one infectious disease represents a long sequence of separate interactions, and that a given animal or *in vitro* model can at best represent only a part of the whole process. For example, the intraperitoneal infection of mice with pneumococci does not closely resemble human pneumonia, yet valuable lessons concerning the interactions of these bacteria with leukocytes can be learned from this model and can be applied to the understanding of similar processes occurring in man. The value of this model depends entirely on our ability to identify the parallels in, say, the functions of the bacterial capsule, or of anticapsular antibodies as they occur in the peritoneal cavity of the mouse and in the lungs of man. Likewise, the finding by Jones and Rutter (1972) that enteropathogenic *E. coli* do not require adhesive mechanisms in order to cause disease in gnotobiotic piglets despite the ability to adhere as a requisite for virulence in the CV animal allowed important general conclusions to be drawn as to the role of bacterial adhesion in pathogenicity. This work was the first demonstration of the fact that adhesion is not *sine qua non* for pathogens that colonize body surfaces but that it gives bacteria an advantage only under certain specific competitive conditions. This insight could be gained in spite of (or, rather, because of) the fact that the monoassociated piglet is very different from its CV counterpart and is therefore not a valid model for diseases of the CV pig.

The early work of Wagner (1959*b*) may be cited as another example. He showed that monoassociation of germfree mice with bacterial species of the mouse indigenous microflora restored the production of "normal" antibodies to the monoassociate. Such normal antibodies are present in the CV but missing in the GF mouse. This study thereby provided an answer to the fundamental question of the origin of normal antibodies. This was in spite of the fact that the monoassociated animals were not true models of what occurs in CV mice; they often had higher antibody titers to the associated bacter.um than to CV mice, and animals associated with lactobacilli developed antibodies to this bacterium that were absent in CV mice.

Numerous other examples could be cited. In general, one can conclude that gnotobiotic animals associated with only one or a few bacteria are poor models of human or animal diseases. The main reason for this shortcoming is the fact that colonization of such animals can be accomplished by most bacteria, and most virulence factors involved in colonization of CV animals or man are not required. By contrast, this very feature of gnotobiotic animals can help identify the nature of factors required for colonization. Moreover, the large bacterial populations that develop in monoassociated animals greatly facilitate the study of virulence factors that come into play after a pathogen has colonized its host, such as penetration of the epithelial barrier (Berg, 1980), toxin production (Klipstein *et al.*, 1979), or induction of ulcerative colitislike lesions (Onderdonk *et al.*, 1981a). Much has been learned about the cariogenic potential and the ability to cause periodontal disease of various bacteria from studies in gnotobiotic animals (e.g., Fitzgerald *et al.*, 1960, 1966; Gibbons *et al.*, 1966; Gibbons and Banghardt, 1968, Llory *et al.*, 1971; Mikx *et al.*, 1976). Gnotobiotic animals certainly provide the only means with which to study synergistic relationships of pathogens with elements of the indigenous microflora (Phillips and Wolfe, 1959; Whipp *et al.*, 1979).

Clearly, most gnotobiotic animals cannot model the entire process of a naturally occurring human or animal disease. Nevertheless, gnotobiotes are excellent models of individual facets of natural disease, i.e., of the functions and interactions of individual virulence and host defense mechanisms. The use of gnotobiotic models does not relieve the investigator of the need for critical evaluation as to whether and where the virulence or host defense mechanisms studied in a gnotobiotic system actually come into play in natural infections of human or animal hosts.

Physiologic Effects

There can be no doubt that the physiologic abnormalities of GF animals are caused by the lack of an indigenous flora and that consequently the normal functions of these physiological parameters are influenced by the flora. There are numerous studies attempting to define a given bacterial species or strain of the flora responsible for a given effect. Few such studies have been conclusive.

For example, the large liquid-filled cecum of GF rodents can be quickly reduced to normal size by conventionalizing the animals. It has also been reported that monoassociation with *Clostridium difficile* (Skelly *et al.*, 1962) or *Salmonella typhimurium* (Wiseman and Gordon, 1965) accomplishes cecal reduction to normal size. These data do not indicate that these two bacterial species are actually responsible for maintaining normal cecal size in CV animals. Both salmonellae and *Cl. difficile* are present at much larger population sizes in monoassociated than in CV animals; it is therefore doubtful that they would have the same effect in both instances. Attempts to reconstitute a functional intestinal microflora under conditions in which population sizes are more nearly normal indicate that reduction of cecal size requires association of the animal with a larger number of different indigenous bacteria (Freter and Abrams, 1972). Nevertheless, studies of cecal reduction by salmonellae have been very useful in identifying various physiologic mechanisms that can bring about reductions in cecal size (Wiseman and Gordon, 1965) and that might therefore be responsible for homeostasis in CV animals and man.

The second example is promotion of growth of young animals by subtherapeutic doses of antibiotics. This effect is widely applied by commercial animal raisers (Burg, 1982). Its mode of action is not clear (Visek, 1978), but it appears that some components of the indigenous microflora have harmful effects on the growth of young animals and that these effects are reversed by feeding antibiotics. In contrast to the previous example, the case for implicating one or a limited number of bacterial species is quite strong here because the low doses of antibiotics used as feed supplements that are effective in enhancing the growth rate of young animals do not cause any noticeable or consistent alteration in the composition of the predominant members of the indigenous flora (reviewed in Visek, 1978). Several laboratories have attempted to identify the bacterial species involved in growth retardation of young animals by monoassociating rats or chicks with individual bacteria of the indigenous flora; testing whether the growth rate of these gnotobiotic animals could be improved by antibiotic supplements to their diets. A considerable number of bacteria have been tested in this manner but, to the knowledge of this author only two were discovered that brought about growth retardation that was reversible by antibiotics: *Clostridium perfringens* (Forbes *et al.*, 1959) and *Streptococcus faecalis* (reviewed by Coates, 1968). Other laboratories did not find similar effects with *Cl. perfringens* (reviewed by Coates, 1968), a discrepancy that might be attributed to strain differences. Here again, the major problem of interpretation is the question whether bacteria that can exert certain physiologic effects when they populate a host animal in high numbers as monoassociates can have the same effect in CV animals in which their numbers are drastically reduced. If the bacterium under test were the only causative agent of antibiotic-reversible growth retardation in CV animals, one would expect this effect to vary in magnitude with the size of the bacterial population. This was not the case in the studies discussed here, i.e., the magnitude of reversible growth retardation in animals monoassociated with *Cl. perfringens* or *S. faecalis* was of the same order as that seen in the CV (Forbes *et al.*, 1959; Coates, 1968) or even somewhat less. Nevertheless, such studies in monoassociated animals have been valuable in exploring the possible mechanisms by which bacteria can cause retardation of their host's growth. For example, several laboratories have reported that cell-free filtrates of intestinal contents simulated or aggravated antibiotic-reversible growth retardation in gnotobiotic and GF animals (Eyssen and DeSomer, 1967; Lee and Dubos, 1968).

The interesting studies of mutagenesis in the gastrointestinal tract (Goldman *et al.*, 1980) may be cited as a final example. Goldman and co-workers monoassociated rats with the Ames his⁻ *Salmonella* mutants and noted an increased concentration of his⁺ revertants when carcinogens were fed. When *Bacteroides vulgatus* was introduced into the gut of rats carrying the *Salmonella* mutants, the revertant response to the carcinogen 2-nitrofluorene was diminished, presumably because the *Bacteroides* reduced the drug to an inactive compound. Although this study does not show that *B. vulgatus* is actually carrying out this conversion in CV animals, the system is an excellent model for studying mechanisms of carcinogen metabolism by elements of the indigenous microflora.

One must conclude that the main value of gnotobiotic animals in studies of the effects of indigenous microflora on host physiology lies in the fact that they facilitate the identification of various mechanisms by which bacteria can affect the host animal. As with the studies of pathogenicity, however, work with gnotobiotic animals is not likely to be able to identify conclusively a given microorganism or group of microorganisms as the cause of physiological reactions that occur in CV animals or man.

Interactions among Microbes of the Indigenous Microflora

Most bacteria (including species not indigenous to intestinal flora) are able to colonize the gut of GF animals, even when these hosts are immunized (Shedlofsky and Freter, 1974), whereas colonization of CV animals or man is difficult to achieve (Freter *et al.*, 1983*b*). It has been generally assumed that the major mechanisms controlling the composition of the indigenous microflora of the large intestine are a consequence of interactions among the numerous microbial species present. One obvious difficulty in studying such interactions is the narrow limits of experimental manipulations that are possible within the CV animal. Consequently, most investigators have resorted to working with *in vitro* models of bacterial interactions or with gnotobiotic animals.

Unfortunately, *in vitro* model systems cannot *a priori* be relied on to reflect mechanisms of *in vivo* interactions. In other words, the interactions among a given group of microorganisms depend to a large extent on the environment in which they take place. Many investigators have reported the lack of correlation between microbial interactions observed among different kinds of bacteria *in vitro* and the interactions of these same bacteria *in vivo*. Some early work (Hentges and Freter, 1962) demonstrated that the interaction between *Shigella flexneri* and a number of other bacteria differed considerably depending on the *in vitro* culture system used for testing these interactions. For example, in a given type of mixed *in vitro* culture (e.g., on nutrient agar plates), some strains of enterobacteria inhibited the growth of *Shigella*, and others did not. Under different culture conditions (e.g., in nutrient broth), there was again a group of inhibitory and noninhibitory enterobacterial strains, but the groups now had a different composition. In other words, it was not possible to predict the inhibitory activity that a given strain would exert against *Shigella* in broth culture from its inhibitory activity on agar plates against the same *Shigella* strain. The inhibitory activity of these strains in the intestine of diassociated mice could not be predicted from their reactions either in broth or on agar media. In order to dramatize the implications of these results, the authors predicted that, for any randomly selected pair of bacterial strains (A and B), they might be able to devise one culture method in which A inhibits B and another culture method in which B inhibits A. Unfortunately, no one has taken up this 20-year-old challenge. Other examples are the studies of Hentges's group (Maier *et al.*, 1972), who have shown that *Shigella flexneri* was strongly inhibited by *Bacteroides* strains *in vitro*, whereas these same strains had no effect on *Shigella* growth in the mouse intestine. One should also mention in this connection the work of Ducluzeau and Raibaud (1974), who found that interactions between *E. coli* and *Shigella flexneri* in gnotobiotic mice differed from those exhibited when the same bacteria were cultured *in vitro*: Some *Shigella* strains capable of colonizing gnotobiotic mice in the presence of *E. coli* were suppressed by the same *E. coli* strain in mixed liquid culture.

Numerous other examples of differences between *in vitro* and *in vivo* interactions among bacteria are known; the existence of these discrepancies is generally accepted among microbial ecologists. For this reason, many workers have turned to an *in vivo* system, the gnotobiotic animal, in the hope that this would be a more relevant model for the study of microbial interactions. Unfortunately, the GF animal differs profoundly from the CV. These differences are especially pronounced in the gastrointestinal tract, the locale of most studies of microbial interactions in animals and man, e.g., the intestinal contents differ, peristalsis is slowed, local and systemic immune mechanisms are absent or reduced, and glycoproteins that often act as receptors for bacterial adhesion (Jones, 1977) are not degraded by indigenous microflora and are therefore of different composition. As has been described by several workers (reviewed in Syed *et al.*, 1970), colonization of GF animals with a single bacterial strain rarely alters the GF abnormalities. Implanting a mixture of several known strains or species of bacteria usually redresses some but rarely all the germfree abnormalities (see also Iwai *et al.*, 1973). A gnotobiotic animal associated with one or a few bacterial strains therefore retains most of these abnormalities. One must conclude that interactions among bacteria in gnotobiotic animals occur in an environment that is unnatural and cannot *a priori* be accepted as being indicative of normal interactions. Examples illustrating this have been described: Two *E. coli* strains interacted strongly (i.e., one strain inhibiting the other) in the intestine of diassociated mice, whereas the same two *E. coli* strains did not interact at all in gnotobiotic mice harboring a more complex indigenous flora (Freter and Abrams, 1972). Likewise, in the studies of Ducluzeau and Raibaud (1974), the interaction of *Shigella flexneri* with *E. coli* strains in gnotobiotic mice differed from that in CV mice in that some *Shigella* strains that were capable of coexisting with *E. coli* in gnotobiotic mice did not grow in CV mice. Also, the rates of plasmid transfer among bacterial donor and recipient strains in the gut of diassociated animals are several orders of magnitude higher than in the CV (Freter *et al.*, 1983*c*).

The gnotobiotic animal associated with one or a few bacteria is therefore no more appropriate as a valid model of bacterial interactions in CV animals than *in vitro* cultures, even though the fact that such animal experiments can be described by the term *in vivo* seems to lend a semantic touch of legitimacy to these undertakings. For example, when two *E. coli* strains (A and B) are implanted as diassociates into gnotobiotic animals, and if strain A suppresses the population level of the other, it cannot be concluded that the better colonizer (strain A) has a higher "intrinsic large intestine colonization potential" (Cohen *et al.*, 1979) or that a higher "relative colonization ability" (Myhal *et al.*, 1982). In other words, the fact that strain A showed greater fitness when competing with strain B in the gnotobiotic animal does not permit the prediction that strain A will also be superior to strain B in competition against B and other indigenous bacteria in the entirely different environment of a CV animal.

Fallacious overinterpretations of data obtained with gnotobiotic animals often stem from the supposition that bacterial strains in the indigenous microflora compete only (or at least predominantly) against other strains of the same species. This has led to *in vivo* studies of certain inhibitory mechanisms, such as the colicins, by which some *E. coli* strain are known to compete successfully *in vitro* with other *E. coli* strains (Ikari *et al.*, 1969) or with closely related genera, e.g., *Shigella* (Halbert, 1948). There are examples showing that this is not necessarily the case; the populations of *E. coli* are controlled at least in part by the strictly anaerobic members of the CV intestinal microflora rather than by closely related species (Meynell, 1963; Freter and Abrams, 1972). Although we do not know the precise mechanisms by which the numerous different bacterial populations of the indigenous microflora compete with each other, these mechanisms are likely to include competition for nutrients and adhesion sites, as well as the production of inhibitory substances (Freter, 1983). All of these are not only different in the gnotobiotic animal as compared with the CV; they almost certainly differ in individual CV animals and individual human beings, depending on the individual strains and genera of which the indigenous flora is composed. It is known that there are wide variations in the composition of the fecal flora among human individuals (Finegold *et al.*, 1974; Holdeman *et al.*, 1976). Consequently, the relative success of a given bacterial strain in colonizing a given host depends not only on the characteristics of this strain itself, but most likely also on the composition of the indigenous microflora in a given individual animal or human host— and a strain that has a relatively strong ability to colonize one individual host may well be relatively poor at colonizing another. Consequently, a universal quality of bacteria such as a "relative colonization potential" is not likely to exist at all. In the event that such a quality were to be defined more narrowly, it could apply only to a certain kind of host, such as the CV mouse, and it would differ almost certainly among CV and gnotobiotic animals of the same species, and in man.

Strictly speaking, only one type of gnotobiotic animal would fulfill all criteria for relevance in studies of microbial interactions among the indigenous microflora, namely, the animal with a "flora minus one." That is to say, the comparison of CV animals with others from which only one bacterial species (or strain) has been removed would permit an unassailable conclusion to be drawn concerning the function of the missing strain in the ecology of the CV host. Unfortunately, such an experiment is technically not feasible at the present time. How close can one come to this ideal? When Corpet and Nicholas (1979) reported that the implantation into gnotobiotic mice or piglets of eight bacteria from the indigenous intestinal flora reduced the population of *Clostridium perfringens* to low levels, does that mean that these eight bacteria are also responsible for the control of *C. perfringens* populations in the CV host? The likelihood for such a conclusion to be correct would increase in the following circumstances: (1) if one could show that all eight of the associated bacteria attained populations in the gnotobiotes that did not exceed their population levels in CV animals, and (2) if the gnotobiotes resembled their CV counterparts physiologically, i.e., if the eight associated bacteria had reversed the GF abnormalities of their host animals.

The above discussion of possible pitfalls in the interpretation of experimental data obtained with gnotobiotic animals has been presented in some detail because the points made therein are not always considered even in the current literature. Other experimental approaches using gnotobiotic animals have yielded valuable data. Here, again, the main value of gnotobiotic studies appears to be the identification of various mechanisms of bacterial interactions that may occur also in CV animals and that, once identified in the gnotobiotic system, can be studied subsequently in CV models. The gnotobiotic technique can indeed furnish invaluable clues as to mechanisms that could profitably be investigated in CV animals. Such clues would often be hard to obtain in any other way. For example, Cohen et al. (1983) reported that the ability of a given *E. coli* strain to compete in diassociated mice against other *E. coli* strains was correlated with the ability to bind a colonic mucus gel protein. As Cohen et al., pointed out, these data open up a number of future avenues of investigation, even if they do not permit any conclusions as to the functions of bacterial mucus gel binding in CV animals. Likewise, the identification of competition for nutrients as a mechanism that controls *E. coli* populations in diassociated mice (Freter, 1962) led to later studies in more complex systems consistent with the conclusion that this mechanism is also of prime importance in the control of intestinal bacteria in CV animals (Freter et al., 1983a); however, there is still no absolute proof of this. Various factors of likely importance in bacterial interactions in CV hosts have been revealed in studies with gnotobiotic animals by Duval-Iflah *et al.* (1981), Onderdonk *et al.* (1981*b*), and many others.

Gnotobiotic animals associated with a limited number of bacteria are also useful in the study of bacterial interactions in areas of the body that normally carry only a limited flora. In such instances, the flora present at the site of interest in the gnotobiotic host animal is likely to resemble in complexity that of the CV counterpart. Consequently, many of the difficulties in interpretation discussed earlier in this section do not apply, such that there is a much greater likelihood that the interactions observed in the gnotobiotic model actually resemble those occurring in the CV animal. Examples are the studies of interaction of yeast and lactobacilli in the mouse stomach (Kotarski and Savage, 1979) and the identification of bacterial adhesion and other factors important in colonization of the chicken crop (Fuller, 1981).

SUMMARY AND CONCLUSIONS

The development of germfree animal techniques is a major technical breakthrough in microbial ecology. The early enthusiasm for this approach has been fully justified, in that comparisons of GF with CV animals have made it possible to describe the numerous interactions between the indigenous microflora and its animal or human host.

More difficult and problematic are studies in gnotobiotic animals involving attempts to identify individual bacterial species or strains responsible for a given function of the indigenous microflora or to identify potentially pathogenic bacteria. The difficulties arise from the fact that gnotobiotic animals which are associated with only one or a few kinds of bacteria retain most of the germfree abnormalities. As a consequence of these abnormalities, bacteria associated with gnotobiotic animals reach much higher population levels than they can attain in CV hosts and so are likely to affect the gnotobiotic animal by mechanisms they cannot deploy in any significant manner at the lower population levels attained in the CV host. Moreover, the gnotobiotic animal represents an environment for bacterial multiplication and colonization which differs profoundly from that of a CV animal; it is therefore likely that competitive interactions among bacteria in gnotobiotes are mediated by mechanisms that differ from those operating in CV animals. By contrast, studies in gnotobiotic animals of pathogenic potential, of effects on host physiology or of competitive bacterial interactions are indeed very useful if they are concerned with the mechanisms underlying these phenomena. The identification in gnotobiotic systems of such mechanisms of microbe–host and microbe–microbe interactions give valuable clues as to the kinds of mechanisms that could profitably be sought out and explored in subsequent studies involving the more complex CV systems.

It is apparent that gnotobiotic animals cannot be expected to model the entire series of microbe–microbe and microbe–host interactions that occur in normal or diseased CV animals or in human beings. By contrast, gnotobiotes are very useful models of the individual mechanisms underlying such interactions. It is also apparent that much ingenuity and hard work will be necessary to design and execute future studies that can conclusively identify specific bacterial species (or groups of bacterial species) responsible for any specific function of the indigenous microflora and to define the mechanisms by which these functions are mediated.

REFERENCES

Aly, R., and Shinefield, H. R., 1982, *Bacterial Interference*, CRC Press, Boca Raton, Florida.

Ashe, W. K., Sherp, H. W., and Fitzgerald, R. J., 1965, Previously unrecognized virus from submaxillary glands of gnotobiotic and conventional rats, *J. Bacteriol.* **90:**1719–1729.

Baker, J. A., and Ferguson, M. S., 1942, Growth of Platyfish (*Platypoecilus maculatus*) free from bacteria and other microorganisms, *Proc. Soc. Exp. Biol. Med.* **51:**116–119.

Berg, R. D., 1980, Mechanisms confining indigenous bacteria to the gastrointestinal tract, *Am. J. Clin. Nutr.* **33:**2472–2484.

Bohnhoff, M., Drake, B. L., and Miller, C. P., 1954, Effect of streptomycin on susceptibility of intestinal tract to experimental Salmonella infection, *Proc. Soc. Exp. Biol. Med.* **86:**132–137.

Bohnhoff, M., Miller, C. P., and Martin, W. R., 1964, Resistance of the mouse's intestinal tract to experimental *Salmonella* infection, *J. Exp. Med.* **120:**817–828.

Burg, R. W., 1982, Fermentation products in animal health. *ASM News* **48:**460–463.

Coates, M. E., 1968, Nutrition and metabolism, in: *The Germfree Animal in Research* (M. E. Coates, H. A. Gordon, and B. S. Wostmann, eds.), pp. 161–179, Academic Press, London.

Coates, M. E., and Fuller, R., 1977, The gnotobiotic animal in the study of gut microbiology, in: *Microbial Ecology of the Gut* (R. T. J. Clarke and T. Bauchop, eds.) pp. 311–346, Academic Press, New York.

Coates, M. E., Gordon, H. A., and Wostmann, B. S., 1968, *The Germfree Animal in Research,* Academic Press, New York.

Cohen, P. S., Pilsucki, R. W., Myhal, M. L., Rosen, C. A., Laux, D. C., and Cabelli, V. J., 1979, Colonization potentials of male and female *E. coli* K12 strains *E. coli* B and human fecal *E. coli* strains in the mouse G.I. tract, *Recombinant DNA Tech. Bull.* **2:**106–113.

Cohen, P. S., Rossoll, R., Cabelli, V. J., Young, S., and Laux, D. C., 1983, Relationship between the mouse colonizing ability of a human fecal *Escherichia coli* strain and its ability to bind a specific mouse colonic mucous gel protein, *Infect. Immun.* **40:**62–69.

Corpet, D., and Nicolas, J. L., 1979, Antagonistic effect of intestinal bacteria from the microflora of holoxenic (conventional) piglets, against *Clostridium perfringens* in the digestive tract of gnotoxenic mice and gnotoxenic piglets, in: *Clinical and Experimental Gnotobiotics* (T. M. Fliedner, H. Heit, D. Niethammer, and H. Pflieger, eds.), pp. 169–174, Gustav Fischer, New York.

Dubos, R. 1963, Staphylococci and infection immunity, *Am. J. Dis. Child.* **105:**643–645.

Ducluzeau, R., and Raibaud, P., 1974, Interaction between *Escherichia coli* and *Shigella flexneri* in the digestive tract of "gnotobiotic" mice, *Infect. Immun.* **9:**730–733.

Duval-Iflah, Y., Raibaud, P., and Rousseau, M., 1981, Antagonisms among isogenic strains of *Escherichia coli* in the digestive tracts of gnotobiotic mice, *Infect. Immun.* **34:**957–969.

Eyssen, H., and DeSomer, P., 1967, Effects of *Streptococcus fecalis* and a filtrable agent on growth and nutrient absorption in gnotobiotic chicks, *Poultry Sci.* **46:**323–333.

Finegold, S. M., Attebery, H. R., and Sutter, V. L., 1974, Effect of diet on human fecal flora: Comparison of Japanese and American diets, *Am. J. Clin. Nutr.* **27:**1456–1469.

Fitzgerald, R. J., Jordan, H. V., Stanley, H. R., Poole, W. L., and Bowler, A., 1960, Experimental caries and gingival pathologic changes in the gnotobiotic rat, *J. Dent. Res.* **39:**923–935.

Fitzgerald, R. J., Jordan, H. V., and Archard, H. O., 1966, Dental caries in gnotobiotic rats infected with a variety of *Lactobacillus acidophilus*, *Arch. Oral Biol.* **11:**473–476.

Fliedner, T., Heit, H., Niethammer, D., and Pflieger, H., 1979, Clinical and experimental gnotobiotics, *Zentralbl. Bakteriol.* **1** (Suppl. 7), 396 pp.

Forbes, M., Park, J. T., and Lev, M., 1959, Role of the intestinal flora in the growth response of chicks to dietary penicillin, *Ann. N.Y. Acad. Sci.* **78**(1):321–327.

Freter, R., 1954, The fatal enteric cholera infection in the guinea pig, *Bacteriol. Proc.* **1954:**56.

Freter, R., 1956, Coproantibody and bacterial antagonism as protective factors in experimental enteric cholera, *J. Exp. Med.* **104:**419–426.

Freter, R., 1962, *In vivo* and *in vitro* antagonism of intestinal bacteria against *Shigella flexneri*. II. The inhibitory mechanism, *J. Infect. Dis.* **110:**38–46.

Freter, R., 1983, Mechanisms that control the microflora in the large intestine, in *Human Intestinal Microflora in Health and Disease* (D. J. Hentges, ed.) pp. 33–54, Academic Press, New York.

Freter, R., and Abrams, G. D., 1972, Function of various intestinal bacteria in converting germfree mice to the normal state, *Infect. Immun.* **6:**119–126.

Freter, R., Brickner, H., Botney, M., Cleven, D., and Aranki, A., 1983a, Mechanisms which control bacterial populations in continuous flow culture models of mouse large intestinal flora, *Infect. Immun.* **39:**676–685.

Freter, R., Brickner, H., Fekete, J., and Vickerman, M. M., 1983b, Survival and implantation of *E. coli* in the intestinal tract, *Infect. Immun.* **39:**686–703.

Freter, R., Freter, R. R., and Brickner, 1983c, Experimental and mathematical models of *E. coli* plasmid transfer *in vitro* and *in vivo*, *Infect. Immun.* **39:**60–84.

Freter, R., Stauffer, E., Cleven, D., Holdeman, L. V., and Moore, W. E. C., 1983d, Continuous flow cultures as *in vitro* models of the ecology of large intestinal flora, *Infect. Immun.* **39:**666–675.

Fubara, E. S., and Freter, R., 1972, Availability of locally synthesized and systemic antibodies in the intestine, *Infect. Immun.* **6:**965–981.

Fuller, R., 1968, The routine microbiological control of germfree isolators, in: *The Germfree Animal in Research* (M. E. Coates, H. A. Gordon, and B. S. Wostmann, eds.), pp. 37–45, Academic Press, New York.

Fuller, R., 1981, Epithelial attachment as a factor controlling bacterial colonization of the gastrointestinal tract, in: *Recent Advances in Germfree Research* (S. Sasaki, A. Ozawa, and K. Hashimoto, eds.), pp. 143–147, Tokai University Press, Tokyo.

Gibbons, R. J., and Banghart, S., 1968, Induction of dental caries in gnotobiotic rats with a levan-forming Streptococcus and a Streptococcus isolated from subacute bacterial endocarditis, *Arch. Oral Biol.* **13:**297–308.

Gibbons, R. J., Socransky, S. S., and Kapsimalis, B., 1964, Establishment of human indigenous bacteria in germ-free mice, *J. Bacteriol.* **88:**1316–1323.

Gibbons, R. J., Berman, K. S., Knoettner, P., and Kapsimalis, B., 1966, Dental caries and alveolar bone loss in gnotobiotic rats infected with capsule forming Streptococci of human origin, *Arch. Oral Biol.* **11:**549–560.

Goldman, P., Carter, J. H., and Wheeler, L. A., 1980, Mutagenesis within the gastrointestinal tract determined by histidine auxotrophs of *Salmonella typhimurium*, *Cancer* **45:**1068–1072.

Gorbach, S. L., and Onderdonk, A., 1979, Experimental animal models and human disease, *Gastroenterology* **76:**643–645.

Gordon, H. A., 1981, Contributions of the microbial flora to the physiologic normality of the animal host, in: *Recent Advances in Germfree Research* (S. Sasaki, A. Ozawa, and K. Hashimoto, eds.), pp. 347–353, Tokai University Press, Tokyo.

Gordon, H. A., and Pesti, L., 1972, The gnotobiotic animal as a tool in the study of host microbial relationships, *Bacteriol. Rev.* **35:**390–429.

Gordon, H. A., and Wostmann, B. S., 1973, Chronic mild diarrhea in germfree rodents: A model portraying host–flora synergism, in: *Germfree Research, Biological Effects of Gnotobiotic Environments* (J. B. Heneghan, ed.), pp. 593–601, Academic Press, New York.

Gustafsson, B. E., 1959, Lightweight stainless steel systems for rearing germfree animals, *Ann. N.Y. Acad. Sci.* **78**:17–28.

Gustafsson, B. E., and Norin, K. E., 1977, Development of germfree animal characteristics in conventional rats by antibiotics, *Acta Pathol. Microbiol. Scand. Sect. B* **85**:1–8.

Halbert, S. P., 1948, The relation of antagonistic coliform organisms to *Shigella* infections. II. Observations in acute infections, *J. Immunol.* **60**:359–381.

Heneghan, J. B., 1973, *Germfree Research, Biological Effects of Gnotobiotic Environments,* Academic Press, New York.

Hentges, D. J., 1970, Enteric pathogen-normal flora interactions, *Am. J. Clin. Nutr.* **23**:1451–1456.

Hentges, D. J., and Freter, R., 1962, *In vivo* and *in vitro* antagonism of intestinal bacteria against *Shigella flexneri.* I. Correlation between various tests, *J. Infect. Dis.* **110**:30–37.

Hill, M. J., 1980, Bacterial metabolism and human carcinogenesis, *Br. Med. Bull.* **36**:89–94.

Holdeman, L. V., Good, I. J., and Moore, W. E. C., 1976, Human fecal flora: variation in bacterial composition within individuals and a possible effect of emotional stress, *Appl. Environ. Microbiol.* **31**:359–375.

Ikari, V. S., Kenton, D. M., and Young, V. M., 1969, Interaction in the germfree mouse intestine of colicinogenic and colicin-sensitive microorganisms, *Proc. Soc. Exp. Biol. Med.* **130**:1280–1284.

Iwai, H., Ishihara, Y., Yamanaka, J., and Ito, T., 1973, Effects of bacterial flora on cecal size and transit rate of intestinal contents in mice, *Jpn. J. Exp. Med.* **43**:297–305.

Jones, G. W., 1977, The attachment of bacteria to the surfaces of animal cells, in: *Microbial Interactions* (J. L. Reissig, ed.), pp. 139–176, Chapman and Hall, London.

Jones, G. W., and Rutter, J. M., 1972, Role of the K88 antigen in the pathogenesis of neonatal diarrhea caused by *Escherichia coli* in piglets, *Infect. Immun.* **6**:918–927.

Kawai, Y., and Morotomi, M., 1978, Intestinal enzyme activities in germfree, conventional, and gnotobiotic rats associated with indigenous microorganisms, *Infect. Immun.* **19**:777–778.

Khoury, K. A., Floch, M. A., and Hersch, T., 1969, Small intestinal mucosal cell proliferation and bacterial flora in the conventionalization of the germfree mouse, *J. Exp. Med.* **130**:659–670.

Kim, Y. B., 1981, Immunology—introductory remarks, in: *Recent Advances in Germfree Research* (S. Sasaki, A. Ozawa, and K. Hashimoto, eds.), pp. 483–491, Tokai University Press, Tokyo.

Klipstein, F. A., Goetsch, C. A., Engert, R. F., Short, H. B., and Schenk, E. A., 1979, Effect of monocontamination of germfree rats by enterotoxigenic coliform bacteria, *Gastroenterology* **76**:341–348.

Kotarski, S. F., and Savage, D. C., 1979, Models for study of the specificity by which indigenous lactobacilli adhere to murine gastric epithelia, *Infect. Immun.* **26**:966–975.

Lattuada, C. P., Norberg, R. M., and Harlan, H. P., 1982, Large isolators for rearing rodents, in: *Recent Advances in Germfree Research* (S. Sasaki, A. Ozawa, and K. Hashimoto, eds.), pp. 53–58, Tokai University Press, Tokyo.

Lee, C. J. and Dubos, R., 1968, Lasting biological effects of early environmental influences. III. Metabolic responses of mice to neonatal infection with a filterable weight-depressing agent, *J. Exp. Med.* **128**:753–762.

Levenson, S. M., Doft, F., Lev, M., and Kan, D., 1969, Influence of microorganisms on oxygen consumption, carbon dioxide production and colonic temperature of rats, *J. Nutri.* **97**:542–552.

Llory, H., Guillo, B., and Frank, R. M., 1971, A cariogenic Actinomyces viscosus—a bacteriological and gnotobiotic study, *Helv. Odont. Acta* **15**:134–138.

Luckey, T. D., 1963, *Germfree Life and Gnotobiology,* Academic Press, New York.

Luckey, T. D., 1965, Effects of microbes on germfree animals, *Adv. Appl. Microbiol.* **7**:169–223.

MacDonald, T. T., Carter, P. B., and Phillips, P. G., 1979, Modulation of cellular immunity by the enteric flora: Delayed type hypersensitivity in the gnotobiotic mouse, in: *Clinical and Experimental Gnotobiotics* (T. M. Fliedner, H. Heit, D. Niethammer, and H. Pflieger, eds.), pp. 221–226, Gustav Fischer, Verlag, Stuttgart.

Maier, B. R., Onderdonk, A. B., Baskett, R. C., and Hentges, D. J., 1972, Shigella, indigenous flora interactions in mice, *J. Clin. Nutr.* **25:**1433–1440.

Meynell, G. G., 1963, Antibacterial mechanisms of the mouse gut, *Br. J. Exp. Pathol.* **44:**209–219.

Mikx, F. H. M., van der Hoeven, J. S., and Walker, G. J., 1976, Microbial symbiosis in dental plaque studied in gnotobiotic rats and in the chemostat, in: *Microbial Aspects of Dental Caries,* Vol. 3 (H. M. Stiles, W. J. Loesche, and T. C. O'Brien, eds.), Special Supplement, Microbiology Abstracts, pp. 763–771, Information Retrieval, Inc., Washington, D. C.

Miller, R. S., and Hoskins, L. C., 1981, Mucin degradation in human colon ecosystems. Fecal population densities of mucin-degrading bacteria estimated by a "most probable number" method, *Gastroenterology* **81:**759–765.

Mirand, E. A., and Back, N., 1968, *Germfree biology, experimental and clinical aspects,* Plenum Press, New York.

Miyakawa, M., 1959, The Miyakawa remote-control germfree rearing unit, *Ann. N.Y. Acad. Sci.* **78:**37–46.

Miyakawa, M., and Luckey, T. D., 1968, *Advances in Germfree Research and Gnotobiology,* CRC Press, Cleveland.

Myhal, M. L., Laux, D. C., and Cohen, P. S., 1982, Relative colonizing abilities of human fecal and K12 strains of *Escherichia coli* in the large intestines of streptomycin-treated mice, *Eur. J. Clin. Microbiol.* **1:**186–192.

National Academy of Sciences, National Research Council, Institute of Laboratory Animal Resources, Subcommittee on Standards for Gnotobiotes, 1970, *Gnotobiotes, Standards and Guidelines for the Breeding, Care and Management of Laboratory Animals,* National Academy of Sciences, Washington, D.C.

Nielsen, E., and Friis, C. W., 1980, Influence of an intestinal microflora on the development of the immunoglobulins IgG1, IgG2a, IgM and IgA in germfree BALB/c mice, *Acta Pathol. Microbiol. Scand.* **88:**121–126.

Onderdonk, A. B., Franklin, M. L., and Cisneros, R. L., 1981a, Production of experimental ulcerative colitis in gnotobiotic guinea pigs with simplified microflora, *Infect. Immun.* **32:**225–231.

Onderdonk, A. B., Marshall, B., and Levy, S. B., 1981b, Competition between congenic *Escherichia coli* K-12 strains *in vivo, Infect. Immun.* **32:**74–79.

Phillips, B. P., and Wolfe, P. A., 1959, The use of germfree guinea pigs in studies on the microbial interrelationships in amoebiasis, *Ann. N.Y. Acad. Sci.* **78**(1)**:**308–314.

Pleasants, J. R., 1974, Gnotobiotics, in: *Handbook of Laboratory Animal Science,* Vol. 1 (E. C. Melby, Jr., and N. H. Altman, eds.), pp. 119–174, CRC Press, Cleveland.

Pleasants, J. R., Schmitz, H. E., and Wostmann, B. S., 1982, Immune responses of germfree C3H mice fed chemically defined ultrafiltered "antigen-free" diet, in: *Recent Advances in Germfree Research* (S. Sasaki, A. Ozawa, and K. Hashimoto, eds.), pp. 535–538, Tokai University Press, Tokyo.

Pollard, M., 1981, Development of model systems for cancer research, in: *Recent Advances in Germfree Research* (S. Sasaki, A. Ozawa, and K. Hashimoto, eds.), pp. 621–625, Tokai University Press, Tokyo.

Pollard, M., and Matsuzawa, T., 1964, Induction of leukemia in germfree mice by x-rays, *Am. J. Pathol.* **44:**17a.

Raibaud, P., Dickinson, A. B., Sacquet, E., Charlier, H., and Mocquot, G., 1966, La microflora due tube digestif du rat. IV. Implantation controlée chez le rat gnotobiotique de differents genres microbiens isoles du rat conventionnel, *Ann. Inst. Pasteur* **111:**193–210.

Reddy, B. S., 1981, Role of gut microflora and nutritional factors in the development of large bowel and breast cancer, in: *Recent Advances in Germfree Research* (S. Sasaki, A. Ozawa, and K. Hashimoto, eds.)., pp. 627–631, Tokai University Press, Tokyo.

Reyniers, J. A., 1959, Germfree vertebrates: Present status, *Ann. N.Y. Acad. Sci.* **78:**1–400.

Sacquet, E., 1969, Equipment design and management, in: *The Germfree Animal in Research* (M. E. Coates, H. A. Gordon, and B. S. Wostmann, eds.), pp. 1–22, Academic Press, London.

Sasaki, S., Ozawa, A., and Hashimoto, K., 1982, *Recent Advances in Germfree Research,* Tokai University Press, Tokyo.

Savage, D. C., 1977, Microbial ecology of the gastrointestinal tract, *Annu. Rev. Microbiol.* **31:**107–133.

Savage, D. C., Siegel, J. E., Snellen, J. E., and Whitt, D. D., 1981, Transit time of epithelial cells in the small intestines of germfree mice and ex-germfree mice associated with indigenous microorganisms, *Appl. Environ. Microb.* **42:**996–1001.

Shedlofsky, S., and Freter, R., 1974, Synergism between ecological and immunological control mechanisms of intestinal flora, *J. Infect. Dis.* **129:**296–303.

Skelly, B. J., Trexler, P. C., and Tanami, J., 1962, Effect of a *Clostridium* species upon cecal size of gnotobiotic mice, *Proc. Soc. Exp. Biol. Med.* **110:**455–458.

Syed, S. A., Abrams, G. D., and Freter, R., 1970, Efficiency of various intestinal bacteria in assuming normal functions of enteric flora after association with germ-free mice, *Infect. Immun.* **2:**376–386.

Trexler, P. C., 1959, The use of plastics in the design of isolator systems, *Ann. N.Y. Acad. Sci.* **78:**29–35.

Trexler, P. C., 1982, Gnotobiotic technology: Introductory remarks, in: *Recent Advances in Germfree Research* (S. Sasaki, A. Ozawa, and K. Hashimoto, eds.), pp. 33–37, Tokai University Press, Tokyo.

van der Waaij, D., Berghuis-deVries, J. M., and Lekkerkerk, J. E. C., 1971, Colonization resistance of the digestive tract in conventional and antibiotic-treated mice, *J. Hyg.* **69:**405–411.

Vieira, E. C., Moraes e Santos, T., Nicoli, J. R., Vieira, L. Q., and Pleasants, J. R., 1982, Diet sterilization for germfree animals in a hospital-type autoclave, in: *Recent Advances in Germfree Research* (S. Sasaki, A. Ozawa, and K. Hashimoto, eds.), pp. 83–87, Tokai University Press, Tokyo.

Visek, W. J., 1978, The mode of growth promotion by antibiotics, *J. Animal Sci.* **46:**1447–1469.

Wagner, M., 1959*a*, Determination of germfree status, *Ann. N.Y. Acad. Sci.* **78:**89–101.

Wagner, M., 1959*b*, Serologic aspects of germfree life, *Ann. N.Y. Acad. Sci.* **78**(1):261–271.

Whipp, S. C., Robinson, I. M., Harris, D. L., Glock, R. D., Matthews, P. J., and Alexander, T. J. L., 1979, Pathogenic synergism between *Treponema hyodysenteriae* and other selected anaerobes in gnotobiotic pigs, *Infect. Immun.* **26:**1042–1047.

Wilkins, T. D., and Appleman, D., 1976, Review of methods for antibiotic susceptibility testing of anaerobic bacteria, *Lab. Med.* **7**(4):12–15.

Wiseman, R. F., and Gordon, H. A., 1965, Reduced levels of a bioactive substance in the cecal content of gnotobiotic rats monoassociated with *Salmonella typhimurium, Nature (Lond.)* **205:**572–573.

Wostmann, B. S., 1981, The germfree animal in nutritional studies, *Annu. Rev. Nutr.* **1:**257–279.

Wostmann, B. S., and Bruckner-Kardoss, E., 1981, Functional characteristics of gnotobiotic rodents, in: *Recent Advances in Germfree Research* (S. Sasaki, A. Ozawa, and K. Hashimoto, eds.), pp. 321–325, Tokai University Press, Tokyo.

Wostmann, B. S., Bruckner-Kardoss, E., and Knight, P. L., Jr., 1968, Cecal enlargement, cardiac output, and O_2 consumption in germfree rats, *Proc. Soc. Exp. Biol. Med.* **128:**137–140.

Wostmann, B. S., Reddy, B. S., Bruckner-Kardoss, E., Gordon, H., and Singh, B., 1973, Causes and possible consequences of cecal enlargement in germfree rats, in: *Germfree Research Biological Effect of Gnotobiotic Environments* (J. B. Heneghan, ed.), pp. 261–270, Academic Press, New York.

Yale, C. E., 1982, Autoclavable lightweight stainless steel isolators for laboratory animals, in: *Recent Advances in Germfree Research* (S. Sasaki, A, Ozawa, and K. Hashimoto, eds.), pp. 45–51, Tokai University Press, Tokyo.

ENRICHMENT CULTURES IN BACTERIAL ECOLOGY

J. S. Poindexter and E. R. Leadbetter

> It has been more than implied that results obtained from the breeding pen, the seed pan, the flower pot and the milk bottle do not apply to evolution in the "open," nature at large, or to wild types. To be consistent, this same objection should be extended to the use of the spectroscope in the study of the evolution of the stars, to the use of the test tube and the balance by the chemist, of the galvanometer by the physicist. All these are the unnatural instruments used to torture Nature's secrets from her.
>
> *T. H. Morgan (1916)*

INTRODUCTION

Morgan (quoted in Clark, 1984) could have included monotypic ("pure") cultures of bacteria in either list. Only in extremely rare, and often then only transient, situations does a single type of bacterium occupy a natural habitat alone—even the "microhabitat" so often mentioned in these chapters. The respective contributions of the various members to the activities of a bacterial community cannot be assessed, however, without knowledge of the potential activities of each type of member. To enable separation of the types for such assessment, the elective culture technique, developed largely by Beijerinck and by Winogradsky, remains an indispensable method in bacterial ecology.

An elective culture not only serves to provide, ultimately, pure populations of distinctive types; the course of population changes during its incubation also reveals the respective influences of physical and chemical factors on the persistence of each type of bacterium present, as well as the influences of the community members on each other.

J. S. Poindexter • Science Division, Long Island University, Brooklyn, Campus, Brooklyn, New York 11201. *E. R. Leadbetter* • Department of Molecular and Cell Biology, The University of Connecticut, Storrs, Connecticut 06268.

The principles, purposes, and limitations of this technique are reviewed in this chapter. Other discussions of elective cultivation are available in the reviews of Veldkamp (1970, 1977), Schlegel and Jannasch (1967), and Meers (1973) and in the collected works of Beijerinck (1921–1940) and of Winogradsky (1949). For detailed procedures, see Chapter 8 of Gerhardt *et al.* (1981).

PURPOSES AND PRINCIPLES

An elective culture, also known as an enrichment culture, is an arrangement consisting of a natural sample introduced into an artificial environment in which conditions are imposed that will change the relative numbers of the types of bacteria present in the sample. One or more types are favored and rise in proportion to the others whose proportions drop because they reproduce more slowly than the favored types, fail to reproduce at all, or die. The conditions imposed in an enrichment procedure result in a population change, presumably a change that could occur *in situ*. The specific conditions that successfully enrich a given type imply the respective activities, capabilities, and sensitivities of the various microorganisms present in the initial sample.

Such treatment of a natural sample serves two principal investigative purposes: (1) to select from among a natural, mixed population a type of bacterium known to exist; and (2) to determine whether there exist bacteria that are favored by a particular set of conditions. In some instances, elective cultures can also be employed to promote the growth of bacteria not yet cultivable as pure populations; it seems reasonable to regard most such instances as examples of bacterial dependence on other microorganisms, whose contributions to the welfare of the dependent type are not yet sufficiently understood to allow investigators to substitute chemical or physical conditions for the presence of the supportive microbes.

In order to serve the first of the two principal purposes, a natural sample, such as soil, water, sediment, decaying plant or animal carcass, or live body part, is mixed with a source of nutrients known to be especially suitable for promoting the growth and reproduction of the desired type. The mixture is then incubated under conditions of appropriate pH, temperature, illumination, atmosphere, and ionic strength. All these factors, as a minimum set, must be considered in the design of conditions for enrichment. Enrichment cultures themselves can be used for identifying conditions most favorable for the development of a population of a given type; for example, Blakemore *et al.* (1979) first determined the appropriate pH, buffer, reducing agents, and Fe and N sources with enrichment cultures and were subsequently able to design a procedure for successful enrichment and isolation of magnetic bacteria.

Even before inoculation, the sample must be appropriately selected, and it may be subjected to treatment that will enhance the probability of the desired enrichment. Basically, the sample should be selected from a habitat where the desired type is likely to thrive. For the first purpose, one would not seek, for example, cyanobacteria from the depths of a compost pile or dioxygen-sensitive bacteria from the surface of a rushing stream. Acidophilic bacteria would be sought from samples such as acid mine wastes, while a marsh sediment redolent of sulfide would be a suitable source of sulfate-reducing anaerobes. Examples of appropriate sources of specific types of bacteria are presented in Table I.

Pretreatment of samples prior to incubation under elective conditions generally consists of physical methods that kill or remove potential competitors or predators likely to be present in the sample. For example, heating (to 60–100°C) will inactivate most non-spore-forming bacteria. Gentle washing of a particulate or solid-surface sample will remove loosely associated microbes and thereby enrich the sample for types that adhere tightly to substrata. In a few instances, relatively unspecialized types may predominate in a sample but compete poorly under contained enrichment conditions; such types are favored by dilution of the sample to the point where minor but highly competitive types are numerically negligible or, preferably, absent. One of the most commonly used pretreatments for water samples is filtration that removes large cells such as algae, fungi, and protozoa (the last especially likely to include bacteriovores) and particles to which bacteria cling. This practice significantly reduces the complexity of the microbial population of the inoculum and favors subsequent development of planktonic bacteria.

TABLE I

Appropriate Sources of Selected Bacterial Physiotypes

Physiotype	Appropriate sources
Chemolithotrophic	Soil, hot springs
Phototrophic (sulfur)	Marine sediments, marsh mats
Phototrophic (nonsulfur)	Soil, freshwater sediments, hot springs
Polymer degradative	Decaying animal, microbial, or plant material
Acidophilic	Coal mine drainage, peat bogs
Alkalophilic	Alkaline soils and springs
Thermophilic	Hot springs, domestic hot-water systems
Psychrophilic	Cold-preserved foods, cold soils
Osmophilic	Dried fruits, honey
Methano- or methylotrophic	Water and air–water interfaces above methanogenic sediments
Luminous	Seawater; surfaces and light organs of marine animals
Dissimilatory nitrate-reducing	Waterlogged soils
Dissimilatory sulfate-reducing	Marine sediments, anoxic freshwater sites
Diazotrophic	Soil, water
Halophilic	Salterns, dried marine material
Methanogenic	Rumen fluid, anoxic sediments

To serve the second principal purpose of elective cultivation, i.e., to ask whether there exists a particular type of bacterium, the enrichment conditions are exploratory. At this point in our understanding of bacterial diversity, probably the most frequent type of exploration is the search for bacteria capable of degrading synthetic organic materials, especially those intended for large-scale distribution in agricultural areas as insecticides or herbicides, or components of petroleum mixtures subject to "spills." Enrichment conditions then provide the test substance as sole organic nutrient, whereas conditions of pH, temperature, availability of O_2, etc., may be varied in order to expand the exploration for aerobes or anaerobes, mesophiles or thermophiles, i.e., to support any physiotype capable of subsisting on the test compound. A mixture of generally suitable organic nutrients may also be included in order to allow growth in the presence of the test substance for a period of adaptation to its utilization. In some cases, the synthetic substance may prove usable only if other nutrients are available, even after adaptation, a phenomenon called *cometabolism* (see Hegeman, 1985).

This sort of use of the enrichment technique has been invaluable in the description of bacterial metabolic and physiologic potentials, in recognition of the microhabitat, and in cataloging the diversity of bacterial niches. From such studies has come the generalization that any naturally occurring organic substance can be utilized as a nutrient by some kind of microorganism: for a water-soluble substance, a bacterium can be found to utilize it; for insoluble organic materials, the participation of fungi is often necessary (see Griffin, 1985). Exploratory use of enrichment cultures in this way is not limited to organic nutrients; the suitability of ammonia as an energy source, of atmospheric dinitrogen as a nitrogen source for aerobic bacteria, and of H_2S as an electron source for photosynthesis was recognized by Winogradsky, Beijerinck, and van Niel, respectively, largely as consequences of the study of enrichment cultures.

Exploratory use of enrichment cultures can also be employed to determine whether there are bacteria with properties other than utilization of a specific substrate. Such cultures have been used, for example, to detect the existence of microorganisms that can grow and multiply under a variety of extreme conditions of heat, pressure, acidity or alkalinity, salinity, or osmotic strength; to detect microbes capable of carrying out specific transformations (e.g., sugar to lactic acid, acetic acid to methane, or CO_2 to organic material without illumination); and to identify the types that can colonize surfaces composed of natural or synthetic materials (see Chapter 2, this volume).

In contrast to sample selection to enrich for a specific type of bacterium, exploratory enrichment cultures are better inoculated with a variety of samples. The presence of a substance may be transient or unsuspected, and/or the bacteria that carry out a given transformation may be so efficient that the process is not detectable in the natural habitat. Consequently, the most suitable source may not be recognizable; on the contrary, it may be a source that seems unlikely.

A second major reason that suitability of a sample as a source of bacteria exhibiting a specific property is not always predictable is the reality of the microhabitat. The scale of the microhabitat is one to two orders of magnitude below the limit of human vision. A single particle of soil or marine detritus, the surface of a leaf, or a bit of animal excrement may house a myriad of microbial types distributed among microhabitats. The desired type may be present and active, but sequestered in a microhabitat *in situ* and not detectable in an undispersed sample. A single natural particle occupied by millions of tenaciously attached highly diverse bacteria would yield only a single colony if plated directly. On the other hand, competitiveness of a given type may depend on the persistence of its microhabitat, which dispersal of the sample could disrupt so that that type could not compete with others in the enrichment culture. Physicochemical methods such as those discussed by Karl and by White (Chapters 3 and 4, this volume) may likewise not be sufficiently sensitive to detect only a few million bacterial cells and, unless the organisms are morphologically distinctive or specific antibody is available, microscopic methods may also be incapable of detecting their presence. As usual, absence of evidence—cultural, chemical, or microscopic—does not constitute evidence of absence; and because of microhabitat occupation, it can also not be taken as evidence of inactivity.

In some cases, highly selective enrichments of only one or a few types of bacteria can be achieved; examples are described below (see Table II, below). Generally, the more extreme or demanding the conditions, the fewer the types that will be enriched. It must be noted, however, that very few sets of conditions result in exclusion of bacterial growth, and most conditions permit either simultaneous or successive populations to develop. This familiar experience of all who have worked with enrichment cultures is attributable to the diversity of metabolic and physiologic capabilities that have evolved among prokaryotes. Very few natural sites are bacteriologically sterile, although presumably some are sites of transient inactive bacterial presence.

BACTERIAL DIVERSITY

The diversity of bacteria is attributable to the very long history of this group of organisms, a history that extends over a period roughly 300 times longer than the history of plants and 500 times longer than that of mammals. Whereas bacterial diversity is understandably a consequence of so long a history, it is perplexing for bacterial taxonomists. Since the technologic achievement of pure cultures in the late nineteenth century (see Clarke, 1985), which provided convincing evidence of hereditary distinctiveness of bacterial types, bacteriologists have attempted to classify bacteria as other major biological groups are classified. Such attempts require designation not only of species, but of higher taxonomic levels, and so interpretation of phylogenetic relationships. The relative facility of phylogenetic inferences regarding eukaryotic organisms provided by evidence of sexual recombination in nature (a common definition of species), of similarities in development from zygote to maturity, and of structural properties in fossilized ancestors is not available for prokaryotes. Instead, phylogenetic inferences have traditionally been based on the extent of similarity among determinable traits of modern isolates— morphologic, physiologic, and/or details of protein or nucleic acid structure. Although superficially cladistic, such inference is based on the presumption that common characteristics reflect common ancestry, with the troubling awareness that common experience of selective pressures may select for similarities in lines of independent descent, just as it does among eukaryotes. Until very recently, however, there was little in the way of practical and convincing means for distinguishing analogous from homologous traits among prokaryotes. Studies of nucleic acid fine structure may provide a means of doing so.

The first characteristics used for classifying bacteria were cell morphology (Cohn, 1872) and physiologic traits (Migula, 1897, 1900). The latter comprised mainly the ability to grow in the presence or absence of air and the identity and relative proportions of fermentation products; both traits are clearly relevant to the ecology of a bacterial group. The history of classification systems for bacteria, the traits they emphasized, and their purposes (determinative or phylogenetic) were recounted in detail by van Niel (1946; see also Floodgate, 1962). At that time, van Niel despaired of arriving at a system that would rationally reflect bacterial phylogeny, although he (with Kluyver) had earlier insisted that "a true reconstruction of the course of evolution is the ideal of every taxonomist" (Kluyver and van Niel, 1936). Nevertheless, van Niel acknowledged several other legitimate uses of classification systems.

Since the 1940s, several additional kinds of characteristics have become available to bacterial taxonomists. Two of these—cell wall chemistry and ultrastructure and nucleic acid composition and homologies—are proving especially useful in inferring probable natural relationships among large clusters of genera, i.e., relationships at suprageneric taxonomic levels. Chief among such characteristics are the presence and detailed composition of the cell wall (murein, pseudomurein, or neither), certain details of lipid composition, the base composition of DNA (expressed as moles% G + C), DNA–DNA and DNA–rRNA homologies, and nucleotide sequences within the rRNA. The last of these has engendered considerable enthusiasm and confidence—especially among its practitioners—that the long-sought means of rationally assessing phylogeny among bacteria has been found.

The technique calculates "similarity values" of one characteristic (rRNA sequence catalogs) (Fox et al., 1980; Stackebrandt and Woese, 1981) rather than several, as employed earlier in bacterial taxonomy (Sneath and Sokal, 1962). Attempts to employ rRNA "distance" (Beers and Lockhart, 1962; Beers et al., 1962) rather than similarity have not been reported. Analysis of rRNA sequences was originally regarded as an approach to bacterial phylogeny free of the distortions that arise from approaches that consider traits upon which natural selection has acted (Stackebrandt and Woese, 1981). More recently, emphasis has shifted to employing as taxonomic criteria only those sequences within certain relatively conserved regions of the rRNA. These subsequences, called signatures, are interpreted as regions that evolved mainly during periods of evolution of major groups, and so are useful in recognizing those groups (Woese *et al.*, 1985*b*). This revised view of the phylogenetic implications of rRNA sequences is consistent with the notion of discontinuous rates of evolution, or "punctuated equilibria" (Eldredge and Gould, 1972; see also Stanley, 1981). Continuous or discontinuous, it would seem naive to expect that nucleic acids, which are currently believed to govern the phenotype of an organism, could evolve independently of the consequences for fitness of that phenotype. The evolution of proteins is demonstrably subject to selective pressures (see, e.g., Dickerson, 1972) and, although a mechanistic explanation has not yet been proposed, the application of rRNA sequence analysis to bacterial taxonomy is providing strong evidence that this is also true for the evolution of rRNA.

The relatively high cost of sequence analysis may be one of the reasons that actual practice of the analysis is not yet widespread (see Stackebrandt and Woese, 1981); nevertheless, as more laboratories engage in this approach, the number of species so analyzed should increase over the 400 or so employed during this first decade of its availability. The approach unfortunately has disadvantages more serious than expense, a major one being its interpretive limitations. For example, the length of branches in phylogenetic trees based on this analysis cannot be interpreted, nor can the order of branching within a phylogenetic line (Fox *et al.*, 1980; Woese *et al.*, 1985*b*). The significance of rRNA sequences seems not yet clear, and the resulting taxonomic scheme is still unsettled. Consequently, generic groups appear in significantly different positions as the emphasis shifts from the use of catalog similarities (Fox *et al.*, 1980) to signatures (Woese *et al.*, 1985*b*). For example, *Leptospira* appeared almost as distant from the spirochetes as from all other eubacteria in the earlier analysis, but was later moved into the same major phylum with the spirochetes; similarly, *Desulfovibrio* moved from a position within the purple photosynthetic bacteria and their relatives to a category accommodating bdellovibrios and myxococci. Such drastic rearrangements imply that the ability to interpret relationships on the basis of rRNA sequences is to some extent still speculative, as is the use of older criteria.

Probably another major reason for the somewhat limited participation in attempts to classify bacteria according to rRNA sequences is that this approach yields a classification system whose sole use is inferring phylogeny. It is not required to yield any other information regarding bacterial characteristics. Nevertheless, it does, and therein may lie its greatest value: the resulting scheme implies an evolutionary heirarchy of bacterial characteristics inferred independently (so far as we know at present) of those traits. It can be anticipated that phylogeny revealed through rRNA analyses, especially of signature regions, will also reveal which bacterial characteristics have varied at crucial points in the descent of bacteria. Identification of points at which major lines of descent have diverged will further imply not only which characteristics were strongly favored and whose appearance established new lines of descent, but also the environmental changes that were the likely pressures that favored establishment of a fundamentally new kind of bacterium. The historical periods of those changes may be available from evidence outside bacteriology. The advent of rRNA analysis as a phylogenetic approach is indeed welcome, for its implications—when translated into characteristics of strucuture, biochemistry and physiology—will help us not simply decipher bacterial phylogeny, but understand it in terms of the history of bacterial interactions with each other, with eukaryotes, and with their evolving environments.

It is interesting in this respect to compare older systems with that which is developing through the newer, molecular biological approaches. In a genus-by-genus comparison of the system proposed by Kluyver and van Niel (1936) and in recent versions of bacterial "phyla" based on rRNA analysis (summarized in Woese et al., 1985b), a remarkable, and reassuring, degree of agreement is apparent. Had Kluyver and van Niel held gram reaction as more significant (a feature we now know to be correlated with structure and chemistry of the cell wall), and had these workers had DNA composition available as a character and known of its role in heredity (before the recognition of DNA as genetic material by Avery et al., 1944), the correspondence would undoubtedly be far higher. The older system used cell morphology as the principal taxonomic criterion because it was the closest parallel among bacteria to structural development, features of which have been invaluable in inferring phylogeny among eukaryotes. Spheroidal morphology was regarded as primitive and as the source of other cell shapes, so that the tribe Micrococceae served as a central and internally heterogeneous group through which other families and tribes were related. Features of metabolism, principally catabolism, were used at lower taxonomic levels. Ten genera were created in order to cluster known isolates, several genera were regarded as only provisionally placed, and many genera, esepecially those of medical importance, were at that time so incompletely characterized either morphologically or physiologically as to preclude any rational positioning.

The comparison is illustrated in Fig. 1, which is based on only those 22 genera (19 eubacterial and three archaebacterial) that appear in both systems. It is evident that the outline of bacterial classification is not significantly different in the two systems. A striking feature is seen among the three genera of methanogens; superimposed on the eubacteria in the older system, they were nevertheless well separated, as they are now, as archaebacteria, in the rRNA system. Both systems acknowledge the close relationship between mycobacteria and actinomycetes, and neither system assigns a position among eubacterial groups for myxobacteria, spirochetes, or cyanobacteria. In contrast, the genus *Alcaligenes*, which Kluyver and van Niel could place only provisionally near gram-positive bacteria with similar catabolic characteristics, has apparently found a place among other gram-negative bacteria in the newer

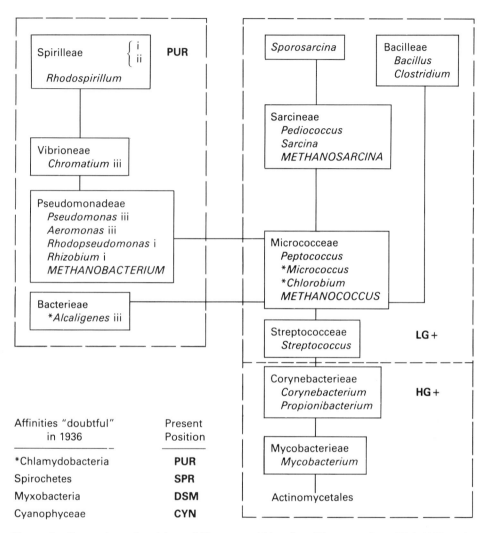

FIGURE 1. Comparison of positions of 22 genera within tribes (Kluyver and van Niel, 1936) and within "phyla" (Woese *et al.*, 1985*b*). Names of "phyla" are in bold face: **PUR** (purple photosynthetic bacteria and relatives; i, ii, iii refer to subgroups within this "phylum"); **LG+** (gram-positive bacteria, low moles % G + C); **HG+** (gram-positive bacteria, high moles % G + C) [**LG+** and **HG+** are "subphyla" of **GM+** (gram-positive bacteria other than radiation-resistance micrococci)]; **SPR** (spirochetes and relatives); **DSM** (sulfur-dependent bacteria and relatives); **CYN** (cyanobacteria). Names of archaebacterial genera are in capital letters. *Positions of eubacterial genera revised on the basis of rRNA analysis: *Alcaligenes* is the only genus shown here in a position different from that in Table II of Kluyver and van Niel (1936); see text. *Micrococcus* spp. are currently assigned to both "subphyla" of **GM+**. *Chlorobium* has been removed to its own "phylum" (**GN**). *Sphaerotilus* (Chlamydobacteria) has been assigned to **PUR**.

classification scheme, as have the sheathed bacteria (Chlamydobacteria). (Kluyver and van Niel required homogeneity with respect to the gram reaction only at the generic level, whereas in the newer system, all the gram-positive bacteria are assigned to one or the other of two phyla.)

It is especially striking that both systems imply several independent origins of photosynthesis. Five of the tribes in the older system included both photosynthetic and nonphotosynthetic genera; a scattering of photosynthetic groups is also inferred from rRNA analysis (Woese *et al.*, 1985a). *Chlorobium*, for example, now separated from all other eubacteria, had been placed by Kluyver and van Niel in the central heterogeneous group, apart from all purple photosynthetic genera. The use of light is not limited to eubacteria; among the archaebacteria, a unique means for utilization of solar energy has evolved, employing bacteriorhodopsin rather than bacteriochlorophylls (Stoeckenius, 1976). Recently recognized photosynthetic types such as *Chloroflexus* and *Heliobacterium* have been distinguished from longer-known types on the basis of both biochemical features of their photosynthesis and rRNA sequences. The implication that the use of extraterrestrial energy has been an adaptive advantage in so wide a variety of morphologic and physiologic settings should impress us yet again with the limitations imposed on living systems by their need for metabolic energy and the inescapable pressures brought to bear on their evolution by environmental conditions that determine the availability of usable energy.

Three eubacterial and two archaebacterial groups currently classified were unknown to, or insufficiently characterized to be considered by Kluyver and van Niel; these are the peptidoglycanless eubacteria (*Planctomyces, Pasteuria*), *Chloroflexus* and *Herpetosiphon*, and *Deinococcus*, none of which is placed by rRNA analysis within a eubacterial phylum containing genera accommodated by Kluyver and van Niel, and the extreme halophiles and thermoacidophiles (or sulfur-dependent archaebacteria) among the archaebacteria. These groups are peculiar in biochemistry, rRNA sequences, and ecology. Their discovery, isolation, and characterization required attention to habitats, behaviors, or morphologies not typical of bacteria. It is interesting to note that most bacterial types discovered during the past 20–30 years and regarded as new by their discoverers are not accommodated in the emerging "phyla" based on rRNA analysis and are being set apart as their own "phyla." They are unique, by everyone's criteria.

Probably the most discouraging result of the rRNA approach is the implication that the origins of major bacterial groups (clusters of genera) are so ancient that even this, the most promising addition to the taxonomist's toolbox, generally fails to detect major relationships that have been undetectable or uninterpretable using other criteria. Despite its helpfulness at lower taxonomic levels and its contribution to the recognition of archaebacteria as more distinct [by morphologic, physiologic, and ecologic, as well as marcromolecular, criteria (Woese and Fox, 1977; Woese *et al.*, 1978)] from eubacteria than any eubacterial groups are from each other, rRNA sequence analysis has served more to justify (both in the natural historian's and in the printer's sense) earlier phylogenetic systems than to revolutionize bacterial classification. It continues to imply that bacteria evolve as organisms, each a constellation of structural, biochemical and physiologic properties that determine its fitness.

Aside from formal classification of bacteria, informal group designations of greater or lesser breadth are widely used in reference to bacterial types, but, unlike "seed plants" or "birds," these terms do not usually refer to discrete recognized taxa. Instead, they usually acknowledge physiotypes and, less commonly, morphotypes.

The physiotype designations are especially relevant to enrichment cultures, all of which in practice favor a range of physiotypes during incubation that is narrower than the range of types in most natural samples. Terms such as lithotroph, organotroph, oligotroph, thermophile, and acidophile are used to refer to organisms with the capacity for lithotrophy, organotrophy, etc. Unfortunately, this usage conveys a connotation of physiologic distinctiveness that is not usually intended; nonparasitic types, in particular, are versatile as well as diverse.

The versatility of bacteria undoubtedly is important to their activities in nature, but it complicates classification (and ecological interpretations) by confusing the question of niche. If, for example, a bacterium can, in pure culture, use both inorganic and organic substances as energy sources and so is regarded as a facultative lithotroph, or mixotroph, which type of substance does it use in nature? Does its versatility enable it to use whichever is available, or do conditions in its usual habitats compel it to use one sort only in order to avoid competition (see Chapter 7, this volume)? Is its niche lithotrophy, organotrophy, or versatility? To some extent, the study of the bacterium's behavior in enrichment cultures will help resolve this question. Nevertheless, its classification will probably be based on its versatility, whether overall properties, biochemical properties, or details of its nucleic acid structure are employed as taxonomic criteria.

Through most of the evolution of bacterial taxonomy, species have been recognized by their inferred niches and, since the work of Winogradsky, considerable taxonomic weight has been given to properties believed to be integrally related to each type of bacterium's role in nature. The recent addition of molecular biological approaches to bacterial taxonomy has not changed this situation. Despite the lack of criteria as discrete (not to be confused with concrete) as those available to students of eukaryotes, lower prokaryote taxa (species and genera) originally recognized on the basis of cultural—both enrichment and pure—characteristics have withstood the challenge of comparative enzymology, serology, nucleic acidology, and computerized analysis, and neither genera nor species are significantly more susceptible to being discarded or repositioned on the basis of such studies than on the basis of criteria longer in use. Distinctive types of bacteria are still more conveniently, usefully, and quite dependably recognized by their behavior under specified environmental conditions, i.e., by traits directly related to their respective niches and to the conditions appropriate for their artificial selection in enrichment cultures.

CONDITIONS FOR ELECTIVE CULTIVATION

Closed-System Liquid Enrichment Cultures

The most widely used type of elective culture consists of a nutrient-containing liquid (the culture medium) inoculated with the sample and incubated, with or without agitation, under a defined set of physical and chemical conditions. This type of culture is suitable when the desired type is able to grow and multiply at the expense of a nutrient(s) not suitable, under the cultural conditions, for most other types, and/or if the desired type is able to multiply under physiochemical conditions not tolerated by other types likely to be present in the sample. It should be emphasized that the nutrient mixture and environmental conditions are not necessarily optimal for growth of the desired type; rather, they are optimal for competition by the desired type with cohabitants of the site sampled.

Besides the identity of the soluble nutrients provided, parameters that will influence microbial multiplication during incubation include the temperature; the presence, intensity, and wavelength of illumination; mechanical agitation; the composition of the gaseous atmosphere; and the concentration of solutes (expressed as pH, salinity, total ionic strength, osmotic strength or water activity, or total organic nutrient concentration).

A small selection of examples of usually dependable enrichment procedures is presented in Table II. The examples discussed in the remainder of this section are intended to illustrate the kinds of conditions that can be varied in order to increase the selectivity of a given enrichment, the procedural steps that can increase the likelihood of success of the enrichment, and how certain ecological implications can be drawn from events observable in enrichment cultures.

Chemolithotrophic Bacteria

The most important considerations in the enrichment of chemolithotrophs are the exclusion of light and of organic compounds. The first of these would permit development of photolithotrophs, and the second would support metabolic activity of organotrophs. When these two potential sources of energy are excluded, particular chemolithotrophs can be favored according to the identities of the electron donor and acceptor provided. Ammonia oxidizers (e.g., *Nitrosomonas, Nitrosococcus*) are favored when NH_3 is provided as the electron donor, nitrite oxidizers (e.g., *Nitrobacter*) by nitrite availability, and *Thiobacillus* spp. when sulfide or thiosulfate is present. Selectivity can be further restricted among the thiobacilli by excluding O_2 and substituting nitrate as respiratory electron acceptor; this selection strongly favors a single species, *Thiobacillus denitrificans*.

Enrichment for chemolithotrophs is usually transient, at best (see, e.g., Kuenen and Tuovinen, 1981), because their growth and multiplication are accompanied by the fixation of CO_2. Organotrophs present in the sample will

TABLE II
Examples of Dependably Selective Enrichment Procedures

| | Conditions | | | | | |
| | Source of | | | | | |
Organism	Aeration	Energy	Carbon	Other nutrients[a]	pH	Other
Nitrobacter winogradskyi	Aerobic	Nitrite	CO_2	Fe	7	25–30°C
Paracoccus denitrificans	Anaerobic	H_2	CO_2	Nitrate (2%)	7.2–7.5	
Thiobacillus denitrificans	Anaerobic	$S_2O_3^{-2}$	CO_2	Nitrate (0.5%)	7	
Thiocapsa pfennigii	Anaerobic	Light, >900 nm	CO_2	S^{2-}, vit. B_{12}	7.2–7.4	
Rhodopseudomonas palustris	Anaerobic	Light, >900 nm	benzoate	NH_4^+, pABA, vit. B_{12}, CO_2	5.2–6.8	
Chlorobium spp.	Anaerobic	Light, 700–750 nm	CO_2	S^{2-}, vit. B_{12}	6.6–6.9	500–2000 lx
Sporocytophaga myxococcoides	Aerobic	Cellulose		Casitone, yeast extract	7.2	Sample sonicated
Bacillus pasteurii	Aerobic	Urea (5%)		Yeast extract	9.0	Sample heated
Bacillus stearothermophilus	Aerobic	Peptone, yeast extract			7	65°C, sample heated
Beijernckia spp.	Aerobic	Glucose		N_2, MoO_4^{-2}	5	30°C
Clostridium pasteurianum	Anaerobic	Glucose or sucrose		N_2, pABA and biotin	7.2	Thioglycolate
Halobacterium spp.	Aerobic	Amino acids		Yeast extract	7	Agitation, NaCl (25%)
Methanosarcina spp.	Anaerobic	Calcium acetate		CO_2	7	37°C, S^{2-}

[a] Nutrients in addition to essential minerals, including sources of N, P, S, major cations, and trace metals.

sooner or later be supported by excretions from the lithotrophs or by the death and decomposition of some of their population. For this reason, enrichment of the lithotrophs may be furthered by removal of a sample shortly after bacterial multiplication is evident and dilution of the sample into a second volume of the same medium. This secondary enrichment culture is thereby inoculated with a preenriched population; multiplication of the desired type in the secondary enrichment may be detectable earlier than it had been in the primary culture, and its numbers may rise higher before organotrophs begin to multiply.

Phototrophic Bacteria

Phototrophic bacteria are of two distinct types with respect to O_2: anoxygenic types, whose phototrophic growth is typically O_2-inhibited; and oxygenic types, whose phototrophic development is not inhibited by aerobic conditions (see Pfennig, 1985; Allen, 1985). Initially oxic conditions will favor the latter types (cyanobacteria), whose oxygenesis will subsequently inhibit photosynthesis by the anoxygenic purple and green bacteria. However, eukaryotic algae will also be favored with the cyanobacteria; their competitiveness with cyanobacteria can be reduced by incubation temperatures above 35°C, although this will limit the diversity of the cyanobacteria enriched (Rippka et al., 1981). Several types of cyanobacteria are capable of N_2 fixation, and for their enrichment the omission of a nitrogen compound from the culture medium will, at least transiently, favor their development relative to eukaryotic phototrophs and other cyanobacteria unable to assimilate N_2.

Enrichment of anoxygenic phototrophic bacteria is facilitated by anaerobic incubation with longer wavelength light than can be used efficiently by oxygenic phototrophs. Among the anoxygenic types, provision of reduced sulfur compounds will favor green or purple sulfur bacteria at shorter and longer wavelengths, respectively, and organothrophic purple bacteria are favored by wavelengths above 800 nm and substitution of nonfermentable organic commpounds for sulfur compounds. However, students of the anoxygenic phototrophs are keenly aware that the diversity of even so narrow (it may seem) a physiotype is greater in natural samples than in their enrichment cultures, and certainly than in their collections of pure cultures. On occasion, the persistence, insight, and imaginativeness of investigators have yielded yet one more type of phototrophic bacterium, invariably with the assistance of some modification of an elective culture procedure.

Chemoorganotrophic Bacteria: Anaerobic Respirers

Besides O_2, two inorganic compounds are used efficiently by certain types of bacteria as inorganic electron acceptors: nitrate and sulfate. The capacity for nitrate respiration occurs among both gram-positive and gram-negative bacteria found in soils and sediments and among various types of aquatic bacteria from both freshwater and oceanic samples. Enrichment of nitrate respirers by anaerobic incubation in the dark with a nonfermentable organic compound as carbon source will favor any of several types of bacteria, depending principally on the identity of the carbon source and to a lesser extent on the source of the inoculum. Probably the most restrictive such carbon source is methanol; methanol-nitrate cultures inoculated with soil or water and incubated anaerobically in the dark regularly yield a nearly pure population of *Hyphomicrobium* (Sperl and Hoare, 1971; Attwood and Harder, 1972). This highly selective enrichment procedure depends on the rather rare combination of methylotrophy and nitrate respiration, a combination recognized with *Hyphomicrobium* isolates obtained indirectly from other types of enrichments and applied to the design of the procedure specific for hyphomicrobia.

Although nitrate respiration is not uncommon among oxidative chemoorganotrophic bacteria, many cannot reduce the product of the first reductive step (nitrite), which may accumulate to toxic levels. Active respirers will reduce nitrite as well as nitrate, and so will not inhibit their own anaerobic development. Nitrate and nitrite respirers are almost invariably also able to respire O_2, and aerobic incubation increases the efficiency—as both rate and yield—of their growth. Nevertheless, under aerobic conditions, they are often poor competitors with other, strictly aerobic types, and elective cultivation of the nitrate respirers from among the aerobic microbes of a given habitat or sample requires that during enrichment they depend on their less efficient, but more competitive, respiration of nitrate. The extent to which bacteria depend on nitrate respiration for their survival in natural communities is implied by the diversity of bacterial groups in which this property occurs.

Sulfate dissimilation, typically to sulfide, permits enrichment of nonfermentative bacteria, most of which are capable of growing (in pure culture) only under anaerobic conditions. For these bacteria, the enrichment procedure employing organic compounds, sulfate or thiosulfate, and anaerobic conditions is presumably a direct reflection of their usual mode of metabolism in nature. In contrast to nitrate respirers, dissimilatory sulfate-reducing bacteria are not distributed in a variety of habitats; nevertheless, they are abundant wherever conditions appropriate to this mode of existence are found, as in marine and estuarine sediments where seawater provides sulfate and decaying plant and algal matter provides the organic material.

Chemoorganotrophic Bacteria: Free-Living Diazotrophic Bacteria

All organisms require a source of nitrogen in their diet, and bacteria are exceptional in this regard only in their peculiar ability to utilize the most abundant form of nitrogen in the biosphere—atmospheric dinitrogen. (Dinitrogen-fixing plants do so only in symbiosis with certain bacteria, which in turn fix N_2 only in association with the plant; see Benson, 1985.) Chemoorganotrophic bacteria capable of fixing N_2 as free-living (nonsymbiotic) organisms occur among several bacterial groups. The longest-known genera are

Azotobacter and *Clostridium,* but since the development of the acetylene-reduction assay for the enzyme dinitrogenase (reviewed in Benson, 1985), several other types have been recognized as possessing this ability. Nevertheless, in elective cultivation of free-living N_2 fixers, particularly from samples of soil, the organisms most likely to arise in significant numbers are azotobacters and clostridia. Development of either is strongly favored by enrichment in media lacking combined nitrogen and provided with an atmosphere containing N_2.

Azotobacter spp. are favored by aerobic conditions and the provision of a carbon source not commonly utilized by aerobic chemoorganotrophs. The traditional and usually successful enrichment medium includes mannitol or benzoate as carbon source. As N_2 is fixed by the azotobacters and some NH_3 or other N-compound is released, the relative advantage of the azotobacters over other chemoorganotrophs continues due to their ability to utilize either of these carbon sources. As in the case of nitrate respirers, the enrichment conditions favor the competitiveness of azotobacters, yet their growth in pure cultures is enhanced by the availability of exogenous NH_3. In both cases, elective cultivation depends on the ability of the desired type to call into use a system (for nitrate respiration or for N_2 fixation, respectively) repressed by conditions under which they grow more efficiently, but where their competitiveness is reduced.

Enrichment of N_2-fixing clostridia is favored by anaerobic conditions, a complex carbohydrate such as starch as the carbon source, and pretreatment of the sample with heat ("pasteurization"). This last condition kills non-spore-forming anaerobes that would begin to develop as the clostridia solubilized the starch. All three conditions prevent the development of azotobacters.

Chemoorganotrophic Bacteria: Fermentative Bacteria

"Fermentative" as a designation of physiotype refers to organisms capable of growing in the absence of an external electron acceptor; the needed acceptor is generated endogenously from an organic compound that serves also as carbon source for growth and as electron donor for energy metabolism. Accordingly, elective cultivation depends on the ability of such types to grow and multiply in the dark, in the absence of O_2, nitrate, or sulfate as electron acceptor, at the expense of a single organic compound, although sometimes with organic micronutrients present. The most widely used fermentable compounds are sugars; elective cultivation of specific types of fermentative bacteria requires either that a nonsugar or a sugar that is not widely fermented be provided or that a condition be imposed that prevents the development of most of the fermentative types.

Lactic acid-producing bacteria, for example, are favored if the initial pH of the enrichment culture is adjusted to pH 5; although many fermentative bacteria excrete organic acids as products of their fermentative metabolism, their own growth is often inhibited by acid accumulation. Many types of lactic acid bacteria are exceptionally tolerant of acidic conditions and are not so inhibited. Again, as in enrichment of anaerobic N_2-fixers, clostridia are favored by pretreatment of the sample with heat to kill vegetative cells and enrich for heat-resistant endospores in the inoculum. Specific carbon sources will also favor clostridia, since many members of this group digest and ferment insoluble carbohydrates such as starch and cellulose, while others are capable of fermenting amino acids—not a common bacterial property. Less restrictive as a carbon source is the sugar lactose; a great variety of bacteria can ferment lactose. However, among lactose-fermenting types are the enteric bacteria, so called because of their wide distribution in animal intestines. Enteric bacteria (and their free-living relatives) generally grow rapidly and do not require growth factors such as vitamins; they are also typically resistant to the action of bile salts. Thus, their enrichment is favored by anaerobic conditions, lactose as sole organic nutrient, and the presence of a bile salt such as desoxycholate. This enrichment depends on a property evoked in these bacteria by the substrate (lactose), although pure cultures grow at least equally well by fermenting glucose.

Bacteria Tolerant of Extreme Conditions

Several groups of bacteria display characteristics of types already discussed, as well as other physiotypes, but also exhibit an uncommon resistance to, or tolerance of, one or another condition that prevents development of metabolically similar types. An incubation temperature of 55–60°C will select for thermophiles of several genera among cyanobacteria, for *Bacillus stearothermophilus* among aerobic spore-forming bacteria, for *Desulfovibrio thermophilus* among sulfate reducers, or for *Thermus* spp. among aquatic organotrophs. An initial pH lower than 5 will favor *Thiobacillus thiooxidans* among sulfur-oxidizing lithotrophs, whereas a pH value greater than 9 will favor certain *Bacillus* spp. among relatively unspecialized oxidative chemoorganotrophs. Aerobic incubation in the dark in a medium with salinity above the equivalent of 20% sodium chloride will permit the development only of some fungi and, among the bacteria, of *Halobacterium* spp. Simply allowing a water sample to stand in the dark, without addition of nutrients, and especially if diluted with clean, sterile water, will favor the slow but dependable development of oligotrophic bacteria such as *Caulobacter* and *Seliberia* spp. Inclusion of specific antibiotics has been found to favor fairly narrow groups of bacteria; the most dependable is rifampicin, used to inhibit most bacteria during the enrichment of spirochetes.

It should be emphasized that none of the examples cited provides a suitable and predictable elective enrichment by dependence on only a single environmental factor; seldom will the identity of an organic substrate or of the electron donor or acceptor, the wavelength of light, the pH or incubation temperature, or any other condition alone suffice to yield a specific enrichment. Only in otherwise specified cultural contexts is a single factor likely to favor one type of bacterium over another (see Table III). This does not so

TABLE III
Single Factors That Influence the Course of Certain Enrichments

Conditions	Factor	Organisms favored
Aerobic, nitrate, 30°C	Carbon source	Succinate: *Pseudomonas fluorescens* or *putida* Muconate: *Pseudomonas acidovorans*
Aerobic, benzoate	Nitrogen source	Nitrate: *Pseudomonas* spp. N_2: *Azotobacter* spp.

Anaerobic, glucose, NH_4^+	Micronutrients	Vitamin and amino acid supplements:
		    Lactic acid bacteria
		    Without supplements: Enteric bacteria

Anaerobic, glucose, yeast extract	pH	pH 5.3: *Lactobacillus* spp. pH 7.0: *Streptococcus lactis*
Aerobic, succinate, nitrate	Temperature	35°C: *Pseudomonas fluorescens* or *putida* 42°C: *Pseudomonas aeruginosa*
Anaerobic, ethanol, biotin	Wavelength of illumination	800–900 nm: *Rhodospirillum rubrum* 1000–1100 nm: *Rhodopseudomonas viridis*

much imply a limitation of elective cultivation methods as it serves as evidence of the multiple environmental influences that, in nature, determine and constantly change the composition of bacterial communities.

Open-System Elective Cultivation

An elective culture constitutes a closed system when it is allowed to develop from a single inoculum in a finite volume of medium. An open-system culture may be "open" in either or both regards: repeated or continuous inoculation or renewal of medium. Such enrichment techniques have been used for elective cultivation from natural samples (Jannasch, 1967; Veldkamp, 1976, 1977; Jannasch and Mateles, 1974; see also Chapter 7, this volume) and in the selection of various types of genetically-altered bacteria (see reviews by Harder *et al.*, 1977; Calcott, 1981; and Dykhuizen and Hartl, 1983).

The usual open-system enrichment culture employs an arrangement in which the enrichment medium is delivered continuously to a constant-volume culture vessel from which liquid leaves by overflow as rapidly as it enters from the medium reservoir.

A major practical and theoretical problem in bacterial ecology and in designing enrichment cultures is identification of the nutrient that limits the size of the population *in situ*. That substance presumably is consumed as rapidly as it appears within the community so that it will escape detection by most types of analyses. That is, the limiting nutrient should be undetectable; a utilizable substance that can accumulate is probably not limiting microbial metabolism. It follows that those inhabitants of a site having the greatest affinity for the limiting nutrient will be favored by its continued scarcity. As soon as the habitat is contained as a consequence of its becoming a sample, the limiting nutrient will be exhausted. The competitiveness of the type(s) best suited for exploitation of environmental conditions limited by that nutrient is thereby reduced; it/they may never appear numerically significant in any sort of contained enrichment culture.

The closest approximation to conditions favoring such bacteria are chemostatlike cultures provided with the ambient liquid (the water of an aquatic habitat, an extract or perfusate of a particulate habitat) as a source of nutrients. Establishing such conditions is not simple in practice, largely because of the problems relating to containment of the liquid itself. If it is to be used without sterilization, but stored in a reservoir, its limiting nutrient will be exhausted in the reservoir, and populations changing in numerical proportions in the reservoir will be introduced continuously into the growth vessel. The latter problem seems relatively insignificant when the bacterial count in the liquid is low and a substantial population develops within the growth vessel (Schlegel and Jannasch, 1967), an observation consistent with the ecological principle of competitive exclusion (see Chapter 9, this volume). If the liquid is sterilized by heat, the limiting nutrient may (1) be destroyed by the sterilization process, (2) sorb to the reservoir, or (3) deteriorate on standing in the reservoir due to spontaneous chemical changes. Ideally, the ambient liquid could be delivered directly from the habitat to the culture vessel via a filter sterilization system, with the filter almost continually replaced to prevent the accumulation of metabolically active microbial cells. The technical difficulties of this ideal are obvious; it is difficult to operate such a culture at a rate sufficient to eliminate each problem, even when the habitat is relatively handy, such as ocean water on board a research vessel.

Open-system enrichment cultures can be especially useful in enriching for slow-growing oligotrophic bacteria when the nutrient liquid is supplied at so low a rate that copiotrophs (organisms whose competitiveness is dependent on an abundance of nutrients) cannot develop efficiently. This method is particularly useful in enrichments for bacteria favored by the initial conditions, but not by the conditions they create as they grow and multiply in a container. This includes any kind of bacterium whose growth is self-limited by its metabolic by-products (e.g., acid producers), as well as those (e.g., N_2 fixers) whose growth improves conditions for other types. Both types appear, in closed-system enrichment cultures, as stages in successions.

An additional advantage of open-system cultures is that conditions within the culture are more amenable to experimental control. For example, the pH can be continuously adjusted, the mixed population can be subjected to alternation of energy sources or atmospheres to select for facultative organisms competitive under two (or more) sets of conditions or under fluctuating conditions (Kuenen et al., 1977; Dykhuizen and Davies, 1980), or the rate of delivery of nutrients can be changed to challenge the multiplying organisms to adjust to quantitative fluctuations in nutrient availability.

Open-system enrichment cultures are not regarded by their users as being nearly natural conditions. However, they offer potential for investigation of population fluctuations that closed-system cultures, which are unidirectional with time, cannot. Thus, they represent a dynamic sort of enrichment culture that could be more fully exploited in studies of bacterial ecology than has yet been realized in practice.

Two-Phase Elective Cultivation

In nature, very few bacteria seem to exist as free-floating cells not associated with any physical phase other than water. Even in the open ocean, a high proportion of the "planktonic" bacteria are associated with particles (ZoBell, 1946; Sieburth, 1979). A variety of bacteria contain gas vacuoles and do not seem to associate with particles; these may migrate vertically in the water column (Walsby, 1981).] In general, rapid multiplication in free suspension does not appear to be the natural habit of the majority of bacteria. It is, however, typically the growth habit required (by bacteriologists) for bacterial cultivation—both during enrichment and in pure cultures. For most types of studies—biochemical, physiological, and genetical—bacteria that adhere to culture vessels or to each other are regarded as unsuitable for experimentation. They live, nevertheless, and they can be electively cultivated as well as studied *in situ* (see Chapters 1 and 2, this volume).

Bacteria capable of establishing themselves at interfaces can be enriched by inclusion of an appropriate interface as a component of the enrichment culture. For example, an immobile solid surface exposed to a stream of nutrients has been used to enrich the surface with populations of *Leptothrix* (Mulder and Deinema, 1981); the wall of the culture vessel also is the site of enrichment of *Sphaerotilus* (Mulder and Deinema, 1981) and of *Gallionella* (Hanert, 1981). Each of these types is morphologically distinctive and can be recognized by microscopic examination of material removed from the surface. Macroscopic masses may also accumulate; in the case of *Gallionella*, in particular, iron oxides accumulate in the extracellular stalks, and the mass becomes characteristically orange in color (illustrated in Hanert, 1981).

At the surface of an unagitated liquid culture, there is a stable gas–liquid interface. Various types of bacteria, by mechanical means or due to positive aerotaxis, tend to accumulate and may be numerically enriched in this artificial neuston even when practically undetectable elsewhere in the culture. The surface biofilm of clean-water samples not amended or amended with only very low concentrations of nutrients is the most dependable enrichment for *Caulobacter* species, as long as care is taken to remove for examination or subcultivation only the film and very little of the bulk aqueous phase (Schmidt, 1981).

Liquid–liquid interfaces can also be employed as enrichment sites. This sort of interface occurs when a water-insoluble substrate such as a petroleum fraction is provided as the principal carbon source. Hydrocarbon-oxidizing bacteria typically establish themselves on the surface of the microdroplets of hydrocarbonaceous liquid dispersed in the aqueous phase (see, e.g., Fig. 1 in Rosenberg and Gutnick, 1981). When agitation of the culture is halted and the droplets coalesce, they may provide a mechanical enrichment from a microhabitat that is allowed to enlarge to practical dimensions.

The most refractory technical problem in the use of two-phase enrichment cultures lies beyond the scope of this discussion. That problem is separation of the population from the interface for isolation of monotypic populations in the absence of the interface. Even when separable from the nonaqueous surface, adhesive bacteria may adhere to each other so tenaciously that suitable methods for separation disrupt the cells as or more efficiently than the clumps. The resulting purified populations may still require an interface for their maintenance and grow poorly or not at all in the favored routine suspended-cell culture.

In a reverse manner, certain motile bacteria are readily enriched by their ability to move along interfaces that pose barriers to most types. Both spirochetes and gliding bacteria can migrate through agar gels and membrane filter pores that impede the movement of bacteria with external flagella. Multiplying as they migrate, their presence becomes discernible as a fog of bacterial cells within the agar away from the site of inoculation (Canale-Parola, 1973) or as colonies arising adjacent to the uninoculated side of a membrane filter (Reichenbach and Dworkin, 1981).

Bacteria that can utilize solid, water-insoluble substrates present a special case of two-phase elective cultivation in which only a thin film of the aqueous phase is involved. (Unlike fungi, many of which can use "dry" insoluble substrates, bacteria are generally capable of only very limited development when water is available only as vapor or as liquid absorbed within the substrate.) A typical enrichment for cellulose-digesting bacteria such as *Cytophaga* spp. consists of sterile paper (cellulose) laid on the surface of an agar medium containing the required mineral nutrients. Cytophagas produce cellulolytic enzymes, solubilize the paper, and assimilate the digestion products (sugars). Subsequent transfer of cells growing on the paper to secondary enrichment cultures with cellulose suspended in an agar overlayer usually leads to successful enrichment for these organisms. Similar arrangements employing chitin, uric acid, starches, or proteins will enrich for other types of hydrolytic bacteria. Development of hydrolytic bacteria results in release of soluble nutrients and, as in natural habitats, converts the substrate into forms available to other types of bacteria.

ELECTIVE CULTURES AS MICROBIAL SUCCESSIONS

Very few closed-system enrichment cultures yield a single predominant population, pure or mixed, that develops to a constant proportion that is maintained indefinitely. On the contrary, the development of the population favored by the initial selective conditions invariably changes conditions within the culture and leads to the rise of one or more other types. The conditions affected will include some or all of the following: concentration of nutrients, identity of solutes (especially organic substances), pH, pCO_2, and availability of O_2. In the special case of two-phase (solid–liquid) cultures, the solid surface may be conditioned by pioneering adhesive bacteria so that it becomes suitable as a substratum for adhesion by other types (Sieburth, 1979; see also Chapter 2, this volume).

As the microbial population develops and changes, the proportion of the desired type may pass a peak; only during this peak may its relative numbers be sufficiently high to allow subsequent isolation. The peak may be recognized by direct examination of the culture for properties associated with the growth and multiplication of the desired organism (e.g., an extreme change in pH; the production of H_2, N_2, O_2, or CH_4; production of a particular organic compound, especially of fermentation products such as ethanol or lactic acid; solubilization of a substrate or deposition of an insoluble product such as elemental sulfur or metal oxide or sulfide; accumulation of pigment in cell masses or diffusing into the medium), or by direct microscopical detection of a distinctive aspect of morphology (endospores; sheathed filaments, cellular or extracellular stalks, capsules) or motility (jerking motions of spirochetes, gliding motility of cytophagas, very rapid swimming of bdellovibrios). Without such indicators, the peak may be detectable only in hindsight, by identification of colonies among those arising on solidified media inoculated periodically from the enrichment culture. However, if other characteristics of the culture were monitored each time a sample was removed to solidified medium, a parameter may be identified that will be a suitable indicator for use in subsequent enrichment for that organism.

In enrichment cultures that promote early development of the desired type, subsequent isolation usually employs media and other conditions that are similar to those provided in the enrichment culture at the time of its inoculation (see, e.g., Veldkamp, 1970). For organisms that arise at a later stage of the succession within the enrichment culture, however, isolation conditions should mimic the conditions as they existed in the culture at or just before the peak development of the desired type. For this reason alone, the changes in the enrichment culture should be monitored as completely as possible. Subsequent enrichments, as well as isolation procedures, may be redesigned so that earlier stages in the succession can be eliminated and the desired type favored earlier and more selectively.

Particularly when it can be demonstrated that the course of a succession can be modified on the basis of such information, useful implications regarding the influence of conditions in the habitat can be drawn.

LIMITATIONS AND PROSPECTS FOR ELECTIVE CULTURES IN BACTERIAL ECOLOGY

Because enrichment cultures result in changes in relative proportions of members of the microbial population in a sample, they are not useful when the desired type is numerically predominant in a sample or when an organism is sufficiently peculiar in physiologic traits that stringently selective isolation conditions can be employed. The first instance is typical of clinical specimens consisting of normally sterile materials such as blood or other body tissues. In such materials, the presence of the organism is manifested by the symptoms typical of its invasion and multiplication in the plant or animal host; a suitable isolation procedure normally requires only direct inoculation of the tissue sample into or onto a growth-supporting medium that mimics the composition of the susceptible tissue and that preferably permits differentiation of the organism sought from other microbes likely to accompany it. This approach is suitable even for the isolation of organisms such as *Salmonella* spp. from stool specimens. Most enteric bacteria are obligate anaerobes and will not multiply on plates incubated aerobically; among the facultative anaerobes, salmonellae become so numerous in hosts exhibiting intestinal disease symptoms of their presence that they are usually detectable on primary plating media incubated aerobically.

In other instances, the desired type may be the usual predominant bacterial type in a given habitat, such as species of *Arthrobacter* in soil samples (Keddie and Jones, 1981), of *Halobacterium* in brines (Larsen, 1981), and of *Thermus* in acidic hot springs (above 70°) (Brock, 1981). For morphologically distinctive types that develop in fairly homogenous, extensive (even macroscopic) masses, such as *Leptothrix* (Mulder and Deinema, 1981) and *Chloroflexus* (Castenholz and Pierson, 1981), the desired type can be removed mechanically from the habitat or sample and the inoculum thereby enriched by manipulation before cultivation.

Such situations are special and limited to a few types of samples or bacteria. Nevertheless, they are instructive with respect to the ecology of those materials and bacteria as examples of unusual habitats to which bacterial access is limited and wherein bacterial multiplication is restricted, and of niches for which bacterial diversity is low. The survival and competitiveness of such bacteria can in many such cases be interpreted with reference to their properties of invasiveness, survival during conditions unsuitable for the growth of most microbes, or morphologic traits associated with the development of cohesive or mutually protective colonies. Thus, the lack of need for enrichment cultivation can itself be a source of information about the ecology of a bacterium.

Enrichment Cultures as Enumeration Techniques: The MPN Method

Most probable number (MPN), or dilution-to-extinction, determinations are a quantitative application of the elective culture technique. These methods consist of diluting a liquid sample or a suspension of a solid sample, such as soil, and using the serial dilutions as inocula for separate enrichment cultures. From the proportion of cultures that eventually exhibit evidence of the presence of the organism sought (actually, from the proportion lacking the desired type; see Gerhardt *et al.*, 1981), an approximation of the number of such organisms in the undiluted sample can be calculated. In essence, it provides an estimation of how far the sample must be diluted in order to yield aliquots free of the desired type.

This method is especially useful when the presence of the organism can be detected readily in the enrichment culture itself. Morphologically distinctive bacteria such as spirilla (Scully and Dondero, 1973), spirochetes (Henry and Johnson, 1978), and prosthecate bacteria (Belyaev, 1967; Staley, 1971) have been enumerated by this technique. Extreme thermophiles can also be so enumerated, as well as enriched (Brock, 1918), since any evidence of bacterial growth in enrichment cultures incubated at 70–75°C is evidence of their presence (Brock and Boylen, 1973). The method has also been applied in enumerating bdellovibrios; after incubation with potential host bacteria as enrichment cultures, samples were plated with the potential host, and the plates were examined for *Bdellovibrio* plaques (Klein and Casida, 1967). In this case, one further cultural step was used after the enrichment culture.

The technique is also used in the detection and estimation of potential pathogens in the analysis of water, and of milk and other foods. For decades, the initial step in the officially prescribed determination of fecal coliform bacteria was essentially an enrichment culture for bacteria capable of producing gas by lactose fermentation—the presumptive test.

Whether the initial sample is liquid or particulate, the MPN procedure can yield only a minimum estimate of bacterial numbers, since clumps of cells are not completely disrupted by the procedures. Nevertheless, dilution procedures have proved invaluable in detecting and estimating numbers of bacteria (1) whose proportions in the native populations are too low to permit their detection by direct plating, (2) about which too little is yet known to allow their isolation, and/or (3) that are not known to be sufficiently specialized physiologically to allow the design either of enrichment procedures that will bring the organism to predominance or of an adequately selective plating medium.

Eukaryotes as Living Media

In some noteworthy cases, living organisms (entire host organisms, tissue slices, or cell cultures) are the only known suitable environment for growth and multiplication of the desired type of bacterium. In addition to *Bdellovibrio*, this situation continues today for most endosymbionts of invertebrate animals and ciliate protozoa, the leprosy bacillus (*M. leprae*), and all but one (and a few questioned) of the actinomycete endosymbionts (*Frankia* spp.) of the N_2-fixing root nodules on nonleguminous plants.

Some once refractory isolates have been obtained through procedures that employ living media for enrichment followed by isolation and cultivation in artificial media. The guinea pig, for example, is a suitable initial medium for isolation of *Legionella pneumophila* from clinical and autopsy specimens

(Balows and Brenner, 1981) because it is a relatively poor host for other bacteria likely to be present, except for rickettsiae. Spleen homogenate from the febrile guinea pig is then further enriched in *L. pneumophilia* by subcultivation in the yolk sacs of embryonated hens' eggs. From eggs in which the embryos die, yolk sac material is inoculated onto a complex but artificial medium. Plant pathogens, as well, may be enriched first by cultivation in healthy plants (for xanthomonads: Starr, 1981) or cultivated tissues (for agrobacteria: Ark and Schroth, 1958).

Monotypic cultures in artificial media are highly desirable for characterization of the properties of symbiotic bacteria, and the environments that are eventually found suitable for their axenic cultivation are instructive with respect to the identity of host factors on which they naturally depend. Nevertheless, their propagation as monoxenic populations (i.e., with the host the only "foreigner") permits the preparation of monotypic bacterial suspensions using physical methods that separate the symbionts from the cells of their host. Such procedures allow certain characteristics of the symbiotic bacteria to be determined; for example, the pigment content of *Prochloron* (Withers *et al.*, 1978a,b), the DNA base composition of *Paramecium* endosymbionts (Preer, 1981), the cell-wall composition of *Frankia* spp. (Becking, 1981), the favorable temperatures for multiplication of *Cristispira* (Kuhn, 1981), and antigenic characteristics of rickettsiae (Philip *et al.*, 1976; Robinson *et al.*, 1976) and chlamydiae (Wang, 1971).

For decades, students of symbionts have expended considerable effort in the establishment of monotypic cultures. Monoxenic populations are helpful, but the ecology (as well as the genetics, biochemistry, and physiology) of a given symbiont can be fully appreciated to the satisfaction of those who know it most intimately only when the formulation of an environment that is an adequate substitute for the natural host has been achieved. Enrichment cultures, with living or artificial media, are one of the principal tools in such endeavors.

Stable Prokaryote Associations

Chapter 8 of this volume is devoted to a discussion of microbial associations. Some of these, often called consortia, exhibit characteristic cell arrangements in which a constant relative number of cells of each partner is maintained throughout reproductive events. Others are known as mixed populations of noncontiguous cells, called cocultures. Certain bacterial groups appear frequently in such associations—green sulfur bacteria in the first case, and anaerobic bacteria that occur with H_2-utilizing methanogens or desulfovibrios in the second. Both members of several known consortia have resisted separate cultivation, even though the majority include one member that is phototrophic (Trüper and Pfennig, 1971). By contrast, most of the organisms that arise as members of cocultures during enrichment procedures have yielded one partner in pure culture, while the second has become familiar as an associate of various bacteria (McInerney *et al.*, 1981). The second organism in such cases is seen as the dependent partner (the syntroph) whose environmental needs can be met by more than one other type of bacterium, but whose needs have not yet been satisfied by an artificial inanimate environment.

In such instances, enrichment cultures can reveal the existence of bacterial types and allow inference of some of their properties even when conditions cannot be devised to favor them sufficiently to permit their existence independently of their natural cohabitants. The information so far available concerning the existence of such bacteria has been obtained solely through studies of their presence and behavior in enrichment cultures.

Unexpected Organisms

Because elective cultures are usually set up with the purpose of favoring a particular type of organism, and certain procedures seem likely to be strongly selective, the appearance of a distinctly different creature may surprise even experienced investigators. In such instances, the pursuit of the unexpected can lead to the recognition of a previously unknown organism. A few examples are cited here as evidence that bacteriologists have acted in accord with Pasteur's observation (promise?) that "chance favors the prepared mind." From the multitude of available examples, the four selected illustrate, in order, an organism repeatedly observed in enrichments for a very different physiotype, an organism that could easily have been ignored as a relatively uninteresting example of the type sought, an organism physiologically similar to the expected organism but exhibiting an unexpected physiological trait, and finally, the most recent of this set, an organism enriched by an unintended change in the composition of the enrichment medium.

The first example is *Hyphomicrobium*, a morphologically unique organism that produces an elongated cellular appendage from which progeny develop by budding. *Hyphomicrobium* was detected by microscopic examination of enrichment cultures for nitrifying bacteria, first by Rullman (1897), later by Stutzer and Hartleb (1898), who named the organism, by Prouty (1929), who insisted it was not a nitrifying organism, and by Kingma-Boltjes (1936), who succeeded in isolating it from a population of nitrifiers enriched from a soil sample. *Hyphomicrobium* was later characterized as a methylotropic bacterium and is now known to be capable of growing at the expense of volatile organic substances common in the air of laboratories. Its specific enrichment, employing methanol, nitrate, and anaerobic conditions, was devised only relatively recently (Sperl and Hoare, 1971; Attwood and Harder, 1972). In hindsight, its frequent appearance as a contaminant in enrichment cultures for aerobic chemolithotrophs was most likely due to its capacity for oligotrophy with respect to organic substrates. Its detection was a consequence of its distinctive morphology.

Bdellovibrios, exceptionally small bacteria that are predators on other bacteria, were discovered by Stolp and Petzold (1962) during the course of isolation of bacteriophage lytic for pseudomonads. A soil suspension had been "sterilized" bacteriologically by passage through a filter with a maximum pore size of 1.35 μm. The filtrate was plated on a nutrient medium heavily seeded with the intended host pseudomonads, but after 1 day of incubation phage plaques were not detectable. For reasons not given in the report, the plate was examined again 2 days later; lytic zones were then discerned. A suspension of material from a lytic zone was diluted and replated, and once again lytic areas were detected that, in contrast to phage plaques, appeared only after 2 days or more of incubation and continued to expand in size for another 5 days; phage plaques on this host normally appeared within 24 hours and did not increase in size on further incubation. Microscopical examination of material from the growing lytic zones revealed the presence of small, rapidly motile bacteria that attached to and lysed the larger pseudomonads. Subsequent studies (Stolp and Petzold, 1962; Stolp and Starr, 1963; Stolp, 1965) found that for the isolation of bdellovibrios from soil, liquid enrichment cultures prior to plating were not an advantage because bacteriophages were enriched by the same procedure. The phage plaques developed earlier when the enrichment cultures were plated, obscuring—even precluding—the development of lytic zones by *Bdellovibrio*. In the initial samples, bdellovibrios tended to outnumber phages, and their isolation was favored by direct plating of the sample.

In an enrichment for soil thiobacilli, London and Rittenberg (1967) employed mineral media containing thiosulfate or sulfur as energy source and monitored the pH of the cultures, a strong acidification accompanying bacterial growth being the usual evidence of multiplication of the sulfate-forming bacteria. Aliquots of the enrichment cultures that had reached pH 2.8 were plated on thiosulfate-mineral salts medium; on such a medium, thiobacilli typically produce colonies with centers that are opaque and yellow-orange due to the deposition of elemental sulfur. In this instance, however, one of the two predominant colony types lacked the distinctive sulfur deposit. Subsequent isolation and characterization of the organism found it to be capable of oxidizing sulfur compounds as energy sources, like other thiobacilli, but not capable of subsisting on CO_2 as its sole source of carbon. The bacterium was designated *T. perometabolis*, "a thiobacillus crippled in its autotrophic potential." On yeast extract-thiosulfate agar, colonies were significantly larger than on mineral salts-thiosulfate agar; on prolonged incubation on the richer medium, a small amount of elemental sulfur did appear within the colonies.

In the cases of *Hyphomicrobium* and *Thiobacillus perometabolis*, atmospheric, inoculum-borne, and agar-contaminating substances apparently supplied nutrients required by the unanticipated organisms. These instances serve as reminders that the recipe for an enrichment or isolation medium is not necessarily a full description of the usable nutrients present. In the last of these examples, the recipe itself was "inadvertently" altered, but whether intentionally or whether this was recognized only during analysis of the results was not indicated in the report.

During a series of enrichments for anoxygenic photosynthetic bacteria, Gest and Favinger (1983) employed strict anaerobic conditions and a medium that included organic substrates and a small amount (0.01%) of an ammonium salt. Abundant growth in such enrichment cultures depends on the ability of the bacteria to fix N_2; the small amount of added combined N was provided to support the initiation of growth. Ordinarily, the salt employed was NH_4Cl, but in this series $(NH_4)_2SO_4$ was substituted; this change resulted in sufficient sulfide production by desulfovibrios in the sample to prevent the multiplication of the expected purple nonsulfur bacteria. Instead, an entirely new type of photosynthetic bacterium, designated *Heliobacterium chlorum*, appeared. On subsequent isolation and characterization, this "new" bacterium has exhibited several properties different from the purple nonsulfur bacteria. As a greenish, nonsulfur, photoheterotrophic bacterium, it exhibits some similarity to *Chloroflexus*, including possible gliding motility and rod shape, but it is distinct from the latter organism by lacking chlorosomes, containing a bacteriochlorophyll not previously known, and being unable to grow aerobically. It does not show any significant rRNA similarity to other photosynthetic bacteria (Woese *et al.*, 1985a). Its distribution in nature is yet to be determined; although it was isolated from "an essentially aerobic" soil locale, it is exceptionally sensitive to O_2 and may prove more dependably isolable from anaerobic materials.

Prospects

The examples cited in the preceding section were selected to illustrate a variety of ways in which enrichments not intended to be exploratory nevertheless uncovered the existence of previously unknown organisms. The selection inadvertently also illustrates that microhabitats are the habitats of microbes: all four enrichment procedures began with soil samples. They were intended to yield, respectively, nitrifying bacteria, bacteriophages, sulfur-oxidizing autotrophs, and purple photosynthetic bacteria; they yielded as well (or, in the last case, instead) budding prosthecate methylotrophic oligotrophs, bacterial predators of bacteria, a nonautotrophic sulfur-oxidizing bacterium, and yet another combination of properties that allows phototrophy. The potentials of prokaryotic existence and the extent of bacterial diversity have not been exhausted by a century of study of even this single "habitat"; and a great deal can yet be learned about bacteria in nature through the attentive observation of enrichment cultures. In addition to the living fruits yet to be gathered from exploratory enrichment cultures, the bacterial ecologist can, by this approach, assess and control conditions within a sample of the microbial world and thereby "carry out ecological studies of a precision and refinement that are far beyond the reach of the botanist or zoologist" (Stanier, 1951).

REFERENCES*

Allen, M. M., 1985, Oxygenic photosynthesis in prokaryotes, in: *Bacteria in Nature*, Vol. 1 (E. R. Leadbetter and J. S. Poindexter, eds.), pp. 133–153, Plenum Press, New York.

Ark, P. A., and Schroth, M. N., 1958, Use of slices of carrot and other fleshy roots to detect crown gall bacteria in soil, *Plant Dis. Rep.* **41**:1279–1281.

Attwood, M. M., and Harder, W., 1972, A rapid and specific enrichment procedure for *Hyphomicrobium* spp., *Ant. van Leeuwenhoek J. Microbiol. Serol.* **38**:369–378.

Avery, D. T., MacLeod, C. M., and McCarty, M., 1944, Studies on the chemical nature of the substance inducing transformation of pneumococcal types. Induction of transformation by a deoxyribonucleic acid fraction isolated from pneumococcus type III, *J. Exp. Med.* **79**:137–158.

Balows, A., and Brenner, D. J., 1981, The genus *Legionella*, in: *The Prokaryotes* (M. P. Starr *et al.*, eds.), pp. 1091–1102, Springer-Verlag, Berlin.

Becking, J. H., 1981, The genus *Frankia*, in: *The Prokaryotes* (M. P. Starr *et al.*, eds.), pp. 1991–2003, Springer-Verlag, Berlin.

Beers, R. J., and Lockhart, W. R., 1962, Experimental methods in computer taxonomy, *J. Gen. Microbiol.* **28**:633–640.

Beers, R. J., Fisher, J., Megraw, S., and Lockhart, W. R., 1962, A comparison of methods for computer taxonomy, *J. Gen. Microbiol.* **28**:641–652.

Beijerinck, M. W., 1921–1940, *Verzamelde Geschriften*, Vols. 1–6, Nijhoff, Den Haag.

Belyaev, S. S., 1967, Distribution of the caulobacter group of bacteria, *Mikrobiologiya* **36**:157–162 (in Russian).

Benson, D. R., 1985, Consumption of atmospheric nitrogen, in: *Bacteria in Nature*, Vol. 1 (E. R. Leadbetter and J. S. Poindexter, eds.), pp. 155–198, Plenum Press, New York.

Blakemore, R. P., Maratea, D., and Wolfe, R. S., 1979, Isolation and pure culture of a freshwater magnetic spirillum in chemically defined medium, *J. Bacteriol.* **140**:720–729.

Brock, T. D., 1981, Extreme thermophiles of the genera *Thermus* and *Sulfolobus*, in: *The Prokaryotes* (M. P. Starr *et al.*, eds.), pp. 978–984, Springer-Verlag, Berlin.

Brock, T. D., and Boylen, K. L., 1973, Presence of thermophilic bacteria in laundry and domestic hot-water heaters, *Appl. Microbiol.* **25**:72–76.

Calcott, P. H., 1981, Genetic studies using continuous cultures, in *Continuous Culture of Cells* (P. H. Calcott, ed.), pp. 127–140, CRC Press, Boca Raton, Florida.

Canale-Parola, E., 1973, Isolation, growth and maintenance of anaerobic free-living spirochetes, in: *Methods in Microbiol.*, Vol. 8 (J. R. Norris and D. W. Ribbons, eds.), pp. 61–73, Academic Press, New York.

Castenholz, R. W., and Pierson, B. K., 1981, Isolation of members of the family Chloroflexaceae, in: *The Prokaryotes* (M. P. Starr *et al.*, eds.), pp. 290–298, Springer-Verlag, Berlin.

Clark, R. W., 1984, *The Survival of Charles Darwin: A Biography of a Man and an Idea*, Random House, New York.

Clarke, P. H., 1985, The scientific study of bacteria, 1780–1980, in: *Bacteria in Nature*, Vol. I (E. R. Leadbetter and J. S. Poindexter, eds.), pp. 1–37, Plenum Press, New York.

Cohn, F., 1872, Untersuchungen über Bacterien, *Beitr. Biol. Pflanz.* **2**:127–224.

Dickerson, R. E., 1972, The structure and history of an ancient protein, *Sci. Am.* **226**:58–72.

Dykhuizen, D., and Davies, M., 1980, An experimental model: bacterial specialists and generalists competing in chemostats, *Ecology* **61**:1213–1227.

---

* Several references cite chapters in M. P. Starr, H. Stolp, H. G. Trüper, A. Balows, and H. G. Schlegel, eds., 1981, *The Prokaryotes, A Handbook on Habitats, Isolation, and Identification of Bacteria*, Springer-Verlag, Berlin. Those chapters contain extensive discussions of enrichment and isolation procedures and provide more complete lists of primary reports than are cited here. *The Prokaryotes* was an outgrowth of a symposium on enrichment cultures, published as H. G. Schlegel, ed., 1965, *Anreicherungskultur und Mutantenauslese*, Gustav Fischer, Stuttgart. The reader is directed to both works as major sources of information concerning enrichment cultivation.

Dykhuizen, D. E., and Hartl, D. L., 1983, Selection in chemostats, *Microbiol. Rev.* **47**:150–168.

Eldredge, N., and Gould, S. J., 1972, Punctuated equilibria: an alternative to phyletic gradualism, in: *Models in Paleobiology* (T. J. M. Schopf, ed.), pp. 82–115, Freeman, Cooper, San Francisco.

Floodgate, G. D., 1962, Some remarks on the theoretical aspects of bacterial taxonomy, *Bacteriol. Rev.* **26**:277–291.

Fox, G. E., Stackebrandt, E., Hespell, R. B., Gibson, J., Maniloff, J., Dyer, T. A., Wolfe, R. S., Balch, W. E., Tanner, R. S., Magrum, L. J., Zahlen, L. B., Blakemore, R., Gupta, R., Bonen, L., Lewis, B. J., Stahl, D. A., Luehrsen, K. R., Chen, K. N., and Woese, C. R., 1980, The phylogeny of prokaryotes, *Science* **209**:457–463.

Gerhardt, P., Murray, R. G. E., Costilow, R. N., Nester, E. W., Wood, W. A., Krieg, N. R., and Phillips, G. B. (eds.), 1981, *Manual of Methods for General Bacteriology*, American Society for Microbiology, Washington, D.C.

Gest, H., and Favinger, J. L., 1983, *Heliobacterium chlorum*, an anoxygenic brownish-green photosynthetic bacterium containing a "new" form of bacteriochlorophyll, *Arch. Microbiol.* **136**:11–16.

Griffin, D. M., 1985, A comparison of the roles of bacteria and fungi, in: *Bacteria in Nature*, Vol. I (E. R. Leadbetter and J. S. Poindexter, eds.), pp. 221–255, Plenum Press, New York.

Hanert, H. H., 1981, The genus *Gallionella*, in: *The Prokaryotes* (M. P. Starr *et al.*, eds.), pp. 509–515, Springer-Verlag, Berlin.

Harder, W., Kuenen, J. G., and Matin, A., 1977, Microbial selection in continuous culture, *J. Appl. Bacteriol.* **43**:1–24.

Hegeman, G., 1985, The mineralization of organic materials under aerobic conditions, in: *Bacteria in Nature*, Vol. I (E. R. Leadbetter and J. S. Poindexter, eds.), pp. 97–112, Plenum Press, New York.

Henry, R. A., and Johnson, R. C., 1978, Distribution of the genus *Leptospira* in soil and water, *Appl. Environ. Microbiol.* **35**:492–499.

Jannasch, H. W., 1967, Enrichment of aquatic bacteria in continuous culture, *Arch. Mikrobiol.* **59**:165–173.

Jannasch, H. W., and Mateles, R. I., 1974, Experimental bacterial ecology studied in continuous culture, *Adv. Microbiol. Physiol.* **11**:165–212.

Keddie, R. M., and Jones, D., 1981, Saprophytic, aerobic coryneform bacteria, in: *The Prokaryotes* (M. P. Starr *et al.*, eds.), pp. 1838–1878, Springer-Verlag, Berlin.

Kingma-Boltjes, T. Y., 1936, Über *Hyphomicrobium vulgare* Stutzer et Hartleb, *Arch. Mikrobiol.* **7**:188–205.

Klein, D. A., and Casida, L. E., 1967, Occurrence and enumeration of *Bdellovibrio bacteriovorus* in soil capable of parasitizing *Escherichia coli* and indigenous soil bacteria, *Can. J. Microbiol.* **13**:1235–1241.

Kluyver, A. J., and van Niel, C. B., 1936, Prospects for a natural system of classification of bacteria, *Zentralbl. Bakteriol.* **94**:369–403.

Kuenen, J. G., Boonstra, J., Schröder, H. G. J., and Veldkamp, H., 1977, Competition for inorganic substrates among chemoorganotrophic and chemolithotrophic bacteria, *Microbial Ecol.* **3**:119–130.

Kuenen, J. G., and Tuovinen, O. H., 1981, The genera *Thiobacillus* and *Thiomicrospira*, in: *The Prokaryotes* (M. P. Starr *et al.*, eds.), pp. 1023–1036, Springer-Verlag, Berlin.

Kuhn, D. A., 1981, The genus *Cristispira*, in: *The Prokaryotes* (M. P. Starr *et al.*, eds.), pp. 555–563, Springer-Verlag, Berlin.

Larson, H., 1981, The family Halobacteriaceae, in: *The Prokaryotes* (M. P. Starr *et al.*, eds.), pp. 985–994, Springer-Verlag, Berlin.

London, J., and Rittenberg, S. C., 1967, *Thiobacillus perometabolis* nov. sp., a non-autotrophic thiobacillus, *Arch. Mikrobiol.* **59**:218–225.

McInerney, M. J., Bryant, M. P., Hespell, R. B., and Costerton, J. W., 1981, *Syntrophomonas wolfei*, gen. nov. sp. nov., an anaerobic syntrophic, fatty acid-oxidizing bacterium, *Appl. Environ. Microbiol.* **41**:1029–1039.

Meers, J. L., 1973, Growth of bacteria in mixed cultures, *CRC Crit. Rev. Microbiol.* **2**:139–184.

Migula, W., 1897, *System der Bakterien*, Vol. I, Gustav Fischer, Jena.

Migula, W., 1900, *Specielle Systematik der Bakterien*, Vol. II, Gustav Fischer, Jena.

Mulder, E. G., and Deinema, M. H., 1981, The sheathed bacteria, in: *The Prokaryotes* (M. P. Starr *et al.*, eds.), pp. 425–440, Springer-Verlag, Berlin.

Pfennig, N., 1985, Stages in the recognition of bacteria using light as a source of energy, in: *Bacteria in Nature*, Vol. I (E. R. Leadbetter and J. S. Poindexter, eds.), pp. 113–131, Plenum Press, New York.

Philip, R. N., Casper, E. A., Ormsbee, R. A., Peacock, M. G., and Bergdorfer, W., 1976, Microimmunofluorescence test for the serological study of Rocky Mountain spotted fever and typhus, *J. Clin. Microbiol.* **3**:51–61.

Preer, L. B., 1981, Prokaryotic symbionts of *Paramecium*, in: *The Prokaryotes* (M. P. Starr *et al.*, eds.), pp. 2127–2136, Springer-Verlag, Berlin.

Prouty, C. C., 1929, The use of dyes in the isolation of a nitrite organism, *Soil Sci.* **28**:125–136.

Reichenbach, H., and Dworkin, M., 1981, The order Cytophagales (with addenda on the genera *Herpetosiphon*, *Saprospira*, and *Flexithrix*), in: *The Prokaryotes* (M. P. Starr *et al.*, eds.), pp. 356–379, Springer-Verlag, Berlin.

Rippka, R., Waterbury, J. B., and Stanier, R. Y., 1981, Isolation and purification of cyanobacteria: some general principles, in: *The Prokaryotes* (M. P. Starr *et al.*, eds.), pp. 212–220, Springer-Verlag, Berlin.

Robinson, D. M., Brown, G., Gan, E., and Huxsoll, D. L., 1976, Adaptation of a microimmunofluorescence test to the study of human *Rickettsia tsutsugamushi* antibody, *Am. J. Trop. Med. Hyg.* **25**:900–905.

Rosenberg, E., and Gutnick, D. L., 1981, The hydrocarbon-oxidizing bacteria, in: *The Prokaryotes* (M. P. Starr *et al.*, eds.), pp. 903–912, Springer-Verlag, Berlin.

Rullman, W., 1897, Über ein Nitrosobakterium mit neuen Wuchsformen, *Zentralbl. Bakteriol. Parasitol.* **3**:228–231.

Schlegel, H. G., and Jannasch, H. W., 1967, Enrichment cultures, *Annu. Rev. Microbiol.* **21**:49–70.

Schmidt, J. M., 1981, The genera *Caulobacter* and *Asticcacaulis*, in: *The Prokaryotes* (M. P. Starr *et al.*, eds.), pp. 446–476, Springer-Verlag, Berlin.

Scully, D. A., and Dondero, N. C., 1973, Estimation with several culture media of spirilla of 11 natural sources, *Can. J. Microbiol.* **19**:983–989.

Sieburth, S. M., 1979, *Sea Microbes*, Oxford University Press, New York.

Sneath, P. H. A., and Sokal, R. R., 1962, Numerical taxonomy, *Nature (Lond.)* **193**:855–860.

Sperl, G. T., and Hoare, D. S., 1971, Denitrification with methanol: A selective enrichment for *Hyphomicrobium* species. *J. Bacteriol.* **108**:733–736.

Stackebrandt, E., and Woese, C. R., 1981, The evolution of prokaryotes. *Symp. Soc. Gen. Microbiol.* **32**:1–31.

Staley, J. T., 1971, Incidence of prosthecate bacteria in a polluted stream, *Appl. Microbiol.* **22**:496–502.

Stanier, R. Y., 1951, The life-work of a founder of bacteriology, *Q. Rev. Biol.* **26**:35–37.

Stanley, S. M., 1981, *The New Evolutionary Timetable Fossils, Genes, and the Origin of Species*, Basic Books, New York.

Starr, M. P., 1981, The genus *Xanthomonas*, in: *The Prokaryotes* (M. P. Starr *et al.*, eds.), pp. 742–763, Springer-Verlag, Berlin.

Stoeckenius, W., 1976, The purple membrane of salt-loving bacteria, *Sci. Am.* **234**:38–46.

Stolp, H., 1965, Isolierung von *Bdellovibrio bacteriovorus*, in: *Anreicherungskultur and Mutantenauslese* (H. G. Schlegel, ed.), pp. 52–56, Gustav Fischer, Stuttgart.

Stolp, H., and Petzold, H., 1962, Untersuchungen über einen obligat parasitischen Mikroorganismus mit lytischer Aktivität für *Pseudomonas*-Bakterien, *Phytopathol. Z.* **45**:364–390.

Stolp, H., and Starr, M. P., 1963, *Bdellovibrio bacteriovorus* gen. et sp. n., a predatory, ectoparasitic, and bacteriolytic microorganism, *Ant. van Leeuwenhoek J. Microbiol. Serol.* **29**:217–248.

Stutzer, A., and Hartleb, R., 1898, Untersuchungen über die bei der Bildung von Saltpeter beobachteten Mikroorganismen. 1, *Abh. Mitt. Landwirt. Inst. Köningl. Univ. Breslau* **1**:75–100.

Trüper, H. G., and Pfennig, N., 1971, Family of phototrophic green sulfur bacateria: *Chlorobiaceae* Copeland, the correct family name; rejection of *Chlorobacterium* Lauterborn; and the taxonomic situation of the consortium-forming species, *Int. J. System. Bacteriol.* **21**:8–10.

van Niel, C. B., 1946, The classification and natural relationships of bacteria, *Cold Spring Harbor Symp. Quant. Biol.* **11:**285–301.

Veldkamp, H., 1970, Enrichment cultures of prokaryotic organisms, in: *Methods in Microbiology,* Vol. 3A (J. R. Norris and D. W. Ribbons, eds.), pp. 305–361, Academic Press, London.

Veldkamp, H., 1976, *Continuous Culture in Microbial Physiology and Ecology,* Meadowfield Press, Ltd., U.K.

Veldkamp, H., 1977, Ecological studies with the chemostat, in: *Advances in Microbial Ecology,* Vol. 1 (M. Alexander, ed.), pp. 59–94, Plenum Press, New York.

Walsby, A. E., 1981, Gas-vacuolate bacteria (apart from cyanobacteria), in: *The Prokaryotes* (M. P. Starr *et al.,* eds.), pp. 441–447, Springer-Verlag, Berlin.

Wang, S. P., 1971, A microimmunofluorescence method. Study of antibody response to TRIC organisms in mice, in: *Trachoma and Related Disorders Caused by Chlamydial Agents* (R. L. Nichols, ed.), pp. 273–288, Excerpta Medica, Amsterdam.

Winogradsky, S., 1949, *Microbiologie du sol. Oeuvres complètes,* Masson, Paris.

Withers, N., Vidaver, W., and Lewin, R. A., 1978a, Pigment composition, photosynthesis and fine structure of a non-blue-green prokaryotic algal symbiont (*Prochloron* sp.) in a didemnid ascidian from Hawaiian waters, *Phycologia* **17:**167–171.

Withers, N. W., Alberte, R. S., Lewin, R. A., Thornber, J. P., Britton, G., and Goodwin, T. W., 1978b, Carotenoids, chlorophyll–protein composition, and photosynthetic unit size of *Prochloron* sp., a prokaryotic green alga, *Proc. Natl. Acad. Sci. U.S.A.* **75:**2301–2305.

Woese, C. R., and Fox, G. E., 1977, Phylogenetic structure of the prokaryotic domain: the primary kingdoms, *Proc. Natl. Acad. Sci. U.S.A.* **74:**5088–5090.

Woese, C. R., Magrum, L. J., and Fox, G. E., 1978, Archaebacteria, *J. Mol. Evol.* **11:**245–252.

Woese, C. R., Debrunner-Vossbrinck, B. A., Oyaizu, H., Stackebrandt, E., and Ludwig, W., 1985a, Gram-positive bacteria: possible photosynthetic ancestry, *Science* **229:**762–765.

Woese, C. R., Stackebrandt, E., Macke, T. J., and Fox, G. E., 1985b, A phylogenetic definition of the major eubacterial taxa, *Syst. Appl. Microbiol.* **6:**133–142.

ZoBell, C. E., 1946, *Marine Microbiology,* Chronica Botanica, Waltham, Massachusetts.

MIXED SUBSTRATE UTILIZATION BY MIXED CULTURES

Jan C. Gottschal

INTRODUCTION

In spite of the vast amount of information already available on the ecology and physiology of microorganisms, the elucidation of the principles underlying microbial associations encountered in virtually all natural habitats still constitutes a major challenge to microbiologists. This is not too surprising if it is realized how in most ecosystems the interplay of a large species diversity and the broad scale of interspecies relationships almost inevitably render any attempt to model such an environment in the laboratory a gross oversimplification.

Yet there is no need for despair, for we recognize that our knowledge of dynamic events in "nature's mixed culture" has increased considerably since the first microorganism was made visible by Antonie van Leeuwenhoek. Most of this knowledge was obtained from laboratory studies on pure cultures, sometimes combined with data from field measurements. It is, however, only relatively recently that much more attention has been paid to the study of deliberately created mixed cultures and that the observed microbial interactions have become the subject of detailed investigations. Certainly it has been this shift toward mixed culture studies that has rapidly increased our understanding of many microbial processes observed in natural habitats. The state of knowledge of this field has been regularly reviewed (Bungay and Bungay, 1968; Veldkamp and Jannasch, 1972; Meers, 1973; Fredrickson, 1977; Harrison, 1978; Haas *et al.*, 1980; Bazin, 1981*b;* Slater, 1981; Bull and Slater, 1982; Slater and Bull, 1982). In recent years, a similar shift in interest from growth of pure cultures on a single principal substrate to growth on multiple substrates has occurred. A considerable amount of information is now available on the regulation of growth on substrate mixtures, both in batch culture

---

Jan C. Gottschal • Department of Microbiology, University of Groningen, 9751 NN Haren, The Netherlands.

and in substrate-limited chemostats. Some overviews of this particular topic are available, e.g., Mateles *et al.,* (1967), Paigen and Williams (1970), Harder and Dijkhuizen (1976, 1982), and Weide (1983).

In view of the abundance of experimental and theoretical work done on mixed cultures and on mixed substrate utilization, the relative scarcity of information on mixed cultures grown in the presence of mixed substrates is somewhat surprising. The aim of this chapter is to bring together those examples that may serve to illustrate the principles governing growth of mixed cultures in which more than one substrate is provided (continuously) by the experimenter. Thus, examples of cultures in which one primary substrate is converted into a number of secondary substrates, resulting in at least an equal number of secondary populations, are not discussed here. The available literature in which such mixed cultures are described in detail has been amply reviewed (Megee *et al.,* 1972; Meers, 1973; Meyer *et al.,* 1975; Fredrickson, 1977; Harrison, 1978; Miura *et al.,* 1980; Linton and Drozd, 1982; Slater and Bull, 1982). This topic is best treated by discussing only those studies in which the participating species were identified and in which the experimental conditions were controlled as rigorously as possible. The experiments in which these requirements were fulfilled had been performed in chemostats in which the mixed populations were competing for various mixtures of substrates present at growth-rate-limiting concentrations. For this reason, a short treatment of the theory of growth under nutrient limitation in continuous culture precedes the sections in which examples of mixed cultures grown under multiple substrate limitation are presented.

THEORY OF GROWTH IN CONTINUOUS CULTURE

The remarkable speed at which many bacteria are able to grow has certainly contributed considerably to making them very attractive subjects for the study of basic properties of the living world. However, this very property can present serious problems to scientists endeavoring to study bacteria under constant environmental conditions because a bacterial population tends to use up the available substrates very rapidly, thereby causing cessation of growth. No wonder that, for more than half a century, microbiologists have considered various means of prolonging the active growth phase of microbial cultures by the continuous addition of fresh medium and continuous harvesting of cells. It was, however, not until the 1950s when the basic theory of the chemostat was formulated (Monod, 1950; Novick and Szilard, 1950; Herbert *et al.,* 1956), that the use of continuous culture techniques began to play a role of importance in microbiological research. Over the past two decades, the chemostat, in one form or another, indeed appeared to be the best tool available for studying (mixed) microbial populations under well-controlled conditions. Knowledge of the basics of chemostat operation is therefore of paramount importance for the appraisal of the implications of (multiple) substrate limitation in mixed continuous cultures. Only a very short and therefore necessarily oversimplified presentation of the theory of the chemostat operating under (mixed) substrate limiting conditions is given here. More detailed information on this subject can be found in Herbert *et al.* (1956), Powell (1958, 1967), Pirt (1975), Taylor and Williams (1975), and Bazin (1981*a*); see also Button (1985).

Substrate-Limited Growth

Under conditions in which all requirements for growth are satisfied, the increase in biomass *dx* during a short time interval *dt* is expected to be proportional to the amount of biomass *x* already present and to the time interval:

$$dx = \mu \cdot x \cdot dt \tag{1}$$

hence

$$\mu = 1/x \cdot (dx/dt) \tag{2}$$

where the parameter μ represents the growth rate per unit of cell mass and is termed the specific growth rate. As long as all required nutrients are present in sufficiently high concentration, constant exponential growth is usually observed (batch culture), the rate of which is virtually unaffected by the substrate concentration over a wide range. However, upon depletion of an essential nutrient, the specific growth rate will decline. In a classic paper, Monod (1942) showed empirically that the relation of specific growth rate to substrate concentration accorded very well with the following expression:

$$\mu = \mu_m \cdot s/(s + K_s) \tag{3}$$

where μ_m represents the maximum specific growth rate attainable, which is approached asymptotically as *s* increases. The saturation constant K_s is the concentration of limiting nutrient required for growth at half the maximum rate.

Beyond doubt, bacteria in the natural environment will usually grow at a submaximal rate because of the typically very low substrate concentrations (Konings and Veldkamp, 1983). In batch cultures, it is virtually impossible to cultivate bacteria at such decreased growth rates in a well-controlled manner. However, it is exactly this mode of growth that can be obtained without difficulty in a chemostat (Herbert *et al.,* 1956; Veldkamp, 1977; Tempest and Neijssel, 1978).

A chemostat can simply be described as a culture into which fresh medium is continuously introduced at a constant rate *F;* the culture volume *V* remains unchanged due to constant removal of culture fluid at that same rate. The value

$$F/V = D \tag{4}$$

is referred to as the dilution rate. The inflowing medium usually contains all nutrients required for growth in such concentrations that, during growth, only one of them becomes limiting. The concentration of the growth-limiting substrate in the medium reservoir is usually termed S_r. After inoculation of the culture vessel, the substrates present will be consumed at a certain rate which depends not only on the physical and chemical conditions prevailing in the culture, but also on the nature of the substrate(s) and the properties of the microorganism(s). The specific substrate consumption rate q is defined as the amount of substrate consumed per unit of cell mass per unit of time:

$$q = 1/x \cdot (ds/dt) \tag{5}$$

For growth to occur, a certain amount of substrate consumed must be converted into cell material. The extent to which this takes place is conveniently defined by the growth yield Y of a microorganism, as

$$Y = dx/ds \tag{6}$$

By making use of these expressions, the values of the concentrations of biomass and limiting substrate can be conveniently described under various conditions. For example, the change in biomass is given by the following balance:

$$\text{Change in biomass} = \text{growth} - \text{output}$$

thus

$$dx/dt = \mu x - Dx = (\mu - D)x \tag{7}$$

If the value of D is lower than the μ_m, a steady state will eventually be obtained (Pirt, 1975), so that $dx/dt = 0$; hence $\mu = D$. Similarly, the balance for the growth-limiting substrate is as follows:

Change in substrate concentration =
$$\text{input} - \text{output} - \text{substrate used for growth}$$

hence

$$ds/dt = D(S_r - s) - \mu x/Y \tag{8}$$

In steady state, identified by the constancy of both cell density and residual concentration of the limiting substrate, that is

$$dx/dt = 0 \quad \text{and} \quad ds/dt = 0$$

it follows (1) from eq. (8) solved for x, that

$$\bar{x} = Y(S_r - s) \tag{9}$$

(2) from Eq. (3) solved for s, that

$$\bar{s} = K_s \cdot D/(\mu_m - D) \tag{10}$$

where $\bar{x}$ represents the steady-state cell density, and $\bar{s}$ represents the steady-state concentration of the limiting substrate within the culture. Equations (9) and (10) clearly illustrate that for a given microorganism and substrate, the steady-state cell density and limiting substrate concentration are determined operationally by the magnitude of S_r and D, respectively. This rule must be applied with some caution, however, because it assumes that $\bar{s} << S_r$ and that Y has a constant value. Indeed, in most cases, $\bar{s}$ has a very low value as long as the steady-state dilution rate (hence the specific growth rate) remains well below μ_m. Moreover, for most substrates, K_s values are in the micromolar range or even lower (Pirt, 1975; Law and Button, 1977; Tilman, 1977; Dykhuizen and Davies, 1980), and steady-state substrate concentrations are consequently very low.

Although the cell yield of a given microorganism growing at the expense of a given substrate is reasonably constant over a range of dilution rates, it has been shown for most microorganisms that the yield drops progressively when the dilution rate is lowered to very low values. This phenomenon has been attributed to the fact that bacteria require a certain constant (small) amount of energy in order to maintain their integrity. Since at very low dilution rates the proportion of the total amount of energy source consumed to satisfy this "maintenance energy requirement" will increase considerably, this will be manifested as a decrease in yield under conditions of energy-source limitation. For more details concerning the bioenergetic aspects of bacterial biomass formation, the reader is referred to Pirt (1965, 1975), Powell (1967), and Stouthamer (1977, 1979).

Multiple Substrate Limitation

In view of the large number of publications dealing with bacterial growth under multiple substrate-limiting conditions (reviewed by Harder and Dijkhuizen, 1976, 1982), it is not surprising that several attempts have been made to describe mathematically the growth patterns observed. This is even less surprising if one appreciates the considerable difficulties in interpreting, in physiologically and ecologically meaningful terms, the results obtained with (mixed) cultures grown on two or more substrates. In such complex systems, it is sometimes hoped that adequate mathematical description of bacterial growth and parameters, such as specific substrate consumption rate and prevailing substrate concentrations, might aid in understanding the behavior of these cultures. Monod-type kinetics (Monod, 1950) were used in most cases. In some instances, however, somewhat different considerations of growth kinetics served as the basis of the mathematical treatment of substrate-limited growth (Contois, 1959; Schulze and Lipe, 1964; Williams, 1967; Dabes *et al.*, 1973; Sinclair and Ryder, 1975; MacDonald, 1976). The less commonly employed kinetic models are not discussed here. Only a brief summary of the more generally used extensions of the Monod-type growth kinetics for the case of multiple substrate limitation in mixed cultures is presented.

There are at least two fundamentally different views regarding the description of the specific growth rate of a culture in which more than one substrate is present at a limiting concentration. Some investigators have postulated that at any moment only one factor can be truly growth limiting (Ryder and Sinclair, 1972; Sykes, 1973; Droop, 1974; Tilman, 1977). On the other hand, in most studies dealing with multiple substrate limitation, it is assumed that various substrates can contribute jointly to the specific growth rate (Fredrickson *et al.*, 1970; Cooney and Mateles, 1971; Megee *et al.*, 1972; Sinclair and Ryder, 1975; Taylor and Williams, 1975; Yoon *et al.*, 1977; Dykhuizen and Davies, 1980; Gottschal and Thingstad, 1982). Although these opposing views confuse the issue of multiple substrate-limited growth, the controversy is less serious than it appears when a distinction is made between two rather different types of combinations of rate-limiting nutrients. Such combinations consist of nutrients required either for distinctly different metabolic purposes (e.g., O_2 + C source; C source + phosphate) or for physiologically similar functions (e.g., glucose + fructose; O_2 + NO_3^-). In the first case, one would expect only one substrate to be growth rate limiting. Nevertheless, the second substrate might be present at a concentration that limited the rate of a specific process whose rate nevertheless did not restrict the overall rate of growth (Bader *et al.*, 1975). That is, the rate of the limited process could still be greater than needed to sustain the growth rate allowed by the availability of the other, the primary limiting nutrient. By contrast, multiple substrate limitation as a result of a limited supply of substrates used similarly in metabolism would not generally appear to pose a conceptual problem, since both would serve to alleviate a shortage of the same growth rate-limiting intracellular requirement (e.g., energy, or carbon, or nitrogen). This latter type of multiple substrate limitation will be termed "mixed substrate limitation" in this discussion.

If one attempts to describe growth under conditions of mixed substrate limitation mathematically, the specific growth rate should be best described as a function of all these substrates, i.e., $\mu = \mu(s_1, s_2, \ldots, s_4)$. If, on the other hand, the substrates involved do not serve similar metabolic functions, a description of μ based on single substrate limitation might be appropriate. Nevertheless, it must be emphasized that even in this latter case, the more general model, with μ being a function of more than one substrate, might still be applied successfully. An example of such a situation can be found in the work of Sinclair and Ryder (1975). These investigators studied the growth of *Candida utilis* under "dual"-limitation of oxygen and glycerol; they compared their experimental results with the predictions generated by two mathematical models. Truly dual substrate limitation was assumed in one of these models and effectively single substrate limitation in the other. Differences between the predictions based on the two models were within the experimental error. Since in such cases there will be little to discriminate experimentally between single or dual substrate limitation (Bader *et al.*, 1975; Bazin, 1981a), it is probably a wise policy to choose the more general, multisubstrate function for the description of μ. An example of such a description of the specific growth rate is presented in some detail in the section on growth of mixed cultures on mixed substrates.

Mixed Cultures

In this section, the theory of growth of mixed cultures both on one and on two or more potentially limiting substrates is considered briefly. Moreover, since a detailed analysis of the growth kinetics of mixed cultures is virtually impossible in closed systems (i.e., batch cultures) because of the constantly changing environmental conditions, the brief treatment of some theoretical aspects of growth of mixed cultures is limited to growth under rigorously controlled conditions as usually obtained in continuous cultures.

Competition for One Growth-Limiting Substrate

When a chemostat has been inoculated with a mixture of bacteria, unrestricted growth of those species able to use the available primary substrate is possible for some time. During this initial period, the specific growth rate of each species will, in theory, approach its maximum specific growth rate (μ_m) under the prevailing conditions. In other words, the term $s/(K_s + s)$, as used in the Monod-type expression $\mu = \mu_m \cdot s/(K_s + s)$ [see Eq. (3)], approaches unity, because $s >> K_s$ during this period. This implies that under these conditions, competition for the available amount of substrate is dominated by the value of μ_m of each individual species. However, it will also be evident that upon reduction of the initially present amount of substrate, the specific growth rate of the individual species will decrease according to their respective μ–s relationships.

This stage of substrate-limited growth will not last long unless new medium is introduced at a rate lower than the maximal rate of growth of at least one member of the mixed culture. Assuming that there are no other interactions among the species other than competition for the available limiting substrate, it will be evident that under these conditions the competitiveness of each species will depend on the interrelationship of its μ_m and K_s on the one hand and the imposed dilution rate D on the other. This is most clearly seen by examining Eq. (10) and by recalling that the species that attains the highest specific growth rate at the limiting substrate concentration [see Eq. (3)] will eventually outnumber its competitors. A more thorough mathematical treatment of this process can be found in the classic paper by Powell (1958) (see also Chapter 8, this volume, for discussion of a similar model). A visualization of an arbitrary μ–s relationship of two organisms A and B is presented in Fig. 1A. At any given substrate concentration, the specific growth rate of organism A will be higher than that of organism B. A different situation obtains if the two μ–s curves intersect, as shown in Fig. 1B. In this case, the specific growth rate of organism A exceeds that of bacterium B to the left of the intersection, whereas organism B exhibits the higher specific growth rate at substrate concentrations to the right of the intersection. Since in a given

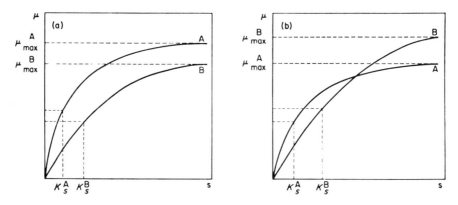

FIGURE 1. μ–s relationship of two organisms A and B. (a) $K_{s(A)} < K_{s(B)}$ and $\mu_{max(A)} >$ $\mu_{max(B)}$; (b) $K_{s(A)} < K_{s(B)}$ and $\mu_{max(A)} < \mu_{max(B)}$. (From Veldkamp, 1970.)

culture at steady state the concentration of the limiting substrate is determined by the dilution rate [Eq. (10)], it is evident that at relatively low D values organism A will predominate but that at higher values organism B will outgrow its competitor. It should be emphasized that the cell yield does not appear in the equation describing the μ–s relationship and therefore does not affect the outcome of competition in such mixed cultures. In practice, relative cell numbers are the values used to assess that outcome, so that "outcompete" means "outnumber in steady state."

In theory, coexistence can occur either if the two μ–s relationships coincide or if the substrate concentration is kept exactly at the value at which the curves intersect. In practice, however, no examples of coexistence based on this principle have yet been reported (Megee *et al.*, 1972; Veldkamp and Jannasch, 1972; Meers, 1973; Harder *et al.*, 1977; Mur *et al.*, 1977; de Freitas and Fredrickson, 1978; Laanbroek *et al.*, 1979; Kuenen and Gottschal, 1982). Although such findings clearly illustrate the principle of competitive exclusion (Hardin, 1960) (see Chapter 9, this volume), an overwhelming number of examples of stable mixed cultures limited by only one primary substrate have been reported. This paradox is partly attributable to the formation of secondary growth substrates. Several such mixed cultures depend on phenomena such as interdependence based on excretion of secondary growth requirements such as vitamins (Megee *et al.*, 1972; Meyer *et al.*, 1975), production of autoinhibitors (de Freitas and Fredrickson, 1978), the presence of a predator (Jost *et al.*, 1973), (selective) adhesion (Bungay and Bungay, 1968), or fluctuations in the value of certain physical parameters (e.g., light) (van Gemerden, 1974) (see Chapter 8, this volume, for several examples). Furthermore, alternate availability of two different substrates, each in turn limiting growth, can lead to coexistence of species (Gottschal *et al.*, 1981). It may also be possible that fluctuations, either random or controlled, in the dilution rate around the value at which the μ–s relationship of two species crosses would lead to species coexistence (Grenney *et al.*, 1973; Stewart and Levin, 1973). In an effort to control such fluctuations, Bazin (1981*b*) suggested on-line monitoring of population sizes followed by appropriate adjustments of dilution rate to achieve a constant average for both species.

Competition for More than One Growth-Limiting Substrate

Only a very limited number of studies has appeared in which a mathematical description has been presented of mixed cultures competing for a mixture of two or more primary substrates; almost invariably, Monod-type growth kinetics form the basis of such models (Megee *et al.*, 1972; Pirt, 1975; Taylor and Williams, 1975; Tilman, 1977; Yoon *et al.*, 1977; Dykhuizen and Davies, 1980; Miura *et al.*, 1980; Gottschal and Thingstad, 1982). To provide an idea of the value of such mathematical exercises, two examples are briefly summarized in this section. In the first example (Tilman, 1977), the growth of a mixed culture of the diatoms *Asterionella formosa* and *Cyclotella meneghiniana* was studied in a semicontinuous culture system with various mixtures of silicate and phosphate as the primary rate-limiting nutrients. Since both nutrients are required for growth, it was assumed that at any one moment only one of them would be rate limiting (see also the section, Multiple Substrate Limitation). Assuming Monod-type kinetics, with the *i*th of *n* total species and the *j*th of *m* different resources, this was expressed as follows:

$$\frac{dN_i}{dt} \cdot \frac{1}{N_i} = \underset{1 \leq j \leq m}{\text{MIN}} \left(\mu_{m(i)} \cdot \frac{s_j}{K_{ij} + s_j} - D \right) \tag{11}$$

$$\frac{ds_j}{dt} = D(S_{r(j)} - s_j) - \sum_{i=1}^{n} N_i \mu_{m(i)} \cdot \frac{s_j}{K_{ij} + s_j} \cdot \frac{1}{Y_{ij}} \tag{12}$$

in which N_i represents the number of cells of the *i*th species, K_{ij} denotes the half-saturation constant of species *i* for substrate *j*, and the other parameters have their usual meaning (see the section, Substrate-Limited Growth) and are related to a given species or substrate by the use of the subscripts *i* and/or *j*, respectively. The term min is used to indicate that the growth rate will be determined only by that one nutrient that is truly rate limiting (Droop, 1974). Obviously, this assumption implies that the changeover from one limitation to another takes place abruptly, which means that the dynamic aspects of the species composition of the mixed culture, in response to varying ratios of the two potentially limiting nutrients, cannot readily be described in this way. Indeed, only steady-state conditions have been analyzed with this model system. The following steady-state equations are obtained from Eqs. (11) and (12) when the time derivatives $dN_i/dt = ds_j/dt$ equal zero:

$$N_{i(j)} = Y_{ij}(S_{r(j)} - s_j) \tag{13}$$

$$s_j = DK_{ij}/(\mu_{m(i)} - D) \tag{14}$$

where $N_{i(j)}$ denotes the steady-state cell number of species *i* when this species alone is limited by substrate j, and s_j represents the steady-state concentration of substrate j when only species i is present. Next, the ratio of the reservoir substrate concentrations at which species i would just switch over from one to another limitation was determined. This would occur, in the case of two potentially limiting substrates, when $N_{i(1)}$ just equaled $N_{i(2)}$. Combining this equality with Eq. (13) yields the following expression:

$$S_{r(1)} = s_1 + (S_{r(2)} - s_2) \cdot Y_{i2}/Y_{i1} \qquad (15)$$

When the same procedure is followed for the second species, three areas can be distinguished with their boundaries defined by the substrate ratios, i.e., $S_{r(1)}/S_{r(2)}$, as calculated from Eq. (15) for the two competing species. There will be one area in which s_1 limits the growth rate of both species, in another region s_2 will be limiting for the two species and, unless both species exhibited identical μ–s relationships and identical yield values for both substrates, a third area will be present in which one species is limited by substrate 1 and the other is limited by substrate 2. Only in this area will stable coexistence of the two species be possible (Stewart and Levin, 1973; Taylor and Williams, 1975; Yoon et al., 1977). In the two other areas, competition between the two species will occur for one common growth-limiting substrate, which, according to the considerations discussed in the section Competition for One Growth-Limiting Substrate, will lead to dominance by one species.

Equation (14) still applies as a description of the steady-state concentration of both substrates when these are supplied at such a ratio that stable coexistence occurs. The contribution of each species to the total cell number in the mixed culture can be found by rearranging equation (12). Assuming that $s_j << S_{r(j)}$, one obtains:

$$N_1 = N_{1(1)} - N_2 \cdot Y_{11}/Y_{21} \qquad (16)$$

$$N_2 = N_{2(2)} - N_1 \cdot Y_{22}/Y_{12} \qquad (17)$$

where N_1 and N_2 are the numbers of cells of species 1 and 2, respectively, per unit of culture volume, and $N_{1(1)}$ and $N_{2(2)}$ have the same meaning as $N_{i(j)}$ in Eq. (13). With this set of equations, it becomes possible to predict the composition of a given mixed culture at steady state and to compare this result with the observed ratio of the two species in the culture. Tilman (1977) made such a comparison and found very close agreement, indeed. Apart from such a comparison with actual experimental results, the outcome of this and similar models (see below) may also be used, perhaps even more fruitfully, to obtain clear-cut information on the relative importance of the various growth parameters for the outcome of competition for limiting resources in mixed cultures. From this particular model, it would follow, for instance, that in sharp contrast with competition for single growth-limiting nutrients, cell yield is of crucial importance when a species is competing for more than one nutrient at the same time. For instance, it can be seen from Eq. (15) that the boundary between the areas in which a species is limited (under given environmental conditions) by one or the other substrate is almost solely determined by the ratio of the yields on each of the two substrates. Moreover, the competitiveness of a species within the area of predicted coexistence, is determined principally by the ratio of its own yield on one of the two substrates to that of its competitor.

A model of the type just described does have limitations; apart from the fact that it oversimplifies reality, this particular one seriously lacks generality because it does not apply to situations in which the two (or more) substrates affect growth rate simultaneously. A brief discussion of a model in which the specific growth rate is assumed to depend on the plural effect of more than one substrate (Taylor and Williams, 1975) follows.

In this model, growth in a chemostat is assumed to be the result of the multiplicative effect of several substrates. In mixed cultures with m possible limiting substrates and n species of organisms, the chemostat growth equations [see Eqs. (7), (8), (11), and (12)], assuming Monod kinetics, may be written as follows:

$$\frac{dx_i}{dt} = \left\{\mu_{m(i)} \prod_{1 \leq j \leq m} \left(\frac{s_j}{s_j + K_{ij}}\right) - D\right\} \cdot x_i \tag{18}$$

$$\frac{ds_j}{dt} = (S_{r(j)} - s_j) D - \sum_{i=1}^{n} \left(\frac{dx_i}{dt} + x_i D\right) \Big/ Y_{ij} \tag{19}$$

$$i = 1, 2, \ldots, n$$

$$j = 1, 2, \ldots, m$$

where Π represents a product.

At steady-state conditions with $x_i \neq 0$, these equations can be written as

$$\mu_{m(i)} \prod_{1 \leq j \leq m} \left(\frac{s_j}{s_j + K_{ij}}\right) = D \tag{20}$$

and

$$\bar{s}_j = S_{r(j)} - \sum_{i=1}^{n} \bar{x}_i / Y_{ij} \tag{21}$$

from which it may be seen that, in the absence of any other interactions, equilibrium values for s_j can only be found for $m \geq n$ (Taylor and Williams, 1975). In other words, in order to sustain a stable mixed culture of a certain number of species, at least the same number of growth-limiting substrates is required. This is only a minimum requirement, since the actual growth parameters of the competing organisms pose further restrictions on both the dilution rate and the relative concentrations of the growth limiting substrates in the medium.

This can be shown conveniently by choosing $n = m = 2$ and then finding the equilibrium substrate concentrations and cell densities at steady state [from Eqs. (20) and (21)], and finally by showing that such an equilibrium is stable.

This somewhat laborious task was performed by Taylor and Williams (1975), who came to the important conclusion that stable coexistence of two species A and B would be possible only if

$$\frac{Y_{a1}}{Y_{a2}} < \frac{S_{r2}}{S_{r1}} < \frac{Y_{b1}}{Y_{b2}} \tag{22}$$

with a and b referring to species A and B, respectively, with the usual assumption of $S_{r(i)} >> s_i$, and with the additional assumption that A is the faster-growing species on limiting amounts of one substrate and B outgrows species A on the other. In other words, this relationship very conveniently expresses the range of input substrate concentrations that will permit stable coexistence of the two species A and B. With a combination of species and substrates in which one species outcompetes the other on both substrates, the same general analysis still holds, but this time Eq. (20) will have either two solutions or none at all. The authors were unable to perform a rigorous stability analysis for the predicted equilibria in this case, but they indicated that only one equilibrium would be stable. This would imply the existence of one range of ratios of input substrate concentrations yielding stable coexistence, but also another range for which establishment of coexistence or dominance by one species would depend on the size of the inoculum. A further complication would occur if the two species exhibited intersecting μ–s relationships for either one or both of the limiting nutrients. In such cases, whether relation (22) is fully applicable will also depend on the actual dilution rate used (below or above the rate at which the curves intersect).

I conclude this section with some cautionary remarks on the use of mathematical models in attempts to understand the behavior of mixed cultures grown on mixed substrates. The two models described above have been presented mainly to illustrate that, in spite of the fundamental difference in the underlying assumption concerning growth rate limitation, both models show very good qualitative agreement in their predictions of the steady-state composition of the mixed cultures. Both models show that, in contrast to the situation under single nutrient limitation, coexistence is to be expected over a range of dilution rates. As to the species composition in the mixed culture, the importance of the ratio of the yields of a given organism from the two substrates was clearly demonstrated. Furthermore, in the respective discussions of the two models, both publications (Taylor and Williams, 1975; Tilman, 1977) emphasized that numerical analysis of the differential equations indicated that, with certain substrate ratios in the input medium, attainment of the final predicted species composition took a surprisingly long time. Moreover, similar results were obtained by Yoon *et al.* (1977) and Gottschal and Thingstad (1982), who predicted on the basis of somewhat different models that, under some conditions, up to 100 volume changes would be required for the final steady-state situation to be reached. This clearly indicates the very limited value of an exact prediction of the steady state culture composition for our understanding of the species composition in a natural habitat. Specific growth rates might often be in the range of 0.01 hr^{-1}, which would mean that by the time a mixed population could have stabilized, theoretically, environmental perturbations of some sort probably would have taken place. Not only does nature often invalidate predictions of the kind described above, it also presents ecologists with the almost insurmountable problem of how to study microorganisms provided with a multitude of (limiting) nutrients. Indeed, both models described above in theory also apply to conditions of real multiple substrate limitation. In practice, however, a full mathematical analysis of a multispecies/multisubstrate-limited culture has not yet been reported, most likely because of the greatly increased experimental and mathematical complexity with every additional species and substrate examined. This latter aspect in particular should be emphasized because even in case such multispecies models are fully analyzed mathematically, their practical value would become very small indeed if experimental techniques were not available to test the predictions generated by such models.

PURE CULTURES, MIXED SUBSTRATES

This section does not present a complete overview of the information currently available on the physiological and ecological aspects of growth on mixtures of various substrates, a subject that has been dealt with extensively in several papers (Harder and Dijkhuizen, 1975, 1982; Bull and Brown, 1979; Weide, 1983). Rather, I have chosen to discuss only a few examples to illustrate some characteristics of growth of pure cultures on mixed substrates that might be of importance in evaluating observations done with mixed cultures grown under similar conditions.

The phenomenon of diauxic growth (Monod, 1942) and the even more general phenomenon of sequential utilization of substrates (sugars in particular) has been studied by many investigators. Several illustrative examples can be found in the works of Edwards (1969), Standing *et al.* (1972), Smith and Bull (1976), Clark and Holms (1976), and Krauel *et al.* (1982). The general practice that emerges from these studies (see Harder and Dijkhuizen, [1976] for more examples) with mixtures of various sugars is that, in batch culture, most microorganisms tested utilized glucose first and consumed other sugars (fructose, lactose, sucrose, galactose, maltose, or xylose) only after the glucose concentration had dropped to a very low level. This very common feature, found with so many microorganisms, has generally been explained by assuming that degradation products of a readily metabolized carbon source (*viz.* glucose) interferes with the synthesis of enzymes required for the uptake or metabolism of other carbon sources. This control mechanism, catabolite repression, has been shown to operate on a wide range of different enzymes (Paigen and Williams, 1970) and is surely not restricted to the metabolism of sugars.

Although a compound such as glucose can prevent the metabolism of other sugars, it does so only at relatively high concentrations. Not surprisingly, therefore, several investigators have compared the pattern of substrate utilization on a mixture of sugars in batch culture and in continuous culture. Mateles *et al.* (1967) showed that at lower dilution rates, both glucose and fructose were completely consumed, whereas at higher rates fructose remained unutilized. This phenomenon was observed not only for pure cultures of *Pseudomonas fluorescens* and *Escherichia coli*, but also for a natural enrichment culture on glucose + lactose, with lactose remaining unutilized at higher growth rates. Somewhat later, Silver and Mateles (1969) repeated a glucose + lactose mixed-substrate experiment using *E. coli* B$_6$ with the same result; however, they also found that β-galactosidase-constitutive mutants were strongly selected for in such cultures and eventually dominated the culture. These mutants, though, continued to utilize both glucose and lactose up to the washout rate (0.9 hr^{-1}).

Smith and Bull (1976) reported on the growth of *Saccharomyces fragilis* on mixtures of glucose, fructose, sucrose, and sorbitol, the major carbohydrate components of coconut water. In batch culture, fructose and glucose were utilized simultaneously with concurrent "inversion" of sucrose into glucose + fructose, whereas sorbitol was consumed only after the sugars had almost disappeared from the culture fluid. When the same experiment was conducted in a continuous culture under carbohydrate limitation at various dilution rates increasing from 0.02 to 0.46 hr^{-1}, it appeared that only at growth rates below 0.05 hr^{-1} were all four compounds utilized completely. Between 0.05 and 0.2 hr^{-1}, sorbitol appeared in increasing concentrations in the culture fluid, sucrose remained unmetabolized to an increasing extent between 0.18 and 0.35 hr^{-1}, and above this dilution rate both fructose and glucose began to accumulate in the culture fluid. This general pattern of sequential utilization of substrates in batch culture, but simultaneous consumption in continuous culture has been further illustrated by the results of studies on the metabolism of a range of microorganisms capable of both heterotrophic and autotrophic growth (Dijkhuizen and Harder, 1979a,b; Gottschal and Kuenen, 1980a; Smith *et al.*, 1980; Perez and Matin, 1982; Wood and Kelly, 1983) or heterotrophic and methylotrophic growth (van Verseveld, 1979; Egli *et al.*, 1982, 1983). In these studies, the so-called mixotrophic metabolism was investigated by supplying the continuous culture with various mixtures of the "autotrophic (methylotrophic) substrate" and the "heterotrophic substrate." Simultaneous metabolism of both of these substrates was indeed observed for all substrate mixtures tested: formate + acetate, formate + oxalate (Dijkhuizen and Harder, 1979a,b), methanol + mannitol, formate + mannitol (van Verseveld, 1979), thiosulfate + acetate (Gottschal and Kuenen, 1980a), thiosulfate + glucose (Smith *et al.*, 1980; Perez and Matin, 1982), formate + glucose (anaerobically) (Wood and Kelly, 1967), and methanol + glucose (Egli *et al.*, 1982, 1983). A most significant conclusion to be drawn from these studies, pertinent to our understanding of the growth of bacteria under mixed-substrate limitation, is that such metabolically flexible bacteria are capable of accurately tuning their enzymic machinery to the nutritional demands. This was most apparent for the oxidation potential of the culture for the substrates supplied, and for the capacity to convert CO_2 or formaldehyde into cellular material in the cases of some methylotrophic organisms. Both activities clearly responded to the ratio of "autotrophic" or "methylotrophic" and "heterotrophic" substrates in the inflowing medium. In Fig. 2, an example of this phenomenon is shown. In this case, *Thiobacillus* A2 was grown on a range of mixtures of acetate + thiosulfate as the carbon- and/or energy-limiting substrates (Gottschal and Kuenen, 1980*a*). An important consequence of this regulatory mechanism is that in some cases it enables the organism to optimize its biomass yield. This aspect revealed itself as an increase in cell density of cultures grown on certain mixtures of the "heterotrophic" and the "autotrophic" substrates compared with the density expected from the sum of the growth on both substrates alone (van Verseveld, 1979; Dijkhuizen and Harder, 1979*a*; Gottschal and Kuenen, 1980*a*). In these particular examples, the observed increase in growth efficiency has been explained by emphasizing the energy-saving effect of the utilization of an organic substrate on the synthesis of biomass as compared

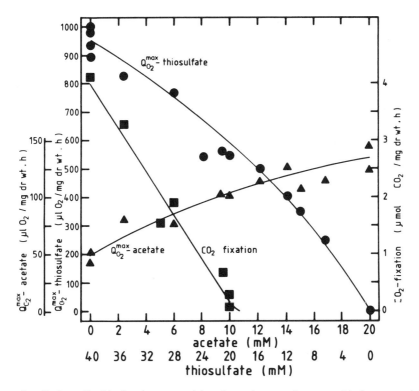

FIGURE 2. Carbon dioxide fixation potential and maximum substrate oxidation potentials of whole cells of *T.* A2 as a function of different acetate and thiosulfate concentrations in the reservoir medium of chemostat cultures. Data were obtained with cells from thiosulfate- + acetate-limited chemostat cultures in steady state at a dilution rate of 0.05 hr[-1]. ($\bullet$ --- $\bullet$) QO_2^{max}-thiosulfate; ($\blacktriangle$ --- $\blacktriangle$) QO_2 ^{max}-acetate; ($\blacksquare$ --- $\blacksquare$) CO_2-fixation potential. (From Gottschal and Kuenen, 1980*a*.)

with biosynthesis on the basis of CO_2 fixation. Obviously, similar positive effects on cell yields can be expected in general when additional substrates are present and if metabolic regulation permits the use of energetically "cheaper" pathways for biomass synthesis. Although cell yield probably does not play a role of any significance in interspecies competition for single rate-limiting substrates, it definitely affects the competitiveness of a species when more than one substrate is rate limiting at the same time.

To conclude this short section on pure culture studies, I offer a few tentative remarks on the possible influence of the presence of one rate-limiting nutrient on the steady state substrate concentration $(\bar{s})$ and possibly the saturation constant (K_s) of the other(s). The problem is that it is usually extremely difficult to measure accurately the residual concentration of the growth-limiting nutrient in chemostats at steady state. This is certainly true for the concentration of thiosulfate in the continuous cultures of *Thiobacillus* A2 grown on various acetate + thiosulfate mixtures (Fig. 2) (Gottschal and Kuenen, 1980*a*). Nevertheless, from the data on the maximum thiosulfate-oxidizing potential in these cultures, it may perhaps be inferred indirectly that the thiosulfate steady state concentration during mixed substrate limitation was lower than during growth limitation by thiosulfate alone. Although the maximum thiosulfate oxidizing potential decreased with decreasing concentration of thiosulfate in the feed, this relationship was not strictly linear (Fig. 2). The available specific overcapacity to oxidize thiosulfate was always slightly higher with mixed feeding than with thiosulfate alone. A similar situation seems to exist in cultures of *Pseudomonas oxalaticus* OX1 with respect to formate-oxidizing potential of cells grown in the presence of a mixture of formate and fructose (Dijkhuizen, personal communication). If, then, the tentative assumption is made that under nutrient limitation the initial substrate-capturing step limits the flux of metabolites through the organism, it might be concluded that the higher the overcapacity for capturing the available rate-limiting substrate, the lower the residual substrate concentration will be. More direct evidence to justify a generalization of such substrate interactions is scarce. In one study, it was reported (Law and Button, 1977) that the steady-state glucose concentration in a carbon- and energy-limited chemostat culture of a *Corynebacterium* sp. was strongly affected by the presence of varying amounts of amino acids in the feed (Table I). However, these data cannot be taken to indicate unequivocally that mixtures of several growth-limiting substrates mutually depress the steady state substrate concentrations because other data were presented showing that steady state arginine concentrations were higher in the presence of glucose than in its absence. Another example of this type of substrate interference was reported by Egli *et al.* (1983). These investigators demonstrated that in carbon-limited chemostat cultures of the yeast *Kloeckera* sp. 2201 with various mixtures of glucose + methanol in the feed, the residual methanol concentration was as much as 50-fold lower than during growth with methanol as the sole growth-limiting substrate. The absolute amounts of glucose and methanol in the feed were adjusted so that as the ratio of these two substrates was varied, the culture density remained fairly constant. Ac-

TABLE I

Effect of Additional Substrate on the Concentration of Glucose at Steady State
(D = 0.03 hr^{-1})^a

Additions to medium (μg/liter)				Steady-state glucose concn. (μg/liter)
Arginine	Glutamate	Amino acids^b	Glucose	
0	0	0	2500	210
1250	0	0	1250	100
800	800	0	800	50
0	0	1000	1700	<10
0	0	1250	1	0.3
0	0	0	1	0.7

^{a} From Law and Button (1977).
^{b} 50 μg of each of the common amino acids per liter.

cordingly, the observed effect on the residual methanol concentration was not simply a consequence of an elevated level of methanol-oxidizing enzyme resulting from the presence of more biomass in the culture. We can only speculate on the nature of the physiological basis for these fascinating substrate interactions. It is not even clear whether the explanation could be described in terms of the familiar Monod-type kinetics of microbial growth or whether more sophisticated enzyme kinetic models must be developed.

Although I have discussed only a few examples of pure cultures grown on mixed substrates, it does perhaps illustrate sufficiently the diversity of the possible metabolic implications of multiple-substrate-limited growth. It is hoped that it also points to the gigantic complexity facing microbiologists when studying mixed cultures grown under mixed substrate limitation. In the next few pages, an attempt will be made to discuss briefly the work of several investigators who have endeavored to analyze such complex cultures.

MIXED CULTURES, MIXED SUBSTRATES

The paucity of reports describing experimental examples of mixed cultures grown under well-defined and controlled conditions with more than one primary growth substrate is not because mixed culture studies are regarded as unimportant; rather, it is indicative of the enormous difficulties associated with the study of such mixed cultures. A brief summary of those studies in which mixed cultures were described in sufficient detail to illustrate how the principles of mixed substrate utilization by pure cultures may be of use to our understanding of mixed cultures grown under similar conditions is provided here.

In their pioneering study on mixed substrate utilization in continuous culture, Mateles *et al.* (1967) reported that a mixed bacterial population, obtained by inoculating a chemostat with river water, behaved very similarly to pure cultures of *Escherichia coli* when grown with glucose + lactose as the limiting carbon- and energy-sources. Both substrates were metabolized simultaneously at relatively low dilution rates, but at higher dilution rates progressively more of the lactose remained unused. Chian and Mateles (1968) extended this work with a river water-inoculated chemostat that was fed with a medium containing glucose + butyrate as limiting substrates; this resulted in a mixed culture of only two bacterial species: a pseudomonad and a coliform (not further identified). Of the two types of bacteria, the coliform was strongly dominant (90% on the basis of cell number) at dilution rates below 0.8 hr^{-1}. Above this value, the pseudomonad became increasingly numerous, relative to the coliform, accounting for up to 75% of the population at a dilution rate of 1.15 hr^{-1}. At dilution rates below 0.8 hr^{-1}, the residual glucose concentration remained undetectably low, whereas it increased rapidly to 550 mg/liter (approx. one-half the glucose concentration in the reservoir medium) at $D = 1.15$ hr^{-1}. In contrast, the residual concentration of butyrate remained very high (800–950 mg/liter; butyrate concentration in the feed was 1000 mg/liter) at dilution rates between 0.4 and 1.15 hr^{-1}. Moreover, acetate was excreted by the mixed population at dilution rates above 0.6 hr^{-1}. Attempts to grow the pseudomonad alone in batch culture on a mixture of glucose + butyrate failed. In the presence of the coliform, however, both species grew well with the pseudomonad in the end accounting for 70–90% of the total population. The μ_m values were not determined for either organism. From these results, it will be clear that the analysis of this seemingly straightforward example of a mixed culture on two growth-limiting substrates is not so simple. On the basis of the available data, it is not possible to decide to what extent each substrate contributed to the growth of the two species and whether butyrate was limiting at any dilution rate. In fact, it is quite possible that the growth of the pseudomonad was limited simply by the availability of a certain growth factor excreted by the coliform.

Another interesting case of coexistence of a pseudomonad (*Ps. aeruginosa*) and a coliform (*Klebsiella aerogenes*) was described by Pirt (1975). In a mixed chemostat culture at 40° and pH 7.2, the *Pseudomonas* species was limited by glucose and *Kl. aerogenes* by *p*-hydroxybenzoate when these two substrates were present simultaneously in the medium reservoir of the chemostat. However, when the temperature was lowered to 37°C or less, *Kl. aerogenes* used both substrates and displaced the pseudomonad. Not enough data were provided to permit detailed analysis of the various possible outcomes of this type of mixed culture. Nevertheless, this relatively simple case nicely illustrates the very important general phenomenon of exclusion of species A when species B is able to metabolize both substrates and species A only one. This is less trivial than it might seem at first glance because under many conditions this will also be true even though species A would outgrow species B if grown on the shared substrate alone. This very important principle follows directly from theoretical considerations (Phillips, 1973; Taylor and Williams, 1975; Yoon *et al.*, 1977; Gottschal and Thingstad, 1982), as confirmed experimentally (Gottschal *et al.*, 1979; Laanbroek *et al.*, 1979; Dykhuizen and Davies, 1980).

During the course of a study on the ecophysiology of glutamate- and aspartate-fermenting bacteria, two different glutamate-fermenting *Clostridia* were isolated (Laanbroek *et al.*, 1979). One of them, isolated from a continuous culture (glutamate + aspartate as limiting substrates), was identified as *Cl. cochlearium* and had a very limited range of potential substrates: glutamate, glutamine, and histidine only. *Cl. tetanomorphum,* another glutamate-fermenting clostridium, was isolated from a batch-culture enrichment; this organism was less restricted in its range of utilizable growth substrates. When these two clostridia were grown together in a glutamate-limited chemostat, *Cl. cochlearium* displaced *Cl. tetanomorphum* from the culture, both at a dilution rate of 0.04 hr^{-1} and at 0.33 hr^{-1}. This result is, however, somewhat difficult to understand in terms of Monod growth kinetics because, as determined from the residual substrate concentrations and the μ_m values of each organism grown in pure culture, the μ–s relationships intersected between a dilution rate of 0.2 and 0.3 hr^{-1}, with *Cl. tetanomorphum* exhibiting the higher specific growth rate at the glutamate concentration prevailing below $D = 0.20$ hr^{-1}. Perhaps interactions of yet unknown nature were responsible for this deviation of the expected outcome of the competition at the rather low dilution rate. Despite this uncertainty concerning the exact growth kinetics at very low dilution rate, the mixed culture of these two bacteria was successfully used to demonstrate the effect of the addition of a second substrate. When glucose was added to the medium, stable coexistence of both species was observed at a dilution rate of 0.04 hr^{-1}. Only *Cl. tetanomorphum* was able to use glucose. When glucose was subsequently omitted from the medium, *Cl. tetanomorphum* was displaced from the culture. From the changes in the size of both populations, it could be inferred that during coexistence, *Cl. tetanomorphum* utilized glutamate and glucose simultaneously. It was further argued that an increase of the glucose concentration in the feed should eventually result in the elimination of *Cl. cochlearum*. It may be noted here that in this particular case the yield on glucose is likely to affect strongly the availability of glutamate for *C. cochlearum*. Although this yield effect does not seem to have been studied experimentally, it has been emphasized in several theoretical studies on mixed cultures (Taylor and Williams, 1975; Yoon *et al.*, 1977; Gottschal and Thingstad, 1982).

　　During our own work on the ecology of the facultatively chemolithotrophic *Thiobacillus* A2 (for the sake of convenience called mixotroph), various mixed chemostat cultures were studied in which the mixotroph was cocultured with *Thiobacillus neapolitanus* (an autotroph), with a heterotrophic *Spirillum*, or with both organisms (Gottschal *et al.*, 1979). The medium fed to these cultures contained various mixtures of thiosulfate and acetate as limiting substrates. The values of μ_m and yield of each species were determined in pure chemostat cultures. While the autotroph grew only on thiosulfate and did not assimilate acetate to a significant extent, the mixotroph could grow on both substrates separately and utilized the two substrates simultaneously when they were provided together at limiting concentrations (Gottschal and Kuenen, 1980a). The heterotroph grew on acetate and many other organic substrates, but did not metabolize thiosulfate.

Competition experiments between the mixotroph and the heterotroph with acetate as the limiting substrate, and between the mixotroph and the autotroph with only thiosulfate in the feed, always resulted in elimination of the mixotroph. The outcome of these experiments became entirely different, though, when both substrates were present. In the mixed culture of the mixotroph and the heterotroph, increasing amounts of thiosulfate in the acetate-containing medium resulted in coexistence, with increasing numbers of the mixotroph and decreasing numbers of the heterotroph (Fig. 3). Indeed, at about 10 mM thiosulfate, the heterotroph had been almost excluded from the culture. A strictly analogous experiment was performed using a mixed culture of the mixotroph and the autotroph. Increasing amounts of acetate or glycolate were added to a thiosulfate-containing medium. In this case, the autotroph approached extinction at an acetate or glycolate concentration of approximately 8–10 mM with 40 mM thiosulfate (Gottschal *et al.*, 1979). The actual number of bacteria of both species (not the percentage of total cell counts) in these mixed cultures changed in a linear fashion with the concentration of the second substrate. This conforms to predictions based on a general mathematical model developed by Gottschal and Thingstad (1982) on the basis of an earlier, more general model described by Phillips (1973). It could also be concluded from such models that an increase in cell yield of

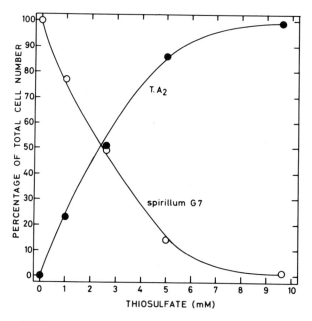

FIGURE 3. Effect of different concentrations of thiosulfate on the outcome of the competition between *T.* A2 and a heterotrophic "specialist" spirillum G7, for acetate. Dilution rate = 0.07 hr^{-1}. The inflowing medium contained acetate (10 mM) together with increasing concentrations of thiosulfate (0 – 10 mM). After steady states had been established, the percentages of *T.* A2 (● --- ●) and spirillum G7 (○ --- ○) were determined. (From Gottschal *et al.*, 1979.)

the mixotroph on the shared substrate should result in the elimination of the specialist at lower concentrations of the additional growth-limiting substrate. This example of elimination of the more "specialized" species by the versatile mixotroph thus clearly illustrates how, by the addition of a second substrate, the "generalist" may outgrow the "specialist" even though the latter organism clearly has a much higher μ_m on the *shared* substrate. Since the aim of these particular mixed culture studies had been to investigate how a "generalist" like *Thiobacillus* A2 could maintain itself amid the various types of "specialists" present in nature, a mixed culture of all three species was studied with different mixtures of thiosulfate and acetate present in the feed. The predicted outcome of such a mixed-culture experiment, based on the model mentioned above (Gottschal and Thingstad, 1982) was as follows: (1) coexistence of the heterotroph + mixotroph at a high acetate/ thiosulfate ratio in the feed; (2) coexistence of the autotroph + mixotroph at a high thiosulfate/acetate ratio; and (3) elimination of the two specialists at intermediate substrate ratios. The actual results of the experiment substantiated the first two predictions, but at intermediary ratios both "specialists" appeared to coexist together with the dominant mixotroph (Fig. 4). Thus, three species would seem to coexist on

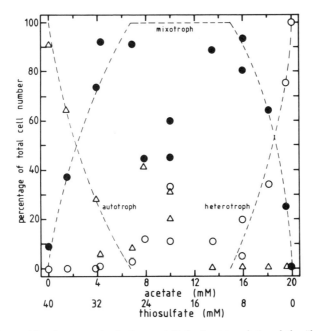

FIGURE 4. Competition between the "mixotroph," the "autotroph," and the "heterotroph" for thiosulfate and acetate as growth-limiting substrates in the chemostat at a dilution rate of 0.075 hr^{-1}. Concentrations of these nutrients in the inflowing medium ranged from 0 to 20 mM acetate and from 40 to 0 mM thiosulfate. Relative cell numbers of the three competing species were determined at steady state. The dashed line represents the theoretical prediction. (● --- ●) mixotroph; (△ --- △) autotroph; (○ --- ○) heterotroph. (From Gottschal and Thingstad, 1982.)

only two limiting substrates. This contrasts with the generally accepted theorem that for stable coexistence to occur between primary consumers competing for nutrients, the number of these nutrients must be greater than or equal to the number of competing species (Levins, 1968; Phillips, 1973). An explanation for this observed deviation from the theory can at present only be speculative. One obvious possibility is that, in addition to the purely competitive interaction, other yet-unrecognized inhibitory or stimulatory interactions played a role (Meyer *et al.*, 1975; de Freitas and Fredrickson, 1978). Another possibility is that the experimental data were in fact taken from chemostat cultures that were not in true steady state. It was pointed out by several investigators (Taylor and Williams, 1975; Yoon *et al.*, 1977; Tilman, 1977; Gottschal and Thingstad, 1982) that under certain mixed-culture conditions, extremely long periods of time (up to 100 volume changes) are needed to reach true steady states.

In an elegant study by Dykhuizen and Davies (1980), the same general problem of coexistence of a "generalist" and a "specialist" was investigated. These workers took as a starting point of their investigations an apparently widely accepted axiom in ecology, which states that a generalist should be less efficient than a specialist; otherwise, specialists simply would become extinct! As the acceptance of this axiom appears to be based mainly on intuition (Cody, 1974), the authors decided to test this statement by conducting competition experiments between more and less efficient specialist and generalist strains of *E. coli* in continuous cultures limited by maltose or maltose + lactose. In this study, an *E. coli* strain that contained a deletion of the lactose operon was termed a specialist. This strain was otherwise identical to the generalist, which was obtained from the specialist by transduction of the lactose operon. So-called inefficient generalists and specialists were obtained by selecting streptomycin-resistant mutants of the two strains, assuming that in such mutants protein synthesis would be less efficient (see Gorini and Davies, 1968; Garrett and Wittman, 1974). The precise meaning of efficiency in this context is perhaps somewhat unclear, since (1) the cell yield of the generalist was unaffected, and its μ_m on lactose and maltose was approximately 10% lower than that of the specialist, and (2) both μ_m and cell yield of the specialist were 5–10% lower than that of the generalist.

In one set of competition experiments, mixed cultures of generalists and specialists were grown with maltose as the sole limiting substrate (Dykhuisen and Davies, 1980). Under such conditions, the generalists were displaced by the specialist regardless of their degree of "efficiency." However, when lactose was provided in addition to maltose, the generalists coexisted or even completely eliminated the specialists, particularly at higher concentrations of lactose relative to maltose (Fig. 5). This situation strongly resembled the one described earlier with mixed cultures of the thiobacilli and the heterotrophic specialist. In these studies, however, the effect of an additional parameter (the relative efficiency of the competing species) was also taken into account. Clearly, the inefficient generalist required more lactose in the feed to attain dominance in the culture than did the efficient strain. The mathematical

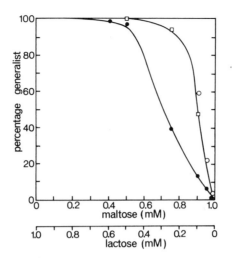

FIGURE 5.   Effect of different ratios of maltose and lactose in the feed on the number of "generalists" relative to the total cell number in mixed cultures with "specialists" in a chemostat at steady state ($D = 0.33$ hr^{-1}. ($\bullet$ --- $\bullet$) efficient specialist + *in*efficient generalist; ($\circ$ --- $\circ$) efficient specialist + efficient generalist; ($\square$ --- $\square$) inefficient specialist + inefficient generalist. (Drawn on the basis of data reported by Dykhuizen and Davies, 1980.)

description of these mixed cultures predicted that an increase of the μ_m of the generalist on the shared substrate (maltose) would result in an increase in the relative number of generalists in the mixed culture at a given ratio of the two substrates. This is the same result as was predicted in the case of competition between the mixotroph and specialists for an increase in the yield of the generalist on the nonshared substrate. This is not surprising, since changes in both of these parameter will affect the substrate-capturing capacity of the generalists (see Chapter 8, this volume, regarding the J parameter). Thus, an increase in yield on the nonshared substrate will lead to the production of more biomass per unit of substrate, which in turn will lead to more total consumption of the shared substrate. Similarly, an increase in μ_m of the generalist on the shared substrate will also lead to increased consumption of this nutrient by that organism, albeit this time because of a higher rate of substrate consumption by the population.

A quite different example of coexistence of two species grown under dual substrate limitation was reported by Tilman (1976, 1977). Two species of freshwater diatoms, *Asterionella formosa* and *Cyclotella meneghiniana*, were chosen as subject organisms in a study of "the paradox of the plankton" (Tilman, 1976; Hutchinson, 1961). This "paradox" is the striking discrepancy between the observed abundance of coexisting diatom species in most freshwater lakes, on the one hand, and on the other, the predicted lack of species diversity when it is considered that the two major potentially limiting nutrients, phosphate and silicate, are shared by all diatoms. Evidently, the cooperative action of several physical and biological parameters may be responsible for this paradox. Tilman chose to concentrate on the extent to which information on the species-specific potential of resource acquisition would allow prediction of the steady-state outcome of competition for two substrates by two species. Two mathematical models were constructed, one based on Monod-type growth kinetics and the other on the physiological model of variable internal stores (Droop, 1974; Lehman *et al.*, 1975; Veldkamp, 1977). A very important underlying assumption of both models was that either phosphate or silicate limited the growth of each individual species. Using the μ_m, K_s and yield values obtained in pure cultures (Tilman and Kilham, 1976), it was calculated that in (semi)continuous cultures of the two species separately, *Asterionella* should be phosphate limited at silicate/phosphate (Si/P) ratios higher than 96 in the feed, and silicate limited below that ratio. For *Cyclotella*, the boundary Si/P value was calculated to be 6. It was then argued that in mixed cultures of these two species, each would be limited by a different nutrient (silicate or fosfate) at intermediate Si/P ratios. According to both models, such a situation should lead to coexistence of these two species under steady-state conditions. The result of a very large number of competition experiments at various Si/P ratios confirmed this view unambiguously; it was further shown that at very high Si/P ratios *Asterionella* outgrew *Cyclotella*, whereas at very low ratios the reverse occurred. Most interestingly, Tilman (1977) also investigated whether these laboratory results would correspond to the field situation. To this end, a large number of samples, obtained from many different locations on Lake Michigan, were analyzed with respect to the Si/P ratio and to diatom species composition. In order to apply the laboratory results to the field situation, it was necessary to convert the silicate and phosphate concentrations in the lake into supply rates ($D \times S_r$) as used in the laboratory. When this was done, fairly good agreement was obtained between field situation and prediction. Coexistence of an *Asterionella* strain and a *Cyclotella* strain was observed in samples with Si/P ratios from about 150 to 10, whereas only one of these strains was found in samples exhibiting much higher or lower Si/P ratios. The results of this work illustrate very well how the knowledge of growth parameters of pure cultures can be used to predict the outcome of mixed culture experiments, and apparently in some cases may serve to provide a plausible explanation for species distribution in natural environments. It must be born in mind, however, that the above data have not brought us much closer to an explanation for Hutchinson's paradox. Tilman (1977) also mentioned that many other factors must influence the natural distribution of phytoplankton, and he mentioned specifically that one very important difference from most laboratory situations is that true steady states probably are never attained in natural environments. As Grenney *et al.* (1973) stated it, the major factor contributing to the observed phenomenon of numerous species coexisting in the same apparently isotropic environment could well be the continual variation in environmental conditions. Most likely, a phytoplankton community tends toward different equilibria at different times.

In another ecophysiological study on phototrophic microorganisms, purple sulfur bacteria in this case, the fact that the natural environment does not normally produce steady-state conditions was incorporated in a laboratory study (van Gemerden, 1974). The two *Chromatium* species, *Chromatium vinosum* and *Chromatium weissei* could not coexist when grown together in a sulfide-limited chemostat in the light: *Chr. vinosum* outgrew its competitor at all dilution rates tested. However, when instead of a continuous supply of light a regimen of light and dark was imposed (6 hr dark–6 hr light or 8 hr dark–4 hr light), the two species coexisted. This result was explained by taking into account the following three observations: (1) sulfide is not metabolized by these organisms in the dark; (2) the maximum rate of sulfide oxidation of *Chr. weissei* is approximately two times higher than that of *Chr. vinosum;* and (3) at elevated sulfide concentrations, *Chr. weissei* accumulated considerably more intracellular sulfur than did *Chr. vinosum.* Thus, when a dark period began, sulfide accumulated in the culture and both populations were washed out at the same rate. Upon reillumination, *Chr. weissei* seized the major part of the available sulfide, stored it as sulfur, and subsequently used it as additional electron donor during the period in which sulfide again became limiting, thus allowing it to grow faster than expected on the basis of its μ–s (sulfide) relationship. The relative abundance of the two species depended on the imposed light–dark rhythm. However, at very short dark periods (relative to the light periods), coexistence was no longer possible, and *Chr. weissei* was again eliminated from the culture.

Recently, van Gemerden and Beeftink (1983) reported two other examples of coexistence of the purple nonsulfur bacterium *Rhodopseudomonas capsulata* and the purple sulfur bacterium *Chr. vinosum*. *Rps. capsulata* exhibited a higher μ_m and a lower K_s for growth on sulfide than did *Chr. vinosum*. However, in mixed cultures of these two species with sulfide as the sole limiting substrate, *Rps. capsulata* did not outgrow *Chr. vinosum*. Quite the opposite appeared to happen: *Chr. vinosum* strongly dominated in the mixed culture with *Rps. capsulata* coexisting at a level of only 5% of the total biomass. This was convincingly explained by pointing out that *Rps. capsulata* oxidizes sulfide in a 2-electron yielding step to elemental sulfur that is deposited extracellularly and is thus available to *Chr. vinosum,* which in addition to a certain fraction of the available sulfide also oxidizes sulfur in a 6-electron yielding step to sulfate. The apparent competitiveness of *Rps. capsulata* was somewhat improved by enabling it to make use of its capacity to grow mixotrophically (Wijbenga and van Gemerden, 1981). To this end, acetate was included in the medium, which indeed resulted in an increase in the size of the *Rps. capsulata* population. The results of these experiments may explain why these latter bacteria are not known to bloom in nature. They also imply ecological consequences of the (in)ability to oxidize sulfide to the level of sulfate. Once more, these results point to the crucial role that mixotrophy may play in the survival of "generalists" in mixed cultures, and probably in nature, as well.

CHEMOSTAT ENRICHMENTS WITH MIXED SUBSTRATES

The answer to the question of to what extent the results of studies on bacteria in the laboratory can be extrapolated to the enormously more complicated mixed culture–mixed substrate situations occurring in nature is perhaps still almost as elusive as it was when Winogradsky studied his famous "Winogradsky columns." This reflects the variety of practical problems in assessing the *in situ* activity of microbial species and microbial communities, in identifying and isolating those microorganisms that mediate a particular biological process in the field, and in characterizing the precise physicochemical nature of the microenvironment in which the microorganisms reside. It is beyond the scope of this chapter to discuss these matters; however, the various aspects of this fascinating area of research have been dealt with in detail in several reviews (Harder *et al.*, 1977; Slater, 1981; Wimpenny, 1981; Fry, 1982; Parkes, 1982).

The outcome of any study on the ecology and/or physiology of bacteria in the laboratory depends to a large degree on the procedure used to isolate the desired bacteria. Often, only one type of enrichment procedure is followed. It is important to note that the classic batch-type enrichment procedure usually leads to the isolation of an entirely different type of microorganisms than when continuous culture enrichment techniques are employed (Jannasch, 1967; Schlegel and Jannasch, 1967; Veldkamp, 1970; Harder *et al.*, 1977; Parkes, 1982). A major drawback of batch-culture enrichments is that only those bacteria that thrive best at high substrate concentrations will become dominant. As a result, all those species that might be especially adapted to growth under low, growth-limiting substrate conditions will generally be overlooked (see Chapter 6, this volume). The type of bacteria termed "generalists" in the previous section will also be overlooked, as their competitive advantage appears to be the simultaneous consumption of two (or more) substrates present at limiting concentrations. An illustration of this principle was encountered during our attempts to isolate selectively thiobacilli able to grow mixotrophically (Gottschal and Kuenen, 1980b). These types of bacteria had never been isolated in batch-enrichment cultures because they were always outgrown by heterotrophic and autotrophic "specialists," which exhibit a substantially higher μ_m on the inorganic sulfur source and the organic substrate, respectively. Previously isolated "mixotrophs" had been obtained by screening bacteria grown on solid media for their capacity to grow both on organic and on sulfur-containing inorganic substrates. However, making use of the present knowledge of the competitive behavior of such "generalists" in mixed-substrate-limited chemostat cultures (Gottschal *et al.*, 1979), chemostat enrichments were performed using various mixtures of thiosulfate + acetate as limiting substrates. Using freshwater samples to inoculate the chemostat, "mixotrophs" indeed became dominant in four out of five different enrichments of this type; in the fifth culture, a heterotroph became dominant that oxidized thiosulfate but was unable to grow autotrophically cf. *Thiobacillus perometabolis* (London and Rittenberg, 1967). Accordingly, chemostat enrichment procedures with mixed substrate limitation might prove a useful technique for the *selective* isolation of "generalists."

The application of multiple-substrate limitation does not always lead to enrichment of one dominant organism that metabolizes the major fraction of both substrates. Chian and Mateles (1968) were probably the first to use mixed-substrate limitation in chemostat enrichments. These workers showed repeatedly that when mixtures of glucose + butyrate were used, mixed cultures of two dominant species, each metabolizing only one substrate, were obtained. A similar result was obtained by Laanbroek *et al.* (1977), who used glutamate + aspartate as limiting substrates in anaerobic chemostat enrichments. These authors found that in such enrichments two species became dominant: a glutamate-fermenting *Clostridium cochlearium* and an aspartate-fermenting *Campylobacter* species.

Furthermore, when samples of marine mud were used to inoculate chemostats in which thiosulfate + acetate were limiting, no mixotrophic thiobacilli came to the fore; rather, a mixed culture of an "autotrophic" thiobacillus and a heterotroph was obtained (Gottschal and Kuenen, 1980*b*). Since with freshwater samples this procedure resulted in dominant cultures of "mixotrophic" thiobacilli, it may be that these organisms are simply not present in marine sediments. Nevertheless, from the results obtained with the glucose + butyrate and with the glutamate + aspartate enrichments, it cannot yet be concluded that "generalists" capable of simultaneously metabolizing these organic substrates are absent from the environments sampled. Rather, one can speculate that such "jack of all trades" do thrive in nature, but that the fierce selection pressure in homogeneous continuous cultures favors the development of faster-growing specialists on the individual substrates. The rationale for this difference between "generalists" growing on mixtures of organic substrates and "generalists" growing on mixtures of organic and inorganic compounds can at present only be surmised. Perhaps the fact that both the specific growth rate and the cell yield can be higher for "mixotrophs" when grown on mixtures of an organic substrate plus an inorganic energy source is attributable to the energy-saving effect that organic substrates would have on biosynthesis by making CO_2-fixation unnecessary. By contrast, such an effect would not be expected with mixtures of functionally similar organic substrates unless they were metabolized via pathways with markedly different energetics. This would obviously strongly diminish the selective advantage of mixed substrate utilization, and selective pressure would probably work in favor of two specialized species, each exhibiting the highest specific growth rate at the prevailing concentration of the respective substrates. More experimental data on various mixed-substrate enrichments are required to evaluate this explanation. Of particular interest should be chemostat enrichments on such mixtures as H_2 or NH_4 + organic compound, methanol + formate (both requiring C_1 assimilation), or H_2S + light + organic substrate.

CONCLUSION

The outcome of competition experiments between microbes of different species grown in continuous culture with only one limiting nutrient often results in a pure culture of a single species. This result very elegantly illustrates

Gause's principle of competitive exclusion (1934): "if two species coexist, they must occupy different niches"; or "complete competitors cannot coexist" (Hardin, 1960).

In this chapter, much attention has been paid to coexistence of two (or more) species in continuous culture with more than one limiting nutrient. We have seen that the presence of additional substrates permits the coexistence of species. To be more precise, if competition for the limiting nutrients is the sole interaction in the mixed cultures, it follows directly from Gause's principle that for N species to coexist there must be present at least n distinguishable nutrients with $N \leq n$ (Levins, 1968; Phillips, 1973). However, I wish to stress here that this is only a minimum requirement and by no means a sufficient one. For stable coexistence to be possible, with $N = n = 2$, the contour plots reflecting the steady-state concentrations of the two substrates in a mixed culture of two species A and B should intersect (Levins, 1968; Taylor and Williams, 1975). Whether this is so will depend mainly on the usual growth parameters of the individual organisms: μ_m, Y(ield) and K_s. In such intersecting contour plots, areas can be found in which one population is able to outgrow all its competitors. Coexistence can then be expected at the points at which these areas touch. However, direct extrapolation of these seemingly straightforward conclusions to less well characterized multispecies–multisubstrate mixed cultures or to the natural environment are not justified, as the possible occurrence of all sorts of additional inhibitory and stimulatory interactions must be expected. Moreover, the importance of (regular) fluctuations in environmental conditions in maintaining the enormous species diversity in most habitats cannot be overemphasized. This implies that, although analysis of increasingly complex mixed cultures in continuous-flow systems at steady state will undoubtedly add considerably to our knowledge of the various strategies of competing microbial species, the study of mixed cultures under nonequilibrium conditions deserves much more attention than it has received to date. Mathematical models, properly describing the population dynamics of mixed cultures under such transient conditions, are badly needed. The use of such models in combination with experimental work on the behavior of mixed cultures under nonequilibrium conditions might alleviate considerably the extraordinarily difficult task faced by microbial ecologists in understanding the fascinating properties of nature's mixed culture.

ACKNOWLEDGMENT.   I am grateful to Dr. L. Dijkhuizen for his critical comments on the manuscript. This chapter includes literature up to June 1983, when the manuscript was first written.

REFERENCES

Bader, F. G., Meyer, H. S., Fredrickson, A. G., and Tsuchiya, H. M., 1975, Comments on microbial growth rate, *Biotechnol. Bioeng.* **17:**279–283.

Bazin, M. J., 1981a, Theory of continuous culture, in: *Continuous Cultures of Cells*, Vol. I (P. H. Calcott, ed.), pp. 27–62, CRC Press, Boca Raton, Florida.

Bazin, M. J., 1981*b*, Mixed culture kinetics, in: *Mixed Culture Fermentations* (M. E. Buskell and J. H. Slater, eds.), pp. 25–51, Academic Press, London.
Bull, A. T., and Brown, C. M., 1979, Continuous culture applications to microbial biochemistry, in: *International Review of Biochemistry*, Vol. 21: Microbial Biochemistry (J. R. Quayle, ed.), pp. 177–226, University Park Press, Baltimore.
Bull, A. T., and Slater, J. H., 1982, Microbial interactions and community structure, in: *Microbial Interactions and Communities*, Vol. 1 (A. T. Bull and J. H. Slater, eds.), pp. 13–44, Academic Press, London.
Bungay, H. R. III, and Bungay, M. L., 1968, Microbial interactions in continuous culture, *Adv. Appl. Microbiol.* **10:**269–290.
Button, D. K., 1985, Kinetics of nutrient-limited transport and microbial growth, *Microbiol Rev.* **49:**270–297.
Chian, S. K., and Mateles, R. I., 1968, Growth of mixed cultures on mixed substrates. I. Continuous culture, *Appl. Microbiol.* **16:**1337–1342.
Clark, B., and Holms, W. H., 1976, Control of the sequential utilization of glucose and fructose by *Escherichia coli*, *J. Gen. Microbiol.* **95:**191–201.
Cody, M. L., 1974, *Competition and the Structure of Bird Communities*, Monographs in Population Biology, No. 7, Princeton University Press, Princeton, New Jersey.
Contois, D. E., 1959, Kinetics of bacterial growth: Relationship between population density and specific growth rate of continuous cultures. *J. Gen. Microbiol.* **21:**40–47.
Cooney, C. L., and Mateles, R. I., 1971, Fermentation kinetics, *Proc. Int. Congr. Microbiol. Stand.* 441–449.
Dabes, J. N., Finn, R. K., and Wilke, C. R., 1973, Equations of substrate limited growth: The case for Blackman kinetics, *Biotechnol. Bioeng.***15:**1159–1177.
de Freitas, M. J., and Fredrickson, A. G., 1978, Inhibition as a factor in the maintenance of the diversity of microbial ecosystems, *J. Gen. Microbiol.* **106:**307–320.
Dijkhuizen, L., and Harder, W., 1979*a*, Regulation of autotrophic and heterotrophic metabolism in *Pseudomonas* OX1. Growth on mixtures of acetate and formate in continuous culture, *Arch. Microbiol.* **123:**47–53.
Dijkhuizen, L., and Harder, W., 1979*b*, Regulation of autotrophic and heterotrophic metabolism in *Pseudomonas* OX1. Growth on mixtures of oxalate and formate in continuous culture, *Arch. Microbiol.* **123:**55–63.
Droop, M. R., 1974, The nutrient status of algal cells in a continous culture, *J. Mar. Biol. Assoc. U.K.* **54:**825–855.
Dykhuizen, D., and Davies, M., 1980, An experimental model: Bacterial specialists and generalists competing in chemostats, *Ecology* **61**(5):1213–1227.
Edwards, V. H., 1969, Correlation of lags in the utilization of mixed sugars in continous fermentation, *Biotechnol. Bioeng.* **11:**99–102.
Egli, T., Käppeli, O., and Fiechter, A., 1982, Regulatory flexibility of methylotrophic yeasts in chemostat cultures: Simultaneous assimilation of glucose and methanol at a fixed dilution rate, *Arch. Microbiol.* **131:**1–7.
Egli, T., Lindley, N. D., and Quayle, J. R., 1983, Regulation of enzyme synthesis and variations of residual methanol concentration during carbon-limited growth of *Kloeckera* sp. 2201 on mixtures of methanol and glucose, *J. Gen. Microbiol.* **129:**1269–1281.
Fredrickson, A. G., 1977, Behaviour of mixed cultures of microorganisms, *Annu. Rev. Microbiol.* **31:**63–87.
Fredrickson, A.G., Megee, R. D., and Tsuchiya, H. M., 1970, Mathematical models for fermentation processes, *Adv. Appl. Microbiol.* **13:**419–465.
Fry, J. C., 1982, The analysis of microbial interactions and communities *in situ*, in: *Microbial Interactions and Communities*, Vol. I (A. T. Bull and J. H. Slater, eds.), pp. 103–152, Academic Press, London.
Garrett, A., and Wittman, H. G., 1974, Structure of bacterial ribosomes, *Adv. Protein Chem.* **27:**277–347.
Gause, G. F., 1934, *The Struggle for Existence*, Williams & Wilkins, Baltimore.

Gorini, L., and Davies, J., 1968, The effect of streptomycin on ribosome function, *Curr. Topics Microbiol. Immunol.* **44:**100–122.

Gottschal, J. C., and Kuenen, J. G., 1980*a*, Mixotrophic growth of *Thiobacillus* A2 on acetate and thiosulfate as growth limiting substrates in the chemostat, *Arch. Microbiol.* **126:**33–42.

Gottschal, J. C., and Kuenen, J. G., 1980*b*, Selective enrichment of facultatively chemolithotrophic thiobacilli and related organisms in continous culture, *FEMS Microbiol. Lett.* **7:**241–247.

Gottschal, J. C., and Thingstad, T. F., 1982, Mathematical description of competition between two and three bacterial species under dual substrate limitation in the chemostat: A comparison with experimental data, *Biotechnol. Bioeng.* **24:**1403–1418.

Gottschal, J. C., De Vries, S., and Kuenen, J. G., 1979, Competition between the facultatively chemolithotrophic *Thiobacillus* A2, an obligately chemolithotrophic *Thiobacillus* and a heterotrophic *Spirillum* for inorganic and organic substrates, *Arch. Microbiol.* **121:**241–249.

Gottschal, J. C., Nanninga, H. J., and Kuenen, J. G., 1981, Growth of *Thiobacillus* A2 under alternating growth conditions in the chemostat, *J. Gen. Microbiol.* **126:**85–96.

Grenney, W. J., Bella, D. A., and Curl, H. C., 1973, A theoretical approach to interspecific competition in phytoplankton communities, *Am. Nat.* **107:**405–425.

Haas, C. N., Bungay, H. R., and Bungay, M. L., 1980, Practical mixed culture processes, *Annu. Rep. Ferment. Proc.* **4:**1–29.

Harder, W., and Dijkhuizen, L., 1976, Mixed substrate utilization, in: *Continuous Culture. 6: Applications and New Fields* (A. C. R. Dean, D. C. Ellwood, C. G. T. Evans, and J. Melling, eds.), pp. 297–314, Ellis Horwood, Chichester, England.

Harder, W., and Dijkhuizen, L., 1982, Strategies of mixed substrate utilization in microorganisms, *Phil. Trans. R. Soc. Lond. B* **297:**459–480.

Harder, W., Kuenen, J. G., and Matin, A., 1977, Microbial selection in continuous culture, *J. Appl. Bacteriol.* **43:**1–24.

Hardin, G., 1960, The competitive exclusion principle, *Science* **131:**1292–1297.

Harrison, D. E. F., 1978, Mixed cultures in industrial fermentation processes, *Adv. Appl. Microbiol.* **24:**129–164.

Herbert, D., Elsworth, R., and Telling, R. C., 1956, The continuous culture of bacteria: A theoretical and experimental study, *J. Gen. Microbiol.* **14:**601–622.

Hutchinson, G. E., 1961, The paradox of the plankton, *Am. Nat.* **95:**137–145.

Jannasch, H. W., 1967, Enrichment of aquatic bacteria in continuous culture, *Arch. Mikrobiol.* **59:**165–173.

Jost, J. L., Drake, J. F., Fredrickson, A. G., and Tsuchiya, H. M., 1973, Interactions of *Tetrahymena pyriformis, Escherichia coli, Azotobacter vinelandii* and glucose in a minimal medium, *J. Bacteriol.* **113:**834–840.

Konings, W. N., and Veldkamp, H., 1983, Energy transduction and solute transport mechanisms in relation to environments occupied by microorganisms, in: *Microbes in their Natural Environments,* Symposium 34: The Society for General Microbiology (J. H. Slater, A. Whittenbury, and J. W. T. Wimpenny, eds.), pp. 153–186, Cambridge University Press, Cambridge.

Krauel, U., Krauel, H. H., and Weide, H., 1982, Abbau von Mischsubstraten durch Hefen. I. Substrate der Sulfitablauge, *Z. Allg. Mikrobiol.* **22:**545–555.

Kuenen, J. G., and Gottschal, J. C., 1982, Competition between chemolithotrophs and methylotrophs and their interactions with heterotrophic bacteria, in: *Microbial Interactions and Communities,* Vol. 1 (A. T. Bull and J. H. Slater, eds.), pp. 153–187, Academic Press, London.

Laanbroek, H. J., Kingma, W., and Veldkamp, H., 1977, Isolation of an aspartate-fermenting, free living *Campylobacter* spec., *FEMS Microbiol. Lett.* **1:**99–102.

Laanbroek, H. J., Smit, A. J., Klein Nulend, G., and Veldkamp, H., 1979, Competition for L-glutamate between specialized and versatile *Clostridium* species, *Arch. Microbiol.* **120:**61–66.

Law, A. T., and Button, D. K., 1977, Multiple-carbon-source-limited growth kinetics of a marine coryneform bacterium, *J. Bacteriol.* **129:**115–123.

Lehman, J. T., Botkin, D. B., and Likens, G. E., 1975, The assumption and rationales of a computer model of phytoplankton population dynamics, *Limnol. Oceanogr.* **20:**343–364.

Levins, R., 1968, *Evolution in Changing Environments. Some Theoretical Explorations,* Princeton University Press, Princeton, New Jersey.

Linton, J. D., and Drozd, J. W., 1982, Microbial Interactions and communities in Biotechnology, in: *Microbial Interactions and Communities,* Vol. 1 (A. T. Bull and J. H. Slater, eds.), pp. 357–406, Academic Press, London.

London, J., and Rittenberg, S. C., 1967, *Thiobacillus perometabolis* nov. sp., a nonautotrophic *Thiobacillus, Arch. Mikrobiol.* **59:**218–225.

MacDonald, N., 1976, Time delay in simple chemostat models, *Biotechnol. Bioeng.* **18:**805–812.

Mateles, R. I., Chian, S. K., and Silver, R., 1967, Continuous culture on mixed substrates, in: *Microbial Physiology and Continuous Culture* (E. O. Powell, C. Evans, R. E. Strange, and D. W. Tempest, eds.), pp. 233–239, Her Majesty's Stationary Office, London.

Meers, J. L., 1973, Growth of bacteria in mixed cultures, *CRC Crit. Rev. Microbiol.* **2:**138–184.

Megee, R. D. III, Drake, J. F., Frederickson, A. G., and Tsuchiya, H. M., 1972, Studies in intermicrobial symbiosis. *Saccharomyces cerevisiae* and *Lactobacillus casei, Can. J. Microbiol.* **18:**1733–1742.

Meyer, J. S., Tsuchiya, H. M., and Frederickson, A. G., 1975, Dynamics of mixed populations having complementary metabolism, *Biotechnol. Bioeng.* **17:**1065–1081.

Miura, Y., Tanaka, H., and Okazaki, M., 1980, Stability analysis of commensal and mutual relations with competitive assimilation in continuous mixed culture, *Biotechnol. Bioeng.* **22:**929–948.

Monod, J., 1942, *Recherches sur la croissance des cultures bactériennes,* Hermann and Cie, Paris.

Monod, J., 1950, La technique de culture continue; théorie et applications, *Ann. Inst. Pasteur Paris* **79:**390–410.

Mur, L. R., Gons, H. J., and van Liere, L., 1977, Some experiments on the competition between green algae and blue-green bacteria in light-limited environments, *FEMS Microbiol. Lett.* **1:**335–338.

Novick, A., and Szilard, L., 1950, Description of the chemostat, *Science* **112:**715–716.

Paigen, K., and Williams, B., 1970, Catabolite repression and other control mechanisms in carbohydrate utilization, *Adv. Microbial Physiol.* **4:**251–324.

Parkes, R. J., 1982, Methods for enriching, isolating, and analysing microbial communities in laboratory systems, in: *Microbial Interactions and Communities,* Vol. 1 (A. T. Bull and J. H. Slater, eds.), pp. 45–102, Academic Press, London.

Perez, R. C., and Matin, A., 1982, Carbon dioxide assimilation by *Thiobacillus novellus* under nutrient-limited mixotrophic conditions, *J. Bacteriol.* **150:**46–51.

Phillips, O. M., 1973, The equilibrium and stability of simple marine biological systems. I. Primary nutrient consumers, *Am. Nat.* **107:**73–93.

Pirt, S. J., 1965, The maintenance energy of bacteria in growing cultures, *Proc. R. Soc. Lond. B* **163:**224–231.

Pirt, S. J., 1975, *Principles of Microbe and Cell Cultivation,* Blackwell, Oxford.

Powell, E. O., 1958, Criteria for the growth of contaminants and mutants in continuous culture, *J. Gen. Microbiol.* **18:**259–268.

Powell, E. O., 1967, The growth rate of microorganisms as a function of substrate concentration, in: *Microbial Physiology and Continuous Culture Proceedings of the Third International Symposium* (E. O. Powell, C. G. T. Evans, R. E. Strange, and D. W. Tempest, eds.), pp. 34–56, Her Majesty's Stationary Office, London.

Ryder, D. N., and Sinclair, C. G., 1972, Model of the growth of aerobic microorganisms under oxygen limiting conditions, *Biotechnol. Bioeng.* **14:**787–798.

Schlegel, H. G., and Jannasch, H. W., 1967, Enrichment cultures, *Annu. Rev. Microbiol.* **21:**49–70.

Schulze, K. L., and Lipe, R. S., 1964, Relationship between substrate concentration, growth rate, and respiration rate of *Escherichia coli* in continuous culture, *Arch. Mikrobiol.* **48:**1–20.

Silver, R. S., and Mateles, R. I., 1969, Control of mixed substrate utilization in continuous cultures of *Escherichia coli, J. Bacteriol.* **97:**535–543.

Sinclair, C. G., and Ryder, D. N., 1975, Models for the continuous culture of microorganisms under both oxygen and carbon limiting conditions, *Biotechnol. Bioeng.* **17:**375–398.

Slater, J. H., 1981, Mixed cultures and microbial communities, in: *Mixed Culture Fermentations* (M. E. Buskell and J. H. Slater, eds.), pp. 1–24, Academic Press, London.

Slater, J. H., and Bull, A. T., 1982, Environmental microbiology: Biodegradation, *Phil. Trans. R. Soc. Lond. B* **297**:575–597.

Smith, M. E., and Bull, A. T., 1976, Studies of the utilization of coconut water waste for the production of the food yeast *Saccharomyces fragilis, J. Appl. Bacteriol.* **41**:81–95.

Smith, A. L., Kelly, D. P., and Wood, A. P., 1980, Metabolism of *Thiobacillus* A2 grown under autotrophic, mixotrophic and heterotrophic conditions in chemostat culture, *J. Gen. Microbiol.* **121**:127–138.

Standing, C. N., Fredrickson, A. G., and Tsuchiya, H. M., 1972, Batch- and continuous culture transients for two substrate systems, *Appl. Microbiol.* **23**:354–359.

Stewart, F. M., and Levin, B. R., 1973, Partitioning of resources and the outcome of interspecific competition: A model and some general considerations, *Am. Nat.* **107**:171–198.

Stouthamer, A. H., 1977, Energetic aspects of the growth of microorganisms, in: *Microbial Energetics*, Twenty-seventh Symposium of the Society of General Microbiology, London (R. A. Haddock and W. A. Hamilton, eds.), pp. 285–315, Cambridge University Press, Cambridge.

Stouthamer, A. H., 1979, The search for correlation between theoretical and experimental growth yields, in: *International Review of Biochemistry*, Vol. 21, Microbial Biochemistry (J. R. Quayle, ed.) pp. 1–47, University Park Press, Baltimore.

Sykes, R. M., 1973, Identification of the limiting nutrient and specific growth rate, *J. Water Pollut. Control Fed.* **45**:888–895.

Taylor, P. A., and Williams, P. J. LeB., 1975, Theoretical studies on the coexistence of competing species under continuous flow conditions, *Can. J. Microbiol.* **21**:90–98.

Tempest, D. W., and Neijssel, O. M., 1978, Ecophysiological aspects of microbial growth in aerobic nutrient-limited environments, in: *Advances in Microbial Ecology*, Vol. 2 (M. Alexander, ed.), pp. 105–153, Plenum Press, New York.

Tilman, D., 1977, Resource competition between planktonic algae: An experimental and theoretical approach, *Ecology* **58**:338–348.

Tilman, D., and Kilham, S. S., 1976, Phosphate and silicate growth and uptake kinetics of the diatoms *Asterionella formosa* and *Cyclotella meneghiniana* in batch and semicontinuous culture, *J. Phycol.* **12**:375–383.

Tilman, D., 1976, Ecological competition between algae: Experimental confirmation of resource-based competition theory, *Science* **192**:463–465.

van Gemerden, H., 1974, Coexistence of organisms competing for the same substrate: An example among the purple sulfur bacteria, *Microb. Ecol.* **1**:104–119.

van Gemerden, H., and Beeftink, H. H., 1983, Ecology of phototrophic bacteria, in: *The Phototrophic Bacteria: Anaerobic Life in the Light*, Studies in Microbiology, Vol. 4 (J. G. Ormerod, ed.), pp. 146–185, Blackwell, Oxford.

van Verseveld, H. W., 1979, Influence of environmental factors on the efficiency of energy conservation in *Paracoccus denitrificans*, thesis, Vrije Universiteit, Amsterdam.

Veldkamp, H., 1970, Enrichment cultures of prokaryotic organisms, in: *Methods in Microbiology*, Vol. 3A (J. R. Norris and D. W. Ribbons, eds.), pp. 305–361, Academic Press, London.

Veldkamp, H., 1977, Ecological studies with the chemostat, in: *Advances in Microbial Ecology* Vol. 1 (M. Alexander, ed.), pp. 59–94, Plenum Press, New York.

Veldkamp, H., and Jannasch, H. W., 1972, Mixed culture studies with the chemostat, *J. Appl. Chem. Biotechnol.* **22**:105–123.

Weide, H., 1983, Mikrobielle Verwertung von Mischsubstraten, *Z. Allg. Mikrobiol.* **23**:37–70.

Wijbenga, D. J., and van Gemerden, H., 1981, The influence of acetate on the oxidation of sulfide by *Rhodopseudomonas capsulata, Arch. Microbiol.* **129**:115–118.

Williams, F. M., 1967, A model of cell growth dynamics, *J. Theor. Biol.* **15**:190–207.

Wimpenny, J. W. T., 1981, Spatial order in microbial ecosystems, *Biol. Rev.* **56**:295–342.

Wood, A. P., and Kelly, D. P., 1983, Autotrophic, mixotrophic and heterotrophic growth with denitrification by *Thiobacillus* A2 under anaerobic conditions, *FEMS Microbiol. Lett.* **16**:363–370.

Yoon, H., Klinzing, G., and Blanch, H. W., 1977, Competition for mixed substrates by microbial populations, *Biotechnol. Bioeng.* **19**:1193–1210.

TRANSIENT AND PERSISTENT ASSOCIATIONS AMONG PROKARYOTES

Michael J. McInerney

INTRODUCTION

Since most, if not all, microbial ecosystems contain a large number of metabolically diverse species, it would be impossible and perhaps confusing to describe in detail all the possible interactions and associations that occur in nature. Such a purely descriptive approach does little to explain the basis of these interactions or their functional importance in the ecosystem. The function of any ecosystem pertains to the flow of energy and the cycling of materials (Odum, 1971). Thus, the analysis of any ecosystem must include the role of individual species in the flow of energy and material cycling as well as the ways in which interactions among species affect these functions. This approach defines the trophic structure of the ecosystem and allows one to visualize more clearly the kinds of interactions that occur within and between trophic levels as well as their importance in the ecosystem.

Microbial interactions are also important in the development of community structure (Bull and Slater, 1982; Atlas and Bartha, 1981; see Chapter 9, this volume). The greater the diversity of species, the greater the complexity of the community. Communities with high species diversity have greater stability and are able to tolerate environmental fluctuations within certain broad limits. Species diversity, analogous to genetic diversity within a population, allows for a more varied response within a dynamic ecosystem. Microbial interactions add to functional diversity of species (Bull and Slater, 1982) and the homeostatic mechanisms that maintain the stability of a community are based on interactions between community members.

Michael J. McInerney • Department of Botany and Microbiology, University of Oklahoma, Norman, Oklahoma 73019.

Thus, the study of microbial interactions must pertain to the function of these interactions in the ecosystem and how these interactions affect species diversity and community stability. This metabolic or biochemical approach to ecology has been successfully applied to the study of the rumen of herbivores (Hungate, 1966), probably making it one of the best-studied microbial ecosystem. This and several other anaerobic ecosystems will serve as models for the discussion of bacterial interactions. The general principles learned from the study of these systems can be applied to other microbial ecosystems.

To obtain a complete understanding of any microbial ecosystem, many different experimental approaches are required (C. J. Smith and Bryant, 1979; R. T. J. Clarke, 1977). Information on the overall reactions of the complete microbial community obtained with *in vivo* or *in vitro* techniques are required. For many environments, *in vivo* studies are relatively easy because samples can be obtained without markedly affecting the overall physiology of the system. However, for certain environments such as the large bowel of man, sampling is a major problem. When using *in vitro* techniques to study mixed microbial communities, special care must be taken to simulate the natural environment accurately. Batch-culture studies of mixed ruminal bacteria that contain high levels of readily degradable carbohydrates often give erroneous results due to the rapid growth and selection of one or a few species (C. J. Smith and Bryant, 1979). Research with pure cultures is essential to understanding microbial metabolism in the ecosystem. However, the metabolic activities of bacteria are often modified by interactions with other species, and data obtained with pure cultures must be interpreted with caution. An example of this is the effect that H_2-using bacteria (such as methanogens) have on H_2-production by fermentative bacteria (Wolin, 1974). This chapter relates how data obtained from both pure cultures and known mixtures of species can be used to visualize the principles involved in establishing individual niches and the biochemical ecology of the ecosystem. The first part discusses bacterial interactions in general terms, and the second part discusses these interactions are related to particular ecosystems.

CLASSIFICATION OF BACTERIAL INTERACTIONS

There have been several attempts to classify the types of interactions that occur among bacterial species and between bacteria and eukaryotic organisms (see Bull and Slater, 1982). Some of these attempts have led to rather complex and confusing systems of classification and nomenclature making the utility of some systems questionable. However, in many cases, these systems can facilitate the description and categorization of the many interactions that occur in microbial communities. I take the position that it is far less important to know whether an interaction is commensalistic or protocooperative than to know how the interaction is important in energy flow or cycling of materials in the ecosystem.

TABLE I
Analysis of Population Interactions[a]

Type of interaction	Species[b] 1	2	General nature of interaction
Neutralism	0	0	Neither population affects the other
Competition	−	−	Inhibition when resource is in short supply
Amensalism	−	0	Population 1 inhibited, 2 not affected
Parasitism	+	−	Population 1, the parasite, usually smaller than 2, the host
Predation	+	−	Population 1, the predator, usually larger than 2, the prey
Commensalism	+	0	Population 1 benefits, while 2 is unaffected
Protocooperation	+	+	Interaction favorable to both but not obligatory
Mutualism	+	+	Interaction favorable to both and obligatory

[a] From Odum (1971).
[b] 0, no significant interaction; +, growth, survival, or other population attribute benefited (positive term added to growth equation); −, population growth or other attribute inhibited (negative term added to growth equation).

One of the more widely used schemes, originally proposed by Odum (1953), is based on the effects of one species on the population size of the second species. The effects are rated as positive (+), neutral (0), or negative (−) depending on whether the population size of the affected species increases, remains the same, or decreases. The effects of species 1 on species 2 and *vice versa* are arranged in a binary matrix that contains all the possible combinations (Table I). As Odum (1971) points out, considering population interactions in this manner avoids the confusion that results when only terms or definitions are employed. The following will provide examples of these kinds of interactions. More comprehensive reviews of these types of interactions are available (Meers, 1973; Fredrickson, 1977; Bull and Slater, 1982). The definitions of the terms are as used by Atlas and Bartha (1981) and Odum (1971). Other definitions that differ from these are noted.

Neutralism

Neutralism implies the lack of interaction between two species or populations where neither population affects the other. Brock (1966) points out that it is rare that two microbial populations do not interact. Indeed, examples of demonstrated neutralism are rare (Bull and Slater, 1982; Meers, 1973). One possible example is provided by the work of Lewis (1967): A *Lactobacillus* species and a *Streptococcus* species were grown in pure culture and in mixed cultures in a chemostat; the population densities of each species in pure culture were almost the same as in the mixed culture, indicating that neither species affected the growth of the other. The significance and occurrence of neutralism in natural environments are unknown.

Commensalism

In a commensalistic relationship, one population benefits from the association while the other population remains unaffected. The most commonly cited examples of commensalistic relationships involve the production by one species of a growth factor essential for the growth of a second species or the removal of an inhibitory compound by one species, which thereby allows another species to grow. Commensalistic relationships of the first type are important in successions of microbial communities, and may also occur in the contamination of laboratory cultures. Several examples of these two types of commensalistic interactions were summarized by Meers (1973).

Other examples of commensalism include the generation of carbon compounds by one species that are used as energy sources by a second species. For example, *Acetobacter suboxydans* oxidized mannitol to fructose, thereby allowing growth of *Saccharomyces carlsbergensis*, which used fructose, but not mannitol, as an energy source (Chao and Reilly, 1972). However, other interactions may also have affected *A. suboxydans* in this study (Chao and Reilly, 1972).

The ability of bacteria to cometabolize compounds allows for the development of commensalistic relationships. In cometabolism, one species metabolizes a (second) substrate that it is unable to use as sole carbon or energy source, but the product of this reaction may be used by another organism. Beam and Perry (1974) showed that when *Mycobacterium vaccae* cometabolize cyclohexane to cyclohexanol when growing on propane as its energy source, the cyclohexanol was used by a *Pseudomonas* species as its source of energy and cell carbon.

Commensalistic relationships may also be based on the modification of the environment by one species in such a way that the growth of other species is allowed or stimulated. The removal of oxygen by facultative anaerobes enables obligate anaerobes to grow. The activity of osmotolerant yeasts reduces the osmolarity, thereby enabling less tolerant species to grow. Other examples include the removal or detoxification of antibiotics, metals, or other compounds by one species, allowing growth of others (Meers, 1973).

Other interactions may complicate studies or commensalism. Megee *et al.* (1972) grew *Saccharomyces cerevisiae* and a riboflavin-requiring strain of *Lactobacillus casei* in continuous culture. When riboflavin was absent from the medium, *L. casei* depended on the yeast to produce riboflavin, but both species competed for the limiting nutrient, glucose. *L. casei* had a competitive advantage over the yeast because of its higher affinity for glucose and a greater death rate of the yeast under starvation conditions. When riboflavin was added to the medium, the yeast was washed out. However, when riboflavin was absent, the commensalistic interaction counteracted the effects of competition, and a stable population of both species coexisted over a wide range of dilution rates.

The ability of one species to alter its physiology to avoid competition may be important in developing commensalistic relations. Both *Lactobacillus plantarum* and *Propionibacterium shermanii* are important in the production of Swiss cheese. Both species use glucose, but *P. shermanii* will use lactate preferentially when both glucose and lactate are present (Lee *et al.*, 1976). The inoculation of both organisms into batch cultures of glucose minimal medium initially results in a competition for glucose. As the fermentation proceeds, competition is reduced as *P. shermanii* uses the lactate produced by *L. plantarum* from glucose (Lee *et al.*, 1976). This avoidance of competition also occurs in continuous culture. If *P. shermanii* did not preferentially use lactate it would be excluded from chemostats operated at high dilution rates, whereas *L. plantarum* would be excluded at low dilution rates. The replacement of competition by commensalism allows both species to coexist in a steady state over a wide range of dilution rates (Lee *et al.*, 1976).

Some examples of commensalism may be an oversimplification of the interactions that occur in nature. The degradation of cellulose by anaerobic bacteria in the rumen produces soluble sugars and fermentation products that can be used by other bacteria as energy sources. This has been cited as an example of commensalism (Atlas and Bartha, 1981). However, many of these cellulolytic bacteria require B vitamins, heme, volatile fatty acids, and other growth factors (Bryant, 1973, 1977) that are supplied by other anaerobic species. Although the removal of oxygen by facultative bacteria allows obligate anaerobes to grow and may seem to be commensalism, the facultative bacteria may actually benefit from the relationship by using the products of anaerobic metabolism as sources of energy or nutrients.

Protocooperation and Mutualism

In protocooperation and mutualism, both interacting populations benefit from the relationship. Protocooperation differs from mutualism in that the interaction is not obligatory while, in mutualism, the interaction is obligatory. The meaning of "obligatory" may depend on the perspective of the scientist. To some, obligatory means only that the interaction is necessary for survival in the habitat, and there is no requirement that the interaction be specific for the two interacting species (Fredrickson, 1977). To others, obligatory means both a specificity for the interacting species and a necessity of the interaction for survival in the habitat (Atlas and Bartha, 1981; Odum, 1971). This latter interpretation more closely agrees with that of Odum (1971) and will be used here. Thus, in protocooperation, both may depend on the interaction to survive in the habitat, but the interaction need not be with a particular species, whereas the mutualistic interaction is necessary for survival and is usually specific for the species involved.

This definition of protocooperation is slightly different from that of Atlas and Bartha (1981) who state that both species can exist in the habitat on their own but the association offers some mutual benefit. One reason for this different interpretation can be illustrated in the following example. The anaerobic cellulolytic bacteria in the rumen produce hydrolysis and fermentation products from cellulose that are used by other ruminal bacteria. The cellulolytic bacteria require B vitamins and other growth factors that are supplied by other bacteria (Bryant, 1973). The interaction is obligate, since the cellulolytic bacteria cannot survive in the rumen without B vitamins, but it is not specific in terms of the species involved. Any species that produces B vitamins would serve. Many biologists would agree that such an interaction is protocooperative, not mutualistic. It is often difficult to ascertain whether a relationship is commensalistic or protocooperative or whether a relationship is obligatory and specific for the individual species involved and should be considered mutualistic. Again, in the present context, understanding the function of the interaction in the ecosystem is more important than deciding which label to assign to it.

One of the most common examples of protocooperation is the cross-feeding of nutrients between species. This type of interaction is also referred to as syntrophy. *Lactobacillus arabinosus* and *Streptococcus faecalis* grow together, but not singly, in a glucose minimal medium (Nurmikko, 1954). *S. faecalis* requires folic acid, which is produced by *L. arabinosus*, and *L. arabinosus* requires phenylalanine, which is produced by *S. faecalis*. An analogous interaction occurs between a nicotinic acid-requiring *Bacillus polymyxa* and a biotin-requiring *Proteus vulgaris* (Yeoh *et al.*, 1968). Other examples of syntrophy include interspecies hydrogen reactions and the cycling of electron donors and acceptors between photosynthetic and sulfur-reducing species. These are discussed more fully later in the sections on syntrophic associations in methanogenic environments and with phototrophs.

Protocooperative interactions may involve the removal of an inhibitory compound. Mixed cultures growing on methane as the sole carbon and energy source grow better and are more stable than pure cultures (Sheehan and Johnson, 1971; Harrison *et al.*, 1975). The instability of pure cultures growing on methane is believed to be due to the accumulation of inhibitory intermediate oxidation products of methane (Wilkinson and Harrison, 1973). One mixed culture contained a methane-oxidizing pseudomonad that was inhibited by methanol (Wilkinson and Harrison, 1973); a *Hyphomicrobium* sp. present in the mixed culture had a high affinity for methanol and prevented its accumulation in the medium. The *Pseudomonas* species could not be grown in pure culture, and the *Pseudomonas-Hyphomicrobium* coculture did not grow so well as a mixed culture that also contained *Flavobacter* and *Acinetobacter* species. Thus, additional, unidentified interactions must also have occurred among these species (Wilkinson and Harrison, 1973; Wilkinson *et al.*, 1974).

An unusual synergistic interaction is the production of an enzyme by an association of species not produced by either species alone. Closely related *Pseudomonas* strains produce portions of the enzyme lecithinase; in co-cultures, an active enzyme is produced, even though no single strain produces an active enzyme (Bates and Lies, 1963). Other examples of protocooperation are cited in other reviews (Meers, 1973; Frederickson, 1977; Atlas and Bartha, 1981; Veldkamp, 1977).

Analysis of food chains and trophic structure of an ecosystem allows one to visualize protocooperative and commensalistic relationships more clearly. Which species metabolizes the primary substrate, and which species utilizes the products from the degradation of the primary substrate? The reason for the great diversity of species in some ecosystems becomes apparent when one recognizes the large number of ways that cross-feeding of nutrients and energy sources between species can occur. This explains why many environments can contain large numbers of nutritionally fastidious organisms.

The terms mutualism and symbiosis are often used synonymously to describe an obligate interaction between two species where both species benefit from the relationship (Atlas and Bartha, 1981). This kind of interaction often requires close physical proximity and is specific for the organisms involved; usually, one member cannot be replaced by another closely related species. This definition differs somewhat from that used by other workers (Meers, 1973; Fredrickson, 1977). The meanings of close physical proximity and of obligate are subject to different interpretations. Many associations among bacteria that are merely food web associations have often been interpreted as symbiosis (Nurmikko, 1954; Khan and Murray, 1982a; Chao and Reilly, 1972).

Most of the classic examples of mutualism involve interaction with eukaryotes and will be discussed in other volumes of this series. Some examples of this kind of interaction are the lichens (Hale, 1974), the nitrogen-fixing bacteria in root nodules (Burns and Hardy, 1975), and the anaerobic cellulose-degrading bacteria in the rumen of herbivores (Hungate, 1966).

Competition

Competition represents a negative interaction between two species in which the population sizes of both species are adversely affected. Resource-based competition occurs when chemicals, particulate matter or, in some cases, light is removed from a common environment resulting in the reduction of the net growth rates of the two populations (Fredrickson and Stephanopoulos, 1981). A resource is any substance that is used for growth and proliferation of the population. Two populations are said to compete for a resource if both populations use, although do not necessarily require the resource, and if the resource has a dynamic effect on at least one of the populations (Fredrickson and Stephanopoulos, 1981). A resource has a dynamic effect on a population if its availability has a significant effect on the net growth rate of the population (Fredrickson and Stephanopoulos, 1981). The outcome of competition is usually the exclusion of one of the competitors from the habitat. This is called the competitive exclusion principle (see Chapter 9, this volume) and, as originally proposed by Hardin (1960), it states, "complete competitors cannot coexist." However, there are competitive situations that can be described as complete where two populations can coexist, sometimes at steady state (Fredrickson and Stephanopoulos, 1981; Fredrickson, 1977; Meers, 1973; Veldkamp, 1977). These situations are discussed more fully below.

The principle of competition for a single resource by two populations where competition is the only interaction between the two populations is called pure and simple competition and has been discussed by others (Veldkamp and Jannasch, 1972; Fredrickson and Stephanopoulos, 1981; Powell, 1958; Harder et al., 1977; Slater and Bull, 1978; see also Gottshal, this volume, for discussion of a similar model). The analysis of competition is an extension of the single-strain growth model of Monod (Herbert *et al.*, 1956), where the outcome of competition depends on the substrate concentration, the specific and maximal growth rates of the individual species, and their respective affinities for the substrate. For two competing strains grown in continuous culture with constant substrate input, the equations for substrate utilization and growth are (Hansen and Hubbel, 1980):

$$\frac{dS}{dt} = (S_R - S)D - \frac{\mu_{m1}}{Y_1}\left(\frac{S \cdot N_1}{K_{s_1} + S}\right) - \frac{\mu_{m2}}{Y_2}\left(\frac{S \cdot N_2}{K_{s_2} + S}\right)$$

$$\frac{dN_1}{dt} = \frac{\mu_{m1}(S \cdot N_1)}{K_{s_1} + S} - D \cdot N_1$$

$$\frac{dN_2}{dt} = \frac{\mu_{m2}(S \cdot N_2)}{K_{s_2} + S} - D \cdot N_2$$

where S is the concentration of the limiting resource in the culture, S_R is the reservoir concentration of the limiting resource, and D is the dilution rate of the culture. For the ith species, N_i is the concentration of the cells in the culture, $-D$ represents the rate of loss due to cell outflow, μ_{m_i} is the maximal specific growth rate, Y_i is the yield of cells (cells per unit of resource) and K_{s_i} is the half-saturation constant (Monod constant) for the limiting resource (Herbert *et al.*, 1956).

The outcome of competition can be predicted by comparing curves relating the specific growth rate μ to the concentration of the resource in the vessel S (Fig. 1) (Veldkamp, 1970; Harder *et al.*, 1977; Fredrickson and Stephanopoulos, 1981). When the intrinsic rate of increase in each population is zero, the two populations can coexist (Fredrickson and Stephanopoulos, 1981). The intrinsic rate of increase of a population in a homogeneous system is the difference between the rate at which new cells are added by reproduction and immigration and the rate at which cells are lost by death and outflow from the culture. If external factors such as temperature are held constant, only one resource is limiting, and competition is the only interaction that occurs between the two populations, then the intrinsic rates of increase of the

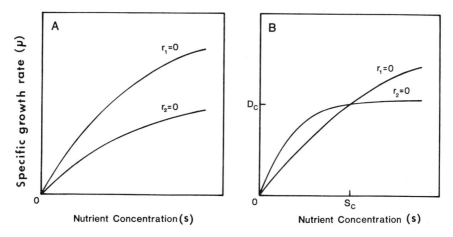

FIGURE 1. Relationship between specific growth rate (μ) and steady-state concentration(s) of limiting substrate for pure and simple competition in a chemostat. (A) Population 1 always grows faster than population 2. (B) Population 1 grows faster than population 2 only at $S < S_c$. The intrinsic rates of increase, r_1 and r_2, for populations 1 and 2 equal 0 along the curves shown. In a chemostat culture, μ is determined operationally and at steady state is equal to the dilution rate D. (After Veldkamp, 1970, and Fredrickson and Stephanopoulos, 1981; see also Fig. 1 of Chapter 7, this volume.)

populations are functions only of S and of the dilution rate, $D = \mu$ in steady state; see Fig. 1). For the two populations represented in Fig. 1A, there are no meaningful values of S and D that satisfy the conditions necessary for coexistence, and population 2 will be excluded from a mixed-population chemostat at all dilution rates. In Fig. 1B, the curves intersect at $S = S_c$ ($\mu = D_c$), and the conditions for coexistence can be satisfied if D is held exactly at D_c. At all other dilution rates, one of the populations will be excluded—population 1 at dilution rates less than D_c and population 2 at dilution rates greater than D_c. The coexistence at $\mu = D_c$ is theoretically stable, but in practice, there are always random variations in D with time (Fredrickson and Stephanopoulos, 1981). Stephanopoulos et al. (1979) modeled the random fluctuations of D as "white noise" and showed that one competitor will be excluded if the intensity of the noise in D and the bias of the mean of D away from D_c are not both zero. If the intensity of the noise and the magnitude of the bias are small, there will be a drift toward the exclusion of one population. Although this drift may be small, it will always occur. Several studies have verified this theoretical analysis of competition (Fredrickson, 1977; Fredrickson and Stephanopoulos, 1981; Meers, 1971, 1973; Jannasch, 1967; Harder and Veldkamp, 1971; Powell, 1958; Kuenen and Gottschal, 1982).

    A more mathematical assessment of competition involves the use of the J parameter (Hansen and Hubbel, 1980), which is essentially the ratio, for a given species, of its half-saturation constant for the limiting resource and its intrinsic rate of increase. Which species survives or whether wash-out of the culture occurs depends on S_R and the J parameter. For the ith species, the J parameter is as follows:

$$J_i = K_{s_i} (D/r_i)$$

where r_i is the intrinsic rate of increase of the ith species, $r_i = (\mu_{m_i} - D)$ and K_{s_i} and D are as before (Hansen and Hubbel, 1980). If the J parameters of n number of species are ordered, $J_1 < J_2 < \ldots < J_n$ then washout will occur of all species if $J_1 > S_R$. However, if $J_1 < S_R$, species 1 will survive and outnumber all other species. The usefulness of the J parameter as a criterion to predict competitive ability is not obvious. Hansen and Hubbel (1980) tested the validity of the J parameter as a criterion to determine the outcome of competition for a single limiting resource (Table II). These workers showed that if two strains had equal r and D values, the strain with the lower K_s value won (experiment 1, Table II), and if two strains had identical K_s and D values, the strain with the higher r value won (experiment 2, Table II). If two strains had different K_s and r values but identical J values, both species coexisted in the chemostat (experiment 3, Table II). Although the outcome of competition was predicted, the time courses of population density changes deviated from the theoretical predictions (Hansen and Hubbel, 1980).

If two species competing for a single limiting resource have μ–S curves as shown in Fig. 1B, then the outcome of the competition will depend on dilution rate (Meers, 1971). These reversals are predicted by theory only if the species with the smaller K_s also has the smaller maximum specific growth rate. Examples of these kinds of situations have been reviewed (Veldkamp, 1977; Kuenen and Gottshal, 1982; Jannasch and Mateles, 1974). The outcome of competition can depend on other environmental factors such as temperature. Harder and Veldkamp (1971) studied the competition for lactate by two marine psychrophilic bacterial species in which both the dilution rate and

TABLE II
Use of the J Parameter to Predict the Outcome of Competition between Two Populations[a]

Experiment	Bacterial strain[b]	K_S (μg/liter)	r (per hr)	J (μg/liter)	D (per hr)	Outcome[c]
1	A	3.0	0.75	0.240	0.06	A
	B	310	0.85	21.90	—	—
2	C	1.6	0.61	0.198	0.075	C
	D	1.6	0.89	0.135	—	—
3[d]	C	1.6	0.61	0.198	0.075	Coexist
	D	0.9	0.34	0.199	—	—

[a] Data are taken from Hansen and Hubbel (1980). Cultures were grown in a 200-ml chemostat at 34°C with tryptophan as the limiting nutrient.
[b] Strain A was *Escherichia coli* strain C-8; strain B was *Pseudomonas aeruginosa* strain PAO 283; strain C was a nalidixic acid-resistant and spectinomycin-sensitive strain of C-8; and strain D was a nalidixic acid-sensitive and spectinomycin-resistant strain of C-8.
[c] Which strain won the competition or whether both strains persisted at steady state is indicated.
[d] Nalidixic acid (0.5 μg/ml) was added to alter the kinetic parameters of strain D.

the temperature were varied. At $-2°C$, the obligate psychrophile excluded the facultative psychrophile at all dilution rates, whereas at 16°C, the facultative psychrophile excluded the obligate psychrophile at all dilution rates. At 4° and at 10°C, however, the outcome of the competition was dependent on dilution rate. At low dilution rates, the obligate psychrophile was excluded and, at high dilution rates, the facultative psychrophile was excluded.

Fredrickson and Stephanopoulus (1981) concluded that pure and simple competition in a homogenous environment leads to the exclusion of one of the competitors if all input into the system is constant with time. However, there are examples wherein coexistence of two or more species does occur. One of these is the so-called paradox of plankton raised by Hutchinson (1961), where many species of phytoplankton, obviously competing for limited supplies of nutrients, persist in relatively homogeneous environments of large bodies of water. One explanation for this paradox is that competitive exclusion might not occur because the ecosystem has periodic, time-varying inputs so that the competitors coexist in limit cycles. Limit cycles are a perpetually transient state in which population densities continually oscillate. This can allow competitors to coexist if the curves of $\mu–S$ are of the type shown in Fig. 1B. However, the periodic disturbance of inputs cannot by itself explain the coexistence of a large number of species on a few limiting resources (Stewart and Levin, 1973). Another explanation to resolve the paradox of phytoplankton is the heterogeneity of the environment where n competing populations can coexist if there are at least n different subenvironments (Fredrickson and Stephanopoulos, 1981; Wimpenny, 1981).

Coexistence of competitors can occur if there is more than one limiting resource (Fredrickson and Stephanopoulos, 1981; Yoon *et al.*, 1977; Yoon and Blanch, 1977; Fredrickson, 1977). The number of species potentially capable of winning a competition depends on the number of limiting resources in the medium and on the distribution of the minimal J parameters for these resources among the competing species (Hansen and Hubbel, 1980). If there are more limiting resources than species and if each species has at least one minimal J parameter for one of these resources, all strains will coexist. Two species of diatoms coexisted in semicontinuous cultures when both phosphorus and silica were limiting; one had a lower J parameter for phosphorus, while the other had a lower J parameter for silica (Tilman, 1977). Gottschal *et al.* (1979) studied the competition for thiosulfate and acetate by the facultative chemolithotroph, *Thiobacillus* strain A2, and the heterotroph *Spirillum* strain G7 in continuous culture. The competition was partial because *Spirillum* strain G7 used only acetate, while *Thiobacillus* strain A2 grew chemolithotrophically with thiosulfate, heterotrophically with acetate, or mixotrophically with both. When the medium contained only acetate, the specialist, *Spirillum* strain G7, outgrew the more versatile *Thiobacillus* strain A2. When the medium contained both thiosulfate and acetate, the two populations coexisted at steady state, with the population of *Thiobacillus* strain A2 increasing as thiosulfate concentration in the reservoir increased. Similar results were obtained when the obligate chemolithotroph, *Thiobacillus neapolitanus*, and *Thiobacillus* strain A2 were grown in continuous culture with limiting amounts of thiosulfate (Gottschal *et al.*, 1979). *T. neapolitanus* outgrew *Thiobacillus* strain A2, but the two populations coexisted since the latter used glycollate excreted by *T. neapolitanus*. As the concentration of acetate or glycollate was increased in the reservoir, the number of *Thiobacillus* strain A2 increased. *T. neapolitanus* was excluded at concentrations of the organic compounds greater than 12 mM. Versalilists like *Thiobacillus* strain A2 have a competitive advantage over specialists such as *Spirillum* strain G7 or *T. neapolitanus* if several resources are limiting and the turnover of inorganic and organic substrates is of the same order of magnitude (Kuenen and Gottschal, 1982).

Other circumstances can lead to the coexistence of two competing populations. Modeling studies (de Freitas and Fredrickson, 1978) showed that two populations competing for a single limiting resource can coexist in an ideal chemostat if at least one of the populations produces an autoinhibitor. Other circumstances include commensalistic and protocooperative interactions among the competing populations (Megee *et al.*, 1972; Gottschal *et al.* 1979; Meyer *et al.*, 1975; and Lee *et al.*, 1976). Additional examples of coexistence are discussed by Fredrickson and Stephanopoulos (1981).

Amensalism (Antagonism)

Amensalism occurs when one population produces a substance inhibitory to other populations. The first population may be unaffected or may gain a competitive edge which is beneficial. Lactic acid bacteria produce organic acids such as lactic acid that lower the pH of the environment and prevent the growth of other bacteria. This is important in the production of various foods such as sauerkraut or yogurt. A similar situation occurs in mine drainage, where growth of *Thiobacillus thiooxidans* lowers the pH to 1–2. Wolin (1969) showed that *Escherichia coli* was inhibited by the volatile fatty acids produced by rumen bacteria; this may explain why *E. coli* is excluded from the rumen. The production of ethanol by yeasts also inhibits the growth of many bacteria in fermenting fruit and vegetable juices. Photosynthesis by cyanobacteria produces oxygen, which inhibits the growth of anaerobic bacteria (Atlas and Bartha, 1981). *Nitrobacter* species are inhibited by high ammonia concentrations, and the production of ammonia by heterotrophs from proteins or amino acids may be an amensalistic interaction in some environments (Meers, 1973).

The production of antibiotics by some microorganisms may also be a mechanism of amensalism. However, the significance of antibiotic production in nature is questionable (Brock, 1966). Nevertheless, a few examples suggest the importance of antibiotic production in nature. *Cephalosporium graminenum* is a pathogen of wheat that survives in dead wheat material between crops. Antibiotic-producing populations of *C. gramineum* are better able to survive than non-antibiotic-producing populations; the former are better able to prevent colonization of dead wheat material by other fungal populations (Bruehl *et al.*, 1969). Cereal grains inoculated with *Bacillus subtilis* or *Chaetomium globosum* are better protected from seed blight caused by *Fusarium roseum* (Baker, 1980), apparently due to antibiotic production in the rhizosphere.

Amensalistic and competitive interactions could occur between two populations and it is possible that these two interactions could balance each other, resulting in coexistence of the two populations. However, analysis of these situations shows that they are unstable and that one population is always excluded (Fredrickson and Stephanopoulos, 1981). Which population is excluded largely depends on the initial concentrations of the populations (Adams *et al.*, 1979).

Predation and Parasitism

Predation and parasitism are interactions where one population benefits to the detriment of the other population. In parasitism, the parasite is usually smaller than the host while, in predation, the predator is usually larger than the prey. Almost all these kinds of interactions involve eukaryotic organisms as either the predator or the host and are outside the context of this chapter. The best studied example of a parasite of bacteria is *Bdellovibrio bacteriovorus* (Starr and Seidler, 1971; Thomashow and Rittenberg, 1979). Bdellovibrios invade the cells of gram-negative bacteria but do not breach the cytoplasmic membrane. For this reason, the term intraperiplasmic growth is used to describe the stages of the life cycle during which bdellivibrios reside inside the host cell. Upon release from the killed host cell, bdellovibrios are free-swimming organisms, seemingly in search of a substrate (host) cell. Because of this, and the fact that bdellovibrios do not use the prey's cellular machinary or energy, bdellovibrios are considered predators (Thomashow and Rittenberg, 1979; Hespell *et al.*, 1974). Later sections of this chapter will discuss the biochemical and molecular aspects of intraperiplasmic growth of bdellovibrios.

Casida (1980*a*) showed that the addition of *Micrococcus luteus* cells to natural soils resulted in relatively rapid death of these cells. Microscopic examination showed that *M. luteus* cells were physically destroyed, and that two different kinds of bacteria were growing in areas where the cells had lysed (Casida, 1980*a,b*). One of these bacteria resembled the streptomycete, *Streptoverticillium*. The streptomycete used slender mycelial filaments to seek out the substrate cells. The mycelial filament either passed through packets of *M. luteus* cells or surrounded them with one or two mycelial filaments, or both (Casida, 1980*b*). The other bacterium was a gram-negative rod that attached to cells of *M. luteus*, resulting in lysis of the cocci (Casida, 1980b). This organism reproduced by budding and was identified as a new genus and species, *Ensifer adhaerens* (Casida, 1982). Neither the *Streptovertcillium* species or *E. adhaerens* was an obligate predator. Both organisms grew in the absence of substrate cells on a variety of media including Noble agar in distilled water, indicating that neither of these organisms was nutritionally fastidious.

Certain microorganisms such as some members of the Myxobacterales obtain nutrients and energy sources by secreting enzymes that lyse and digest bacterial cells. These interactions are beneficial for one population and detrimental to the other and are in the category that includes predation or parasitism. However, these terms are usually reserved for direct interactions involving cell contact. Fredrickson (1977) used the phrase "indirect parasitism" to describe positive–negative interactions that do not involve direct cell contact.

SYNTROPHIC ASSOCIATIONS IN METHANOGENIC ECOSYSTEMS

Microbial Interactions in the Rumen

The microbiology of the rumen and its importance to the nutrition of the ruminant have been extensively reviewed (Hungate, 1966; Bryant, 1959, 1977; Wolin, 1974, 1979) and are not discussed in detail here. This section focuses on how microbial interactions are involved in the production of the main fermentation products, the volatile fatty acids, methane, and carbon dioxide. The general characteristics of the microbial community and the food-chain relationships between different populations are outlined. Several available reviews discuss in greater detail the importance of microbial interactions in the rumen fermentation (Wolin, 1974, 1975, 1976, 1979; Bryant, 1977; Bryant and Wolin, 1975).

Certain herbivorous ungulates (the Ruminantia) possess an enlarged specialized section of the alimentary tract (the rumen) where the delayed passage of food material allows time for the anaerobic fermentation of this material. The fermentation is semicontinuous with a turnover rate of about 0.7 day (Hungate, 1966). The nutrients required for growth of the microbial community are supplied in the feed. The ruminant also maintains appropriate environmental conditions, such as a pH of 6.5 and a temperature of 39°C and controls the dynamics of turnover of the rumen contents (Hungate, 1966). The ruminant benefits from providing an excellent environment for the rumen microorganisms, whose activities and products are essential to the animal. These include microbial enzymes that degrade the major dietary plant polysaccharides, cellulose and hemicellulose, and large amounts of acetic, propionic, and butyric acids, which are used as carbon and energy sources by the animal. Microbially synthesized proteins and B vitamins are used by the animal, making it possible for the ruminant to sustain itself on a diet free of amino acids and vitamins that it does not itself synthesize. More detailed discussion of these topics is available in Hungate (1966) and Bryant (1977).

The microbial community consists of a large number of interacting microorganisms that include diverse species of non-spore-forming, strictly anaerobic bacteria and anaerobic, cilliated protozoa (Bryant, 1959; Hungate, 1966). This discussion focuses on the interactions that occur among the bacterial populations. However, the protozoan populations are intimately involved in metabolic interactions through the engulfment of starch grains and other food particles and their subsequent conversion to products such as hydrogen and lactate, which are further metabolized by bacteria as well as by feeding on the bacteria themselves.

Fermentation Interactions

Figure 2 shows the intermediates and products produced during the fermentation. Cellulose and hemicellulose are the major sources of energy for ruminants except those that are raised as beef cattle, which receive large amounts of starch instead of cellulose. The importance of interactions in the rumen can be realized by considering that even when cellulose is the only energy source added to the diet, the cellulolytic bacteria comprise at most 25% of the viable bacteria (Slyter *et al.*, 1967). Table III shows the energy sources used and the products formed *in vitro* by some representative species in the rumen. Most of these organisms ferment one or more of the major plant polysaccharides present in the diet. *Butyrivibrio fibrisolvens* and *Bacteroides ruminicola* are quite versatile, fermenting many different kinds of carbohydrates (Bryant, 1959, 1977). Other species, such as the ruminococci, are highly specialized and ferment cellulose and pentosan. Selenomonads and *Megasphaera elsdenii* ferment lactic acid, which is not present in the ruminant's diet, but is produced by other rumen microbes. Methane is produced by a very specialized group of bacteria (Balch *et al.*, 1979); the rumen methanogens use primarily hydrogen and formate as energy sources.

In pure culture, representative species produce products such as hydrogen, formate, lactate, ethanol, and succinate that are not normally detected

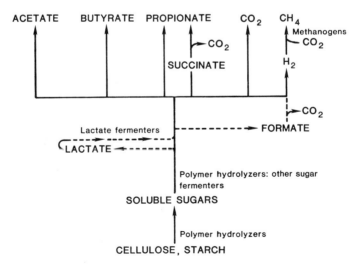

FIGURE 2. Intermediates and products of the rumen fermentation of plant polysaccharides. (– – –) Minor pathways. (From Wolin, 1979.)

TABLE III
Energy Sources and Fermentation Products of Some Representative Rumen Bacteria[a]

Species	Energy sources[b]					Fermentation products[c]
	Cellulose	Xylan	Starch	Other carbohydrates	Other compounds	
Bacteroides						
succinogenes	+	−	∓	+	−	S, A, F
B. ruminicola	−	±	±	+ + +	−	S, A, F
B. amylophilus	−	−	+	−	−	S, A, F
Ruminococcus						
albus	±	±	−	+	−	A, F, E, H$_2$, CO$_2$
R. flavefaciens	±	±	−	+	−	S, A, F, H$_2$
Succinovibrio						
dextrinosolvens	−	−	−	+	−	A, S
Butyrivibrio						
fibrisolvens	∓	±	±	+ + +	−	B, L, F, A, H$_2$, CO$_2$
Eubacterium						
ruminantium	−	±	−	+ +	−	B, F, A, L, CO$_2$, H$_2$
Lachnospira						
multiparus	−	−	−	+ +	Pectin	L, A, F, E, H$_2$, CO$_2$
Selenomonas						
ruminantium	−	−	±	+ + +	Lactate	L, P, A, CO$_2$, H$_2$
Megasphaera						
elsdenii	−	−	−	+	Lactate	B, V, P, A, C, H$_2$, CO$_2$
Methanobrevibacter						
ruminantium	−	−	−	−	H$_2$/CO$_2$ Formate	CH$_4$

[a] Adapted from Bryant (1977).
[b] −, Not used by any strains; ∓, used by a few strains; ±, used by many strains; +, few compounds are attacked; + + +, many compounds are attacked.
[c] A, acetate; B, butyrate; C, caproate; E, ethanol; F, formate; L, lactate; P, propionate; V, valerate.

in *in vivo* fermentation. These products could be extracellular intermediates in the fermentation that are rapidly consumed *in vivo,* or they could be metabolic by-products that are made only *in vitro.* Studies show that both possibilities occur.

The degradation of plant polysaccharides involves attachment of the microorganisms to this material (Baker *et al.,* 1951; Akin and Amos, 1975; Akin *et al.,* 1974). It may be possible for an organism to avoid competition for a commonly used substrate such as cellulose by attaching to different plant cell wall types. Latham *et al.* (1978) observed that *Ruminococcus flavefaciens* predominated on epidermis, phloem, and sclerenchyma cell walls and that *Bacteroides succinogenes* predominated on the mesophyll of the ryegrass used in their study. However, Akin (1980) found that, although the proportions of these two bacteria attached to plant material differed, there was not a significant difference in the type of plant cell wall to which each type of bacterium attached. Electron microscopic studies revealed progressive colonization of the plant material during its digestion (Cheng *et al.*, 1980). Initially, only a few morphotypes were attached to the plant material. Later, a morphologically more diverse population was found attached. This suggested that the degradation of plant material initially involves a few pioneer species. As digestion proceeds, hydrolysis products from the plant polysaccharides and fermentation products of the pioneer species, as well as exudates from the plant cells, become available for use by other bacterial species.

Many of the carbohydrate-fermenting bacteria in the rumen cannot degrade complex plant polysaccharides (Bryant, 1977). These organisms depend on the polysaccharide-hydrolyzing bacteria to produce soluble sugars and oligosaccharides. This interaction seems very important in the rumen, but little is known about the nature of the hydrolysis products or the permeability of the bacteria to them. Thus, whether competition occurs for these products is uncertain. Bryant (1952) observed that a rumen treponeme grew in mixed culture near colonies of cellulolytic bacteria. The best-studied example of this kind of interaction involves the growth of *Selenomonas ruminantium* in coculture with the cellulolytic bacterium, *B. succinogenes* (Scheifinger and Wolin, 1973). In a defined medium with cellulose as the sole energy source and containing all the necessary growth factors, *S. ruminantium* grew well in coculture with *B. succinogenes,* indicating that the hydrolysis products of cellulose degradation were available for use by *S. ruminantium.* Similar crossfeeding relationships must also occur during the degradation of proteins and lipids, but little is known about these interactions.

There is considerable overlapping of substrate utilization capabilities among the carbohydrate-fermenting bacteria in the rumen (Table III) (Bryant, 1973, 1974, 1977). This would lead one to conclude that there must be a considerable amount of competition among these species for the available carbohydrates. However, Russell and Baldwin (1978, 1979) have shown that these species have different substrate affinities, preferences, and sequential utilization patterns for the various carbohydrates. This suggests that each species has a different strategy of substrate utilization, which allows them to occupy different niches in the rumen.

In most ruminants, lactate is not a significant intermediate, but in animals fed high starch diets it can be important (Mackie and Gilchrist, 1979). *Streptococcus bovis* and *Selenomonas ruminantium* are starch-fermenting species that produce considerable amounts of lactate. *Megasphaera elsdenii* and *Veillonella alcalescens* ferment lactate produced by other species to propionate. Other lactate fermenters include *S. ruminantium,* which ferments lactate to propionate by a different pathway.

Blackburn and Hungate (1963) showed that succinate is an important intermediate in the overall fermentation; it is decarboxylated to propionate and carbon dioxide as rapidly as it is produced. Many of the important bacterial species in the rumen produce succinate in pure culture, but very few produce propionate (Table III) (Bryant, 1977). The propionate-producing species include *S. ruminantium* and *M. elsdenii*. The latter organism produces propionate from lactate and is only important in animals fed a high starch diet (Bryant, 1977). *S. ruminantium* uses either carbohydrates or lactate to produce propionate by the randomizing pathway similar to that found in the propionic acid bacteria. *S. ruminantium* also decarboxylates extracellular succinate to propionate and carbon dioxide. In cocultures with *B. succinogenes* or *R. flavifaciens*, *S. ruminantium* grew on the carbohydrates produced from cellulose by either of the other organisms. *S. ruminantium* produced propionate from these carbohydrates, or from succinate produced by the other species. The net result was that the coculture produced propionate instead of succinate as the end product (Scheifinger and Wolin, 1973). *S. ruminantium* apparently does not benefit from the decarboxylation. However, Konings and Veldkamp (1980) found that the excretion of products generates a proton-motive force across the cytoplasmic membrane of some anaerobes, so it may be possible that *S. ruminantium* benefits from this process. Recently, Schink and Pfennig (1982*a*) isolated an anaerobic bacterium, *Propionigenium modestum*, that grows with succinate as its sole source of carbon and energy and produces propionate and carbon dioxide as the end products. Experiments conducted by Miller and Wolin (1979) suggest that the interactions necessary for propionate production in the rumen may not occur in the large intestine of humans. The predominant bacteria in the large intestine, *Bacteroides* species, produce succinate as a major fermentation product in pure culture. When these cultures are supplemented with vitamin B_{12}, less succinate and more propionate is produced. This was not observed when the major succinate-producing rumen bacteria were so tested (Miller and Wolin, 1979).

Interspecies Hydrogen Transfer

Methane is one of the major end products of the rumen fermentation; it is produced by a very specialized group of bacteria called the methanogens (Balch *et al.*, 1979). *Methanobrevibacter ruminantium* is the most numerous methanogen in the rumen (Smith and Hungate, 1958), but *Methanomicrobium mobilis* (Paynter and Hungate, 1968) and *Methanosarcina* species (McInerney *et al.*, 1981) are also present. Methane is produced mainly by the reduction of CO_2 (Kleiber, 1953) with H_2 being a principal electron donor (Hungate *et al.*, 1970). This is confirmed by the fact that the most numerous methanogens isolated from the rumen use only H_2 or formate as energy sources (Hungate, 1966; Bryant, 1977; Smith and Hungate, 1958). Opperman *et al.* (1961) showed that only a small amount of methane could be derived from the methyl group of acetate.

Studies on the metabolic interactions among fermentative bacteria and H_2-using bacteria show that not only do these interactions permit growth of the hydrogen user, they also dramatically affect the balance of bacterial fermentation products (Wolin, 1974, 1982). Hungate (1966) observed that many of the saccharolytic rumen bacteria produce ethanol and lactate in pure culture, yet no ethanol and very little lactate are detected in the rumen. The reason for this is the maintenance of a very low H_2 concentration in the rumen by methanogens allows the fermentative bacteria to excrete more of the electrons generated in glycolysis and pyruvate oxidation as H_2 rather than as reduced products such as ethanol and lactate (Wolin, 1974, 1982). Central to this is the fact that H_2 is produced using electrons generated in the oxidation of reduced pyridine nucleotides (Table IV). The equilibrium of this reaction favors H_2 production only when the concentration of H_2 is very low, as it is when methanogens are effectively using H_2 (Wolin, 1974). H_2 production from electrons generated in the oxidation of pyruvate or acetaldehyde is favorable at one atm of H_2 (Thauer et al., 1977); at low H_2 concentrations, the flow of electrons (NADH) generated in glycolysis is toward the reduction of protons, resulting in H_2 formation. This permits more pyruvate to be degraded to acetate, CO_2, and H_2. Thus, glucose is degraded mainly to acetate, CO_2, and CH_4 (Table IV). At high hydrogen concentrations, as occurs during growth of the fermentative bacteria in pure culture, the flow of electrons from NADH shifts from use in H_2 production to use for reduction of pyruvate and its metabolites, resulting in products such as ethanol, lactate, propionate and butyrate. This results in the degradation of glucose to reduced organic products, as shown in Table IV. Several studies have shown that the concentration of H_2 in the rumen is very low, about 1 μM (Hungate, 1967; Robinson et al., 1981; Robinson and Teidje, 1982); this would favor the flow of electrons from NADH to H_2. Although the degradation of glucose to acetate, CO_2 and H_2 is thermodynamically favorable, the thermodynamic efficiency of ATP synthesis would be too high and incompatible with the entropy requirements for this type of metabolism (Thauer et al., 1977).

The effect of H_2 removal by methanogens on the products formed by other bacteria was first demonstrated during the studies on S organisms isolated from M. omelianskii (Reddy et al., 1972). In pure culture, the S organism fermented pyruvate to ethanol, acetate, CO_2 and a trace of H_2. In a coculture containing S organism and a methanogen, acetate, CO_2, and CH_4, but not ethanol, were formed. Electrons generated in pyruvate oxidation were used for H_2 production, hence CH_4 formation, instead of for ethanol production.

Iannoti et al. (1973) grew Ruminococcus albus alone or in coculture with Vibrio succinogenes in glucose-limited continuous cultures. V. succinogenes uses H_2 to reduce fumarate to succinate in its energy metabolism. In pure culture, R. albus produced acetate, ethanol, CO_2 and H_2 (Table V). In coculture with V. succinogenes, R. albus did not produce ethanol; stoichiometric increases in the amounts of acetate and H_2 occurred. In coculture, all NADH generated in glycolysis by R. albus was reoxidized to NAD^+ by H_2 production (Table V). Theoretically, this would result in more efficient growth of R. albus, since 21% more moles of ATP per mole of glucose fermented could be formed when pyruvate is catabolized to acetate, CO_2 and H_2 rather than being used for ethanol production (Table V). Whether growth of R. albus is more efficient in coculture than in pure culture has not been experimentally verified. Weimer

TABLE IV
Idealized Reactions of Fermentative Bacteria in the Presence and Absence of Methanogens[a]

Equation	$\Delta G^{o\prime}$ (kJ)
I. Hydrogen use by methanogens	
1. $4\ H_2 + HCO_3^- + H^+ \leftrightarrow CH_4 + 3\ H_2O$	−135.6
II. Some partial reactions of fermentative bacteria	
2. Glucose + 2 $NAD^+ \leftrightarrow 2$ pyruvate$^-$ + 2 NADH + 4 H^+	−148.1
3. Pyruvate$^-$ + 2 NADH + 2 $H^+ \leftrightarrow$ propionate$^-$ + 2 NAD^+ + H_2O	−87.0
4. Pyruvate$^-$ + acetate$^-$ + NADH + $H^+ \leftrightarrow$ butyrate$^-$ + NAD^+ + HCO_3^-	−77.4
5. Pyruvate$^-$ + HCO_3^- + 2 NADH + 2 $H^+ \leftrightarrow$ succinate^{2-} + 2 H_2O + 2 NAD^+	−66.9
6. Pyruvate$^-$ + 2 $H_2O \leftrightarrow$ acetate$^-$ + HCO_3^- + H^+ + H^2	−47.3
7. Pyruvate$^-$ + H_2O + NADH + $H^+ \leftrightarrow$ ethanol + HCO_3^- + NAD^+	−38.9
8. Pyruvate$^-$ + NADH + $H^+ \leftrightarrow$ lactate$^-$ + NAD^+	−25.1
9. NADH + $H^+ \leftrightarrow NAD^+$ + H_2	+18.0
III. Some idealized reactions of fermentative bacteria	
A. Without H_2-using methanogens	
10. Glucose + 2 acetate$^- \leftrightarrow 2$ butyrate$^-$ + 2 HCO_3^- + 2 H^+	−302.9
11. Glucose + $H_2O \leftrightarrow$ propionate$^-$ + acetate$^-$ + HCO_3^- + H_2 + 3 H^+	−282.4
12. Glucose $\leftrightarrow$ succinate^{2-} + acetate$^-$ + 3 H^+ + H_2	−262.3
13. Glucose + 2 $H_2O \leftrightarrow 2$ ethanol + 2 HCO_3^- + 2 H^+	−225.9
14. Glucose + 2 $H_2O \leftrightarrow 2$ lactate$^-$ + 2 H^+	−198.3
15. Glucose + 4 $H_2O \leftrightarrow 2$ acetate$^-$ + 2 HCO_3^- + 4 H^+	−191.9
B. With H_2-using methanogens	
16. glucose + $H_2O \leftrightarrow 2$ acetate$^-$ + HCO_3^- + CH_4 + 3 H^+	−342.3

[a] Data are from Thauer et al. (1977) or are calculated from data therein. Calculations are based on H_2 and CH_1 in the gaseous state; all other substances in aqueous solution at 1 mole/kg activity.

and Zeikus (1977) showed that ethanol production from cellulose or cellobiose decreased, and acetate production increased when *Clostridium thermocellum* was grown in coculture with *Methanobacterium thermoautotrophicum* than when grown in pure culture. Similar effects of H_2-using methanogens on lowering the proportions of fermentation products such as succinate, lactate, propionate and butyrate produced by fermentative bacteria are known (Latham and Wolin, 1977; Scheifinger et al., 1975; Chen and Wolin, 1977; Chung, 1976; Ben-Basset et al., 1981; Winter and Wolfe, 1980). The reader is referred to excellent reviews by Mah (1982) and Wolin (1982) for further discussions on this subject.

Formate is produced by some rumen anaerobes (Table III). The formate that is produced by pyridine nucleotide-linked and the ferredoxin-linked CO_2 reductase activity (Thauer et al., 1975) would serve as a mechanism whereby fermentative bacteria that lacked hydrogenase could get rid of electrons generated during glycolysis. The extracellular formate could be catabolized to CO_2 and H_2 by a second organism or used directly by methanogens. This might explain why so many methanogens that use H_2 use formate as well.

TABLE V

Fermentation Balance of Ruminococcus albus Grown in Glucose-Limited Chemostat in Pure Culture or in Coculture with Vibrio succinogenes[a]

Products of R. albus	Moles per 100 moles of glucose fermented	
	Pure culture	Coculture
Ethanol	70	0
Acetate	130	200
H₂	260	400
From pyruvate	200	200
From glycolysis	60	200
CO₂	200	200
Theoretical ATP formation		
Glucose to pyruvate	200	200
Pyruvate to acetate	130	200
Total	330	400

[a] Data from Iannotti *et al.* (1973) reproduced from Wolin (1974).

Nutritional Interactions

Nutritional interactions seem very important in the rumen. Many fermentative anaerobes, and *M. ruminantium*, require one or more volatile fatty acids such as *n*-valeric, isobutyric, D-2-methylbutyric, and/or isovaleric acids, and high levels of acetate may be required or stimulatory (Bryant, 1973; Bryant and Robinson, 1962; Bryant *et al.*, 1971). Acetate and valerate are fermentation products produced by carbohydrate metabolism. The branched-chain volatile acids are produced from branched-chain amino acids present in proteins by organisms such as *Bacteroides ruminicola* and the ciliated protozoa (Pittman and Bryant, 1964; Allison, 1970). Few rumen bacteria use amino acid or peptide nitrogen (Bryant and Robinson, 1962) and are dependent on the ammonia produced by other bacteria or protozoa as the nitrogen source. In addition to ammonia released from proteins, its production from urea by ureoloytic species such as *S. ruminantium* is important (John *et al.*, 1974). Bryant and Wolin (1975) showed that the cellulolytic bacterium *R. albus* and the protein catabolizer *B. ruminicola* grew in defined medium with cellulose as the sole energy source and casein as the sole nitrogen source. In this experiment, *B. ruminicola* depended on *R. albus* to supply sugars and oligosaccharides needed by *B. ruminicola* as energy sources. *B. ruminicola* produced ammonia and isobutyrate or 2-methylbutyrate required for growth of *R. albus*. Similar nutritional interdependence among *Bacteroides amylophilus*, *M. elsdenii*, and *R. albus* has been shown (Muira *et al.*, 1980). The nitrogen metabolism in the ruminant has been reviewed (Armstrong and Weeks, 1983).

Many rumen bacteria require one or more B vitamins for growth; these are produced by other rumen bacteria (Bryant, 1977). Many *Bacteroides* species require heme or a related tetrapyrole that is supplied by other bacteria (Caldwell *et al.*, 1965). *M. ruminantium* requires coenzyme M (Taylor *et al.*, 1974) (2-mercaptoethanesulfonic acid), a unique cofactor found only in methanogens (Balch and Wolfe, 1979). It is presumed that coenzyme M is produced by other rumen methanogens such as *Methanosarcina* species, which contain high levels of this cofactor (Balch and Wolfe, 1979). *M. mobilis* requires an unknown factor present in ruminal fluid (Paynter and Hungate, 1968).

Studies on the nutrition of fermentative bacteria indicate that these organisms are closely adapted to their environment (Bryant, 1973, 1974). For example, the main soluble nitrogen compound is ammonia, not amino acids, and rumen fluid often contains large amounts of branched-chain acids. Thus, it is not surprising that rumen bacteria do not have the ability to effectively assimilate preformed cell monomers such as amino acids and require relatively large amounts of ammonia and volatile acids for growth.

Microbial Interactions in Other Methanogenic Environments

Unlike the rumen and large bowel of many animals, in which acetate, propionate, and butyrate are major end products of fermentation, in many environments, organic matter is completely converted to the gaseous products CH_4 and CO_2. These environments include sludge digestors and aquatic sediments where oxygen, nitrate, or sulfate are not readily available. In these environments, the turnover of materials is much slower than in the rumen; this allows time for complete conversion of organic matter to CO_2 and CH_4. The microbiology of these systems has been reviewed (Bryant, 1979; Mah *et al.*, 1977; Hobson *et al.*, 1974; Zeikus, 1977, 1980; Zehnder, 1978; Torrien and Hattingh, 1969; McInerney and Bryant, 1981*a*) and is not discussed extensively here. Rather, the differences in the microbiology between these ecosystems and the rumen are compared to explain the differences observed in the end products of the fermentations.

Until recently, the complete anaerobic degradation of organic matter to CH_4 and CO_2 was thought to involve two major metabolic groups of bacteria, each responsible for a stage of the fermentation—the acid-forming stage or the methane-forming stage (Barker, 1956). However, it is now accepted that methanogens per se do not degrade alcohols other than methanol, or fatty acids other than formate or acetate (Bryant, 1979; McInerney and Bryant, 1981*a*). The three-stage scheme as illustrated in Fig. 3 best describes the current views of the methane fermentation. As found in the rumen, the first stage involves the fermentative bacteria that hydrolyze the major substrates such as cellulose, proteins, and lipids and produce volatile fatty acids, alcohol, CO_2, H_2, ammonia, and sulfide. Propionate and longer-chain fatty acids, alcohol, and some aromatic acids are degraded by a second group of bacteria called the obligate H_2-producing (proton-reducing) acetogenic bacteria (Bryant, 1979). The third group of bacteria are the methanogens that rapidly use the H_2 produced by other bacteria to reduce CO_2 to CH_4 and some species that

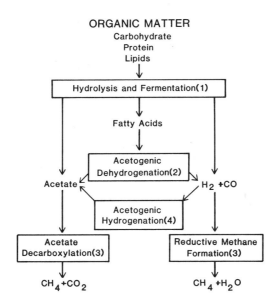

FIGURE 3.   Three-stage scheme for the complete anaerobic degradation of organic matter showing the general pathways and the three major metabolic groups of bacteria. 1, Fermentative
bacteria; 2, hydrogen-producing acetogenic bacteria; 3, methanogenic bacteria. Acetate and
sometimes other acids may be produced from H_2 and CO_2 by a fourth group of bacteria.
(From McInerney and Bryant, 1981a.)

cleave acetate to CO_2 and CH_4 (Mah *et al.*, 1977). In contrast to the rumen,
where methane is formed almost exclusively from the reduction of CO_2 by
H_2 and acetate accumulates, in the complete methane fermentation, 60–90%
of the methane is derived from the methyl group of acetate (Smith and Mah,
1965; Jerris and McCarty, 1965; Cappenberg and Prins, 1974; Mountfort and
Asher, 1977; Mackie and Bryant, 1981). *Methanosarcina* species (Mah *et al.*,
1978) and *Methanothrix soehngenii* (Zehnder *et al.*, 1980) degrade acetate to
CO_2 and CH_4. A fourth group of bacteria called the H_2-using acetogenic
bacteria reduces CO_2 to acetate and sometime butyrate, using H_2 as the electron donor (Balch *et al.*, 1977; Rode *et al.*, 1981; Sharak-Genthner *et al.*, 1981;
Zeikus *et al.*, 1980). These bacteria also use CO and methanol as energy
sources. Bache and Pfennig (1981) showed that one of these bacteria, *Acetobacterium woodii*, demethoxylates methoxylated aromatic acids to the corresponding aromatic acid and produces acetate from the liberated methanol.
Thus, their role in these environments may involve the demethoxylation of
lignins, pectins and methanol fermentation rather than H_2 utilization.

The nutrition, physiology, and metabolism of the major fermentative
bacteria in ecosystems such as waste digestors and aquatic sediments have not
been studied in as much depth as that of the major rumen and intestinal
fermentative bacteria. From the limited information available, it seems that
such bacteria from different ecosystems will have many similarities to their

counterparts obtained from the rumen. Many of the same genera found in the rumen are also found in environments in which complete methane fermentation occurs (McInerney and Bryant, 1981a). In addition, species of *Clostridium* are often found among the predominant organisms (Molongoski and Klug, 1976; Maki, 1954). Ljungdahl (1979) recently reviewed the information regarding anaerobic thermophilic species. The first stage of the fermentation is very similar to that occurring in the rumen. The nutritional interactions and interspecies hydrogen transfer reactions found in the rumen also occur among the bacteria found in these ecosystems (Weimer and Zeikus, 1977; Chung, 1976; Ward *et al.*, 1977; Ben-Bassat *et al.*, 1981; Khan, 1977; Khan *et al.*, 1979; Khan and Murray, 1982a,b). In comparison to the diets of herbivores and to municipal refuse, the organic wastes present in digestors and sediments often have larger amounts of biodegradable protein and fats (Chynoweth and Mah, 1971). Proteins are hydrolyzed to peptides and amino acids, which are fermented mainly to volatile fatty acids, ammonia, and sulfide (Mead, 1971). Branched-chain volatile fatty acids are produced from the corresponding branched-chain amino acid (Bryant, 1979; McInerney and Bryant, 1981a). Aromatic organic acids are produced by various clostridial species from the corresponding aromatic amino acids (Elsden *et al.*,1976). Glycerides, phospholipids, and other fats are hydrolyzed with the release of long-chain fatty acids and other products such as glycerol and galactose. The latter products are fermented, but the long-chain fatty acids are not further degraded by the fermentative bacteria.

The methanogens are the terminal group in the complete methane fermentation. They are essential to the fermentation because they are the only organisms able to catabolize the end products of other bacteria (acetate, CO_2, and H_2) to gaseous products in the absence of light or exogenous electron acceptors other than CO_2. In addition to H_2/CO_2 and acetate, *Methanosarcina* species also use methanol, methylamine, and trimethylamine as energy sources for methanogenesis (Nagle and Wolfe, 1984). Methanogenesis of the latter compound is important in the degradation of phosphatidylcholine, a major constitutent of eukaryotic membranes (Frebig and Gottschalk, 1983; Neill *et al.*, 1978). The nutrition, biochemistry and evolution of this interesting group of bacteria have been reviewed (Bryant *et al.*, 1971; Zeikus, 1977; Mah *et al.*, 1977; Balch *et al.*, 1979; Nagle and Wolfe, 1984).

One of the major differences between the rumen fermentation and the complete methane fermentation is that acetate and longer-chain fatty acids do not accumulate in the latter. As mentioned above, acetate is degraded to CH_4 and CO_2 by certain species of methanogens, and this is quantitatively the most important reaction in methane formation. Only recently have the bacteria that degrade propionate and longer-chained fatty acids been isolated and characterized. These bacteria will now be discussed, as will the importance of effective H_2 removal by methanogens in the degradation of these compounds.

Anaerobic Degradation of Alcohol and Fatty Acids

The isolation of the S organism from *Methanobacillus omelianskii* was the first documentation of a H_2-producing acetogenic bacterium (Bryant *et al.*, 1967). The methanogen was thought to oxidize ethanol to acetate and reduce CO_2 to CH_4 (Barker, 1941), but this fermentation was shown to be carried out by a syntrophic association of two bacterial species. The S organism oxidizes ethanol to acetate and the electrons are used to reduce protons to H_2, while the methanogen uses the H_2 to reduce CO_2 to Ch_4 (Table VI). The formation of H_2 and acetate from ethanol is thermodynamically unfavorable unless H_2 is used to reduce CO_2 to CH_4. Hydrogen inhibits the growth of S organism on ethanol, and good growth occurs only when the H_2-using organism is present (Bryant *et al.*, 1967; Reddy *et al.*, 1972).

Several strains of *Desulfovibrio* produce H_2 from lactate or ethanol when grown without sulfate in the presence of H_2-using methanogens (Table VI) (Bryant *et al.*, 1977). Sulfate is the preferred electron acceptor by *Desulfovibrio* but, in its absence, proton reduction can occur—albeit only via interspecies hydrogen transfer in association with methanogens or other H_2-using bacteria. The complete degradation of lactate or ethanol to CH_4 and CO_2 was observed

TABLE VI

Some Reactions of the Hydrogen-Producing (Proton-Reducing) Acetogenic Bacteria[a]

Equation	$\Delta G^{o\prime}(kJ)$
I. H_2 use by methanogens	
1. $4 H_2 + HCO_3^- + H^+ \leftrightarrow CH_4 + 3 H_2O$	-135.6
II. Some reactions of the hydrogen-producing acetogenic bacteria	
A. Without H_2-using methanogens	
2. $Lactate^- + 2 H_2O \leftrightarrow acetate^- + HCO_3^- + H^+ + 2 H_2$	-4.2
3. $Ethanol + H_2O \leftrightarrow acetate^- + H^+ + 2 H_2$	$+9.6$
4. $Butyrate^- + H_2O \leftrightarrow acetate^- + H^+ + 2 H_2$	-48.1
5. $Valerate^- + H_2O \leftrightarrow acetate^- + propionate^- + H^+ + 2 H_2$	$+48.1$
6. $Propionate^- + 3 H_2O \leftrightarrow acetate^- + HCO_3^- + 3 H_2$	$+76.1$
7. $Palmitate + 14 H_2O \leftrightarrow 8 acetate^+ + 7 H^+ + 14 H_2$	$+351.5$
B. With H_2-using methanogens	
8. $2 Lactate^- + H_2O \leftrightarrow 2 acetate^- + HCO_3^- + CH_4$	-143.6
9. $2 Ethanol + HCO_3^- \leftrightarrow 2 acetate^- + CH_4 + H_2O + H^+$	-116.4
10. $2 Butyrate^- + HCO_3^- + H_2O \leftrightarrow 4 acetate^- + CH_4 + H^+$	-39.4
11. $2 Valerate^- + HCO_3^- + H_2O \leftrightarrow 2 acetate^- + 2 propionate^- + CH_4 + H^+$	-39.4
12. $4 Propionate^- + 3 H_2O \leftrightarrow 4 acetate^- + 3 CH_4 + H^+ + HCO_3^-$	-102.4
13. $2 Palmitate^- + 7 HCO_3^- + 7 H_2O \leftrightarrow 16 acetate^- + 7 CH_4 + 7 H^+$	-246.2

[a] Data are from Thauer *et al.* (1977) or calculated from data therein. See footnote to Table IV for additional comments.

when *Desulfovibrio desulfuricans* was cocultured with *Methanosarcina barkeri* which uses both H_2/CO_2 and acetate for methanogenesis (McInerney and Bryant, 1981b). Other bacteria that oxidize ethanol in syntrophic association with H_2-using bacteria include *Thermoanaerobium brockii* (Ben-Bassat *et al.*, 1981) and *Pelobacter venetianus* (Schink and Stieb, 1983). Methanogenic metabolism of higher alcohols such as *n*-propanol, *n*-butanol, isobutanol, and *n*-pentanol occurred when S organism was cocultured with a H_2-using methanogen (Reddy *et al.*, 1972). As discussed regarding the rumen fermentation, ethanol and lactate are usually not important intermediates in the process, since interspecies hydrogen transfer reactions allow the fermentative bacteria to reoxidize NADH to NAD^+ via proton reduction and H_2 formation. However, under certain circumstances, such as when large amounts of readily degradable carbohydrates are available, these compounds may be produced.

Propionate and longer-chain fatty acids are important intermediates in the complete methane fermentation (Mackie and Bryant, 1981; Lovely and Klug, 1982), and as much as 50 percent of the theoretical chemical oxygen demand of the organic substrates flows through propionate and longer-chain fatty acids (Kasper and Wuhrmann, 1978). Chynoweth and Mah (1971) found that the degradation of long-chain fatty acids such as palmitate and stearate accounts for much of the acetate produced in digestor sludge. As discussed by Bryant (1977) and by McInerney and Bryant (1981a), it is believed that these compounds are degraded by H_2-producing acetogenic bacteria.

The isolation of bacteria that degrade propionate or butyrate anaerobically required that a H_2-using bacterium be included in the isolation media, since these reactions are thermodynamically unfavorable unless the H_2 concentration is very low (Table VI). The inoculum could then be serially diluted, and a single cell of the fatty acid-degrading bacteria would grow, in association with the H_2-using bacterium, and produce a single colony. Monoxenic (two-membered) cultures could be obtained by picking well-isolated colonies. This procedure was first applied in the isolation of the anaerobic fatty acid-oxidizing bacterium, *Syntrophomonas wolfei* (McInerney *et al.*, 1979, 1981b). Monoxenic cultures of this bacterium were obtained with H_2-using methanogens or *Desulfovibrio* species. By β-oxidation, *S. wolfei* β-oxidizes the even-numbered carbon fatty acids such as butyrate, caproate, and caprylate to acetate and H_2, the odd-numbered carbon fatty acids such as valerate and heptanoate to propionate, acetate and H_2, and isoheptanoate to acetate, isovalerate and H_2 (Table VI). It is unable to use any common bacterial energy source or combination of electron donor and acceptor that would allow it to grow without the H_2-using bacterium. Thus, it occupies a very specialized niche, and its growth and metabolism are obligately dependent on H_2 use by other bacteria such as methanogens. A thermophilic butyrate-degrading bacterium has been isolated in coculture with *M. thermoautotrophicum* (Henson and Smith, 1983). Propionate is degraded to the methanogenic intermediates, acetate, CO_2 and H_2 by another H_2-producing acetogenic bacterium called *Syntrophobacter wolinii* (Boone and Bryant, 1980). This bacterium has only been cocultured, either with a H_2-using *Desulfovibrio* species or in a three-membered culture that also contains *Methanospirillum hungatei*. This latter culture was used to demonstrate methanogenesis from propionate via interspecies hydrogen transfer reactions. The bacterium that degrades long-chain fatty acids such as stearate has not been isolated (Lorowitz and Bryant, 1983).

Acetate is cleaved to CH_4 and CO_2 by certain methanogens; this is the major route of acetate degradation in most methanogenic ecosystems. In this reaction, all of the hydrogens on the methyl moiety remain and are found in methane (Pine and Barker, 1956). However, Zinder and Koch (1983) recently isolated a thermophilic nonmethanogenic, acetate-oxidizing bacterium in co-culture with *M. thermoautotrophicum*. This bacterium oxidizes acetate to CO_2 and H_2, which are then used by the methanogen. The significance of these reactions in nature is not clear at this time but this reaction may be important in some thermophilic ecosystems.

Figure 4 shows the importance of H_2 concentration in the energetics of ethanol, propionate and butyrate degradation. Ethanol degradation becomes favorable at partial pressures of H_2 less than 0.15 atm. The degradation of butyrate and propionate is not energetically favorable unless the partial pressure of H_2 is less than about 2×10^{-3} atm or 9×10^{-5} atm, respectively. Methane formation from H_2 and CO_2 is still favorable at partial pressures of H_2 less than 10^{-5} atm. Any slight increase in H_2 partial pressures will stop the degradation of these compounds, with propionate degradation being the first to be affected. Thus, the activity of methanogens in these environments not only regulates the end products by fermentative bacteria, but also regulates the degradation of these compounds by H_2-producing acetogenic bacteria.

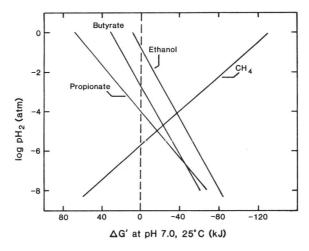

FIGURE 4. Effects of the partial pressure of H_2 (pH 2) in the change in free energy ($\Delta G'$) for reactions involving ethanol, propionate, and butyrate degradation and methane formation (see Table VI). (From McInerney and Bryant, 1981a.)

Anaerobic Degradation of Aromatic Compounds

Many studies have shown that benzoate and certain other aromatic compounds are completely degraded to CO_2 and CH_4 in the absence of light and exogenous electron acceptors other than CO_2 by consortia of microorganisms (Tarvin and Buswell, 1934; Clark and Fina, 1952; Nottingham and Hungate, 1969; Ferry and Wolfe, 1976; Shlomi *et al.*, 1978; Healy and Young, 1978, 1979). The major groups of bacteria observed in these enrichments include benzoate degraders and H_2-using and acetate-using methanogens. Fatty acids other than acetate have been implicated as intermediates (Shlomi *et al.*, 1978), so fatty acid degraders such as *S. wolfei* or *S. wolinii* may also be involved. The degradation of benzoate is thermodynamically unfavorable unless the H_2 concentration is maintained at very low levels (Table VII). *Syntrophus buswellii* is a newly isolated benzoate degrader that grows in coculture with H_2-using *Desulfovibrio* species (Mountfort and Bryant, 1982; Bryant, personal communication) and produces acetate and H_2 from benzoate. Grbić-Galić and Young (1982) reported the isolation in pure culture of a different benzoate degrader that apparently does not require the presence of a H_2-using bacterium. The fermentation products produced from benzoate by this bacterium were not reported, so it is not known why it does not require the presence of a H_2-using bacterium.

For aromatic compounds such as benzene, phenol, *p*-hydroxybenzoate, and benzoate, the concerted action of ring-cleaving bacteria and H_2-using

TABLE VII

Reactions Involved in the Anaerobic Degradation of Aromatic Compounds[a]

Equation	$\Delta G^{o\prime}$ (kJ)
I. Hydrogen use by methanogens	
1. $4 H_2 + HCO_3^- + H^+ \leftrightarrow CH_4 + 3 H_2O$	-135.6
II. Degradation of aromatic compounds by H_2-producing acetogenic bacteria	
A. Without H_2-using methanogens	
2. Benzoate$^-$ + $7 H_2O \leftrightarrow 3$ acetate$^-$ + HCO_3^- + $3 H^+$ + $3 H_2$	$+70.6$
3. Phenol + $5 H_2O \leftrightarrow 3$ acetate$^-$p + $3 H^+$ + $2 H_2$	$+6.6$
4. *p*-Hydroxybenzoate$^-$ + $6 H_2O \leftrightarrow 3$ acetate$^-$ + HCO_3^- + $3 H^+$ + $2 H_2^+$	$+5.4$
B. With H_2-using methanogens	
5. 4 Benzoate$^-$ + $19 H_2O \leftrightarrow 12$ acetate$^-$ + $3 CH_4$ + $1 HCO_3^-$ + $9 H^+$	-124.6
6. 2 Phenol + $7 H_2O + HCO_3^- \leftrightarrow 6$ acetate$^-$ + CH_4 + $5 H^+$	-122.5
7. 2 *p*-Hydroxybenzoate$^-$ + $9 H_2O \leftrightarrow 6$ acetate$^-$ + CH_4 + 5 H^+ + HCO_3^-	-124.8
III. Fermentative degradation of aromatic compounds	
8. Pyrogallol + $3 H_2O \leftrightarrow 3$ acetate$^+$ + $3 H^+$	-158.3
9. Gallate + $4 H_2O \leftrightarrow 3$ acetate$^+$ + HCO_3^- + $3 H^+$	-160.0

[a] Data are from Kaiser and Hanselmann (1982) and Thauer *et al.* (1977) or calculated from data therein. See Table IV for additional comments.

bacteria is required to keep the ring fission reaction exergonic (Table VII) (Kaiser and Hanselmann, 1982). However, this is not true for other kinds of aromatic acids. The addition of oxygen-containing substituents as found in compounds such as pyrogallol or syringic acid upsets the symmetry of the benzene nucleus and causes a substantial lowering of the Gibbs free energies of formation (Kaiser and Hanselmann, 1982). The degradation of these compounds to acetate is exergenic (Table VII; Kaiser and Hanselmann, 1982). Schink and Pfennig (1982b) isolated a pure culture of *Pelobacter acidigallici*, which degrades trihydroxybenzenes such as gallic acid, pyrogallol, 2,4,6-trihydroxybenzoic acid, and phloroglucinol stoichiometrically to 3 moles of acetate (and 1 mole of CO_2). Barik *et al.* (1983) found that phenylacetate-degrading methanogenic enrichments did not contain large numbers of H_2-using methanogens and were able to isolate the phenylacetate-degrading bacterium in pure culture. The degradation of these substituted aromatic compounds requires only food-chain associations, not interspecies hydrogen transfer reactions. In the case of the degradation of syringic acid to CH_4 and CO_2 (Kaiser and Hanselmann, 1982), *A. woodii* demethoxylates syringic acid to gallic acid and produces acetate from the methoxyl moiety. Gallic acid is degraded to acetate by a bacterium morphologically similar to *P. acidigallici*, and acetate is degraded to CH_4 and CO_2 by an acetate-using methanogen.

The anaerobic biodegradation of haloaromatic compounds involves an interesting association of microorganisms. The intitial degradative event involves the loss of the aryl halide without the alteration of the aromatic ring (Suflita *et al.*, 1982). The benzoate produced by dehalogenation is not further metabolized by the dehalogenating-bacterium, but can be further degraded to CO_2 and CH_4 by consortia similar to those described above.

Complete Degradation of Carbohydrates to Gaseous Products

Because of the increased interest in the use of alternative energy sources, the use of known mixtures of bacteria that completely convert cellulose or other carbohydrates to CH_4 and CO_2 has been studied. Examples of these systems has recently been reviewed (Mah, 1982). These systems include one or more fermentative bacteria to degrade the carbohydrate substrate and H_2-using methanogens that participate in interspecies hydrogen transfer reactions. With some systems, *Desulfovibrio* species are added to degrade the ethanol or lactate that accumulates due to inefficient coupling between the fermentative bacterium and the methanogen. Only when an acetate-using methanogen such as *M. barkeri* is included is the conversion of the substrate to CH_4 and CO_2 complete. It is interesting to note that, in the fermentation of glucose by *R. albus*, for instance, about 90% of the glucose energy still remains in the fermentation products. By means of acetogenic and interspecies hydrogen-transfer reactions, these end products are completely converted to CH_4 and CO_2; 86% of the energy originally present in glucose is retained in CH_4. Thus, as Mah (1982) points out, "for a paltry increase in 4–5% of the available energy, anaerobic chemoheterotrophic nonmethanogenic and methanogenic bacteria have developed many types of partnerships in order to eke out a livelihood."

ASSOCIATIONS WITH PHOTOTROPHS

The green sulfur bacteria (*Chorobiaceae*) form the lowest layer of phototrophic organism in ponds, lakes, and estuaries where vertical gradients of light and hydrogen sulfide exist (Pfennig, 1978). These bacteria are nonmotile, obligately phototrophic, and sulfide dependent and cannot store the electron donor in the form of elemental sulfur inside the cell. The inability of the *Chlorobiaceae* to store the electron donor as elemental sulfur inside the cell allows these bacteria to participate in interspecies sulfide and sulfur transfer. The *Chlorobiaceae* grow particularly well in syntrophic association with anaerobic organotrophic bacteria, a feature also observed in the *Chloroflexaceae* (Pfennig, 1978). In addition, *Chlorobium* cultures excrete large amounts of organic compounds into the medium; about 15–25% of the assimilated CO_2 can be recovered in organic compounds in the medium. In natural environments, about 30% of the assimilated CO_2 was excreted (Czeczuga and Gradzki, 1973).

Syntrophic Associations

The interactions between phototrophic bacteria and sulfur and sulfate-reducing bacteria is important in the cycling of sulfur-containing compounds in anaerobic environments (Pfennig, 1980; Nedwell, 1982). The reduction of sulfate to sulfide by sulfate-reducing bacteria has long been recognized as an important aspect of the sulfur cycle. Van Gemerdon (as discussed in Pfennig, 1980) quantitatively studied the bacterial sulfur cycle using cocultures of *Chromatium vinosum* and *Desulfovibrio* species. Formate and sulfate served as electron donor and acceptor for the *Desulfovibrio* species, and yeast extract was added as a carbon source. *C. vinosum* grew autotrophically on the bicarbonate and sulfide formed by the *Desulfovibrio* species. The growth rate of *C. vinosum* was dependent on the rate of sulfate reduction by the *Desulfovibrio* species so that the yield of *C. vinosum* corresponded to the amount of formate added as the electron donor for sulfate reduction. When the sulfate concentration was low, the coculture grew as well as when the sulfate concentration was high because of a rapid turnover of the sulfur compound. Biebl and Pfennig (1978) performed similar experiments using cocultures of *Chlorobium limicola* and *Desulfovibrio* species.

Although the importance of biological sulfate reduction to the sulfur cycle is recognized, the importance of sulfur reduction was questionable since many organisms reduce sulfur by mechanisms that had been regarded as physiologically unimportant (Wolfe and Pfennig, 1977). The first sulfur-reducing anaerobe isolated was *Desulfuromonas acetoxidans* (Pfennig and Biebl, 1976), obtained from a "*Chloropseudomonas*" culture and from other environments. *D. acetoxidans* oxidizes acetate, ethanol or propanol to CO_2 with elemental sulfur as the electron acceptor being reduced to sulfide (Pfennig and Biebl, 1976). In coculture with green sulfur bacteria, sulfide was reoxidized by the phototroph to elemental sulfur. Only a small initial amount of sulfide was needed to obtain good growth of the coculture (Biebl and Pfennig, 1978). Several facultative sulfur-reducing bacteria have been described that also form syntrophic associations with green sulfur bacteria; these include certain *Desulfovibrio* species (Biebl and Pfennig, 1977) and *Campylobacter* species (Wolfe and Pfennig, 1977; Pfennig and Widdel, 1981).

Symbiotic Associations

Several symbiotic consortia of phototrophs and organotrophs were described by Lauterborn and Buder in the early 1900s (described in Pfennig, 1980). These consortia contain a colorless bacterium attached to a pigmented bacterium. Although binomial names were given to these consortia, the bacteriological code of nomenclature does not allow their use to name symbiotic consortia. These names can only be used as trivial names, not as taxa. "Chlochromatium aggregatum" consists of a colorless, motile, rod covered by green, rod shaped cells of the "Chlorobium limicola" type. Growth and cell division of this consortium occur synchronously, indicating a high degree of metabolic interdependence. The consortium is phototactic and, thus, the symbiosis allows the nonmotile green sulfur bacterium to respond to variations in light. "Pelochromatium roseum" consists of a motile colorless bacterium surrounded by *Chlorobium phaeobacteroides*-like cells. Other motile consortia are "Chlorochromatium glebulum" and "Pelochromatium roseoviridie." The latter consortium includes a motile, colorless bacterium surrounded by an inner layer of *Chlorobium phaeobacteroides*-like cells and and outer layer of *Pelodictyon luteolum*-like cells.

Two other nonmotile ectosymbiotic consortia have been described. "Cylinodrogloca bacterifera" consists of a filamentous, colorless bacterium that produces a slime layer in which green sulfur bacteria are imbedded. "Chloroplana vaculata" consists of platelets of alternating rows of *Pelodictyon luteolum*-like cells and a gas-vacuolated colorless rod.

Mechsner (as referenced in Pfennig, 1980) was able to isolate a green sulfur bacterium from the "Chlorochromatium" consortium, but the motile central bacterium did not survive. Given the physiological properties of excretion of large amounts of organic compounds and release of elemental sulfur into the medium by green sulfur bacteria, it is believed that the basis of the symbiotic consortia may be the cycling of sulfur between the sulfide and elemental sulfur oxidation states (Pfennig, 1978). It is possible that the colorless bacterium is obligately dependent on green sulfur bacteria as the electron acceptor (Pfennig, 1980), as *S. wolfei* is dependent on H_2-using bacteria as the electron acceptor (McInerney *et al.,* 1979).

COMPETITION BETWEEN SULFATE REDUCERS AND METHANOGENS

In anaerobic marine or estuarine sediments, sulfate reduction rather than methanogenesis is the dominant terminal process (Nedwell, 1982), with acetate, propionate, butyrate as well as H_2 as the major electron donors for sulfate reduction (Smith and Klug, 1981a; Balba and Nedwell, 1982; Sørenson *et al.,* 1981; Laanbroek and Pfennig, 1981). Widdel and Pfennig (1977, 1981, 1982) isolated a variety of new species of sulfate reducing bacteria that participate in these reactions (Table VIII). Species such as *Desulfobulbus propionicus* and *Desulfovibrio caporans* degrade propionate and C_4 to C_{18} fatty acids, respectively, to acetate and reduce sulfate to sulfide. Other species oxidize substrates such as acetate and longer-chain fatty acids or benzoate completely to CO_2. H_2-using sulfate-reducing bacteria have also been isolated (Badziong *et al.,* 1978). These latter bacteria can participate in interspecies hydrogen transfer reaction with fermentative bacteria. Banat and Nedwell (1983) showed that the degradation of fatty acids and aromatic acids such as benzoate did not involve interspecies hydrogen transfer reactions as is the case in methanogenic environments. As a group, the sulfate-reducing bacteria perform functions provided by the H_2-producing acetogenic bacteria and the methanogens in methanogenic environments. Thus, only two major metabolic groups, the fermentative bacteria and the sulfate-reducing bacteria, function in anaerobic environments in which sulfate levels are high.

Although methanogenesis is not the dominant process in marine and estuarine environments, it does occur and may become dominant in deeper sediments where sulfate is depleted (Nedwell, 1982; Martens and Berner, 1974; Mountfort *et al.,* 1980; Mountfort and Asher, 1981; Oremland and Taylor, 1978). In freshwater sediments or in anaerobic digestors, sulfate reducers are present, and the addition of sulfate to these environments inhibits methanogenesis (Abram and Nedwell, 1978b; Winfrey and Zeikus, 1977; Smith and Klug, 1981a,b; Lovely *et al.,* 1982). This also occurs in mixed cultures containing sulfate reducers and methanogens (Abrams and Nedwell, 1978a).

Three possible explanations of sulfate inhibition of methanogenesis have been proposed. First, the thermodynamics of sulfate reduction with H_2 or acetate as electron donors is energetically more favorable than methanogenesis from these compounds (Claypool and Kaplan, 1974; Martens and Berner, 1974; Winfrey and Zeikus, 1977). However, both processes are thermodynamically favorable, so methanogenic reactions cannot be excluded on this basis (McCarty, 1972; Thauer *et al.,* 1977; Lovely *et al.,* 1982). The second possibility is that methanogens are inhibited by toxic products, sulfide in

TABLE VIII

Recently Isolated Sulfate-Reducing Bacteria That Degrade Fatty and Aromatic Acids[a]

Species	Electron donor
Incomplete oxidation	
Desulfobulbus propionicus	Propionate
Desulfovibrio sapovorans	C_4-C_{18} fatty acids[b]
Complete oxidation	
Desulfobacter postgatei	Acetate
Desulfotomaculum acetoxidans	Acetate, butyrate, and valerate
Desulfococcus multivorans	C_1-C_{14} fatty acids and aromatic acids[c]
Desulfosarcina variabilis	H_2/CO_2, C_1-C_{14} fatty acids and aromatic acids
Desulfonema limicola	H_2/CO_2, C_1-C_{12} fatty acids
Desulfonema magnum	C_2-C_{10} fatty acids and aromatic acids
Desulfovibrio baarsii	C_1-C_{18} fatty acids

[a] From Pfennig and Widdel (1981).
[b] C_n indicates the chain length of the fatty acid used.
[c] Aromatic acids used are benzoate, phenylacetic, and phenylpropionic acids.

particular, produced by the sulfate reducers (Cappenberg, 1974a,b). Although sulfide inhibition of methanogenesis may occur in some cases, it cannot be used as a general explanation for the phenomenon, since many methanogens tolerate high levels of sulfide (Bryant et al., 1977; Scherer and Sahm, 1981; Wellinger and Wuhrmann, 1977). The third possibility is that sulfate reducers outcompete methanogens for energy sources such as hydrogen and acetate (Adram and Nedwell, 1978a,b; Bryant et al., 1977; Martens and Berner, 1974; Mountfort and Asher, 1981; Oremland and Taylor, 1978; Winfrey and Zeikus, 1977; Lovely et al., 1982); this is the currently accepted explanation. Several studies have shown that sulfate reducers have higher affinities for hydrogen and acetate, enabling the sulfate reducers to maintain the pools of these substrates at concentrations too low for the methanogens to effectively use them when sulfate is not limiting for the sulfate-reducing bacteria (Lovely et al., 1982; Robinson and Teidje, 1982; Kristjansson et al., 1982; Schönheit et al., 1982). Even in environments where sulfate concentration is low, sulfate reducers can effectively outcompete methanogens if the input of organic matter is low. In an oligotrophic lake, 30 to 80% of the total amount of organic degradation was attributed to sulfate reduction (Lovely and Klug, 1983).

The maximum potential rate of substrate uptake is as important as the affinity for the substrate in determining the outcome of competition (Lovely et al., 1982). When Wintergreen Lake sediments were amended with 20 mM sulfate, the inhibition of methanogenesis was slow (Lovely et al., 1982). This was due to the small initial potential for hydrogen uptake by sulfate reducers, which was three times lower than the rate of hydrogen uptake by methanogens. Hydrogen uptake by sulfate reducers accounted for only 10% of the total hydrogen turnover rate. The population size of the sulfate reducers was low and limited by sulfate availability. Thus, the maximum potential rate of sulfate reduction was low, and methanogenesis was the dominant process.

Certain methanogens such as *M. barkeri* can use methanol and trimethylamine as methanogenic substrates (Balch *et al.*, 1979). These compounds are not known to be substrates for sulfate reducers. Oremland and Polcin (1982) showed that methanogenesis and sulfate reduction can coexist if there is an adequate supply of these noncompetitive substrates. The importance of these compounds in nature is not well understood, but their availability could explain why methanogenesis and sulfate reduction occur simultaneously in certain environments.

DEGRADATION OF XENOBIOTIC COMPOUNDS

The degradation of xenobiotic compounds and other complex organic compounds often involves the concerted action of several microbial species. The microbial interactions involved in xenobiotic degradation were recently reviewed (Slater and Somervill, 1979; Bull, 1980), and only some examples of these interactions are discussed here.

The degradation of the insecticide Diazinon (0,0-diethyl-0-2-isopropyl-4-methyl-6-pyrimidylthiophosphate) involves a synergistic relationship between *Arthrobacter* and *Streptomyces* species (Gunner and Zurckerman, 1968). Alone, the *Arthrobacter* species could degrade the ethyl substituents, but neither species could metabolize the pyrimidine ring. In mixed culture, Diazinon was completely degraded after 21 days; however, the mechanism has not been elucidated. A similar situation was observed for the degradation of linear alkyl benzenesulphanates (LAS). A bacterial community that degraded LAS was enriched in a chemostat culture with LAS as the sole carbon and energy source (Johanides and Hršak, 1976). The dominant species were representatives of *Pseudomonas* and *Alcaligenes*. Individually, none of the isolates degraded LAS, although the complete community did. The nature of the interaction and the mechanism of degradation are unknown. *Nocardia* species metabolize the ethoxylated side chain of nonionic alkylphenol surfactants leaving the ethoxylated phenylether (Baggi *et al.*, 1978). This latter compound is completely degraded by *Cylindrocarpin* species.

Cometabolism may permit the complete degradation of a compound by a microbial community. An example of this was mentioned above, in which *M. vaccae* cometabolized cyclohexane to cyclohexanol, which was then used by *Pseudomonas* species (Beam and Perry, 1974). A microbial community that degrades Parathion has been selected using continuous cultivation (Munnecke and Hsieh, 1974). Four organisms were isolated that alone did not degrade Parathion, but did so when cultivated together. *Pseudomonas stutzeri* metabolizes Parathion to diethyl thiophosphate and p-nitrophenol; the latter compound inhibits the growth of *P. stutzeri*. This inhibition is relieved by *Pseudomonas aeruginosa,* which uses p-nitrophenol as a carbon and energy source.

P. aeruginosa excretes metabolites that are used by P. stutzeri and two other species for growth. The role of the latter two species in the degradation of Parthion is not clear. DDT (1,1-bis (p-chlorophenyl)2,2,2-trichloroethane) may be degraded by microbial communities. For example, Wedemyer (1967) showed that DDT was dechlorinated to p,p'-dichlorodiphenylmethane anaerobically by Enterobacter aerogenes, and aerobically, a Hydrogenomonas species could co-metabolize the product in the presence of diphenylmethane. Whether this sequence occurs in nature is unclear, since each of these organisms would presumably be active in different environments.

INTRAPERIPLASMIC GROWTH OF BDELLOVIBRIOS

The bdellovibrios exhibit a two-phase life cycle: a free-swimming, non-reproductive stage, and a growth stage spent within the cell of a gram-negative bacterium. Several reviews are available on the biology of bdellovibrios (Shilo, 1969; Starr and Seidler, 1971; Thomashow and Rittenberg, 1979), the last of which was the source of much of the information presented here. Bdellovibrios are very actively motile (10 times faster than E. coli), chemoorganotrophic aerobes. B. bacteriovorus actively respires acetate, amino acids, and pentoses (Hespell et al., 1975; Hespell and Martens, 1978; Hespell and Odelson, 1978). It has the enzymes required to polymerize monomeric units into cell polymers such as RNA, DNA and protein, as well as the synthetic and degradative enzymes needed to obtain these monomers (Hespell et al., 1975; Kuenen and Rittenberg, 1975; Martin and Rittenberg, 1972; Rosson and Rittenberg, 1979). However, growth of bdellovibrios occurs only after it invades its prey or in a complex medium supplemented with concentrated extract from a potential prey. Thus, in nature, it appears to be an obligate predator.

The attack of B. bacteriovorus on the prey cell is a violent collision that is reversible; the predator can detach from the prey and swim away (Thomashow and Rittenberg, 1979). After effective attachment, B. bacteriovorus moves through the outer membrane and peptidoglycan into the periplasm of the prey cell. This is followed by damage of the cytoplasmic membrane making it permeable to small ions and hydrophobic molecules, and inhibition of respiration and of synthesis of RNA, DNA and protein by the prey cell. Unlike bacteriophage and obligate intracellular parasitic bacteria, B. bacteriovorus does not use the prey's metabolic machinery or cellular energy. The peptidoglycan of the prey is altered and the cell rounds up becoming a bdelloplast, which is sealed in a manner that prevents the loss of the cellular components of the prey (Thomashow and Rittenberg, 1978a–c). The regulated degradation of the prey's macromolecular components begins (Martin and Rittenberg, 1972; Hespell et al., 1975; Rosson and Rittenberg, 1979); degradative enzymes apparently generate monomeric units as needed for biosynthesis, since low-molecular-weight fragments of these macromolecules do not accumulate.

Growth of B. bacteriovorus begins about 45–60 min after the initial attack.

B. bacteriovorus elongates into a long helical cell; then fragments and synthesizes flagella. The bdelloplast lyses, and the free-swimming *Bdellovibrio* cells are released. An interesting aspect of intraperiplasmic growth is that it is highly efficient, two to three times more efficient than other bacteria (Rittenberg and Hespell, 1975). The basis of this efficiency is the use of preformed monomeric units such as amino acids, fatty acids, and nucleoside monophosphates obtained from the prey cell. The high-energy phosphate ester bonds in the latter compound are conserved (Rittenberg and Langley, 1975; Kuenen and Rittenberg, 1975). In addition, energy generation is tightly coupled to energy use in biosynthesis (Rittenberg and Hespell, 1975).

Although information is available on how bdellovibrios grow within the bdelloplast, it does not explain why growth inside the bdelloplast is obligatory. Spontaneous mutants occur that are capable of axenic growth (Seidler and Starr, 1969). These occur at a frequency of 10^{-6}–10^{-7}, suggesting that a single gene is involved. The hypothesis is that a specific signal required for growth initiation is supplied by the prey. This signal derepresses a vital gene function(s) in *B. bacteriovorus* and triggers the metabolic events leading to cell division (Rittenberg, 1983). The work of Ruby and Rittenberg (1983) using prematurely-lysed bdelloplasts supports this hypothesis and implies that the initiation of DNA synthesis is the regulated function. Thus, in addition to obtaining food and shelter from the prey cell, the bdellovibrios apparently also obtain a signal needed to initiate growth.

SUMMARY

Evidence has been reviewed in this chapter that prokaryote populations interact in all seven of the effective ways (2–8 in Table I) recognized in ecology. In some instances, notably syntrophic associations among anaerobes, considerable progress has been made in elucidating the metabolic bases for population interactions. Competitive interactions such as for access to one or more limiting substrates (treated extensively in Chapter 7, this volume) are susceptible to theoretical (and mathematical) analysis, but not so easily demonstrated experimentally. In all probability, the difficulties are due not to faulty theory, but to two other causes: first, to the assumption, required for mathematical predictions, that access to a limiting substrate is the sole basis of competition; and second, to the high probability that competition for one resource is transient in natural habitats. Each type of competition is probably periodically relieved, allowing potential competitors to coexist even in microenvironments. When exclusion does occur among prokaryotes in natural habitats, it is more often attributable to antagonistic than to competitive interactions. In contrast to the likely transitory character of both competition and antagonism, the levels of interdependence of some cooperative associations between prokaryote populations have developed to the level of mutualism, so that some types of bacteria are found in nature only in communities or consortia that include their physiologic partners.

Laboratory studies have been essential to unraveling the relationships within mixed bacterial populations. The feasibility of the extension of laboratory approaches to field studies has been admirably demonstrated by the studies of rumen populations, but less constant natural environments still present a formidable challenge to the ingenuity of bacterial ecologists.

REFERENCES

Abram, J. W., and Nedwell, D. B., 1978a, Inhibition of methanogenesis by sulfate-reducing bacteria competing for transferred hydrogen, *Arch. Microbiol.* **117**:89–92.

Abram, J. W., and Nedwell, D. B., 1978b, Hydrogen as a substrate for methanogenesis and sulfate reduction in anaerobe saltmarsh sediments. *Arch. Microbiol.* **117**:93–97.

Adams, J., Kinney, T., Thompson, S., Rubin, L., and Hellin, R. B., 1979, Frequency-dependent selection for plasmid-containing cells of *Escherichia coli, Genetics* **91**:627–637.

Akin, D. E., 1980, Evaluation by electron microscopy and anaerobic culture of types of rumen bacteria associated with digestion of forage cell walls, *Appl. Environ. Microbiol.* **39**:242–252.

Akin, D. E., and Amos, H. E., 1975, Rumen bacterial degradation of forage cell walls investigated by electron microscopy, *Appl. Microbiol.* **27**:1149–1156.

Akin, D. E., Burdick, D., and Michaels, G. E., 1974, Rumen bacterial interrelationships with plant tissue during degradation revealed by transmission electron microscopy, *Appl. Microbiol.* **27**:1149–1156.

Allison, M. J., 1970, Nitrogen metabolism in rumen microorganisms in: *Physiology of Digestion and Metabolism in the Ruminant* (A. T. Philipson, ed.), pp. 456–473, Oriel, Newcastle-upon-Tyne.

Armstrong, D. C., and Weeks, T. E. C., 1983, Recent advances in ruminant biochemistry: Nitrogen digestion and metabolism, *Int. J. Biochem.* **15**:261–266.

Atlas, R. M., and Bartha, R., 1981, *Microbial Ecology, Fundamentals and Applications*, Addison-Wesley, Reading, Massachusetts.

Bache, R., and Pfennig, N., 1981, Selective isolation of *Acetobacterium woodii* on methoxylated aromatic acids and determination of growth yields, *Arch. Microbiol.* **130**:255–261.

Badziong, W., Thauer, R. K., and Zeikus, J. G., 1978, Isolation and characterization of *Desulfovibrio* growing on hydrogen plus sulfate as sole energy source. *Arch. Microbiol.* **116**:41–49.

Baggi, G., Beretta, L., Galli, E., Scolastico, C., and Treccani, V., 1978, Biodegradation of polyoxyethylene alkylphenols, in: *The Oil Industry and Microbial Ecosystems* (K. W. A. Chater, and H. J. Somerville, eds.), pp. 129–136, Heyden, London.

Baker, F., Nasr, H., Morrice, F., and Bruce, J., 1951, Bacterial breakdown of structural starch and starch products in the digestive tract of ruminant and non-ruminant mammals, *J. Path. Bact.* **62**:617–638.

Baker, K. F., 1980, Microbial antagonism—the potential for biological control, in: *Contemporary Microbial Ecology* (D. Ellwood, J. N. Hedger, M. J. Latham, J. M. Lynch, and J. H. Slater, eds.), pp. 327–347, Academic Press, London.

Balba, M. T., and Nedwell, D. B., 1982, Microbial metabolism of acetate, propionate and butyrate in anoxic sediment from the Colne Point saltmarsh, Essex, U.K., *J. Gen. Microbiol.* **128**:1415–1422.

Balch, W. E., and Wolfe, R. S., 1979, Specificity and biological distribution of Coenzyme M (2-mercaptoethanesulfonic acid), *J. Bacteriol.* **137**:256–263.

Balch, W. E., Fox, G. E., Magnum, L. J., Woese, C. R., and Wolfe, R. S., 1979, Methanogens: Reevaluation of a unique biological group, *Microbiol. Rev.* **43**:260–296.

Balch, W. E., Schoberth, S., Tanner, R. S., and Wolfe, R. S., 1977, *Acetobacterium*, a new genus of hydrogen-oxidizing carbon dioxide-reducing, anaerobic bacteria, *Inter. J. Syst. Bacteriol.* **27:**355–361.

Banat, I. M., and Nedwell, D. B., 1983, Mechanism of turnover of C_2-C_4 fatty acids in high-sulfate and low-sulfate anaerobic sediments, *FEMS Microbiol. Lett.* **17:**107–110.

Barik, S., Brulla, W. J., and Bryant, M. P., 1983, Methanogenic enrichments from sewage digestor catabolizing benzoate, phenylacetate and phenol. *Abst. Annu. Mtg. Am. Soc. Microbiol.* **1983:**147.

Barker, H. A., 1941, Studies on the methane fermentation. V. Biochemical activities of *Methanobacillus omelianskii, J. Biol. Chem.* **137:**153–167.

Barker, H. A., 1956, *Bacterial Fermentations*, Wiley, New York.

Bates, J. L., and Liu, P. V., 1963, Complementation of lecithinase activities by closely related pseudomonads: Its taxonomic implications, *J. Bacteriol.* **86:**585–592.

Beam, H. W., and Perry, J. J., 1974, Microbial degradation of cyclo-paraffinic hydrocarbons via cometabolism and commensalism, *J. Gen. Microbiol.* **82:**163–169.

Ben-Bassat, A., Lamed, R., and Zeikus, J. G., 1981, Ethanol production by thermophilic bacteria: Metabolic control of end product formation in *Thermoanaerobium brockii, J. Bacteriol.* **146:**192–199.

Biebl, H., and Pfennig, N., 1977, Growth of sulfate-reducing bacteria with sulfur as electron acceptor, *Arch. Microbiol.* **112:**115–117.

Biebl, H., and Pfennig, N., 1978, Growth yields of green sulfur bacteria in mixed cultures with sulfur and sulfate-reducing bacteria, *Arch. Microbiol.* **117:**9–16.

Blackburn, T. H., and Hungate, R. E., 1963, Succinic acid turnover and propionate production in the bovine rumen, *Appl. Microbiol.* **11:**132–135.

Boone, D. R., and Bryant, M. P., 1980, Propionate-degrading bacterium *Syntrophobacter wolinii* sp. nov. gen. nov., from methanogenic ecosystems, *Appl. Environ. Microbiol.* **33:**1162–1169.

Brock, T. D., 1966, *Principles of Microbial Ecology*, Prentice-Hall, Englewood Cliffs, New Jersey.

Bruehl, G. W., Miller, R. L., and Cunfer, B., 1969, Significance of antibiotic production by *Cephalosporium graminenum* to its saprophytic survival, *Can. J. Plant Sci.* **49:**235–246.

Bryant, M. P., 1952, Isolation and characteristics of a spirochete from the bovine rumen, *J. Bacteriol.* **64:**325–335.

Bryant, M. P., 1959, Bacterial species of the rumen, *Bacteriol. Rev.* **23:**125–153.

Bryant, M. P., 1973, Nutritional requirements of the predominant rumen cellulolytic bacteria, *Fed. Proc.* **32:**1809–1813.

Bryant, M. P., 1974, Nutritional features and ecology of predominant anaerobic bacteria of the intestinal tract, *Am. J. Clin. Nutr.* **27:**1313–1319.

Bryant, M. P., 1977, Microbiology of the rumen, in: *Duke's Physiology of Domestic Animals*, 9th ed. (M. J. Sevenson, ed.), pp. 287–304, Cornell University Press, Ithaca, New York.

Bryant, M. P., 1979, Microbial methane production—Theoretical aspects, *J. Anim. Sci.* **48:**193–201.

Bryant, M. P., and Robinson, I. M., 1962, Some nutritional characteristics of predominant culturable ruminal bacteria, *J. Bacteriol.* **84:**605–614.

Bryant, M. P., and Wolin, M. J., 1975, Rumen bacteria and their metabolic interactions, in: *Proceedings of the First Intersectional Congress of the International Association of Microbiological Societies*, Vol. 2 (T. Hasegawa, ed.), pp. 297–306, Science Council of Japan, Tokyo.

Bryant, M. P., Wolin, E. A., Wolin, M. J., and Wolfe, R. S., 1967, *Methanobacillus omelianskii*, a symbiotic association of two species of bacteria, *Arch. Mikrobiol.* **59:**20–31.

Bryant, M. P., Tzeng, S. F., Robinson, I. M., and Joyner, A. E., 1971, Nutritional requirements of methanogenic bacteria, *Adv. Chem. Ser.* **105:**23–40.

Bryant, M. P., Campbell, L. L., Reddy, C. A., and Crabill, M. R., 1977, Growth of *Desulfovibrio* in lactate or ethanol media low in sulfate in association with H_2-utilizing methanogenic bacteria, *Appl. Environ. Microbiol.* **33:**1162–1169.

Bull, A. T., 1980, Biodegradation: Some attitudes and strategies of microorganisms and microbiologists, in: *Contemporary Microbial Ecology* (D. C. Ellwood, J. N. Hedger, M. J. Latham, J. M. Lynch, and J. H. Slater, eds.), pp. 107–136, Academic Press, London.

Bull, A. T., and Slater, J. H., 1982, Microbial interactions and community structure, in: *Microbial Interactions and Communities*, Vol. 1 (A. T. Bull and J. H. Slater, eds.), pp. 13–44, Academic Press, London.

Burns, R. C., and Hardy, R. W. F., 1975, *Nitrogen Fixation in Bacteria and Higher Plants*, Springer-Verlag, New York.

Caldwell, D. R., White, D. C., Bryant, M. P., and Doetsch, R. N., 1965, Specificity of the heme requirement for growth of *Bacteriodes ruminicola, J. Bacteriol.* **90**:1645–1654.

Cappenberg, T. E., 1974a, Interrelationships between sulfate-reducing and methane-producing bacteria in bottom deposits of a fresh water lake. I. Field observations. *Antonie van Leeuwenhoek J. Microbiol. Serol.* **40**:285–295.

Cappenberg, T. E., 1974b, Interrelationships between sulfate-reducing and methane-producing bacteria in bottom deposits of a fresh water lake. II. Inhibition experiments, *Antonie van Leeuwenhoek J. Microbiol. Serol.* **40**:297–306.

Cappenberg, T., and Prins, R. A., 1974, Interrelations between sulfate-reducing and methane-producing bacteria in bottom deposits of a fresh water lake. III. Experiments with ^{14}C-labeled substrates, *Antonie van Leeuwenhoek J. Microbiol. Serol.* **40**:457–469.

Casida, L. E., 1980a, Death of *Micrococcus luteus* in soil, *Appl. Environ. Microbiol.* **39**:1031–1034.

Casida, L. E., 1980b, Bacterial predators of *Micrococcus luteus* in soil, *Appl. Environ. Microbiol.* **39**:1035–1041.

Casida, L. E., 1982, *Ensifer adherens* gen. nov., sp. nov.: A bacterial predator of bacteria in soil, *Int. J. Syst. Bacteriol.* **32**:339–345.

Chao, C.-C., and Reilly, P. J., 1972, Symbiotic growth of *Acetobacter suboxydans* and *Saccharomyces carlsbergensis* in a chemostat, *Biotechnol. Bioeng.* **14**:75–92.

Chen, M., and Wolin, M. J., 1977, Influence of CH_4 production by *Methanobacterium ruminantium* on the fermentation of glucose and lactate by *Selenomonas ruminantium, Appl. Environ. Microbiol.* **34**:756–759.

Cheng, K.-J., Fay, J. P., Howarth, R. E., and Costerton, J. W., 1980, Sequence of events in the digestion of fresh legume leaves by rumen bacteria, *Appl. Environ. Microbiol.* **40**:613–625.

Chung, K.-T., 1976, Inhibitory effects of H_2 on growth of *Clostridium cellobioparum, Appl. Environ. Microbiol.* **31**:342–348.

Chynoweth, D. P., and Mah, R. A., 1971, Volatile acid formation in sludge digestion, *Adv. Chem. Ser.* **105**:41–54.

Clark, F. M., and Fina, L. R., 1952, The anaerobic decomposition of benzoic acid during methane fermentation, *Arch. Biochem.* **36**:26–32.

Clarke, R. T. J., 1977, Methods for studying gut microbes, in: *Microbial Ecology of the Gut* (R. T. J. Clarke and T. Bauchop, eds.), pp. 1–33, Academic Press, London.

Claypool, G. E., and Kaplan, I. R., 1974, The origin and distribution of methane in marine sediments, in: *Natural Gases in Marine Sediments* (I. R. Kaplan, ed.), pp. 99–140, Plenum Press, New York.

Czeczuga, B., and Gradzki, F., 1973, Relationship between extracellular and cellular production in the sulphuric green bacterium *Chlorobium limicola* Nads. as compared to primary production of phytoplankton, *Hydrobiologia* **42**:85–95.

de Freitas, M. J., and Fredrickson, A. G., 1978, Inhibition as a factor in the maintenance of diversity of microbial ecosystems, *J. Gen. Microbiol.* **106**:307–320.

Elsden, S. R., Hilton, M. G., and Waller, J. M., 1976, The end products of the metabolism of aromatic amino acids by clostridia, *Arch. Microbiol.* **107**:283–288.

Ferry, J. G., and Wolfe, R. S., 1976, Anaerobic degradation of benzoate to methane by a microbial consortium, *Arch. Microbiol.* **107**:33–40.

Fiebig, K., and Gottshalk, G., 1983, Methanogenesis from choline by a coculture of *Desulfovibrio* sp. and *Methanosarcina barkeri, Appl. Environ. Microbiol.* **45**:161–168.

Fredrickson, A. G., 1977, Behavior of mixed cultures of microorganisms, *Annu. Rev. Microbiol.* **31**:63–87.

Fredrickson, A. G., and Stephanopoulos, G., 1981, Microbial competition, *Science* **213**:972–979.

Gottschal, J. C., deVries, S., and Kuenen, J. G., 1979, Competition between the facultative chemolithotrophic *Thiobacillus* A2, an obligately chemolithotrophic *Thiobacillus* and a heterotrophic spirillum for inorganic and organic substrates, *Arch. Microbiol.* **121**:241–249.

Grbić-Galić, D. E., and Young, L. Y., 1982, Anaerobic degradation pathways for ferulate and benzoate and the isolation of a pure culture, *Abst. Annu. Mtg. Am. Soc. Microbiol.* **1982**:199.

Gunner, H. B., and Zuckerman, B. M., 1968, Degradation of Diazinon by synergistic microbial action, *Nature (Lond.)* **217:**1183–1184.

Hale, M. E., 1974, *The Biology of Lichens*, 2nd ed., Edward Arnold, London.

Hansen, S. R., and Hubbell, S. P., 1980, Single-nutrient microbial competition: Qualitative agreement between experimental and theoretically forecast outcomes, *Science* **207:**1491–1493.

Harder, W., and Veldkamp, H., 1971, Competition of marine psychrophilic bacteria at low temperatures, *Antonie van Leeuwenhoek J. Microbiol. Serol.* **37:**51–63.

Harder, W., Kuenen, J. G., and Martin, A., 1977, Microbial selection in continuous culture, *J. Appl. Bacteriol.* **43:**1–14.

Hardin, G., 1960, The competitive exclusion principle, *Science* **131:**1292–1297.

Harrison, D. E. F., Wilkinson, T. G., Wren, S. J., and Harwood, J. H., 1975, Mixed bacterial cultures as a basis for continuous production of single cell protein from C_1 compounds *Continuous Culture 6. Applications and New Fields* (A. C. R. Dean, D. C. Elwood, C. G. T. Evans, and J. M. Melling, eds.), pp. 122–134, Ellis Horwood, Chichester.

Healy, J. B., and Young, L. Y., 1978, Catechol and phenol degradation by a methanogenic population of bacteria, *Appl. Environ. Microbiol.* **35:**216–218.

Healy, J. B., and Young, L. Y., 1979, Anaerobic biodegradation of eleven aromatic compounds to methane, *Appl. Environ. Microbiol.* **38:**84–89.

Henson, J. M., and Smith, P. H., 1983, Isolation of an anaerobic thermophilic, butyrate-utilizing bacterium in coculture with *Methanobacterium thermoautotrophicum*, *Abst. Annu. Mtg. Am. Soc. Microbiol.* **1983:**147.

Herbert, D., Elsworth, R., and Telling, R. C., 1956, The continuous culture of bacteria: A theoretical and experimental study, *J. Gen. Microbiol.* **14:**601–622.

Hespell, R. B., and Mertens, M., 1978, Effects of nucleic acid compounds on viability and cell composition of *Bdellovibrio bacteriovorus* during starvation, *Arch. Microbiol.* **116:**151–159.

Hespell, R. B., and Odelson, D. A., 1978, Metabolism of RNA-ribose by *Bdellovibrio bacteriovorus* during intraperiplasmic growth on *Escherichia coli*, *J. Bacteriol.* **136:**936–946.

Hespell, R. B., Miozzari, G. F., and Rittenberg, S. C., 1975, Ribonucleic acid destruction and synthesis during intraperiplasmic growth of *Bdellovibrio bacteriovorus*, *J. Bacteriol.* **123:**481–491.

Hespell, R. B., Thomashow, M. F., and Rittenberg, S. C., 1974, Changes in cell composition and viability of *Bdellovibrio bacteriovorus* during starvation, *Arch. Microbiol.* **97:**313–327.

Hobson, P. N., Bousfield, S., and Summers, R., 1974, Anaerobic digestion of organic matter, *CRC Crit. Rev. Environ. Control* **4:**131–191.

Hungate, R. E., 1966, *The Rumen and Its Microbes*, Academic Press, New York.

Hungate, R. E., 1967, Hydrogen as an intermediate in the rumen fermentation. *Arch. Microbiol.* **59:**158–164.

Hungate, R. E., Smith, W., Bauchop, T., Yu, J., and Rabinowitz, J. C., 1970, Formate as an intermediate in the bovine rumen fermentation, *J. Bacteriol.* **102:**389–397.

Hutchinson, G. E., 1961, The paradox of the plankton, *Am. Nat.* **95:**137–145.

Iannotti, E. L., Kafkewitz, D., Wolin, M. J., and Bryant, M. P., 1973, Glucose fermentation products of *Ruminococcus albus* grown in continuous culture with *Vibrio succinogenes:* Changes caused by interspecies transfer of H_2, *J. Bacteriol.* **114:**1231–1240.

Jannasch, H. W., 1967, Enrichments of aquatic bacteria in continuous culture, *Arch. Microbiol.* **59:**165–173.

Jannasch, H. W., and Mateles, R. I., 1974, Experimental bacterial ecology studied in continuous culture, *Adv. Microb. Physiol.* **11:**165–212.

Jerris, J. S., and McCarty, P. L., 1965, The biochemistry of methane fermentation using ^{14}C tracers, *J. Water Poll. Contr. Fed.* **37:**178–192.

Johanides, V., and Hršak, D., 1976, Changes in mixed bacterial cultures during linear alkybenzenesulphonate (LAS) biodegradation, in: *Fifth International Symposium on Fermentation* (H. Dellweg, ed.), p. 426, Westkreuz, Berlin.

John, A., Isaacson, H. R., and Bryant, M. P., 1974, Isolation and characteristics of a ureolytic strain of *Selenomonas ruminantium*, *J. Dairy Sci.* **57:**1003–1014.

Kaiser, J.-P., and Hanselmann, K. W., 1982, Fermentative metabolism of substituted monoaromatic compounds by a bacterial community from anaerobic sediments, *Arch. Microbiol.* **133:**185–194.

Kasper, H. F., and Wuhrmann, K., 1978, Kinetic parameters and relative turnovers of some important catabolic reactions in digesting sludge. *Appl. Environ. Microbiol.* **36:**1–7.

Khan, A. W., 1977, Anaerobic degradation of cellulose by mixed culture, *Can. J. Microbiol.* **23:**1700–1705.

Khan, A. W., and Murray, W. D., 1982a, Isolation of a symbiotic culture of two species of bacteria capable of converting cellulose to ethanol and acetic acid, *FEMS Microbiol. Lett.* **13:**377:–381.

Khan, A. W., and Murray, W. D., 1982b, Influence of *Clostridium saccarolyticum* on cellulose degradation by *Acetivibrio cellulolyticus, J. Appl. Bacteriol.* **53:**379–383.

Khan, A. W., Trottier, T. M., Patel, G. B., and Martin, S. M., 1979, Nutrient requirement for the degradation of cellulose to methane by a mixed population of anaerobes, *J. Gen. Microbiol.* **112:**365–372.

Kleiber, M., 1953, Biosynthesis of milk constituents by intact dairy cows studied with C^{14} as tracer, in: *Atomic Energy Agricultural Research* (C. L. Comar, ed.), p. 253, TID5115, Oak Ridge, Tennessee.

Konings, W. N., and Veldkamp, H., 1980, Phenotype responses to environmental change, in: *Contemporary Microbial Ecology* (D. C. Ellwood, J. N. Hedger, M. J. Latham, J. M. Lynch, and J. H. Slater, eds.), pp. 161–191, Academic Press, London.

Kristjansson, J. K., Schönheit, P., and Thauer, R. K., 1982, Different Ks values for hydrogen of methanogenic bacteria and sulfate-reducing bacteria: An explanation for apparent inhibition of methanogenesis by sulfate, *Arch. Microbiol.* **131:**278–282.

Kuenen, J. G., and Gottschal, J. E., 1982, Competition among chemolithotrophs and methylotrophs and their interactions with heterotrophic bacteria, in: *Microbial Interactions and Communities*, Vol. 1 (A. T. Bull and J. H. Slater, eds.), pp. 153–187, Academic Press, London.

Kuenen, J. G., and Rittenberg, S. C., 1975, Incorporation of long-chain fatty acids of the substrate organism by *Bdellovibrio bacteriovorus* during intraperiplasmic growth, *J. Bacteriol.* **121:**1145–1157.

Laanbroek, H. J., and Pfennig, N., 1981, Oxidation of short-chain fatty acids by sulfate-reducing bacteria in freshwater and in marine sediments, *Arch. Microbiol.* **128:**330–335.

Latham, M. J., and Wolin, M. J., 1977, Fermentation of cellulose by *Ruminococcus flavefaciens* in the presence and absence of *Methanobacterium ruminantium, Appl. Environ. Microbiol.* **34:**297–301.

Latham, M. J., Brooker, B. E., Pettipher, G. L., and Harris, P. J., 1978, Adhesion of *Bacteroides succinogenes* in pure culture and in the presence of *Ruminococcus flavefaciens* to cell walls in leaves of perennial ryegrass (*Lolium perene*), *Appl. Environ. Microbiol.* **35:**1166–1173.

Lee, I. H., Fredrickson, A. G., and Tsuchiya, H. M., 1976, Dynamics of mixed cultures of *Lactobacillus plantarum* and *Propionibacterium shermanii, Biotech. Bioeng.* **18:**513–526.

Lewis, D. H., 1967, A note on the continuous flow culture of mixed populations of lactobacilli and streptococci, *J. Appl. Bacteriol.* **30:**406–409.

Ljundahl, L. G., 1979, Physiology of thermophilic bacteria, *Adv. Microbiol. Physiol.* **19:**149–243.

Lorowitz, W. H., and Bryant, M. P., 1983, Methanogenic stearate enrichment cultures, *Abst. Annu. Mtg. Am. Soc. Microbiol.* **1983:**147.

Lovley, D. R., and Klug, M. J., 1982, Intermediary metabolism of organic matter in the sediments of a eutrophic lake, *Appl. Environ. Microbiol.* **43:**552–560.

Lovley, D. R., and Klug, M. J., 1983, Sulfate reducers can outcompete methanogens at freshwater sulfate concentrations, *Appl Environ. Microbiol.* **45:**187–192.

Lovley, D. R., Dwyer, D. F., and Klug, M. J., 1982, Kinetic analysis of competition between sulfate reducers and methanogens for hydrogen in sediments, *Appl. Environ. Microbiol.* **43:**1373–1379.

Mackie, R. I., and Bryant, M. P., 1981, Metabolic activity of fatty acid-oxidizing bacteria and the contribution of acetate, propionate, butyrate and CO_2 to methanogenesis in cattle waste at 40 and 60°C, *Appl. Environ. Microbiol.* **41:**1363–1373.

Mackie, R. I., and Gilchrist, F. M. C., 1979, Changes in lactate-producing and lactate-utilizing bacteria in relation to pH in the rumen of sheep during stepwise adaption to a high-concentrate diet, *Appl. Environ. Microbiol.* **38:**422–430.

Mah, R. A., 1982, Methanogenesis and methanogenic partnerships, *Phil. Trans. R. Soc. Lond. B* **297:**599–616.

Mah, R. A., Ward, D. M., Baresi, L., and Glass, T. L., 1977, Biogenesis of methane, *Annu. Rev. Microbiol.* **31:**309–342.

Mah, R. A., Smith, M. R., and Baresi, L., 1978, Studies on acetate-fermenting strain of *Methanosarcina*, *Appl. Environ. Microbiol.* **35:**1174–1184.

Maki, L. R., 1954, Experiments on the microbiology of cellulose decomposition in municipal sewage plant, *Antonie van Leeuwenhoek J. Microbiol. Serol.* **20:**185.

Martens, C. S., and Berner, R. A., 1974, Methane production in the interstitial waters of sulfate-depleted marine sediments, *Science* **185:**1167–1169.

Martin, A., and Rittenberg, S. C., 1972, Kinetics of deoxyribonucleic acid destruction and synthesis during growth of *Bdellovibrio bacteriovorus* strain 109D on *Pseudomonas putida* and *Escherichia coli*, *J. Bacteriol.* **111:**664–673.

McCarty, P. L., 1972, Energetics of organic matter degradation, in: *Water Pollution Microbiology* (R. Mitchell, ed.), pp. 91–113, Wiley–Interscience, New York.

McInerney, M. J., and Bryant, M. P., 1981a, Basic principles of bioconversion, in: *Anaerobic Digestion and Methanogenesis, Biomass Conversion Processes for Energy and Fuels* (S. S. Sofer and O. R. Zaborsky, eds.), pp. 277–296, Plenum, New York.

McInerney, M. J., and Bryant, M. P., 1981b, Anaerobic degradation of lactate by syntrophic association of *Methanosarcina barkeri* and *Desulfovibrio* species and effect of H_2 on acetate degradation, *Appl. Environ. Microbiol.* **41:**346–354.

McInerney, M. J., Bryant, M. P., and Pfennig, N., 1979, Anaerobic bacterium that degrades fatty acids in syntrophic association with methanogens, *Arch. Microbiol.* **122:**129–135.

McInerney, M. J., Mackie, R. I. and Bryant, M. P., 1981a, Syntrophic association of a butyrate-degrading bacterium and *Methanosarcina* enriched from bovine rumen fluid, *Appl. Environ. Microbiol.* **41:**826–828.

McInerney, M. J., Bryant, M. P., Hespell, R. B., and Costerton, J. W., 1981b, *Syntrophomonas wolfei* gen. nov., sp. nov., an anaerobic, syntrophic, fatty acid-oxidizing bacterium, *Appl. Environ. Microbiol.* **41:**1029–1039.

Mead, G. C., 1971, The amino acid-fermenting clostridia, *J. Gen. Microbiol.* **67:**47–56.

Meers, J. L., 1971, Effect of dilution rate on the outcome of chemostat mixed culture experiments, *J. Gen. Microbiol.* **67:**359–361.

Meers, J. L., 1973, Growth of bacteria in mixed culture, *CRC Crit. Rev. Microbiol.* **2:**139–225.

Megee, R. D. III, Drake, J. F., Fredrickson, A. G., and Tsuchiya, H. M., 1972, Studies in intermicrobial symbiosis. *Saccharomyces cerevisiae* and *Lactobacillus casei*, *Can. J. Microbiol.* **18:**1733–1742.

Meyer, J. S., Tsuchiya, H. M., and Fredrickson, A. G., 1975, Dynamics of mixed populations having complementary metabolism, *Biotech. Bioeng.* **17:**1065–1081.

Miller, T. L., and Wolin, M. J., 1979, Fermentation of saccharolytic intestinal bacteria, *Am. J. Clin. Nutr.* **32:**164–172.

Molongoski, J. J., and Klug, M. J., 1976, Characterization of anaerobic heterotrophic bacteria isolated from freshwater lake sediments, *Appl. Microbiol.* **31:**83–90.

Mountfort, D. O., and Asher, R. A., 1978, Changes in proportions of acetate and carbon dioxide used as methane precursors during anaerobic digestion of bovine waste, *Appl. Environ. Microbiol.* **35:**648–654.

Mountfort, D. O., and Asher, R. A., 1981, Role of sulfate reduction versus methanogenesis in terminal carbon flow in polluted intertidal sediment of Maimea inlet, Nelson, New Zealand, *Appl. Environ. Microbiol.* **42:**252–258.

Mountfort, D. O., and Bryant, M. P., 1982, Isolation and characterization of an anaerobic syntrophic benzoate-degrading bacterium from sewage sludge, *Arch. Microbiol.* **133:**249–256.

Mountfort, D. O., Asher, R. A., Mays, E. L., and Teidje, J. M., 1980, Carbon and electron flow in mud and sandflat intertidal sediments of Delaware inlet, Nelson, New Zealand, *Appl. Environ. Microbiol.* **39:**686–694.

Muira, H., Horiguchi, M., and Matsumoto, T., 1980, Nutritional interdependence among rumen bacteria, *Bacteroides amylophilus*, *Megasphaera elsdenii* and *Ruminococcus albus*, *Appl. Environ. Microbiol.* **40:**294–300.

Munnecke, D. M., and Hsieh, D. P. H., 1974, Microbial decontamination of Parathion and p-nitrophenol in aqueous medium, *Appl. Microbiol.* **28:**212–217.

Nagle, D., and Wolfe, R. S., 1983, Methanogenesis, in: *Comprehensive Biotechnology*, Vol. 1 (M. Moo-Young, ed.), pp. 425–438, Pergamon, London.

Nedwell, D. B., 1982, The cycling of sulphur in marine and freshwater sediments, in: *Sediment Microbiology* (D. B. Nedwell and C. M. Brown, eds.), pp. 53–106, Academic Press, London.

Neill, A. R., Grime, D. W., and Dawson, R. M. C., 1978, Conversion of choline methyl groups through trimethylamine into methane in the rumen, *Biochem J.* **170:**529–535.

Nottingham, P. M., and Hungate, R. E., 1969, Methanogenic fermentation of benzoate, *J. Bacteriol.* **98:**1170–1172.

Nurmikko, V., 1954, Symbiosis experiments concerning the production and biosynthesis of certain amino acids and vitamins in associations of lactic acid bacteria, *Ann. Acad. Sci. Fenn.* **54:**7–58.

Odum, E. P., 1953, *Fundamentals of Ecology*, W. B. Saunders, Philadelphia.

Odum, E. P., 1971, *Fundamentals of Ecology*, 3rd ed., W. B. Saunders, Philadelphia.

Opperman, R. A., Nelson, W. O., and Brown, R. E., 1961, *In vivo* studies of methanogenesis in the bovine rumen: Dissimilation of acetate, *J. Gen. Microbiol.* **25:**103–111.

Oremland, R. S., and Polcin, S., 1982, Methanogenesis and sulfate reduction: Competitive and non-competitive substrates in estuarine sediments, *Appl. Environ. Microbiol.* **44:**1270–1276.

Oremland, R. S., and Taylor, B. F., 1978, Sulfate reduction and methanogenesis in marine sediments, *Geochim. Cosmochim. Acta* **42:**209–214.

Paynter, M. J. B., and Hungate, R. E., 1968, Characterization of *Methanobacterium mobilis*, sp. n., isolated from the bovine rumen, *J. Bacteriol.* **95:**1943–1951.

Pfennig, N., 1978, General physiology and ecology of photosynthetic bacteria, in: *The Photosynthetic Bacteria* (R. K. Clayton and W. R. Sistrom, eds.), pp. 3–18, Plenum Press, New York.

Pfennig, N., 1980, Syntrophic mixed cultures and symbiotic consortia with phototrophic bacteria: A review, in: *Anaerobes and Anaerobic Infections* (G. Gottschalk, N. Pfennig, and H. Werner, eds.), pp. 127–131, Gustav Fisher, Stuttgart.

Pfennig, N., and Biebl, H., 1976, *Desulfuromonas acetoxidans* gen. nov., and sp. nov., a new anaerobic sulfur-reducing, acetate-oxidizing bacterium, *Arch. Microbiol.* **110:**1–12.

Pfennig, N., and Widdel, F., 1981, Ecology and physiology of some anaerobes from the microbial sulfur cycle, in: *Biology of Inorganic Nitrogen and Sulfur* (H. Bothe and A. Trebest, eds.), pp. 169–177, Springer-Verlag, Berlin.

Pine, M. J., and Barker, H. A., 1956, Studies on the methane fermentation. XII. The pathway of hydrogen in the acetate fermentation, *J. Bacteriol.* **71:**644–648.

Pittman, K. A., and Bryant, M. P., 1964, Peptides and other nitrogen sources for growth of *Bacteroides ruminicola*, *J. Bacteriol.* **88:**401–410.

Powell, E. O., 1958, Criteria for the growth of contaminants and mutants in continuous culture, *J. Gen. Microbiol.* **18:**259–268.

Reddy, C. A., Bryant, M. P., and Wolin, M. J., 1972, Characteristics of S organism isolated from *Methanobacillus omelianskii*, *J. Bacteriol.* **109:**539–545.

Rittenberg, S. C., 1983, Bdellovibrio: Attack, Penetration, and growth on its prey, *ASM News* **49:**435–439.

Rittenberg, S. C., and Hespell, R. B., 1975, Energy efficiency of intraperiplasmic growth of *Bdellovibrio bacteriovorus*, *J. Bacteriol.* **121:**1158–1165.

Rittenberg, S. C., and Langley, D., 1975, Utilization of nucleoside monophosphates *per se* for intraperiplasmic growth of *Bdellovibrio bacteriovorus*, *J. Bacteriol.* **121:**1137–1144.

Robinson, J. A., and Tiedje, J. M., 1982, Kinetics of hydrogen consumption by rumen fluid, anaerobic digestor sludge, and sediment, *Appl. Environ. Microbiol.* **44:**1374–1384.

Robinson, J. A., Strayer, R. F., and Teidje, J. M., 1981, Method for measuring dissolved hydrogen in anaerobic ecosystems: Application to the rumen, *Appl. Environ. Microbiol.* **41:**545–548.

Rode, L. M., Sharak Genthner, B. R., and Bryant, M. P., 1981, Syntrophic association by cocultures of the methanol and H_2-CO_2-utilizing species *Eubacterium limosum* and pectin-fermenting *Lachnospira multiparus* during growth in a pectin medium, *Appl. Environ. Microbiol.* **42**:20–22.

Rosson, R. A., and Rittenberg, S. C., 1979, Regulated breakdown of *Escherichia coli* deoxyribonucleic acid during intraperiplasmic growth of *Bdellovibrio bacteriovorus* 109J, *J. Bacteriol.* **140**:620–633.

Ruby, E. G., and Rittenberg, S. C., 1983, Differentiation after premature release of intraperiplasmically growing *Bdellovibrio bacteriovorus*, *J. Bacteriol.* **154**:32–40.

Russell, J. B., and Baldwin, R. L., 1978, Substrate preferences in rumen bacteria: Evidence of catabolite regulatory mechanisms, *Appl. Environ. Microbiol.* **36**:319–329.

Russell, J. B., and Baldwin, R. L., 1979, Comparison of substrate affinities among several rumen bacteria: A possible determinant of rumen bacterial competition, *Appl. Environ. Microbiol.* **37**:531–536.

Scheifinger, C. C., and Wolin, M. J., 1973, Propionate formation from cellulose and soluble sugars by combined cultures of *Bacteroides succinogenes* and *Selenomonas ruminatium*, *Appl. Microbiol.* **26**:789–795.

Scheifinger, C. C., Linehan, B., and Wolin, M. J., 1975, H_2 production by *Selenomonas ruminantium* in the absence and presence of methanogenic bacteria, *Appl. Microbiol.* **29**:480–483.

Scherer, P., and Sahm, H., 1981, Influence of sulfur-containing compounds on the growth of *Methanosarcina barkeri* in a defined medium, *Eur. J. Appl. Microbiol. Biotechnol.* **12**:28–35.

Schink, B., and Pfennig, N., 1982a, *Propionigenium modestum* gen. nov. sp. nov. a new, strictly anaerobic, nonsporing bacterium growing on succinate. *Arch. Microbiol.* **133**:209–216.

Schink, B., and Pfennig, N., 1982b, Fermentation of trihydroxybenzenes by *Pelobacter acidigallici* gen. nov., sp. nov., a new strictly anaerobic, non-spore forming bacterium, *Arch. Microbiol.* **133**:195–201.

Schink, B., and Stieb, M., 1983, Fermentative degradation of polyethylene glycol by a strictly anaerobic, gram-negative, nonsporeforming bacterium, *Pelobacter venetianus*, sp. nov., *Appl. Environ. Microbiol.* **45**:1905–1913.

Schönheit, P., Kristjansson, J. K., and Thauer, R. K., 1982, Kinetic mechanism for the ability of sulfate reducers to out-compete methanogens for acetate, *Arch. Microbiol.* **132**:285–288.

Seidler, R. J., and Starr, M., 1969, Isolation and characterization of host-independent bdellovibrios, *J. Bacteriol.* **100**:769–785.

Sharak Genthner, B. R., Davies, C. L., and Bryant, M. P., 1981, Features of rumen and sludge strains of *Eubacterium limosum*, a methanol- and H_2-CO_2-utilizing species,*Appl. Environ. Microbiol.* **42**:12–19.

Sheehan, B. T., and Johnson, M. J., 1971, Production of bacterial cells from methane, *Appl. Microbiol.* **15**:1473–1478.

Shilo, M., 1969, Morphological and physiological aspects of the interaction of bdellovibrios with host bacteria, *Curr. Top. Microbiol. Immunol.* **50**:174–204.

Shlomi, E. R., Lankhorst, A., and Prins, R. A., 1978, Methanogenic fermentation of benzoate in an enrichment culture, *Microb. Ecol.* **4**:249–261.

Slater, J. H., and Bull, A. T., 1978, Interactions between microbial populations, in: *Companion to Microbiology* (A. T. Bull and P. M. Meadow, eds.), pp. 181–206, Longman, London.

Slater, J. H., and Somerville, H. J., 1979, Microbial aspects of wastewater treatment with particular attention to degradation of organic compounds, *Sym. Soc. Gen. Microbiol.* **29**:221–261.

Slyter, L. L., Weaver, J. M., Oltjen, R. R., and Putman, P. A., 1967, Cellulolytic bacteria from cattle fed purified diets, *J. Anim. Sci.* **26**:880–881.

Smith, C. J., and Bryant, M. P., 1979, Introduction to metabolic activities of intestinal bacteria, *Am. J. Clin. Nutr.* **32**:149–157.

Smith, P. H., and Hungate, R. E., 1958, Isolation and characterization of *Methanobacterium ruminantium*, n. sp., *J. Bacteriol.* **75**:713–718.

Smith, P. H., and Mah, R. A., 1966, Kinetics of acetate metabolism during sludge digestion, *Appl. Microbiol.* **14**:368–371.

Smith, R. L., and Klug, M. J., 1981a, Electron donors utilized by sulfate-reducing bacteria in eutrophic lake sediments, *Appl. Environ. Microbiol.* **42**:116–121.

Smith, R. L., and Klug, M. J., 1981b, Reduction of sulfur compounds in sediments of a eutrophic lake basin, *Appl. Environ. Microbiol.* **41**:1230–1237.

Sørenson, J., Christensen, D., and Jørgensen, B. B., 1981, Volatile fatty acids and hydrogen as substrates for sulfate-reducing bacteria in anaerobic marine sediments, *Appl. Environ. Microbiol.* **42**:5–11.

Starr, M. P., and Seilder, R. J., 1971, The bdellovibrios, *Annu. Rev. Microbiol.* **25**:649–678.

Stephanopoulos, G. N., Aris, R., and Fredrickson, A. G., 1979, A stochastic analysis of the growth of competing microbial populations in a continuous biochemical reactor, *Math. Biosci.* **45**:99–136.

Stewart, F. M., and Levin, B. R., 1973, Partitioning of resources and the outcome of interspecific competition: A model and some general considerations, *Am. Nat.* **107**:171–198.

Suflita, J. M., Horowitz, A., Shelton, D. R., and Teidje, J. M., 1982, Dehalogenation: A novel pathway for the anaerobic biodegradation of haloaromatic compounds, *Science* **218**:1115–1117.

Tarvin, D., and Buswell, A. M., 1934, The methane fermentation of organic acids and carbohydrates, *J. Am. Chem. Soc.* **56**:1751–1755.

Taylor, C. D., McBride, B. C., Wolfe, R. S., and Bryant, M. P., 1974, Coenzyme M, essential for growth of a rumen strain of *Methanobacterium ruminantium*, *J. Bacteriol.* **120**:974–975.

Thauer, R. K., Käufer, B., and Fuchs, G., 1975, The active species of "CO_2" utilized by reduced ferredoxin: CO_2 oxidoreductase from *Clostridium pasteurianum*, *Eur. J. Biochem.* **27**:282–290.

Thauer, R. K., Jungermann, K., and Decker, K., 1977, Energy conservation in chemotrophic anaerobic bacteria, *Bacteriol. Rev.* **41**:100–180.

Thomashow, M. F., and Rittenberg, S. C., 1978a, Intraperiplasmic growth of *Bdellovibrio bacteriovorus* 109J: Solubilization of *Escherichia coli* peptidoglycan, *J. Bacteriol.* **135**:998–1007.

Thomashow, M. F., and Rittenberg, S. C., 1978b, Intraperiplasmic growth of *Bdellovibrio bacteriovorus* 109J: N-deacetylation of *Escherichia coli* peptidoglycan amino sugars, *J. Bacteriol.* **135**:1008–1014.

Thomashow, M. F., and Rittenberg, S. C., 1978c, Intraperiplasmic growth of *Bdellovibrio bacteriovorus* 109J: Attachment of long-chain fatty acids to *Escherichia coli* peptidoglycan, *J. Bacteriol.* **135**:1015–1023.

Thomashow, M. F., and Rittenberg, S. C., 1979, The intraperiplasmic growth cycle—the life of the bdellovibrios, in: *Developmental Biology of Procaryotes* (J. H. Parrish, ed.), pp. 115–138, Blackwell, Oxford.

Tilman, D., 1977, Resource competition between planktonic algae: An experimental and theoretical approach, *Ecology* **58**:338–348.

Toerien, D. F., and Hattingh, W. H. J., 1969, Anaerobic digestion. I. The microbiology of anaerobic digestion, *Water Res.* **3**:385–416.

Veldkamp, H., 1970, Enrichment cultures of prokaryotic organisms, in: *Methods in Microbiology*, Vol. 3A (J. R. Norris and D. W. Ribbons, eds.), pp. 305–361, Academic Press, London.

Veldkamp, H., 1977, Ecological studies with the chemostat, in: *Advances in Microbial Ecology*, Vol. 1 (M. Alexander, ed.), pp. 59–94, Plenum Press, New York.

Veldkamp, H., and Jannasch, H. W., 1972, Mixed culture studies with the chemostat, in: *Environmental Control of Cell Synthesis and Function*, Proceedings of the Fifth International Symposium on the Continuous Culture of Microorganisms (A. C. R. Dean, S. J. Pirt, and D. W. Tempest, eds.), pp. 105–123, Academic Press, London.

Ward, D. M., Mah, R. A., and Kaplan, I. R., 1978, Methanogenesis from acetate: A non-methanogenic bacterium from an anaerobic acetate enrichment, *Appl. Environ. Microbiol.* **35**:1185–1192.

Wedemeyer, G., 1967, Dechlorination of 1,1,1,-trichloro-2,2-*bis*(p-chlorophenyl)ethane by *Aerobacter aerogenes*, *Appl. Microbiol.* **15**:569–574.

Weimer, P. J., and Zeikus, J. G., 1977, Fermentation of cellulose and cellobiose by *Clostridium thermocellum* in the absence and presence of *Methanobacterium thermoautotrophicum*, *Appl. Environ. Microbiol.* **33**:289–297.

Wellinger, A., and Wuhrmann, K., 1977, Influence of sulfide compounds on the metabolism of *Methanobacterium* strain AZ, *Arch. Microbiol.* **115**:13–17.

Widdel, F., and Pfennig, N., 1977, A new anaerobic, sporing, acetate-oxidizing, sulfate-reducing bacterium, *Desulfotomaculum* (emend.) *acetoxidans*, *Arch. Microbiol.* **122**:119–122.

Widdel, F., and Pfennig, N., 1981, Studies on dissimilatory sulfate-reducing bacteria that decompose fatty acids. I. Isolation of new sulfate-reducing bacteria enriched with acetate from saline environments. Description of *Desulfobacter postgatei* gen. nov., sp. nov., *Arch. Microbiol.* **129:**395–400.

Widdel, F., and Pfennig, N., 1982, Studies on dissimilatory sulfate-reducing bacteria that decompose fatty acids. II. Incomplete oxidation of propionate by *Desulfobulbus piopionicus* gen. nov., sp. nov., *Arch. Microbiol.* **131:**360–365.

Wilkinson, T. G., and Harrison, D. E. F., 1973, The affinity for methane and methanol of mixed cultures grown on methane in continuous culture, *J. Appl. Bacteriol.* **36:**309–313.

Wilkinson, T. G., Topiwala, H. H., and Hamer, G., 1974, Interactions in a mixed bacterial population growing on methane in continuous culture, *Biotech. Bioeng.* **16:**41–49.

Wimpenny, J. W. T., 1981, Spatial order in microbial ecosystems, *Biol. Rev.* **56:**295–342.

Winfrey, M. R., and Zeikus, J. G., 1977, Effect of sulfate on carbon and electron flow during microbial methanogenesis in freshwater sediments, *Appl. Environ. Microbiol.* **33:**275–281.

Winter, J. U., and Wolfe, R. S., 1980, Methane formation from fructose by syntrophic associations of *Acetobacterium woodii* and different strains of methanogens, *Arch. Microbiol.* **124:**73–79.

Wolfe, R. S., and Pfennig, N., 1977, Reduction of sulfur by spirillum 5175 and syntrophism with *Chlorobium*, *Appl. Environ. Microbiol.* **33:**427–433.

Wolin, M. J., 1969, Volatile fatty acids and the inhibition of *Escherichia coli* growth by rumen fluid, *Appl. Microbiol.* **17:**83–87.

Wolin, M. J., 1974, Metabolic interactions among intestinal microorganisms, *Am. J. Clin. Nutr.* **27:**1320–1328.

Wolin, M. J., 1975, Interactions betwen bacterial species of the rumen, in: *Digestion and Metabolism in the Ruminant* (I. W. McDonald and A. C. I. Warner, eds.), pp. 134–148, University of New England Publishing Unit, Armidale, Australia.

Wolin, M. J., 1976, Interactions between H₂-producing and methane-producing species, in: *Microbial Formation and Utilization of Gases (H₂, CH₄, CO₂)* (H. G. Schlegel, G. Gottschalk, and N. Pfennig, eds.), pp. 141–150, Goltze KG, Göttingen.

Wolin, M. J., 1979, The rumen fermentation: A model for microbial interactions in anaerobic ecosystems, in: *Advances in Microbial Ecology*, Vol. 3 (M. Alexander, ed.), pp. 49–78, Plenum Press, New York.

Wolin, M. J., 1982, Hydrogen transfer in microbial communities, in: *Microbial Interactions and Communities*, Vol. 1 (A. T. Bull and J. H. Slater, eds.), pp. 323–356, Academic Press, London.

Yeoh, H. T., Bungay, H. R., and Krieg, N. R., 1968, A microbial interaction involving a combined mutualism and inhibition, *Can. J. Microbiol.* **14:**491–492.

Yoon, H., and Blanch, H. W., 1977, Competition for double growth-limiting nutrients in continuous culture, *J. Appl. Chem. Biotechnol.* **27:**260–268.

Yoon, H., Klinzing, G., and Blanch, H. W., 1977, Competition for mixed substrates by microbial populations, *Biotech. Bioeng.* **19:**1193–1210.

Zehnder, A. J. B., 1978, Ecology of methane formation in: *Water Pollution Microbiology*, Vol. 2 (R. Mitchell, ed.), pp. 349–376, Wiley, New York.

Zehnder, A. J. B., Huser, B. A., Brock, T. D., and Wuhrmann, K., 1980, Characterization of an acetate-decarboxylating non-hydrogen-oxidizing methane bacterium, *Arch. Microbiol.* **124:**1–11.

Zeikus, J. G., 1977, The biology of methanogenic bacteria, *Bacteriol. Rev.* **41:**514–541.

Zeikus, J. G., 1980, Microbial populations in digestors, in: *Anaerobic Digestion* (D. A. Stafford, B. I. Wheatley, and D. E. Hughes), pp. 61–89, Applied Science Publishers, London.

Zeikus, J. G., Lynd, L. H., Thompson, T. E., Krzycki, J. A., Weimer, P. J., and Hegge, D. W., 1980, Isolation and characterization of a new, methylotrophic acetogenic anaerobe, the Marburg strain, *Curr. Microbiol.* **3:**381–386.

Zinder, S. H., and Koch, M., 1983, Acetate oxidation by a thermophilic methanogenic syntrophic coculture, *Abst. Annu. Mtg. Am. Soc. Microbiol.* **1983:**147.

APPLICABILITY OF GENERAL ECOLOGICAL PRINCIPLES TO MICROBIAL ECOLOGY

Ronald M. Atlas

ECOSYSTEMS AND MICROBIAL POPULATIONS

The ecosystem represents the fundamental ecological unit (Evans, 1956; Odum, 1971; Whittaker, 1975; Kormondy, 1976). It is a complete self-sustaining unit; within an ecosystem the biological community interacts with its abiotic surroundings in a manner that results in energy flow, trophic structure, and materials cycling (Odum, 1969, 1971). General ecological principles aim at explaining how ecosystems function, describing the bases for interactions between the living components of the ecosystem and their abiotic surroundings, the nature of the energy transfers within the system that establish energy flow, the nature of the interactions among populations that establish trophic structure, and the nature of material transfers within the system. Microorganisms are essential components of all ecosystems; the metabolic activities of microorganisms are critical in the flow of energy, material cycling, and establishment of trophic structure within the ecosystem (Brock, 1966; Alexander, 1971; Lynch and Poole, 1979; Atlas and Bartha, 1981). Without microorganisms, ecosystems could not be self-sustaining, and life on earth as we know it could not exist. Furthermore, the various interactions between microorganisms and their abiotic and biotic surroundings provide excellent models for studying ecological interrelationships and the functioning of ecosystems; our understanding of many general ecological principles comes from studies employing microorganisms.

   In considering the applicability of general ecological principles to microbial ecology we need recognize a basic dilemma: the dependence on pure culture studies to determine the physiological properties of microorganisms,

Ronald M. Atlas • Department of Biology, University of Louisville, Louisville, Kentucky 40292.

but the inability of pure culture studies to yield meaningful information about the ecological functioning of microorganisms. This is a problem not faced by macroecologists who can see what an animal or plant is doing within an ecosystem. The pure culture approach, as traditionally used by microbiologists, in many ways is incompatible with the elucidation of ecologically meaningful data. Whereas many microbiologists would agree with the adage that "work with impure cultures yields nothing but nonsense and *Penicillium glaucum*," microbes in natural ecosystems virtually never occur as pure, monospecific populations. Effectively, microbial ecologists must use pure culture studies to define the potential ecological role of a microorganism and then infer the actual functional roles (niches) of microorganisms that are occupied by those microbes within an ecosystem. Thus the methodological approaches to the study of macroecology often are quite different from the approaches to the study of microbial ecology. However, the theoretical framework, that is the principles of ecological function, should apply equally to micro- and macroorganisms.

ENERGY FLOW THROUGH ECOSYSTEMS

The capture, transformation, and degradation of energy by members of the biological community establish a flow of energy through the ecosystem (H. T. Odum, 1957; Teal, 1962; Phillipson, 1966; E. P. Odum, 1968, 1971; Turner, 1968; Hutchinson, 1970; Woodwell, 1970; Golley, 1972). The energy that supports life on earth originates from the sun (Oort, 1970). Within ecosystems the flow of energy obeys the laws of thermodynamics, i.e., energy is neither created nor destroyed (energy is transferred and transformed), and the system flows in the direction of increasing entropy (less usable energy); thus, there is a unidirectional flow of energy through the ecosystem with the amount of usable energy continuously declining (Woodwell, 1970; Kormondy, 1976). The efficiency of an ecosystem depends on the abilities of the living organisms to capture energy, transform this energy into useful chemical energy, and transfer the captured energy through the system (E. P. Odum, 1968, 1971; Whittaker, 1975). Some ecosystems are able to capture energy and export stored energy as organic matter to other systems, whereas others depend on the import of stored energy as organic matter; i.e., ecosystems vary in their energy flow and hence in their structure (H. T. Odum and E. P. Odum, 1981).

The initial capturing of solar energy to support an ecosystem generally depends on photoautotrophs, the organisms that can convert light energy to the biochemically useful form of energy, ATP, and that can convert inorganic carbon dioxide to organic biomass (Rabinowitch and Govindjee, 1969; Govindjee, 1975; Bolin *et al.*, 1980). In recent years, however, our understanding of chemolithotrophs (microorganisms that can oxidize inorganic compounds to generate ATP) has forced a modification of our basic understanding of ecosystem functioning (Kelly, 1971, 1978; Jannasch and Wirsen, 1979; Karl *et al.*, 1980; Jannasch and Mottl, 1985). We now must conclude that while ecosystems depend on the activities of primary producers to initiate the flow of energy through the system, the initial capturing of energy for the system may either be the result of the conversion of light energy (photoautotrophy) or the oxidation of inorganic compounds (chemolithotrophy).

Photoautotrophy

Rates of Primary Productivity

The rates of primary production by photoautotrophs determine the amount of energy available to most ecosystems (Westlake, 1963; Strickland, 1965; Goldman, 1966; Phillipson, 1966; Kozlovsky, 1968; Woodwell and Whittaker, 1968; Ryther, 1969; Woodwell, 1970; E. P. Odum, 1971; Whittaker and Likens, 1973; Helmut and Whittaker, 1975; National Academy of Sciences, 1975). The relative importance of microbial photoautotrophs versus plants varies with the habitat. Generally, plants are the most important primary producers in terrestrial habitats and photosynthetic microorganisms—algae and bacteria—are most important in aquatic habitats (Russel-Hunter, 1970; Whittaker, 1975; Raymont, 1980). The annual primary production by autotrophic microorganisms in Lake Mendota, a representative freshwater lake, has been found to be 428 gcal/cm^2 per year (Juday, 1940); in Cedar Bog Lake, Minnesota, autotrophic productivity has been reported as 111 gcal/cm^2 per year (Lindeman, 1942); and in the Silver Springs, Florida, river ecosystem, primary productivity has been found to be 2081 gcal/cm^2 per year (H. Odum, 1957). The mean primary productivity in the oceans is in the range of 152–365 g organic carbon per square meter per year (Ryther, 1969; Koblentz-Mishke *et al.*, 1970; Russel-Hunter, 1970; Lieth and Whittaker, 1977; Morris, 1980, 1982; Raymont, 1980) and in most lakes and streams is 100–300 g organic carbon per square meter per year (Lieth and Whittaker, 1977); in these aquatic ecosystems microorganisms are responsible for the primary productivity. By comparison, the primary production in a corn field during a typical growing season (100 days) is 800 gcal/cm^2 (Transeau, 1926). The mean terrestrial primary productivity, which includes inland aquatic ecosystems, is 773 g organic carbon per square meter per year (Lieth and Whittaker, 1977).

Photosynthetic Microorganisms

The photosynthetic microorganisms include the algae, cyanobacteria, prochlorales, and the anaerobic phototrophic bacteria. These microorganisms contain various photosynthetic pigments that allow them to trap light energy. The photosynthetic pigments of different microbial species absorb light at different wavelengths. Because different wavelengths of light have different penetrating powers, microorganisms exhibit zonation within habitats; i.e., photosynthetic microorganisms are spatially separated with respect to depth in relation to their ability to absorb particular wavelengths of light energy (Hutchinson, 1957; Wood, 1965, 1967; Pfennig, 1967; Kormondy, 1969; Odum, 1971; Goldberg *et al.*, 1974; Govindjee and Govindjee, 1974; Person, 1974; Wetzel, 1975; Caldwell, 1977; Kriss *et al.*, 1967; Stanier and Cohen-Bazire, 1977; Bold and Wynne, 1978; Lieth, 1978; Rheinheimer, 1981; Round, 1981; Stanier *et al.*, 1981; Truper and Pfennig, 1981). This spatial separation of habitats allows photosynthetic organisms to coexist in an ecosystem without competing directly for the identical resource; thus, although total primary productivity depends on the quantity of light energy available, different populations use separate portions of the light energy.

The two primary factors affecting the distribution of photosynthetic microorganisms in aquatic ecosystems are the intensity (quantity) and wavelength (quality) of light that penetrates to a particular depth in the water column. The algae and cyanobacteria, which contain chlorophyll *a*, occupy the upper portion of the water column; the anaerobic phototrophic bacteria, whose primary photosynthetic pigments absorb light at longer wavelengths, occur at greater depths. The cyanobacteria exhibit two noteworthy adaptations that allow them to utilize the available light energy efficiently. Many species of cyanobacteria contain gas vacuoles that enable them to migrate up and down in the water column in response to changes in light intensity so that they can carry out primary productivity efficiently at a depth that is appropriate for the amount of available light (Walsby, 1975, 1977, 1981). In addition, many cyanobacteria are capable of anoxygenic photosynthesis; i.e., when the light intensity is low, they are able to shift from their normal oxygen-evolving mode of photosynthesis in which water serves as the electron donor to anaerobic photosynthesis in which hydrogen sulfide serves as the electron donor (Cohen *et al.*, 1975; Garlick *et al.*, 1977; Oren and Padan, 1978; Stanier *et al.*, 1981).

This distribution of photosynthetic microorganisms is based on a shared resource—the available light energy. The stratification pattern exhibited by aquatic photosynthetic microorganisms is a widespread phenomenon among primary producers and is seen also in the pattern of plants found in terrestrial ecosystems; low-lying shrubs utilize the light reaching the forest floor that has penetrated through the forest canopy. The plants growing at the lower levels of the forest utilize the limited light resources that reach them; generally, the shrubs and herbs that are shaded from direct light have accessory pigments that allow them to use the wavelengths of light that have not been filtered out by the leaves of the upper canopy. In tropical forests, there are few, if any, shrubs because very little light passes through the dense forest canopy.

The stratified distribution of primary producers is paralleled by the consumer populations. This is true in both macro- and microbial ecology. Certain birds occupy the overlying forest canopy, whereas other animals feed upon the underlying vegetative growth. Similarly, dense populations of fish occur in regions of extensive algal and cyanobacterial growth. Heterotrophic bacteria, fungi, and protozoa are also abundant in regions where there are high numbers of primary producers (Rheinheimer, 1981). The distribution of these consumers depends upon the availability of food, and areas of high primary productivity are almost always also areas of intense consumption.

Chemolithotrophy

Prior to the discovery of productive thermal rift areas in the deep ocean near the Galapagos Islands (Ballard, 1977; Corliss *et al.*, 1979), the productivity of chemolithotrophs was largely ignored with respect to the flow of energy through ecosystems. The finding of huge clams and wormlike poganophores in the dark depths of the oceans (Edmond and von Damm, 1983) presented a paradox; typically, the ocean depths are characterized by low productivity and biomass because of the lack of light and photosynthetic primary productivity (Jannasch and Wirsen, 1977; Jannasch, 1979). The rich and productive biological communities of the thermal rift areas could not be supported by an influx of biomass from primary producers in the photic zone. Rather, the energy flow in these deep thermal rift ecosystems originates with sulfur-oxidizing chemolithotrophs that obtain their requirements from the heated mineral-rich waters of the thermal vents (Jannasch and Wirsen, 1979; Karl *et al.*, 1980; Edmond *et al.*, 1982; Edmond and von Damm, 1983; Jannasch and Mottl, 1985). Significant primary productivity by chemolithotrophic sulfur oxidizers has been found within the guts of the poganophores and in the vent water (Jannasch and Wirsen, 1979, 1981; Karl *et al.*, 1980). The discovery of the role of chemolithotrophic bacteria in initiating the flow of energy through the thermal rift ecosystems forced an expansion of the general ecological concept that the productivity of all ecosystems was directly tied to the rates of photosynthesis or the rates of influx of organic matter produced by photoautotrophs. It should be noted, however, that the thermal rift areas are quite specialized cases and an exception to the concept that ecological productivity depends directly on photosynthesis.

Heterotrophy

The organic compounds produced by the primary producers provide the energy sources needed by other heterotrophic organisms that require preformed organic matter as a source of energy and carbon. Heterotrophic microorganisms extract their energy needs by metabolizing organic compounds, as do animal populations. Microorganisms can use a wide variety of organic compounds under diverse environmental conditions, carrying out heterotrophic metabolism under conditions where animal populations cannot. Microorganisms can extract energy from organic compounds both anaerobically by fermentation and anaerobic respiration and aerobically by aerobic respiration (Anderson and Wood, 1969; Dawes and Sutherland, 1976; Eriksson and Johnsrud, 1982; Mandelstam *et al.*, 1982; Atlas, 1984). Thus, microorganisms are able to carry out heterotrophic metabolism under anoxic con-

ditions that preclude metabolic activities by animals. This simply is an extension of where, rather than what, metabolic activities microorganisms can perform
as compared to macroorganisms. Under aerobic conditions, macro- and microorganisms share the ability to degrade simple organic nutrients and some
biopolymers, but microorganisms are unique in their capacity to carry out
anaerobic (fermentative) degradation of organic matter. They also are responsible for the recycling of most of the very abundant but difficult-to-digest
biopolymers, such as cellulose and lignin (Imshenetsky, 1967; Kirk, 1971;
Dagley, 1975; Alexander, 1977; Evans, 1977; Ander and Eriksson, 1978;
Hayes, 1979; Crawford and Crawford, 1980; Kirk *et al.*, 1980; Zeikus, 1981;
Eriksson and Johnsrud, 1982). The ability to extract energy through the
degradation of humic materials, waxes, and many man-made synthetics is also
virtually unique to microorganisms (Alexander, 1965; Horvath, 1972; National Academy of Sciences, 1972). It is actually microbial degradation of
cellulose in a cow's rumen that provides the nutrients that the cow metabolizes
for energy (Hungate, 1975).

The activities of microorganisms affect the accessibility of carbon and
energy of organic compounds to the biological community. Some transformations of organic carbon, e.g., the production of polymers such as humic
acids in soil, tend to reduce the rate of cycling or to immobilize that portion
of the carbon and stored energy (Alexander, 1977). Other transformations,
such as the anaerobic degradation of cellulose, mobilize the stored carbon
and energy by producing simpler organic compounds which can be more
readily utilized by the biological community. Transformations that change
the physical state, e.g., production of gaseous compounds, such as CO_2 or
CH_4, from liquids or solids, or transformations which alter the solubility, e.g.,
the production of glucose from cellulose, have a major effect on the mobility
and availability of carbon to the biological community within the habitat.

There is one aspect of microbial heterotrophic activity that merits further
consideration. Some bacterial populations are able to utilize organic compounds at very dilute concentrations. Such bacteria have been termed oligotrophs or low-nutrient bacteria in contrast to copiotrophic bacteria which
grow only at high nutrient concentrations (Jannasch, 1967; Poindexter, 1981*a,b*;
Kuznetsov *et al.*, 1979; Button, 1985). Oligotrophic bacteria are extremely
important in retaining energy within aquatic habitats where nutrient concentrations are frequently very low. Effectively oligotrophic bacteria are secondary producers, capturing organic compounds that would otherwise be excluded from the energy flow within the ecosystem. As such, these microbial
components of ecosystems enhance the efficiency of energy flow through the
ecosystem.

Food Webs

The transfer of energy stored in organic compounds from one organism
to another establishes a food chain; the transfer occurs in steps, each of which

constitutes a trophic level (E. P. Odum, 1962, 1971; Kozlovsky, 1968; Turner, 1968; Woodwell, 1970; Whittaker, 1975; Kormondy, 1976; Atlas and Bartha, 1981). The relationships among organisms of different trophic levels establish the food web. Within a food web, energy is transferred among the biological populations that comprise the community of that ecosystem. Individual food chains, e.g., the detritus food chain, may be based on allochthonous material and not directly on primary producers in the same ecosystem. Organisms that feed directly on primary producers constitute the trophic level of grazers; grazers are preyed upon by predators, and the latter may be preyed upon by a trophic level of larger predators. In reality, many consumers feed on more than one trophic level and on dissolved organic matter. Energy is lost during each transfer between trophic levels. In general, only 10–15% of the biomass from each trophic level is passed on to the next higher trophic level; 85–90% is consumed by respiration or enters the decay portion of the food web (Phillipson, 1966; E. P. Odum, 1971). Consequently, the higher a trophic level, the smaller is its biomass. In a classic example, H. Odum (1957) examined the standing biomass and energy-flow pyramids for the Silver Springs, Florida, ecosystem. The biomass values were 809, 37, 11, and 1.5 kcal/m^2 for the primary producers, primary consumers, secondary consumers, and top consumers, respectively. In addition, 5 kcal/m^2 was stored in the saprotrophic bacteria and fungi, which were not considered part of the direct flow of energy from primary producers to consumers. The energy flow in the same system was 20,810, 3368, 383, and 21 kcal/m^2 per year for the primary producers, primary consumers, secondary consumers, and top consumers, respectively, with an additional 5060 kcal/m^2 per year flowing into the saprotrophic bacteria and fungi.

In most terrestrial and shallow water environments, the predominant primary producers are higher plants, and the dominant grazers are invertebrate or vertebrate herbivores; microbial primary producers and grazers are usually present, but their role is quantitatively less significant than that of higher organisms (Crisp, 1964). In the limnetic zone of deep lakes and in the pelagic portion of the oceans, however, where no autochthonous higher plants exist, the entire food web is based on microbial primary producers, predominantly unicellular planktonic algae and cyanobacteria. In these environments, a substantial portion of the grazers is also microbial (planktonic protozoa), though, of course, they share this role with smaller representatives of the invertebrate zooplankton (Brooks and Dodson, 1965; Curds, 1977; Fenchel and Jorgensen, 1977; Fenchel, 1978; Stout, 1980; Rheinheimer, 1981).

The decay portion of the food web, which represents a sink through which energy is lost from the ecosystem, is dominated by microbial forms in aquatic and terrestrial environments alike (Atlas and Bartha, 1981). The decay portion of the food web involves the degradation of incompletely digested organic matter (e.g., fecal material and urea) and decomposition of dead but not consumed plants and animals. The proportion of the biomass subject to decay rather than consumption varies greatly within different types of habitats, but in forest and saltmarsh habitats decay may account for 80–90% of the total energy flow; at the other extreme, in pelagic and limnetic habitats, the bulk of the primary production is rapidly consumed by grazers and relatively little biomass is channeled through the decay route (E. P. Odum, 1962).

Part of the microbial biomass accumulated during decomposition can be cycled back into the food web through detrital food webs; detritus-based food chains based on the consumption of microbial biomass by predators are particularly important in aquatic ecosystems (Fenchel and Jorgensen, 1977). A detritus food chain is a transfer of energy by which the energy contained in detrital carbon (organic carbon that represents excretions or secretions that are lost from a trophic level) becomes available to the biological community (Wetzel, 1975). Although Kormondy (1976) equated the unidirectional flow of energy through the ecosystem with the one-way movement of carbon (energy) to progressively higher trophic levels, which he called the cardinal principle of ecosystem functioning, it is important to recognize that the recapturing of energy by detrital food webs, as well as the secondary productivity of oligotrophs, does not alter the fundamental principle of unidirectional flow of energy through an ecosystem because energy is still lost during each transfer. The transfer of energy back to higher trophic levels simply represents a mechanism for enhanced efficiency of energy utilization within an ecosystem.

BIOGEOCHEMICAL CYCLING WITHIN ECOSYSTEMS

Whereas energy flows unidirectionally through ecosystems, elements are continuously recycled (McLaren and Peterson, 1967; Cox, 1969; Delwiche, 1970; Pomeroy, 1970, 1974; E. P. Odum, 1971; Likens and Bornmann, 1972; Kormondy, 1976; Sörderlund and Svensson, 1976; Rheinheimer, 1981). Microorganisms are essential to the biogeochemical cycling of elements that are needed to support ecosystems (Starkey, 1964; Brock, 1966; Cox, 1969; Alexander, 1971, 1977; Mishustin and Shil'nikova, 1971; Payne, 1973; Stewart, 1973; Dalton, 1974; Burns and Hardy, 1975; Sörderlund and Svensson, 1976; Campbell, 1977; Cosgrove, 1977; Focht and Verstraete, 1977; Gibson, 1977; Bremmer and Steele, 1978; Lynch and Poole, 1979; Atlas and Bartha, 1981).

Cycling of Carbon

The biogeochemical cycling of carbon primarily represents the transfer of carbon between inorganic carbon dioxide and organic carbon compounds (Bolin, 1970). The global cycling of carbon depends on the primary productivity and the respiration of microbes and of higher organisms (Woodwell and Pecan, 1973; Pomeroy, 1974). Besides their involvement in carbon dioxide production and fixation, microorganisms are important in the movement of carbon within the organic pool of compounds contained within an ecosystem. Additionally, various microorganisms have the ability to produce and utilize gaseous carbon-containing compounds other than carbon dioxide, including methane and carbon monoxide (Hirsch, 1968; Quale, 1972; Mah *et al.*, 1977; Zeikus, 1977; Nozhevnikova and Yurkanov, 1978; Colby *et al.*, 1979; Mah and Smith, 1981; Whittenbury and Dalton, 1981). The cycling of carbon through methane represents an important shunt that removes carbon from the pool of carbon-containing compounds that are accessible to most members of the biological community.

Some of the metabolic activities of microorganisms enable them to carry out unique aspects of the utilization of carbon compounds, such as anaerobic decomposition and the mineralization of complex polymers. Because substantial organic deposits accumulate only under conditions inimical even to microbial activities, microorganisms have been considered to be capable of degrading all naturally occurring organic compounds. This represents the principle of microbial infallibility. Empirical experience indicates that no natural organic compound is protected from biodegradation provided that environmental conditions are favorable. However, there are various synthetic organic compounds that are xenobiotic (foreign to biological systems) and have molecular structures and chemical bond sequences that are not recognized by existing degradative enzymes; some such xenobiotic compounds are resistant to microbial attack (recalcitrant) or are metabolized incompletely, with the result that some man-made organic compounds or their partial-degraded derivatives accumulate in the environment (Alexander, 1965, 1981; National Academy of Sciences, 1972). Thus, while microorganisms normally contribute to the biogeochemical cycling of carbon, there are instances where they fail, and the interruption of the normal biogeochemical cycling of carbon can disrupt the normal functioning of the ecosystem.

Cycling of Nitrogen

Whereas many ecosystems depend on plants for a supply of organic carbon that can be used as a source of energy, all ecosystems are dependent on the bacterial fixation of atmospheric nitrogen (or on human intervention through the synthetic production of nitrogen fertilizers) for providing a supply of fixed nitrogen (Delwiche, 1970; Pomeroy, 1974; Sörderlund and Svensson, 1976). Nitrogen is an essential component of proteins, nucleic acids, and other cell constituents, but the vast supply of molecular nitrogen that occurs in the atmosphere is inaccessible to most biological systems. Plants, animals, and most microorganisms require combined forms of nitrogen for incorporation into cellular biomass, but the ability to fix atmospheric nitrogen is restricted to a limited number of bacteria (Delwiche, 1970; Mishustin and Shil'nikova, 1971; Bennemann and Valentine, 1972; Dalton and Mortenson, 1972; Stewart, 1973; Dalton, 1974; Quispel, 1974; Brill, 1975, 1979; Burns and Hardy, 1975; Winter and Burris, 1976; Gibson, 1977, Becking, 1981; Gordon, 1981; Vincent, 1981; Benson, 1985). Plants could not continue their photosynthetic metabolism without the availability of fixed forms of nitrogen provided by microorganisms or by synthetic fertilizers. As such, the biogeochemical cycling of nitrogen is one process in which microorganisms play an essential function not paralleled among the higher plants and animals; biogeochemical cycling is one aspect of ecology where macroecologists have long had to acknowledge the paramount role of microbial activity (Pomeroy, 1974).

The biological fixation of molecular nitrogen is carried out by several free-living bacterial genera, some of which may be rhizosphere associated, and by several bacteria genera that form mutualistic associations with plants (Becking, 1981; Gordon, 1981; Vincent, 1981). One estimate of global biological nitrogen fixation assumes that 1.7×10^8 metric tons of nitrogen is fixed annually (Burns and Hardy, 1975). Of this, 3.5×10^7 metric tons/year is estimated to be fixed in meadows and grasslands; 4.0×10^7 metric tons/year in forest and woodlands; and an additional 3.6×10^7 metric tons/year in marine habitats.

Microorganisms perform additional critical transformations on fixed forms of nitrogen within the biogeochemical cycle and are particularly responsible for transformations that affect the mobility and accessibility of fixed nitrogen to various terrestrial and aquatic ecosystems (Delwiche, 1970; Alexander, 1977). Nitrogen in living and dead organic matter occurs predominately in the reduced amino form. Many plants, animals, and microorganisms are capable of ammonification, a process in which organic nitrogen is converted to ammonia. Ammonium ions can be assimilated by numerous plants and many microorganisms and are incorporated into amino acids and other nitrogen-containing biochemicals. The nitrogen-containing organic compounds produced by one organism can be transferred to and assimilated by other organisms. Transformations of organic nitrogen-containing compounds are not restricted to microorganisms. Animals, for example, produce nitrogenous wastes, such as uric acid, from the metabolism of nitrogen-containing organic compounds.

The process of nitrification appears to be limited for the most part to a restricted number of autotrophic bacteria (Watson et al., 1981). In nitrification, ammonia or ammonium ions are oxidized to nitrite ions and then to nitrate ions (Wallace and Nicholas, 1969; Aleem, 1970; Painter, 1970; Focht and Verstraete, 1977). The two steps of nitrification, i.e., the formation of nitrite and the formation of nitrate, are carried out by different microbial populations (Walker, 1975; Belser, 1979; Watson et al., 1981). The process of nitrification is extremely important especially in soils because the transformation of ammonium to nitrite and nitrate ions results in a change in charge from positive to negative (Alexander, 1977; Belser and Schmidt, 1978). Positively charged ions tend to be bound by negatively charged clay particles in soil, while negatively charged ions freely migrate in the soil water. Nitrification must therefore be viewed as a mobilization process within soil habitats. In terrestrial ecosystems, ammonia normally is rapidly oxidized by nitrifying bacteria, and plants readily take up nitrate ions into the roots for assimilation into organic compounds. Nitrate and nitrite ions, however, can be leached from the soil column into the groundwater. This is a critical process, since it represents a loss of fixed forms of nitrogen from the soil where these compounds could be utilized by plants for production of organic matter.

Nitrate ions can be converted by a variety of organisms to organic matter through assimilatory nitrate reduction (Payne, 1973). A heterogeneous group of microorganisms, including many bacterial, fungal, and algal species, is capable of assimilatory nitrate reduction. The process of assimilatory nitrate reduction involves several enzyme systems including nitrate and nitrite reductases to form ammonia, which can be incorporated into amino acids.

Microorganisms are also responsible for the return of molecular nitrogen to the atmosphere through denitrification (Delwiche and Bryan, 1976). In the absence of oxygen, nitrate ions can act as terminal electron acceptors; the process is known as nitrate or anaerobic respiration; nitrate ions can also be converted to molecular nitrogen through denitrification (Payne, 1973; Focht and Verstraete, 1977; Hall, 1978; Jeter and Ingraham, 1981). When nitrate ions serve as terminal electron acceptors, the oxidation state of the nitrogen is reduced, leading to dissimilatory nitrate reduction. Dissimilatory nitrate reduction through nitrite can result in the formation of ammonia (nitrate ammonification) or in the production of gaseous forms of nitrogen (denitrification). The initial enzymes involved in these processes are dissimilatory nitrate and nitrite reductase systems. Dissimilatory nitrate reductases are membrane bound, competitively inhibited by oxygen, and not inhibited by ammonia, in contrast to assimilatory nitrate reductases, which are soluble, inhibited by ammonia, and not substantially inhibited by oxygen. Denitrification generally occurs under anaerobic conditions or under conditions of reduced oxygen tension. Although favored by anaerobic conditions, denitrification has been detected in the presence of oxygen; in these cases, the denitrification may have been occurring within anoxic microhabitats. The interface between aerobic and anaerobic layers in soil and sediment is an active zone of denitrification because of the favorable reduced oxygen tensions and the supply of nitrate ions from the overlying aerobic zone. Reflecting the reduced oxygen tension, denitrification is more common in standing waters than in running streams and rivers; denitrification rates are higher in the hypolimnion of eutrophic lakes during summer and winter stratification than during fall and spring turnover (Rheinheimer, 1981).

Because different environmental conditions favor different processes involved in the biogeochemical cycling of nitrogen, there is a spatial zonation of cycling processes (Alexander, 1971, 1977; Atlas and Bartha, 1981). Fixation of nitrogen occurs in both surface and subsurface habitats. Nitrification occurs in aerobic habitats. Denitrification predominates in waterlogged soils and within the sediments of aquatic habitats. The cycling of nitrogen within a given habitat also exhibits seasonal fluctuations; during spring and fall blooms of cyanobacteria, for example, rates of nitrogen fixation in aquatic habitats usually are high, reflecting population fluctuations, as well as availability of needed energy and mineral nutrients for fixation of molecular nitrogen.

Cycling of Sulfur

The element sulfur can exist in a variety of oxidation states within organic and inorganic compounds; microorganisms catalyze the oxidation and reduction of different forms of sulfur, establishing a sulfur cycle (Starkey, 1964; Pomeroy, 1974; Bremmer and Steele, 1978). Sulfur is an essential component of living systems because it is contained in certain amino acids in the form of sulfhydryl (-SH) groups. The sulfhydryl groups are often the reactive sites of enzymes. Sulfur is also an essential component of various coenzymes. Inorganic sulfur-containing compounds serve as sources of sulfur for plants and other macroorganisms, and some are also important because of their toxicity to biological systems (hydrogen sulfide, sulfur dioxide, and sulfuric acid).

Microorganisms perform most of the reactions that cycle sulfur through ecosystems (Kuenen and Tuovinen, 1981; Pfennig *et al.*, 1981; La Riviere and Schmidt, 1981). Organic sulfur-containing compounds can be desulfurized by a variety of microorganisms. Under aerobic conditions, the final inorganic sulfur-containing compound produced from the decomposition of organic sulfur compounds normally is sulfate. Under anaerobic conditions, the final sulfur-containing product normally is hydrogen sulfide. A variety of mercaptans are also formed during the anaerobic decomposition of organic sulfur-containing compounds. Hydrogen sulfide is a highly reactive compound subject to biological as well as nonbiological oxidation. In the presence of oxygen, hydrogen sulfide can be nonbiologically transformed to elemental sulfur and thiosulfate. Biologically, hydrogen sulfide can be oxidized to sulfur and sulfate under aerobic, as well as anaerobic, conditions. Under anaerobic conditions, hydrogen sulfide can serve as an electron donor for bacterial metabolism, and as a result is oxidized to elemental sulfur.

Other Cycling Activities

In addition to their major roles in the cycling of carbon, nitrogen, and sulfur, microorganisms are involved in the biogeochemical cycling of various other elements including oxygen, hydrogen, iron, calcium, silicon, phosphorus, and mercury (Silverman and Ehrlich, 1964; Bolin, 1970; Cloud and Gibor, 1970; Deevey, 1970; Pomeroy, 1974; Cosgrove, 1977; Ridley *et al.*, 1977; Summers and Silver, 1978; National Academy of Sciences, 1978; Atlas and Bartha, 1981). The nature of these microbial activities alters the chemical forms of these elements, affecting their fluxes through ecosystems. Virtually all microbial transformations of inorganic and organic compounds affect the fluxes of elements through ecosystems. Some microbial transformations increase the rate of flow of materials through an ecosystem, sometimes mediating the transfer of materials from one ecosystem to another; other transformations immobilize or retain substances within a particular ecosystem. The overall consequence of microbial metabolism is the cycling of elements among the biological populations of the ecosystem and their abiotic surroundings.

The mobilization of minerals such as phosphate and iron, the oxidation-reduction transformations of nitrogen and sulfur-containing compounds, and the production of carbon dioxide are critical factors in determining productivity. The productivity of many lakes is limited by the availability of mineral nutrients in forms that can be utilized by primary producers, particularly during seasons in which the intensity of light is adequate to support photosynthesis. The relationship between primary productivity and the availability of mineral nutrients means that the extent of primary production is as dependent on the non-phototrophic activities of microorganisms as is it upon the abilities of trees and photosynthetic microorganisms to convert solar energy to ATP. Clearly, the principle that materials cycle within ecosystems is valid, and microorganisms are essential mediators of the cycling processes.

INTERACTIONS OF POPULATIONS WITH THEIR ABIOTIC SURROUNDINGS

Factors Controlling Populations

The transformations of elements by macro- and microorganisms reflect the fact that living organisms require chemical elements in distinct proportions. There are two ecological principles of prime importance that describe the interactions of biological systems and their abiotic surroundings: Liebig's *law of the minimum* (Liebig, 1841), and Shelford's *law of tolerance* (Shelford, 1911). Liebig's law states that the total yield or biomass of any organism will be determined by the nutrient that is present in the lowest (minimum) concentration in relation to the requirements of the given organism. For example, crop yield cannot be increased by addition of excess phosphorus if the crop is suffering from a shortage of nitrogen. This law of the minimum applies to microorganisms just as it applies to plants and animals. In a given ecosystem, the growth of one microbial population may be limited by concentrations of available phosphorus and, within the same ecosystem, growth of another microbial population may be limited by concentrations of available nitrogen. An increase in availability of the particular nutrient limiting a microbial population allows that population to grow and reproduce until another factor becomes limiting.

Shelford's law of tolerance is broader in concept than Liebig's law, stating that for an organism to survive and grow, certain conditions must remain within the tolerance ranges of that given organism. If any condition exceeds the minimum or maximum tolerance of the organism, the organism will fail to thrive and may be eliminated. This principle applies to microorganisms as well as plants and animals, although some microorganisms have adaptations that permit them to exist in extreme natural ecosystems. (Alexander, 1971; Friedman and Galum, 1974; Gray and Postgate, 1976; Baross and Morita, 1978; Ehrlich, 1978; Kushner, 1978; Brock, 1978; Smith, 1982). Nevertheless, neither psychrophilic bacteria nor polar bears will fare well in a tropical rain-forest ecosystem.

E. P. Odum (1971) suggested that a combined concept of Liebig's and Shelford's laws is ecologically most meaningful. Odum's combined law states that the presence and success of an organism or a group of organisms depends upon a complex set of conditions; any condition which approaches or exceeds the limits of tolerance represents the limiting condition or limiting factor. Most organisms are controlled in an ecosystem by the quantity and variability of materials for which there is a minimum requirement, by physical factors that are critical, and by the limits of tolerance of the organisms themselves to these and other components of the environment.

It should be recognized, however, that because of their extremely small size, microorganisms can live in microenvironments. Most measures of environmental determinants fail to consider the existence of microhabitats. For microorganisms, the abiotic surroundings within their microhabitats are what is important. Thus, while the principles of limiting factors and tolerance ranges apply to microorganisms, the scale of interaction is sufficiently different to force consideration of the properties of the micro- rather than the macro-habitat when considering environmental constraints on microbial populations within an ecosystem.

This consideration is important relative to the applicability of Allee's principle to microorganisms. According to this principle, the degree of aggregation, as well as overall density, that results in optimal population growth and survival varies with species and composition—undercrowding (lack of aggregation) as well as overcrowding (excessive aggregation) may be limiting (Allee, 1931; Allee *et al.*, 1949). Clumps of plants and herds of animals reflect both the cooperative and competitive aspects of intrapopulation competition. Too many members of the group overexploit the food and energy resources, but having an appropriate-sized group enhances overall group survival. In macroecology, for example, colonial birds do not successfully reproduce if the population becomes too small, schools of fish are more tolerant to toxins than individuals of the same species, and groups of plants withstand wind stress better than individual plants. However, the benefits of aggregation are outweighed if the group becomes too large; for example, too many plants in a group means that no individual plant receives adequate light and nutrients. Similarly, among microorganisms an aggregated group has better survival properties than individuals, as long as the size of the group is within certain bounds. The formation of microcolonies is probably an adaptation based upon mutual protection or assistance within the population; even motile bacteria that could move away from each other often remain as aggregates. As in the case of fish, aggregation may protect individuals within the group from the adverse effects of toxicants. Dense (aggregated) populations provide a means of closing the system so that materials lost from one cell can be captured by neighboring microorganisms. Aggregation may be advantageous at low nutrient concentrations where the efficiency of substrate acquisition is especially important; many bacteria growing at low nutrient concentrations form rosettes or other relatively small groups (Schmidt, 1981). Cooperative aggregations among microorganisms are probably particularly important when the available substrate is insoluble and exoenzymes are used to convert it to nutrients that can be assimilated. In such cases, not only does aggregation increase the intensity of enzymic attack on the substrate, but the aggregate is also able to capture much of the nutrient resource that otherwise would be lost due to diffusion. For example, myxobacteria utilizing cellulose as a substrate occur in groups that migrate together; the aggregated mass of bacteria exploits the available resource more efficiently than individuals could (Reichenbach and Dworkin, 1981).

Adaptation in Microbial Populations

Like more complex organisms, microorganisms possess features that make them fit for survival in particular ecosystems, and their adaptive features contribute to both change and stability within biological communities. Adaptation in microbes involves changes in both structural and physiological features; it can also involve changes in reproductive or behavioral strategies (Sournia, 1974; Chet and Mitchell, 1976; Kushner, 1978; Wu, 1978). In comparison with higher organisms, where changes subject to natural selection occur rather slowly, the high reproductive rates of microorganisms allow relatively rapid change. Thus, the speed of adaptive change, like the range of tolerable environmental conditions, is greater for microorganisms than for populations of higher organisms.

Within ecosystems, populations have adapted various strategies for survival. Studies on population dynamics of plants and animals have produced a classification system of r or K strategists (Pianka, 1970). Populations that exhibit r strategies have high reproductive rates, but few other competitive adaptations. They tend to prevail in situations that are not resource limited, and in which high reproductive rates outweigh the advantages of other competitive adaptations. K populations depend on the carrying capacity of the environment. Populations with K strategies reproduce more slowly, but have many competitive adaptations. They tend to be successful in resource-limited situations. Populations of r strategists are subject to extreme fluctuations; populations of K strategists tend to be more stable.

This concept of r and K strategist can be meaningfully applied to microbial populations (Atlas and Bartha, 1981; Andrews, 1984). An r strategist microorganism is one that, through rapid growth rates, takes over and dominates situations in which resources are temporarily abundant, e.g., an algal bloom in a lake after addition of phosphate. Winogradsky's "zymogenous" (opportunistic) soil populations (Winogradsky, 1925) closely correspond to the concept of an r strategist. Most microorganisms may appear to be r strategists because of their rapid reproductive capacity. However, other microbial populations, such as Winogradsky's "autochthonous" (humus-degrading) populations, correspond to the concept of K strategists; similarly, oligotrophic bacteria would be examples of K strategists. These organisms grow slowly and continuously at rates that are dictated by the limiting nutrient supply of the ecosystem. Thus, microorganisms exhibit the same basic adaptive strategies as higher organisms for utilizing the available resources of the ecosystem.

The $r–K$ strategy does not, however, fully accommodate the gradient of stress-related adaptations exhibited by living organisms. More recent ecological theory has emphasized disturbance in addition to stress as a major factor in determining the ecological success of a population within an ecosystem. This view leads to the recognition of three kinds of species: opportunistic, stress-tolerant, and biotically tolerant (Van Valen, 1971; Grime, 1977; Vermeij, 1978). The biotically competent organisms are somewhat equivalent to the K strategists and the opportunistic organisms the r strategists. Low stress-low disturbance favors biotically tolerant (competitive) organisms; high stress-low disturbance favors stress-tolerant organisms; and low stress-high disturbance favors opportunistic organisms. These three strategies appear to apply to plants, animals, and microorganisms, and to represent a more meaningful classification system than the one-dimensional $r–K$ system.

The applicability of this three-strategy system to fungi has been considered by Grime (1977). Among the fungi, opportunistic strategies are characteristic of the Mucorales, which frequently are ephemeral colonists of organic substrates; initially, mycelial growth is rapid as the rich supply of carbohydrates is utilized, but as the readily useable substrates diminish mycelial growth ceases and sporulation becomes profuse. Many *Pseudomonas* and certain other bacterial species that can exploit temporarily available abundant resources likewise are opportunistic colonizers. The stress tolerant strategy is apparent among the symbiotic fungi of lichens and mycorrhizas and the free living basidiomycetes that slowly but persistently produce mycelia under environmental conditions that are growth limiting. A similar stress tolerant strategy appears to be applicable to the oligotrophic bacteria of aquatic habitats that grow slowly under nutrient limiting conditions. The competitive strategy is apparent in *Armillaria mellea,* a fungus that successfully inhabits timber for extended periods, producing a consolidated mycelium that can extend rapidly through the production of rhizomorphs. Many of the bacterial populations found within the digestive tracts of animals similarly must exhibit a competitive strategy. The introduction of life cycles adds another dimension of complexity to the view of competitive strategies; for example, fungi characteristically exhibit life cycles and may appear to commit their resources to reproduction during one phase of the life cycle, but to growth-maintenance during another phase (Swift, 1976). Thus, it is overly simplistic to simply view microorganisms as being r or K strategists.

INTERACTIONS BETWEEN DIVERSE POPULATIONS

In addition to interacting with abiotic determinants, the biological populations of an ecosystem interact with each other. The interactions between two different populations can be classified according to whether both populations are unaffected by the interaction, one or both populations benefit from the interactions, or one or both populations are negatively affected by the interaction (Brock, 1966; Hazen, 1970; Alexander, 1971; E. P. Odum, 1971; Krebs, 1972; Colinvaux, 1973; Whittaker, 1975; Atlas and Bartha, 1981). The categories used to describe these interactions constitute an artificial classification system; many specific cases can arbitrarily be placed into one or another category. Possible interactions between microbial populations can be recognized as negative interactions (competition and amensalism); positive interactions (commensalism, synergism, and mutualism); and interactions that are positive for one, but negative for the other population (parasitism and predation). Within a complex natural biological community, it is likely that all of these possible interactions will occur between different populations. The negative interactions between populations act as feedback mechanisms that limit population densities. Positive interactions enhance the abilities of some populations to survive as a community within a particular habitat. In established communities, positive interactions among autochthonous populations are likely to be more developed than in newly established communities, but invaders of established communities are likely to encounter severe negative interactions with autochthonous populations.

The development of positive interactions in some cases permits microorganisms to use available resources more efficiently and/or to occupy habitats that otherwise could not be inhabited. Mutualistic relationships between microbial populations create essentially new organism-systems capable of occupying niches that could not be occupied by either organism alone (Henry, 1966; Society for Experimental Biology, 1975). Positive interactions between microbial populations are based on combined metabolic capabilities and/or combined physiologic capabilities that enhance growth and/or survival rates. Interactions between microbial populations tend to dampen environmental stress. The negative feedback interactions that limit population densities are a self-regulation mechanism that is beneficial to the overall population in the long term by preventing overpopulation (and consequently intensified competition), destruction of the habitat and/or self-destruction (Mitchell, 1971; E. P. Odum, 1971). These interactions between populations are a driving force in the evolution of community structure (MacArthur, 1955; Paine, 1966; Margalef, 1967; Saunders, 1968; Woodwell and Smith, 1969; Whittaker, 1972), and they contribute to the stability of the community (May, 1971, 1973; Holling, 1973; Maynard Smith, 1974).

Competitive Exclusion

A consequence of the environmental limitations expressed by Liebig's and Shelford's Laws is that organisms compete for the resources available to the community. Competition, in its broad sense, includes acquisition of a limiting nutrient, amensalism (production of antagonistic substances), predation, and parasitism. Within any ecosystem, a limited number of niches can be occupied by populations (subsets of species) present within the community, and assuming that an equilibrium state is ever achieved only one population can occupy each niche; competition leads to the exclusion of other populations. This "principle of competitive exclusion" applies to microorganisms as well as to macroorganisms (Hardin, 1960; DeBach, 1966; MacArthur, 1968; Whittaker *et al.*, 1973). Among microorganisms, the winner of direct competition for a limiting nutrient may depend strictly on relative reproductive rates; in chemostats, at least, the population with the highest growth rate becomes established, while others are eliminated (Veldkamp, 1977; Slater, 1980). In natural ecosystems, elimination of particular populations can also be observed (Preston, 1962a,b; MacArthur and Wilson, 1963, 1967; Whittaker, 1975; May, 1976).

There are, however, cases in which the principle fails, or appears to fail (Fredrickson and Stephanopoulos, 1981), forcing further examination of the concept of niche and the idea that communities are at equilibrium (Huston, 1979). For example, phytoplankton seem to present a paradox regarding this principle (Gause, 1934; Hutchinson, 1959, 1961; Patten, 1961). The vast diversity of phytoplankton observed in many aquatic environments appears to present a contradiction to the principle because all phytoplankton compete for the same basic resources; since the euphotic zones of most natural waters are relatively homogeneous, such coexisting plankters may be simultaneously occupying the same niche. Several hypotheses have been proposed to explain this paradox, including contemporaneous disequilibrium (Richerson *et al.*, 1970), i.e., that patches of diverse and monospecific plankton assemblages exist at the same time (contemporaneously), but are spatially separated in the same body of water; also, that physical turbulence in an aquatic system can moderate pressures between plankton populations and permit the coexistence of species competing for the same resources (Kemp and Mitsch, 1979). It appears that such occurrences represent niche sharing based on temporal or spatial separation, a concept that applies broadly to microbial populations, which by their very nature can occupy microhabitats.

By separating populations in time and space, the problem of complete competitive exclusion is overcome, and populations with similar ecological functions can coexist in an ecosystem (Pittendrigh, 1961; Whittaker, 1965; May and MacArthur, 1972; Pianka, 1974). By the same mechanism, microorganisms can coexist if they have differing microhabitats in which they find refuge (Roper and Marshall, 1978). In cases where competitive exclusion does not occur, there may be regular population fluctuations among competing, predator–prey, or host–parasite populations (Utida, 1967; Krebs, 1972; May, 1972, 1973; Luckinbill, 1973; Van den Ende, 1973; Van Gemerden, 1974; Whittaker, 1975); the classical model systems used to demonstrate these regular population oscillations employ protozoan populations (Gause, 1934). The occurrence of regularly timed population fluctuations raises the concept of a temporal niche; microorganisms may occupy a niche in a habitat at one time, but not at another. The existence of temporal niches diminishes competition between populations in some cases, and temporal niches probably allow the coexistence of some populations within the same spatial habitat. In some cases, protozoan populations feeding on the same bacterial populations coexist because one protozoan grows faster than the other in one range of prey resource density and the opposite occurs in another range of resource density (Baltzis and Fredrickson, 1984).

Perhaps a more fundamental concept is that communities are rarely at equilibrium and therefore that populations coexist because they do not, in fact, compete for niches within the ecosystem (Lewin, 1983a,b; Wiens, 1983). The question of equilibrium versus nonequilibrium is one that concerns the central role of competition, as opposed to the role of predation and disturbance, on the structural composition of the community. Clearly, disturbances such as volcanoes and hurricanes and grazing by predators reduce the extent of competition that can lead to displacement from a niche. According to Huston (1979), the diversity of communities can be explained when there is a state of nonequilibrium and when competitive exclusion is prevented by periodic population reductions and environmental fluctuations, by assuming the establishment of a dynamic balance between the rate of competitive exclusion and the frequency of population reduction. This same model may explain the apparent stability of the assemblages of diverse microorganisms in apparent contradiction of the principle of competitive exclusion. Central to this explanation is the view that competition, productivity, and predation act together as an interactive mechanism for regulating community structure.

Cooperative Relationships

In contrast to competitive exclusion, cooperation among biological populations with different functional capacities leads to the development of communities containing multiple populations (Henry, 1966; Margulis, 1971; Society for Experimental Biology, 1975). Cooperation may involve loose relationships, such as commensalism and protocooperation, or obligate mutualistic relationships. In many cases, organisms coexist within a community only through their cooperative relationships. For example, lichens, the result of the mutualistic relationship between fungi and photosynthetic microbial partners, grow in dry habitats where other organisms cannot survive (Ahmadjian, 1967; Hale, 1969). Other consortia of cooperating microorganisms similarly occupy the niches of ecosystems that could not be filled by individual populations alone; for example, methanogens depend on non-methane-producing bacteria to provide low-molecular-weight fatty acid substrates and to reduce the oxygen tension (Zeikus, 1977).

The cooperation of microorganisms with plants is important to the growth of many plants as well as to the establishment of habitats on the plant surfaces for the growth of microorganisms; the establishment of myccorrhizae, the growth of bacteria in the rhizosphere, and the nitrogen-fixing symbioses are all beneficial for both plants and microbes (Katznelson, 1965; Rovira, 1965; Lange, 1966; Meyer, 1966; Gray and Parkinson, 1968; Hartley, 1968; Lynch,

1976; Cooke, 1977; Balandreau and Knowles, 1978). Similarly, the cooperative interactions between animal and microbial populations are mutually beneficial, the microbial contribution to animal digestion that occurs in the fungal gardens of ants in the rumen symbioses, and the symbiotic production of light in the flashlight fish (Buchner, 1965; Hungate, 1966, 1975; Batra and Batra, 1967; Nealson and Hastings, 1972; Weber, 1979) are examples. Interpopulation interactions involving microorganisms (microbe–microbe; microbe–plant; microbe–animal), thus operate on the same bases as those between animal and plant populations. Overall, these interactions act to regulate levels of biological populations.

COMMUNITY STRUCTURE

Diversity, Succession, and Stability

Community Diversity

Communities, including those of microbes, evolve by-passing through a series of successional stages until a stable community structure is achieved at a certain level of diversity assuming that the process is not interrupted by disturbance (Clements, 1936; Blum, 1956; Margalef, 1963; Cooke, 1967; E. P. Odum, 1969; Whittaker, 1975). It is widely accepted that species diversity is a community parameter that relates the stability of that community, and that stable biological communities maintain a minimum level of diversity. Communities of very low or very high diversity are regarded as unstable and subject to repeated or catastrophic change.

In the study of ecology, the term diversity is used to describe the assemblage of species within a community, and it is in this restricted sense that ecologists synonomously describe ecological and species diversity (Shannon and Weaver, 1949; Pielou, 1966a,b, 1975; Margalef, 1968, 1979; May, 1976; Peet, 1974; Whittaker, 1975). Species diversity is a measure of entropy (disorder or randomness) of the community; an index of diversity measures the degree of uncertainty that an individual picked at random from a multispecies assemblage will belong to a particular species (Legendre and Legendre, 1982). The greater the heterogeneity of the assemblage of populations and of individuals within those populations, the greater the diversity of the community; thus, diversity is a measure of the species composition of an ecosystem in terms of the number, distribution, and relative abundances of the species. The interpretation of the meaning of diversity is often difficult because of the problems inherent in the measures (indices) used to describe diversity (Pielou, 1966a,b, 1969, 1975; Woodwell and Smith, 1969; Hurlbert, 1971; Peet, 1974; Margalef, 1979; Atlas, 1983). Although MacArthur (1955) showed that theoretically the greater the diversity, the greater the stability of the community, a direct cause-and-effect relationship between diversity and stability has not yet been established.

In natural communities, changes in diversity are related to changes in the physiologic functions of the populations within that community; in marine ecosystems, for example, the diversity of the bacterial populations is closely related to the available organic substrates and therefore to phytoplankton primary productivity (Kaneko et al., 1977; Martin and Bianchi, 1980; Hauxhurst et al., 1981). Swift (1976) proposed that the distribution of decomposer fungi on leaves supports the classical concepts of island community structure put forth by MacArthur and Wilson (1963, 1967), with the number of species (as an index of diversity) being directly proportional to the area available for colonization; in this special case, species diversity declines as the area is consumed by the decomposers.

The population (species) interactions that lead to the establishment of a defined community of stable diversity are assumed to be based on various physiological interactions. The functional roles (niches) of specific populations within the communities of certain ecosystems have been defined, e.g., for the rumen ecosystem (Hungate, 1975); the bases for interspecific population relationships within such communities also have been defined, and for this system, there is now a relatively complete understanding of community structure and ecosystem function. Work using chemostats has elucidated some of the interactions between microbial populations that lead to the establishment of a stable community structure in an aquatic ecosystem (Slater, 1978, 1979, 1980; Slater and Goodwin, 1980). In chemostat studies, it is often found that stability is achieved when several interacting populations cooperate to best exploit the available resources. In some cases, just two species can constitute a stable community structure, whereas in other experiments, more member populations are needed before the community stabilizes. Hairston et al. (1968) attempted to demonstrate experimentally that higher diversity brings about greater stability using a model system of several bacterial and protozoan populations; they found that increasing the diversity of the bacterial populations enhanced community stability (measured as the persistence of all populations and the evenness of distribution of species abundances within the community), but that two species of Paramecium were more stable than three species; they concluded that more information was needed to establish a functional relationship between stability and diversity. Stability of predatory protozoan populations depends on the specific prey species and not just on the diversity of the species present (Luckinbill, 1979).

Successional Changes

Diversity is a dynamic parameter and changes with successional stages in a community. Theoretically, diversity should increase from the low diversity of the pioneer populations to the higher and stable diversity of the climax community. Using submerged grids and electron microscopic observations, Jordan and Staley (1976) observed this increase in diversity for the periphyton community of Lake Washington. The diversity, measured with the Shannon index, increased during a 10-day period; H' increased from 3.1 at day 1 to 4.2 at day 3 and continued to rise until it reached a maximum value of 4.8 on day 10. Bacteria were the dominant components during the pioneer stage. As the succession proceeded, some pioneer populations disappeared, and the relative proportions of biomass shifted from heterotrophic eubacteria to algae and cyanobacteria.

Eventually, successional changes lead to the establishment of a stable climax community. However, it is often difficult to apply the concept of climax to microbial communities because the short generation times of many microorganisms lead to large population fluctuations, and seemingly minor environmental changes often preclude orderly succession. Initial random events may determine which microorganisms fill the niches of the ecosystem and consequently determine the sequence of successional events that follow. Succession to a climax microbial community does occur in some habitats under certain conditions and, for example, climax algal communities, analogous to climax plant communities, have been recognized (Blum, 1956). Sometimes, the occurrence of regular successional population changes of microorganisms leads to a relatively stable microbial community (Patrick, 1963, 1976; Kaneko et al., 1977). These situations are equivalent, if not identical, to the development of a climax community. In these cases, the successional process follows a repeatable and predictable sequence of population changes; the causative factors responsible for the orderly sequence, however, are often unknown.

Successional processes may be autotrophic or heterotrophic. Autotrophic succession occurs when there is a nonlimiting supply of solar energy. In heterotrophic succession, the energy flow through the system decreases with time; the input of organic matter is insufficient, and the community gradually dissipates the stored energy. Heterotrophic succession occurs in cases of organic pollution. Heterotrophic succession is always temporary because it culminates in disappearance of the community when the stored energy supply is exhausted.

The climax community that results from successional processes tends to maintain steady state conditions. Homeostasis of the climax community is equated with ecological balance and community stability. Changes are damped by the abilities of the indigenous populations to tolerate environmental fluctuations. Except for catastrophic events, the climax community withstands normal environmental disturbances. When acute pollution events disrupt community structure, diversity is altered (Mills and Wassel, 1980; Atlas, 1983); however, after removal of the ecological stress, diversity returns to normal. Homeostasis effectively acts to prevent invasion by foreign (allochthonous) populations and to ensure the persistence of the autochthonous community; i.e., stable communities retain their diversity. Thus, diversification of function (diversity of populations) and homeostasis (community stability) are reflections of the driving ecological principles that establish the characteristics of the climax communities of ecosystems.

SUMMARY

The principles of ecology apply to all biological members of ecosystems, including microorganisms. These principles explain the flow of energy, cycling of elements, interactions of populations with their abiotic surroundings, and interactions among the diverse biological members of the community that lead to the dynamic functioning and stability of ecosystems. Microorganisms are critical components of ecosystems; within ecosystems, microorganisms have essential functions in the flow of energy and cycling of elements. In some cases, new methodological approaches that take into account the fact that microorganisms live in microhabitats are required to establish ecological relationships involving microbial populations. Studies employing microorganisms have helped extend our understanding of ecosystems and undoubtedly will continue to do so. In recent years, the finding of primary production by chemolithotrophic bacteria in the deep sea thermal vents regions of the oceans has expanded our concept of energy flow through ecosystems; biochemical-ecological studies have expanded the recognition of the importance of microbial biogeochemical cycling of elements in the biosphere; and microbiological studies have helped examine the fundamental ecological questions concerning the functional relationship between ecological diversity and stability.

REFERENCES

Ahmadjian, V., 1967, *The Lichen Symbiosis*, Blaisdell, Waltham, Massachusetts.

Aleem, M. I. H., 1970, Oxidation of inorganic nitrogen compounds, *Annu. Rev. Plant Physiol.* **21**:67–90.

Alexander, M., 1965, Biodegradation: Problems of molecular racalcitrance and microbial fallibility, *Adv. Appl. Microbiol.* **7**:35–80.

Alexander, M., 1971, *Microbial Ecology*, Wiley, New York.

Alexander, M., 1977, *Introduction to Soil Microbiology*, Wiley, New York.

Alexander, M., 1981, Biodegradation of chemicals of environmental concern, *Science* **211**:132–138.

Allee, W. C., 1931, *Animal Aggregations: A Study in General Sociology*, University of Chicago Press, Chicago.

Allee, W. C., Emerson, A. E., Park, O., Park, T., and Schmidt, K. P., 1949, *Principles of Animal Ecology*, W. B. Saunders, Philadelphia.

Ander, P., and Eriksson, E-K., 1978, Lignin degradation and utilization by microorganisms, *Prog. Ind. Microbiol.* **14**:1–58.

Anderson, R. L., and Wood, W. A., 1969, Carbohydrate metabolism in microorganisms, *Annu. Rev. Microbiol.* **33**:539–575.

Andrews, J. H., 1984, Relevance of *r*- and *K*- theory to the ecology of plant pathogens, in: *Current Perspectives in Microbial Ecology* (M. J. Klug and C. A. Reddy, eds.), pp. 1–7, American Society for Microbiology, Washington, D.C.

Atlas, R. M., 1983, Diversity of microbial communities, in: *Advances in Microbial Ecology*, Vol. 7 (K. C. Marshall, ed.), pp. 1–47, Plenum Press, New York.

Atlas, R. M., 1984, *Microbiology: Fundamentals and Applications*, Macmillan, New York.

Atlas, R. M., and Bartha, R., 1981, *Microbial Ecology: Fundamentals and Applications*, Addison Wesley, Reading, Massachusetts.

Balandreau, J., and Knowles, R., 1978, The rihizosphere, in: *Interactions Between Non-pathogenic Soil Microorganisms and Plants* (Y. R. Dommergues and S. V. Krupa, eds.), pp. 243–268, Elsevier, Amsterdam.

Ballard, R. D., 1977, Notes on a major oceanographic find, *Oceanus* **20:**35–44.

Baltzis, B. C., and Fredrickson, A. G., 1984, Competition of two suspension-feeding protozoan populations for a growing bacterial population, *Microb. Ecol.* **10:**61–68.

Barross, J. A., and Morita, R. Y., 1978, Microbial life at low temperatures, in: *Microbial Life in Extreme Environments* (D. J. Kushner, ed.) pp. 9–71, Academic Press, London.

Batra, S. W. T., and Batra, L. R., 1967, The fungus gardens of insects, *Sci. Am.* **217**(5):112–120.

Becking, J-H., 1981, The family Azotobacteriaceae, in: *The Prokaryotes* (M. P. Starr, H. Stolp, H. G. Truper, A. Ballows, and H. G. Schlegel, eds.), pp. 795–817, Springer-Verlag, Berlin.

Belser, L. W., 1979, Population ecology of nitrifying bacteria, *Annu. Rev. Microbiol.* **33:**309–333.

Belser, L. W., and Schmidt, E. L., 1978, Nitrification in soils, in: *Microbiology—1978* (D. Schlesinger, ed.), pp. 348–351, American Society for Microbiology, Washington, D.C.

Bennemann, J. R., and Valentine, R. C., 1972, The pathways of nitrogen fixation, *Adv. Microb. Physiol.* **8:**59–104.

Benson, D. R., 1985, Consumption of atmospheric nitrogen, in: *Bacteria in Nature*, Vol. 1 (E. R. Leadbetter and J. S. Poindexter, eds.), pp. 155–198, Plenum Press, New York.

Blum, J. L., 1956, The application of the climax community concept to algal communities of streams, *Ecology* **37:**603–604.

Bold, H. C., and Wynne, M. J., 1978, *Introduction to the Algae*, Prentice-Hall, Englewood Cliffs, New Jersey.

Bolin, B., 1970, The carbon cycle, *Sci. Am.* **223**(3):124–135.

Bolin, B., Degens, E. T., Kempe, S., and Klepner, P., 1980, *The Global Carbon Cycle*, Wiley, New York.

Bremmer, J. M., and Steele, C. G., 1978, Role of microorganisms in the atmospheric sulfur cycle, in: *Advances in Microbial Ecology*, Vol. 2 (M. Alexander, ed.), pp. 155–201, Plenum Press, New York.

Brill, W. J., 1975, Regulation and genetics of bacterial nitrogen fixation, *Annu. Rev. Microbiol.* **29:**109–129.

Brill, W. J., 1979, Nitrogen fixation: Basic to applied, *Am. Sci.* **67:**458–466.

Brock, T., 1966, *Principles of Microbial Ecology*, Prentice-Hall, Englewood Cliffs, New Jersey.

Brock, T. D., 1978, *Thermophilic Microorganisms and Life at High Temperatures*, Springer-Verlag, New York.

Brooks, J. L., and Dodson, S. I., 1965, Predation, body size, and composition of plankton, *Science* **150:**28–35.

Buchner, P., 1965, *Endosymbiosis of Animals with Plant Microorganisms*, Wiley, New York.

Burns, R. C., and Hardy, R. W. F., 1975, *Nitrogen Fixation in Bacteria and Higher Plants*, Springer-Verlag, New York.

Button, D. K., 1985, Kinetics of nutrient-limited transport and microbial growth, *Microb. Rev.* **49:**270–297.

Caldwell, D., 1977, The planktonic microflora of lakes, *Crit. Rev. Microbiol.* **5:**305–370.

Campbell, R., 1977, *Microbial Ecology*, Blackwell Scientific, Oxford, England.

Chet, I., and Mitchell, R., 1976, Ecological aspects of microbial chemotactic behavior, *Annu. Rev. Microbiol.* **31:**221–239.

Clements, F. E., 1936, Nature and structure of the climax, *J. Ecol.* **24:**252–284.

Cloud, P., and Gibor, A., 1970, The oxygen cycle, *Sci. Am.* **223**(3):110–123.

Cohen, Y., Jorgensen, B. B., Padan, E., and Shilo, M., 1975, Sulphide-dependent anoxygenic photosynthesis in the cyanobacterium *Oscillatoria limnetica*, *Nature (Lond.)* **257:**489–492.

Colby, J., Dalton, H., and Whittenbury, R., 1979, Biological and biochemical aspects of microbial growth on C1-compounds, *Annu. Rev. Microbiol.* **33:**481–517.

Colinvaux, P. A., 1973, *Introduction to Ecology*, Wiley, New York.

Cooke, G. D., 1967, The pattern of autotrophic succession in laboratory microecosystems, *BioScience* **17:**717–721.

Cooke, R., 1977, *Biology of Symbiotic Fungi*, Wiley, New York.

Corliss, J. B., Dymond, J., Gordon, L. I., Edmont, J. M., von Herzen, R. P., Ballard, R. D., Green, K., Williams, D., Brainbridge, A., Crane, K., and van Andel, T. H., 1979, Submarine thermal springs of the Galapagos Rift, *Science* **203:**1073–1083.

Cosgrove, D. J., 1977, Microbial transformations in the phosphorus cycle, in: *Advances in Microbial Ecology*, Vol. 1 (M. Alexander, ed.), pp. 95–134, Plenum Press, New York.

Cox, G. W., 1969, *Readings in Conservation Ecology*, Appleton-Century-Crofts, E. Norwalk, Connecticut.

Crawford, D. L., and Crawford, R. L., 1980, Microbial degradation of lignin, *Enzyme Microbiol. Technol.* **2:**11–22.

Crisp, D. J. (ed.), 1964, *Grazing in Terrestrial and Marine Environments*, Blackwell Scientific, Oxford, England.

Curds, C. R., 1977, Microbial interactions involving protozoa, in: *Aquatic Microbiology* (F. A. Skinner and J. M. Shewan, eds.), pp. 69–105, Applied Bacteriology Symposium series No. 6, Academic Press, London.

Dagley S., 1975, Microbial degradation of organic compounds in the biosphere, *Am. Sci.* **63:**681–689.

Dalton, H., 1974, Fixation of dinitrogen by free-living microorganisms, *CRC Crit. Rev. Microbiol.* **3:**183–220.

Dalton, H., and Mortenson, L. E., 1972, Dinitrogen (N_2) fixation (with as biochemical emphasis), *Bacteriol. Rev.* **36:**231–260.

Dawes, I., and Sutherland, I. W., 1976, *Microbial Physiology*, Blackwell Scientific, Oxford, England.

DeBach, P., 1966, The competitive displacement and coexistence principles, *Annu. Rev. Ent.* **11:**183–212.

Deevey, E. S., 1970, Mineral cycles, *Sci. Am.* **223**(3):148–159.

Delwiche, C. C., 1970, The nitrogen cycle, *Sci. Am.* **223**(3):137–146.

Delwiche, C. C., and Bryan, B. A., 1976, Denitrification, *Annu. Rev. Microbiol.* **30:**241–262.

Edmond, J. M., and von Damm, K., 1983, Hot springs on the ocean floor, *Sci. Am.* **248**(4):78–93.

Edmond, J. M., von Damm, K. L., McDuff, R. E., and Measures, C. I., 1982, Chemistry of hot springs on the East Pacific Rise and their effluent dispersal, *Nature (Lond.)* **297:**187–191.

Ehrlich, H. L., 1978, How microbes cope with heavy metals, arsenic and antimony in their environment, in: *Microbial Life in Extreme Environments* (D. J. Kushner, ed.), pp. 381–408, Academic Press, London.

Eriksson, K., and Johnsrud, S. C., 1982, Mineralisation of carbon, in:*Experimental Microbial Ecology* (R. G. Burns and J. H. Slaters, eds.), pp. 239–252, Blackwell Scientific, Oxford, England.

Evans, F. C., 1956, Ecosystem as the basic unit in ecology, *Science* **123:**1127–1128.

Evans, N. C., 1977, Biochemistry of the bacterial catabolism of aromatic compounds in anaerobic environments, *Nature (Lond.)* **270:**17–22.

Fenchel, T. M., 1978, The ecology of micro- and meiobenthos, *Annual Review of Ecology and Systematics* **9:**99–121.

Fenchel, T. M., and Jorgensen, B. B., 1977, Detritus food chains of aquatic ecosystems: The role of bacteria, in: *Advances in Microbial Ecology*, Vol. 1 (M. Alexander, ed.), pp. 1–58, Plenum Press, New York.

Focht, D. D., and Verstraete, W., 1977, Biochemical ecology of nitrification and denitrification, in: *Advances in Microbial Ecology*, Vol. 1 (M. Alexander, ed.), pp. 135–214, Plenum Press, New York.

Fredrickson, A. G., and Stephanopoulos, G., 1981, Microbial competition, *Science* **213:**972–979.

Friedman, E. I., and Galum, M., 1974, Desert algae, lichens, and fungi, in: *Desert Biology* (G. W. Brown, ed.), pp. 165–212, Academic Press, New York.

Garlick, S., Oren, A., and Padan, E., 1977, Occurrence of facultative anoxygenic photosynthesis among filamentous and unicellular cyanobacteria, *J. Bacteriol.* **129:**623–629.

Gause, G. F., 1934, *The Struggle of Existence,* Williams & Wilkins, Baltimore.

Gibson, A. H. (ed.), 1977, *A Treatise of Dinitrogen Fixation,* Section IV: *Agronomy and Ecology,* Wiley, New York.

Goldberg, E. D., McCave, I. N., O'Brien, J. J., and Steele, J. H., 1974, *The Sea,* Wiley, New York.

Goldman, C. R. (ed.), 1966, *Primary Productivity in Aquatic Ecosystems,* University of California Press, Berkeley.

Golley, F. B., 1972, Energy flux in ecosystems, in: *Ecosystem Structure and Function,* Vol. 31 (J. A. Wiens, ed.), pp. 69–90, Oregon State University Annual Biology Colloquia, Oregon State University Press, Corvallis, Oregon.

Gordon, J. K., 1981, Introduction to the nitrogen-fixing prokaryotes, in: *The Prokaryotes* (M. P. Starr, H. Stolp, H. G. Truper, A. Ballows, and H. G. Schlegel, eds.), pp. 781–784, Springer-Verlag, Berlin.

Govindjee, 1975, *Bioenergetics of Photosynthesis,* Academic Press, New York.

Govindjee, and Govindjee, R., 1974, The absorption of light in photosynthesis, *Sci. Am.* **231**(6):68–82.

Gray, T. R. G., and Parkinson, D. (eds.), 1968, *The Ecology of Soil Bacteria,* University of Toronto Press, Toronto.

Gray, T. R. G., and Postgate, J. R., 1976, *The Survival of Vegetative Microbes: Symposium of the Society of General Microbiology,* Vol. 26, Cambridge University Press, Cambridge.

Grime, J. P., 1977, Evidence for the existence of three primary strategies in plants and its relevance to ecological and evolutionary theory, *Am. Nat.* **111**:1169–1194.

Hairston, N. G., Allan, J. D., Colwell, R. K., Futuyama, D. J., Howell, J. Lubin, M. D., Mathias, J., and Vandermeer, J. J., 1968, The relationship between species diversity and stability: An experimental approach with protozoa and bacteria, *Ecology* **49**:1091–1101.

Hale, M. E., 1969, *The Lichens,* W. C. Brown, Dubuque, Iowa.

Hall, J. B., 1978, Nitrate-reducing bacteria, in: *Microbiology—1978* (D. Schlesinger, ed.), pp. 348–351, American Society for Microbiology, Washington, D. C.

Hardin, G., 1960, The competitive exclusion principle, *Science* **131**:1292–1297.

Hartley, J. L., 1968, Mycorrhiza, in: *The Fungi,* Vol. III: *The Fungal Population* (G. C. Ainsworth and A. S. Sussman, eds.), pp. 139–178, Academic Press, New York.

Hauxhurst, J. D., Kaneko, T., and Atlas, R. M., 1981, Characteristics of bacterial communities in the Gulf of Alaska USA, *Microb. Ecol.* **7**:167–182.

Hayes, A. J., 1979, The microbiology of plant litter decomposition, *Sci. Prog.* **66**:25–42.

Hazen, W. E. (ed.), 1970, *Readings in Population and Community Ecology,* W. B. Saunders, Philadelphia.

Helmut, L., and Whittaker, R. H. (eds.), 1975, *The Primary Production of the Biosphere,* Springer-Verlag, New York.

Henry, S. M. (ed.), 1966, *Symbiosis,* Academic Press, New York.

Hirsch, P., 1968, Photosynthetic bacterium growing under carbon monoxide, *Nature (Lond.)* **217**:555–556.

Holling, C. W., 1973, Resilience and stability of ecological systems, *Annu. Rev. Ecol. Syst.* **4**:1–23.

Horvath, R. S., 1972, Microbial co-metabolism and the degradation of organic compounds in nature, *Bacteriol. Rev.* **36**:146–155.

Hungate, R. E., 1966, *The Rumen and Its Microbes,* Academic Press, New York.

Hungate, R. E., 1975, The rumen microbial ecosystem, *Annu. Rev. Microbiol.* **29**:39–66.

Hurlbert, S. H., 1971, The nonconcept of species diversity: A critique and alternative parameters, *Ecology* **52**:577–586.

Huston, M., 1979, A general hypothesis of species diversity, *Am. Nat.* **113**:81–101.

Hutchinson, G. E., 1957, *A Treatise on Limnology,* Wiley, New York.

Hutchinson, G. E., 1959, Homage to Santa Rosalia, or Why are there so many kinds of animals?: *Am. Nat.* **93**:145–159.

Hutchinson, G. E., 1961, The paradox of the plankton, *Am. Nat.* **95**:137–145.

Hutchinson, G. E., 1970, The biosphere, *Sci. Am.* **223**(3):44–53.

Imshenetsky, A. A., 1967, Decomposition of cellulose in the soil, in: *The Ecology of Soil Bacteria* (T. R. G. Gray and D. Parkinson, eds.), pp. 256–296, Liverpool University Press, Liverpool, England.

Jannasch, H. W., 1967, Growth of marine bacteria at limiting concentrations of organic carbon in seawater, *Limnol. Oceanogr.* **12**:264–271.

Jannasch, H. W., 1979, Microbial turnover of organic matter in the deep sea, *BioScience* **29**:228–232.

Jannasch, H. W., and Mottl, M. J., 1985, Geomicrobiology of deep-sea hydrothermal vents, *Science* **229**:717–725.

Jannasch, H. W., and Wirsen, C. O., 1977, Microbial life in the deep sea, *Sci. Am.* **236**:42–52.

Jannasch, H. W., and Wirsen, C. O., 1979, Chemosynthetic primary production at East Pacific sea floor spreading centers, *BioScience* **29**:592–598.

Jannasch, H. W., and Wirsen, C. O., 1981, Morphological survey of microbial mats near deep-sea thermal vents, *Appl. Environ. Microbiol.* **41**:528–538.

Jeter, R. M., and Ingraham, J. L., 1981, The denitifying prokaryotes, in: *The Prokaryotes* (M. P. Starr, H. Stolp, H. G. Truper, A. Ballows, and H. G. Schlegel, eds.), pp. 913–925, Springer-Verlag, Berlin.

Jordan, T. L., and Staley, J. T., 1976, Electron microscopic study of succession in the periphyton community of Lake Washington, *Microb. Ecol.* **2**:241–276.

Juday, C., 1940, The annual energy budget of an inland lake, *Ecology* **21**:438–450.

Kaneko, T., Atlas, R. M., and Krichevsky, M., 1977, Diversity of bacterial populations in the Beaufort Sea, *Nature (Lond.)* **270**:596–599.

Karl, D. M., Wirsen, C. O., and Jannasch, H. W., 1980, Deep-sea primary production at the Galapagos hydrothermal vents, *Science* **207**:1345–1347.

Katznelson, H., 1965, Nature and importance of the rhizosphere, in: *Ecology of Soil-borne Plant Pathogens* (K. F. Baker and W. C. Snyder, eds.), pp. 187–207, University of California Press, Berkeley.

Kelly, D. P., 1971, Autotrophy: Concepts of lithotrophic bacteria and their organic metabolism, *Annu. Rev. Microbiol.* **25**:177–210.

Kelly, D. P., 1978, Bioenergetics of chemolithotrophic bacteria, in: *Companion to Microbiology* (A. T. Bull and P. M. Meadow, eds.), pp. 363–386, Longman, London.

Kemp, W. M., and Mitsch, W. J., 1979, Turbulence and phytoplankton diversity: A general model of the paradox of plankton, *Ecol. Mod.* **7**:201–222.

Kirk, T. K., 1971, Effects of microorganisms on lignins, *Annu. Rev. Phytopathol.* **9**:185–210.

Kirk, T. K., Higuchi, T., and Chang, H-M., 1980, *Lignin Biodegradation: Microbiology, Chemistry and Potential Applications* CRC Press, Boca Raton, Florida.

Koblentz-Mishke, O. J., Valkovinsky, V. V., and Kabanova, J. G., 1970, Plankton primary productivity of the world ocean, in: *Scientific Exploration of the South Pacific* (W. S. Wooster, ed.), National Academy of Sciences, Washington, D. C.

Kormondy, E. J., 1976, *Concepts of Ecology*, Prentice-Hall, Englewood Cliffs, New Jersey.

Kozlovsky, D. G., 1968, A critical evaluation of the trophic level concept. I. Ecological efficiencies, *Ecology* **49**:48–60.

Krebs, C. J., 1972, *Ecology: The Experimental Analysis of Distribution and Abundance*, Harper & Row, New York.

Kriss, A. E., Mishustina, I. E., Mitskerich, M., and Zemetsora, E. U., 1967, *Microbial Population of Oceans and Seas*, Edward Arnold, London.

Kuenen, J. G., and Tuovinen, O., 1981, The genera *Thiobacillus* and *Thiomicrospira*, in: *The Prokaryotes* (M. P. Starr, H. Stolp, H. G. Truper, A. Ballows, and H. G. Schlegel, eds.), pp. 1023–1036, Springer-Verlag, Berlin.

Kushner, D. J. (ed.), 1978, *Microbial Life in Extreme Environments*, Academic Press, New York.

Kuznetsov, S. I., Dubinina, G. A., and Lapteva, N. A., 1979, Biology of oligotrophic bacteria, *Annu. Rev. Microbiol.* **33**:377–387.

Lange, R. T., 1966, Bacterial symbiosis with plants, in: *Symbiosis*, Vol. I (S. M. Henry, ed.), pp. 99–170, Academic Press, New York.

La Riviere, J. W. M., and Schmidt, K., 1981, Morphologically conspicuous sulfur-oxidizing eubacteria, in: *The Prokaryotes* (M. P. Starr, H. Stolp, H. G. Truper, A. Ballows, and H. G. Schlegel, eds.), pp. 1037–1048, Springer-Verlag, Berlin.

Legendre, L., and Legendre, P., 1982, *Numerical Ecology*, Elsevier, Amsterdam.

Lewin, R., 1983a, Santa Rosalia was a goat, *Science* **221**:636–639.

Lewin, R., 1983b, Predators and hurricanes change ecology, *Science* **221**:737–740.

Liebig, J., 1841, Organic chemistry in its application to vegetable physiology and agriculture, in: *Readings in Ecology* (E. J. Kormondy, ed.), pp. 12–16, Prentice-Hall, Englewood Cliffs, New Jersey.

Lieth, H. F. H. (ed.), 1978, *Patterns of Primary Production in the Biosphere—Benchmark Papers in Ecology*, Vol. VIII, Dowden, Hutchinson, and Ross, Stroudsburg, Pennsylvania.

Lieth, M., and Whittaker, R. H., 1977, *Primary Production in the Biosphere*, Springer-Verlag, Berlin.

Likens, G. E., and Bormann, F. H., 1972, Nutrient cycling in ecosystems, in: *Ecosystem Structure and Function* (J. A. Wien, ed.), Oregon State University Press, Corvallis, Oregon.

Lindeman, R., 1942, The trophic dynamic aspect of ecology, *Ecology* **23**:399–418.

Luckinbill, L. S., 1973, Coexistence in laboratory populations of *Paramecium aurelia* and its predator *Didinium nasutum, Ecology* **54**:1320–1327.

Luckinbill, L. S., 1979, Regulation, stability, and diversity in a model experimental microcosm, *Ecology* **60**:1098–1102.

Lynch, J. M., 1976, Products of soil microorganisms in relation to plant growth, *CRC Crit. Rev. Microbiol.* **5**:67–107.

Lynch, J. M., and Poole, N. J., 1979, *Microbial Ecology—A Conceptual Approach*, Blackwell Scientific, Oxford, England.

MacArthur, R., 1955, Fluctuations of animal populations, and a measure of community stability, *Ecology* **36**:533–556.

MacArthur, R. H., 1968, The theory of the niche, in, *Population Biology and Evolution*, pp. 159–176, Syracuse University Press, Syracuse, New York.

MacArthur, R. H., and Wilson, E. O., 1963, An equilibrium theory of insular zoogeography, *Evolution* **17**:373–387.

MacArthur, R. H., and Wilson, E. O., 1967, *The Theory of Island Biogeography*, Princeton University Press, Princeton, New Jersey.

Mah, R. A., and Smith, M. R., 1981, The methanogenic bacteria, in: *The Prokaryotes* (M. P. Starr, H. Stolp, H. G. Truper, A. Ballows, and H. G. Schlegel, eds.), pp. 948–977, Springer-Verlag, Berlin.

Mah, R. A., Ward, D. M., Baresi, L., and Glass, T. L., 1977, Biogenesis of methane, *Annu. Rev. Microbiol.* **31**:309–341.

Mandelstam, J., McQuillen, K., and Dawes, I., 1982, *Biochemistry of Bacterial Growth*, Wiley, New York.

Major, J., 1969, Historical development of the ecosystem concept, in: *The Ecosystem Concept in Natural Resource Management* (G. VanDyne, ed.), Academic Press, New York.

Margalef, R., 1963, On certain unifying principles in ecology, *Am. Nat.* **97**:357–374.

Margalef, R., 1967, Some concepts relative to the organization of plankton, *Oceanogr. Mar. Biol. Annu. Rev.* **5**:257–289.

Margalef, R., 1968, *Perspectives in Ecological Theory*, University of Chicago Press, Chicago.

Margalef, R., 1979, Diversity, in: *Monographs on Oceanographic Methodology* (A. Sournia, ed.), pp. 251–260, UNESCO, Paris.

Margulis, L., 1971, Symbiosis and evolution, *Sci. Am.* **225**(2):48–57.

Martin, Y. P., and Bianchi, M. A., 1980, Structure, diversity, and catabolic potentialities of aerobic heterotrophic bacterial populations associated with continuous cultures of natural marine phytoplankton, *Microb. Ecol.* **5**:265–280.

May, R. M., 1971, Stability in multi-species community models, *Math. Biosci.* **12**:59–79.

May, R. M., 1972, Limit cycles in predator prey communities, *Science*, **177**:900–902.

May, R. M., 1973, *Stability and Complexity in Model Ecosystems*, Princeton University Press, Princeton, New Jersey.

May, R. M., 1976, *Theoretical Ecology Principles and Applications,* W. B. Saunders, Philadelphia.

May, R. M., and MacArthur, R. H., 1972, Niche overlap as a function of environmental variability, *Proc. Natl. Acad. Sci. U.S.A.* **69:**1109–1113.

Maynard, Smith, J., 1974, *Models in Ecology,* Cambridge University Press, Cambridge, England.

McLaren, A. D., and Peterson, G. H. (eds.), 1967, *Soil Biochemistry,* Marcel Dekker, New York.

Meyer, F. H., 1966, Mycorrhiza and other plant symbioses, in: *Symbiosis,* Vol. I (S. M. Henry, ed.), pp. 171–255, Academic Press, New York.

Mills, A. L., and Wassel, R. A., 1980, Aspects of diversity measurement of microbial communities, *Appl. Environ. Microbiol.* **40:**578–586.

Mishustin, E. N., and Shil'nikova, V. K., 1971, *Biological Fixation of Atmospheric Nitrogen,* The Pennsylvania State University Press, Uuniversity Park.

Mitchell, R., 1971, Role of predators in the reversal of imbalances in microbial ecosystems, *Nature (Lond.)* **230:**257–258.

Morris, I., 1980, *The Physiological Ecology of Phytoplankton,* Blackwell Scientific, Oxford, England.

Morris, I., 1982, Primary production of the oceans, in: *Experimental Microbial Ecology* (R. G. Burns and J. H. Slaters, eds.), pp. 239–252, Blackwell Scientific, Oxford, England.

National Academy of Sciences, 1972, *Degradation of Synthetic Organic Molecules in the Biosphere,* National Academy of Sciences, Washington, D. C.

National Academy of Sciences, 1975, *Productivity of World Ecosystems,* National Academy of Sciences, Washington, D. C.

National Academy of Sciences, 1978, *An Assessment of Mercury in the Environment,* National Academy of Sciences, Washington, D. C.

Nealson, K. H., and Hastings, J. W., 1979, Bacterial bioluminescence: Its control and ecological significance, *Microbiol. Rev.* **43:**496–518.

Nozhevnikova, A. N., and Yurkanov, L. N., 1978, Microbiological aspects of regulating the carbon monoxide content of the Earth's atmosphere, *Adv. Microb. Ecol.* **2:**203–244.

Odum, E. P., 1962, Relationship between structure and function in ecosystems, *Jpn. J. Ecol.* **12:**108–118.

Odum, E. P., 1968, Energy flow in ecosystems: A historical review, *Am. Zool.* **8:**11–18.

Odum, E. P., 1969, The strategy of ecosystem development, *Science* **164:**262–270.

Odum, E. P., 1971, *Fundamentals of Ecology,* W. B. Saunders, Philadelphia.

Odum, H., 1957, Trophic structure and productivity of Silver Springs, Florida, *Ecol. Monog.* **27:**55–112.

Odum, H. T., and Odum, E. C., 1981, *Energy Basis for Man and Nature,* McGraw-Hill, New York.

Oort, A. H., 1970, The energy cycle of the Earth, *Sci. Am.* **223**(3):54–63.

Oren, A., and Padan, E., 1978, Induction of anaerobic, photoautotrophic growth in the cyanobacterium *Oscillatoria limnetica,* J. Bacteriol. **133:**558–563.

Paine, R. T., 1966, Food web complexity and species diversity, *Am. Nat.* **100:**65–75.

Painter, H. A., 1970, A review of literature on inorganic nitrogen metabolism in microorganisms, *Water Res.* **4:**393–450.

Patrick, R., 1963, The structure of diatom communities under varying ecological conditions, *Ann. N. Y. Acad. Sci.* **108:**359–365.

Patrick, R., 1976, The formation and maintenance of benthic diatom communities, *Proc. Natl. Acad. Sci. U.S.A.* **120:**475–484.

Pattern, B. C., 1961, Competitive exclusion, *Science* **134:**1599–1601.

Payne, W. J., 1973, Reduction of nitrogenous oxides by microorganisms, *Bacteriol. Rev.* **37:**409–452.

Pearson, W. W., 1974, Bacterial photosynthesis, *Annu. Rev. Microbiol.* **28:**41–59.

Peet, R. K., 1974, The measurement of species diversity, *Annul. Rev. Ecol. Syst.* **5:**285–308.

Pfennig, M., 1967, Photosynthetic bacteria, *Annu. Rev. Microbiol.* **21:**285–324.

Pfennig, N., Widdel, F., and Truper, H. G., 1981, The dissimilatory sulfate-reducing bacteria, in: *The Prokaryotes* (M. P. Starr, H. Stolp, H. G. Truper, A. Ballows, and H. G. Schlegel, eds.), pp. 926–940.

Phillipson, J., 1966, *Ecological Energetics,* Arnold, London.

Pianka, E. R., 1970, On *r*- and *K*-selection, *Am. Nat.* **104:**592–597.

Pianka, E. R., 1974, Niche overlap and diffuse competition, *Proc. Natl. Acad. Sci. U.S.A.* **71:**2141–2145.

Pielou, E. C., 1966a, Shannon's formula as a measure of species diversity: Its use and misuse, *Am. Nat.* **100:**463–465.

Pielou, E. C., 1966b, The measurement of diversity in different types of biological collections, *J. Theor. Biol.* **13:**131–144.

Pielou, E. C., 1969, *An Introduction to Mathematical Ecology*, Wiley, New York.

Pielou, E. C., 1975, *Ecological Diversity*, Wiley, New York.

Pittendrigh, C., 1961, Temporal organization in living systems, *Harvey Lect.* **56:**93–125.

Poindexter, J. S., 1981a, Oligotrophy: Fast and famine existence, in: *Advances in Microbial Ecology*, Vol. 5 (M. Alexander, ed.), pp. 63–89, Plenum Press, New York.

Poindexter, J. S., 1981b, The caulobacters: Ubiquitous unusual bacteria, *Microbiol. Rev.* **45:**123–179.

Pomeroy, L. R., 1970, The strategy of mineral cycling, *Annu. Rev. Ecol. Syst.* **1:**171–190.

Pomeroy, L. R. (ed.)., 1974, *Cycles of Elements—Benchmark Papers in Ecology*, Vol. VI, Dowden, Hutchinson, and Ross, Stroudsburg, Pennsylvania.

Preston, F. W., 1962a, The canonical distribution of commonness and rarity. I, *Ecology* **43:**185–215.

Preston, F. W., 1962b, The canonical distribution of commonness and rarity. II, *Ecology* **43:**410–432.

Quale, J. R., 1972, The metabolism of one-carbon compounds by microorganisms, *Annu. Rev. Microb. Physiol.* **7:**119–203.

Quispel, A. (ed.), 1974, *The Biology of Nitrogen Fixation*, North-Holland, Amsterdam.

Rabinowitch, E., and Govindjee, 1969, *Photosynthesis*, Wiley, New York.

Raymont, J. E. G., 1980, *Plankton and Productivity in the Oceans*, Pergamon Press, Oxford, England.

Reichenbach, H., and Dworkin, M, 1981, The order myxobacteriales, in: *The Prokaryotes* (M. P. Starr, H. Stolp, H. G. Truper, A. Ballows, and H. G. Schlegel, eds.), pp. 315–328, Springer-Verlag, Berlin.

Rheinheimer, G., 1981, *Aquatic Microbiology*, Wiley, New York.

Richerson, P., Armstrong, R., and Goldman, C. R., 1970, Contemporaneous disequilibrium, a new hypothesis to explain the "paradox of the plankton," *Proc. Natl. Acad. Sci. U.S.A.* **67:**1710–1714.

Ridley, W. P., Dizikes, L. J., and Wood, J. M., 1977, Biomethylation of toxic elements in the environment, *Science* **197:**329–332.

Roper, M. M., and Marshall, K. C., 1978, Effects of a clay mineral on microbial predation and parasitism on *Escherichia, Microb. Ecol.* **4:**279–290.

Round, F. E., 1981, *The Ecology of Algae*, Cambridge University Press, New York.

Rovira, A. D., 1965, Interactions between plant roots and soil microorganisms, *Annu. Rev. Microbiol.* **19:**241–266.

Russel-Hunter, W. D., 1970, *Aquatic Productivity*, Macmillan, London.

Ryther, J. H., 1969, Photosynthesis and fish production in the sea, *Science* **166:**72–76.

Saunders, H. L., 1968, Marine benthic diversity: A comparative study, *Am. Nat.* **102:**243–282.

Schmidt, J., 1981, The genera *Caulobacter* and *Asticcacaulis*, in: *The Prokaryotes* (M. P. Starr, H. Stolp, H. G. Truper, A. Ballows, and H. G. Schlegel, eds.), pp. 466–476, Springer-Verlag, Berlin.

Shannon, C. E., and Weaver, W., 1949, *The Mathematical Theory of Communications*, University of Illinois Press, Urbana.

Shelford, V. E., 1911, Physiological animal geography, *J. Morphol.* **22:**551–618.

Silverman, M. P., and Ehrlich, H. L., 1964, Microbial formation and degradation of minerals, *Adv. Appl. Microbiol.* **6:**153–206.

Slater, J. H., 1978, The role of microbial communities in the natural environment, in: *The Oil Industry and Microbial Ecosystems* (K. W. A. Chater and H. J. Somerville, eds.), pp. 137–154, Heyden and Son Ltd., London.

Slater, J. H., 1979, Population and community dynamics, in: *Microbial Ecology—A Conceptual Approach* (J. M. Lynch and N. J. Poole, eds.), pp. 45–63, Blackwell Scientific, Oxford, England.

Slater, J. H., 1980, Physiological and genetic implications of mixed populations and microbial community growth, in: *Microbiology—1980* (D. Schlessinger, ed.), pp. 314–316, American Society for Microbiology, Washington, D. C.

Slater, J. H., and Goodwin, D., 1980, Microbial adaptation and selection, in: *Contemporary Microbial Ecology* (D. C. Ellwood, J. N. Hedger, M. J. Latham, J. M. Lynch, and J. H. Slater, eds.), pp. 137–160, Academic Press, London.

Smith, D. W., 1982, Extreme natural environments, in: *Experimental Microbial Ecology* (R. G. Burns and J. H. Slater, eds.), pp. 555–574, Blackwell Scientific, Oxford, England.

Society for Experimental Biology, 1975, *Symbiosis*, Cambridge University Press, Cambridge, England.

Sörderlund, R., and Svensson, B. H., 1976, The global nitrogen cycle, *Bull. Ecol. Res. Comm. (Stockh.)* **22:**23–73.

Sournia, A., 1974, Circadian periodicities in natural populations of marine phytoplankton, *Adv. Mar. Biol.* **12:**326–386.

Stanier, R. Y., and Cohen-Bazire, G., 1977, Phototrophic prokaryotes: The cyanobacteria, *Annu. Rev. Microbiol.* **31:**225–274.

Stanier, R. Y., Pfennig, N., and Truper, H. G., 1981, Introduction to the phototrophic prokaryotes, in: *The Prokaryotes* (M. P. Starr, H. Stolp, H. G. Truper, A. Ballows, and H. G. Schlegel, eds.), pp. 197–211, Springer-Verlag, Berlin.

Starkey, R. L., 1964, Microbial transformations of some organic sulfur compounds, in: *Principles and Applications in Aquatic Microbiology* (H. Heukelekian and N. Dondero, eds), pp. 405–429, Wiley, New York.

Stewart, W. D. P., 1973, Nitrogen fixation by photosynthetic microorganisms, *Annu. Rev. Microbiol.* **27:**283–316.

Stout, J. D., 1980, Protozoa in nutrient cycling and energy flow, in: *Advances in Microbial Ecology*, Vol. 4 (M. Alexander, ed.), pp. 1–50, Plenum Press, New York.

Strickland, J. D. H., 1965, Production of organic matter in the primary stages of the marine food chain, in: *Chemical Oceanography*, Vol. 1 (J. P. Riley and G. Skirrow eds.), pp. 477–610, Academic Press, New York.

Summers, A. O., and Silver, S., 1978, Microbial transformation of metals, *Annu. Rev. Microbiol.* **32:**637–672.

Swift, M. J., 1976, Species diversity and the structure of microbial communities, in: *The Role of Aquatic and Terrestrial Organisms in Decomposition Processes* (J. M. Anderson and A. MacFayden eds.), pp. 185–122, Blackwell Scientific, Oxford, England.

Teal, J. M., 1962, Energy flow in the salt marsh ecosystem of Georgia, *Ecology* **43:**614–624.

Transeau, E., 1926, The accumulation of energy by plants, *Ohio J. Sci.* **26:**1–10.

Truper, H. G., and Pfennig, N., 1981, Characterization and identification of the anoxygenic phototrophic bacteria, in: *The Prokaryotes* (M. P. Starr, H. Stolp, H. G. Truper, A. Ballows, and H. G. Schlegel, eds.), pp. 299–312, Springer-Verlag, Berlin.

Turner, F. B., 1968, Energy flow and ecological systems, *Am. Zool.* **8:**10–69.

Utida, S., 1967, Damped oscillation of population density at equilibrium, *Res. Pop. Ecol.* **9:**1–9.

Van den Ende, P., 1973, Predator–prey interactions in continuous culture, *Science* **181:**562–564.

Van Gemerden, H., 1974, Coexistence of organisms competing for the same substrate: An example among the purple sulfur bacteria, *Microb. Ecol.* **1:**104–119.

Van Valen, L., 1971, Group selection and the evolution of dispersal, *Evolution* **25:**591–598.

Veldkamp, H., 1977, Ecological studies with a chemostat, in: *Advances in Microbial Ecology*, Vol. 1 (M. Alexander, ed.), pp. 59–94, Plenum Press, New York.

Vermeij, G. J., 1978, *Biogeography and Adaptation: Patterns of Marine Life*, Harvard University Press, Cambridge, Massachusetts.

Vincent, J. M., 1981, The genus *Rhizobium*, in: *The Prokaryotes* (M. P. Starr, H. Stolp, H. G. Truper, A. Ballows, and H. G. Schlegel, eds.), pp. 818–841, Springer-Verlag, Berlin.

Walker, N., 1975, Nitrification and nitryifying bacteria, in: *Soil Microbiology* (N. Walker, ed.), pp. 133–146, Butterworths, London.

Wallace, W., and Nicholas, D. J. D., 1969, The biochemistry of nitrifying organisms, *Biol. Rev.* **44:**359–391.

Walsby, A. E., 1975, Gas vesicles, *Annu. Rev. Plant Physiol.* **26:**427–439.

Walsby, A. E., 1977, The gas vacuoles of blue green algae, *Sci. Am.* **237:**90–97.

Walsby, A. E., 1981, Cyanobacteria: planktonic gas-vacuolate forms, in: *The Prokaryotes* (M. P. Starr, H. Stolp, H. G. Truper, A. Ballows, and H. G. Schlegel, eds.), pp. 224–235, Springer-Verlag, Berlin.

Watson, S. W., Valois, F. W., and Waterbury, J. B., 1981, in: *The Prokaryotes* (M. P. Starr, H. Stolp, H. G. Truper, A. Ballows, and H. G. Schlegel, eds.), pp. 1005–1023, Springer-Verlag, Berlin.

Weber, N. A., 1972, The fungus-culturing behavior of ants, *Am. Zool.* **12:**577–587.

Westlake, D. F., 1963, Comparisons of plant productivity, *Biol. Rev.* **38:**385–425.

Wetzel, R. G., 1975, *Limnology*, W. B. Saunders, Philadelphia.

Whittaker, R. H., 1965, Dominance and diversity in land plant communities, *Science* **147:**250–260.

Whittaker, R. H., 1972, Evolution and measurement of species diversity, *Taxon* **21:**213–251.

Whittaker, R. H., 1975, *Communities and Ecosystems*, Macmillan, New York.

Whittaker, R. H., and Likens, G. E. (eds.), 1973, The primary production of the biosphere, *Hum. Ecol.* **1:**299–369.

Whittenbury, R., and Dalton, H., 1981, The methylotrophic bacteria, in: *The Prokaryotes* (M. P. Starr, H. Stolp, H. G. Truper, A. Ballows, and H. G. Schlegel, eds.), pp. 894–902, Springer-Verlag, Berlin.

Wiens, J. A., 1983, Competition or peaceful existence, *Nat. Hist.* **92**(3):30–34.

Winogradsky, S., 1925, Etudes sur la microbiologie du sol, *Ann. Inst. Pasteur* **39:**299–354.

Winter, H. C., and Burris, R. H., 1976, Nitrogenase, *Annu. Rev. Biochem.* **45:**409–426.

Wood, E. J. F., 1965, *Marine Microbial Ecology*, Reinhold, New York.

Wood, E. J. F., 1967, *Microbiology of Oceans and Estuaries*, Elsevier, New York.

Woodwell, G. M., 1970, The energy cycle of the biosphere, *Sci. Am.* **223**(3):64–74.

Woodwell, G. M., and Pecan, E. V. (eds.), 1973, *Carbon and the Biosphere—Brookhaven Symposia in Biology*, Vol. 24, National Technical Information Service, Springfield, Virginia.

Woodwell, G. M., and Smith, H. H. (eds.), 1969, *Diversity and Stability in Ecological Systems*, Brookhaven Symposia in Biology No. 22, Brookhaven National Laboratory, Upton, New York.

Woodwell, G. M., and Whittaker, R. H., 1968, Primary production in terrestrial communities, *Am. Zool.* **8:**19–30.

Wu, T. T., 1978, Environmental evolution in bacteria, *Crit. Rev. Microbiol.* **6:**33–52.

Zeikus, J. G., 1977, The biology of the methanogenic bacteria, *Bacteriol. Rev.* **41:**514–541.

Zeikus, J. G., 1981, Lignin metabolism and the carbon cycle: Polymer biosynthesis, biodegradation, and environmental recalcitrance, in: *Advance in Microbial Ecology*, Vol. 5 (M. Alexander, ed.), pp. 211–243, Plenum Press, New York.

INDEX

Acetobacter suboxydans, 296
Acetobacterium woodii, 315
Acetogenic bacterium, 314–317
Acidophilic bacteria, 231
Acridine orange direct count (AODC)
      method, 3–9, *see also* Epifluorescence
      techniques; Exogenous fluorochromes
  as active cell indicator, 30
  and biofilm activity detection, 75
  in Chesapeake Bay comparative study,
    23–26
  and ETS activity, 20
  with living versus nonliving cells, 104
  and microhabitats, 183
  as viability measure, 121
Actinomycetes, 35
Active cellular synthesis, 21–22
Adaptive strategies, 353–354
[³H]Adenine method, 153–155
Adenine nucleotide pool turnover, 155
Adenosine : ATP ratio, 128
Adenylate energy charge, 127–128
Adhesion property
  of bacteria, 180
  of biofilm, 57, 59–60, 62, 70, 74
Aerobic conditions, and respiratory quinones,
  193
Aerobic species, cultivation of, 71
Aerobic zone, measurement of, 183
Aggregates, and direct microscopy, 101
Aggregation, benefits of, 352–353
Akin, D. E., 308
Alcohol, anaerobic degradation of, 317–319
Aldehyde fixation
  in AODC method, 4–5, 7
  in direct microscopy, 101–102
  in nalidixic acid method, 24
Alexander, M., 18
Algae, *see also* Eukaryotes; Photosynthesis
  and biomass of, 106, 112–113, 115–116
  dark ³⁵SO₄²⁻ uptake in, 147
  distribution of, 342

Algae (*Cont.*)
  physiological state of populations, 129–130
  primary productivity by, 341
  separation from bacteria, 99, 143
Alkalophilic bacteria, 231
Allee's principle, 352–353
Alternative energy sources, 321–322
Amensalism, 295, 304–305
Ammerman, J. W., 15
Ammonia, and nitrification process, 348–349
Anabolic reduction charge, 127–128
Anaerobic bacteria, *see also* Photosynthesis
  distribution of, 342
  phosphosphingolipids in, 193
  and primary productivity, 341
  signature lipids of, 192–193
Anaerobic conditions, and respiratory
  quinones, 193
Anaerobic degradation, three-stage scheme
  for, 314–315
Anaerobic respiration, 343–344, 349
Anaerobic sediments: *see* Sediments
Anderson, J. R., 115
Anionic polymers, bacterial secretion of, 181
Antagonism: *see* Amensalism
Antibiotics
  animals treated with, 208–209, *see also*
    Gnotobiotic animals
  and bacterial retardation of growth, 217
  and partitioning of taxonomic categories,
    99–100
  production of, in nature, 304–305
  sensitivity of biofilm bacteria to, 80
Antibody stabilization technique, 54–55, 60,
  63
Anticapsular (K30) antiserum, with TEM, 55
Antiseptics, and biofilms, 72
AODC method: *see* Acridine orange direct
  count (AODC) method
Aquatic ecosystems; *see also* Sediments
  and AODC method, 3–7
  assimilatory sulfur metabolism in, 146

Aquatic ecosystems (*Cont.*)
  cell counts in samples from, 102
  cyanobacteria in, 342
  and detrital food webs, 346
  dominant grazers in, 345
  ETS activity in, 20
  and FA combined techniques, 36
  freshwater
    bacterial count in, 4–5, 6, 9
    individual cell activity in, 20
  growth rates in, 146, 148
  and laser-based flow cytometry, 105
  and *Legionella pneumophila*, 35–36
  metabolic activity in, 134–136
  nucleic acid synthesis in, 149–155
  oxygen concentration in, 133
  photosynthetic microorganisms in, 342
  primary producers in, 345
  productivity determination in, 15–17
  seawater
    bacterial count in, 4–5, 6, 9
    sampling of, 92
    use of isotopic tracers with, 96–97
    viable bacteria in, 123
  thermal rifts, 343
ARC: *see* Anabolic reduction charge
Archaebacteria, 238–240
Aromatic compounds, anaerobic degradation
  of, 320–321
Armillaria mellea, 354
Ashe, W. K., 207
Assemblages, bacterial: *see* Standing crop
Associated animals, 206
Asterionella formosa, 269–271, 283–284
Atlas, R. M., 295, 297–298
ATP biomass technique, 107–109,
  153–156
Auger electron spectroscopy (AES),
  187–188
Austin, K. E., 134
Autochthonous populations, 353
Autofluorescence
  and AODC method, 6–7
  and bacterial counts, 6–7, 9
  as problem, 34–35, 103
Autoradiographic analysis: *see*
  Microautoradiography
Axenic animals: *see* Germfree animals
Azam, F., 14, 29, 99, 135–136, 150, 157

Babiuk, L. A., 8, 118
Bache, R., 315
Bacillus pasteurii, 234
Bacillus polymyxa, 298
Bacillus stearothermophilus, 234, 245

Bacterial interactions, classification of,
  294–306
Bacteroides, 307–309
Baker, J. A., 206
Baker, K. H., 20, 36
Bakken, L. R., 105–106
Baldwin, R. L., 309
Banat, I. M., 324
Banse, K., 113
Banwell, J. G., 64
Barik, S., 321
Bartha, R., 295, 298
Bates, S. S., 147
Bazin, M. J., 268
Bdellovibrio
  as bacterial parasite, 305
  discovery of, 254–255
  enrichment media for, 252
  growth of, 327–328
Beam, H. W., 296
Beeftink, H. H., 285
Beijerinck, M. W., 229, 232
Beijerinckia, 234
Berman, T., 94
Biebl, H., 322
Biochemical methods, 106–116
Biofilm bacteria, 50
  cultivation of, 70–74
  problems in sampling of, 64–70
  quantitation of populations, 64, 69–70
  and Robbins device, 71–74
  SEM processing and loss of, 57, 59–64
Biofilms
  coherence of, in SEM technique, 59–60
  detection of activity within, 75–80
  in industrial systems, 73–74
  monitoring of, 77
  in natural ecosystems, 74
  sessile bacterial count in, 69
Biogeochemical cycling, 346–351
Biomass
  of algae: *see* Algae
  and AODC method, 7–8
  and biochemical methods, 106–116
  change in, as growth measure, 156–158
  and community structure, 107–110,
    190–193
  direct measurement of, 184–185
  and dry matter content, 105–106
  of fungi: *see* Fungi
  in natural samples, 89
  of prokaryotes: *see* Prokaryotes
  and total cell numbers, 104–106
Biomass carbon, and biovolume, 105–106
Biopolymers, microbial degradation of, 344

Biosphere, microbial modification of, 182
Biosynthesis, 93
Biotically tolerant species, 354
Biovolume, 7–8, 105–106
Bishop, P. L., 77
Bivalent specific antibodies as TEM stabilizers, 54
Blackburn, T. H., 137, 309
Blakemore, R. P., 230
Bland, P. T., 122
Blending method, and epifluorescent counting, 184
Bohlool, B. B., 32, 35–36
Booth, C. R., 107
Borsheim, K. Y., 18
Bossard, P., 149, 155–156
Bottle effect, 93, *see also* Incubation
Bovine serum albumin (BSA), 34
Bracke, J. W., 32
Brassica napus L., 195–196
Bratbak, G., 105
Brock, T. D., 18, 101, 148, 295
Brlansky, R. H., 35
Bryant, M. P., 306, 309, 313, 318
Bull, A. T., 274
Burnison, B. K., 136
Butyrate degradation, 318–319
Butyrivibrio fibrisolvens, 307–308

Capsules, microhabitat modification and, 180–182
Carbohydrates, degradation of, 321–322
Carbon
  biochemical indicators of, 107
  ^{14}C-radiolabeled organic compounds, *see also* Carbon dioxide; Radioisotopic assay
    and biofilm activity, 79–80
    impurities of ^{14}C]sodium carbonate, 96
    and metabolic activity, 134–138
    and phytoplankton growth, 131 132
  C : S ratio, as variable, 146–147
  in situ mass flux of, 137–138
  and microbial transformations, 344
Carbon cycling, 8, 89, 346–347
Carbon dioxide
  assimilation of, 142–144
  [14]CO$_2$ tracer techniques
    for algal biomass, 115–116
    and carbon fixation, 141–144
  fluxes in, 132
Carcinogen metabolism, 218
Carlucci, A. F., 15

Casida, L. E., 305
C : ATP ratio, 108, *see also* ATP biomass technique
C :A$_T$ ratio, 108–109; *see also* ATP biomass technique
Catabolism, 127–129
Caulobacter, 248
Cells
  division of, 88–89, 91
  morphology of, 235
  number of, and growth, 156–158
  processes in, 88
  size of, and biomass, 104–106
  volume of, 6
Cellulose, anaerobic degradation of, 344
Cephalosporium gramineum, 304–305
Chemical inhibition, of grazing, 16–17
Chemolithotrophic bacteria
  and enrichment cultures, 241–245
  and Romanenko procedure, 143–144
  sources of samples, 231
Chemolithotrophy
  production of organic matter by, 140, 143–144
  and thermal rift ecosystems, 343
Chemoorgantrophic bacteria, enrichment cultures for, 242–245
Chemostats
  and ecosystem stability, 359
  under mixed substrate limiting conditions, 262–265
Cherry, W. B., 35
Chian, S. K., 278, 287
Chitin measurements, and total mycelial biomass, 113–114
Chlorobium, 322–323, 334
"Chlorobium limicola," 323
"Chlorochromatium aggregatum," 323
"Chlorochromatium glebulum," 323
Chloroflexaceae, 322
Chloroform-methanol extraction, 190
Chlorophyll *a*
  and labeling method for algal biomass, 112–113, 115–116
  and photosynthesis, 342
  and phytoplankton biomass, 142
"Chloroplana vaculata," 323
Christensen, D., 137
Christensen, J. P., 127
Christian, R. R., 10, 12–13, 145–146
Chromatium vinosum, 285, 322
Chromatium weissei, 285
Chrzanowski, T. H., 118
Chynoweth, D. P., 318
Climax concept, and succession, 360

Closed ecosystems, and measurement of
    growth, 140
Clostridia
    and cecal reduction, 216–217
    and coexistence in mixed substrates, 279
    enrichment procedure for, 234
    and interactions among indigenous
        microflora, 221
    and reversible growth retardation, 217
C : N : P ratios, and growth rate, 131
Coexistence
    of amensalistic populations, 305
    of *Chromatium* species, 285
    of competitors, 303
    of mixed cultures, 268
    necessary conditions for, 301
    of sulfur and nonsulfur bacteria, 285
Cohen, P. S., 221
Coleman, A. W., 9
Coliform, 278
Cometabolism, and compound degradation,
    326–327
Commensalism, 295–297
Community structure, *see also* Populations
    diversity in, 358–359
    in nature, 87–88
    techniques for analysis of, 190–193
Competition, 295, 299–304, 355–357
Competitive exclusion principle, 278–279,
    299–300, 303
Computer-assisted image microanalysis, 105
Contaminated animals, 206
Conventional animals, 206, 209–213
Conventionalized animals, 206
Coons, A. H., 31
Cooperation, and diverse populations,
    357–358
Corpet, D., 221
[14]C-Radiolabeled organic compounds: *see*
    Carbon
Crawford, C. C., 77
Critical point drying technique, 57, 59–60,
    62
Cross-feeding, 298–299, *see also*
    Protocooperation
Cross-reactions, and FA method, 34
Cuhel, R. L., 147–148
Cullen, J. J., 129
Culture techniques
    for biomass estimation, 100–101
    and microorganisms in nature, 88
    see also Enrichment cultures; Mixed
        cultures; Pure cultures
CV animals: *see* Conventional animals

Cyanobacteria, *see also* Photosynthesis
    and AODC method, 6–7
    distribution of, in aquatic ecosystems, 342
    and enrichment cultures, 245
    and nitrogen fixation, 349
    and primary productivity, 341
    sources of samples, 231
Cyclotella meneghiniana, 269–271, 283–284
"Cylinodrogloca bacterifera," 323

Daley, R. J., 3
DAPI: *see* 4'6-Diamidino-2-phenylindole
Davies, M., 282
Davis, W. M., 128
DCMU-enhanced fluorescence, 129–130
DDT, degradation of, 327
Decay, 345–346
Dehydration, of biofilms, 54, 57
Denitrification, 192, 349
Density values, bacterial dry mass, 8
Desulfobulbus propionicus, 324
Desulfovibrio, 317–318, 323–324
Desulfuromonas acetoxidans, 323
Detrital food webs, 345–346
4'6-Diamidino-2-phenylindole (DAPI), 9
    in epifluorescent microscopy, 103
    and microhabitats, 183–184
Diassociated animals, 206
Diatoms, 283–284, *see also* Autofluorescence
Diazinon, degradation of, 326
Diazotrophic bacteria, 231
Dietz, A. S., 136
Diffusion, and bacterial motion, 179–182
Dilution, and grazing prevention, 16–17
Direct microscopy
    and microorganisms in nature, 88
    power of, 1–2
    subjectivity of, 3, 13, 101
Disease, and gnotobiotic animals, 213–216
Disruption methods, in biofilm sampling,
    69–70
Dissimilatory nitrate-reducing bacteria, 231
Dissimilatory sulfate-reducing bacteria, 231
DiTullio, G., 131, 148
Diversity, *see also* Coexistence
    and ecological niches, 232
    of populations, and enzymatic activity,
        124–125
    and successional changes, 358–360
DNA-specific dyes: *see* 4'6-Diamidino-2-
        phenylindole, Hoechst dye No. 33258
DNA synthesis: *see* Nucleic acid synthesis
Dobrynin, E. G., 105
Domsch, K. H., 3, 30, 115

Dortch, Q., 125
Double-labeled molecules, 94, *see also* Isotopic tracers
Dowdle, W. R., 33
Dual laser flow cytometry, 130
Ducklow, H. W., 10
Ducluzeau, R., 219–220
Dunaliella tertiolecta, 147
Dundas, I., 105
Duval-Iflah, Y., 222
Dye-exclusion test, and viability, 118
Dyes, and UFOs in IL procedures, 35
Dykhuizen, D., 282

EC$_A$: *see* Adenylate energy charge
Ecosystems, *see also specific ecosystems*
  biogeochemical cycling within, 346–351
  energy flow through, 340–346
  and pure culture studies, 339–340
Egli, T., 276
Eighmy, T. T., 77
Elective cultures: *see* Enrichment cultures
Electrochemical methods, and biomass, 116
Electron-beam microscopy
  and microhabitats, 185–187
  versus conventional optical microscopy, 102
Electron particle counters, and biovolume, 104
Electron transport system (ETS) activity
  activators, 25, 28
  in Chesapeake Bay comparative study, 23, 25
  direct determinations of, 20
  and enzyme activity, 126–127
  in individual bacterial cells, 19–20
  and INT-reduction method, 25
Ellery, W. N., 7
Energy, *see also* Chemolithotrophy; Photoautotrophy
  aerobic versus anaerobic extraction of, 343–344
  capture of, 340–343
  direction of flow of, 340, 346
  transfer of, and food webs, 343–346
Energy dispersive spectrometry (EDS), 187–188
England, J. M., 120
Enrichment cultures
  closed-system liquid, 241–246
  as enumeration techniques, 251–252
  exploratory use of, 232
  limitations of, 250–256
  and living media, 252–253

Enrichment cultures (*Cont.*)
  as microbial successions, 249–250
  and mixed substrates, 285–287
  open-system, 246–248
  and physiotype designations, 240
  prospects for, 256
  purposes for, 230–233
  and selectivity, 233–234
  and stable prokaryote associations, 253–254
  two-phase, 248–249
  and unexpected organisms, 254–256
Ensifer adhaerens, as predator, 305
Entamoeba histolytica, 214
Environment modification by microbes, 179–182
Enzymes
  activity of, and populations, 124–125
  concentration of, in GF and CV animals, 209–210
Epifluorescence techniques, 3–9, *see also* Fluorescent antibody techniques
  and active cell determination, 30
  and active substrate uptake measures, 21
  and cell count, 102–104
  comparison of, for determination of individual cell activity, 22–30
  europium-chelate/fluorescent brightener in, 9
  and microhabitats, 183–185
  and planktonic microorganisms, 104
Eppley, R. W., 113, 125
Equilibrium, in population interactions, 356–357
Erythrosine method, 2
Escherichia coli
  and amensalism, 304
  generalist versus specialist strains of, 282
  generalizability of models based on, 139, 143
  growth of
    and dark CO_2 assimilation, 143
    on mixed substrates, 274
    and SO_4^{2-} uptake, 146
  population levels in CV animals, 212
  propulsion system of, 179
  and *Shigella flexneri,* 219–220
  and TEM study, 53–55, 58
Ethanol degradation, 319
Ethidium bromide, in bacterial count techniques, 9, 30
Eubacterium ruminantium, 308
Euchrysine-2GNX: *see* Epifluorescent techniques; Exogenous fluorochromes

Eukaryotes
  activity of, compared to prokaryotes,
    193–194
  and AODC method, 6–7
  and habitat properties, 190–191
  as living enrichment media, 252
  marine distribution of phototrophic, 99
Evans, E. A., 98
Exclusion principle: see Competitive exclusion
    principle
Ex-germfree animals, 206
Exogenous fluorochromes, and natural
    environments, 103

FAINT technique, 36
Fallon, R. D., 7, 12, 146
FA methods: see Fluorescent antibody
    techniques
Fatty acids, anaerobic degradation of,
    317–319
Favinger, J. H., 256
FDA method: see Fluorescein diacetate
FDB method: see Fluorescin dibutyrate
FDC method: see Frequency of dividing cells
    method
Ferguson, M. S., 206
Ferguson, R. L., 14–15, 106, 157
Fermentative bacteria, 192, 314–316, see also
    Methane fermentation
Field-collected data, reliability of, 91
Filtration, and grazing prevention, 16–17
Findlay, R. H., 194–195
Firefly bioluminescence assay, 108
FITC: see Fluorescein isothiocyanate (FITC)
Flavin nucleotide level, as total biomass
    indicator, 109
Fliermans, C. B., 34–36
Fluorescein diacetate (FDA), 30, 117–118; see
    also Fluorescent antibody techniques
Fluorescein dibutyrate (FDB), 117–118
Fluorescein isothiocyanate (FITC), 8–9, 31,
    33; see also Fluorescent antibody
    techniques
Fluorescein-labeled antibodies: see
    Fluorescent antibody (FA) techniques
Fluorescence microscopy: see Epifluorescence
    techniques
Fluorescent antibody (FA) techniques, 31–37,
    see also specific antibodies
  combined, and active-inactive cell
    differentiation, 36
  direct versus indirect, 33
  limitations of, 104
  phage-antiphage system, 33

Fluorochrome methods: see Epifluorescence
    techniques
FMN level, 109
Food webs, 344–346
Formaldehyde: see Aldehyde fixation
Formate, 312
Forsdyke, D. R., 136
Fourier-transforming infrared spectrometry,
    and microhabitats, 188–189
Fractionation, 99
Francisella tularensis, 192
Frankland, J. C., 114
Fredrickson, A. G., 303, 306
Frequency of dividing cells (FDC) method,
    10–14, 144–146
Frictional resistance (f_F), and biofilm
    formation, 72–73, 75
FT/IR spectroscopy, 188–189
Fuhrman, J. A., 6, 8, 14, 15, 18, 29, 150, 157
Fumigation/respiration method, for total soil
    microbial biomass, 114–115
Fungi, see also Eukaryotes
  biomass of
    and biochemical methods, 113–116
    and dry matter content, 105–106
  survival strategies and, 354
  and universal acceptor problem, 35

Gas chromatography/mass spectrometry (GC-
    MS), 189–190
Gas vacuoles, 187
Gehron, M. J., 129
Generalist/specialist coexistence, 281–283,
    286–287, 303–304
Germfree (GF) animals, see also Gnotobiotic
    animals
  data interpretation with, 211–213
  defined, 206
  exogenous contamination of, 207–208
  and fastidious microorganisms, 207
  and indigenous microflora, 209–213
  isolation methods, 207–208
  techniques with, 206–209, 213
Gest, H., 256
GF animals: see Germfree (GF) animals
Glycocalyx polysaccharides, staining of,
    53–54
Glynn, A. A., 180
Gnotobiotic animals, 206, 211–213
Gnotoxenique animals: see Gnotobiotic
    animals
Gocke, K., 137
Goldman, J. C., 131
Goldman, P., 218

Gorbach, S. L., 214
Gottschal, J. C., 272, 280, 303
Govan, R. W., 71
Grain density autoradiography, and viability, 121
Gram-negative bacteria, 191, 193
Gram-positive bacteria, 191
Grazing, *see also* Predation
  assay methods for, and productivity determination, 17–19
  and filtration, 16–17
  pressures from, and microbiota metabolic activity, 191
  as trophic level, 345
Grbić-Galić, D. E., 320
Grenney, W. J., 284
Grime, J. P., 354
Growth, *see also* Growth rate (μ)
  diauxic, 273
  distinguished from cell division, 89
  methods for estimation of, 91, 140
  of mixed cultures on mixed substrates, 277–285
  substrate-limited, 262–265, 267
Growth rate (μ)
  and autoecological approach, 146, 148
  and frequency of dividing cells method, 10–13
  and GTP : ATP ratio, 128
  and mean cell volume, 14
  and productivity measures with incubation, 15
  and Redfield ratio, 131

Haas, L. W., 9, 104
Habitats, *see also* Microhabitats
  diversity of, 86–87
  scale of, 86–87
  zonation within, 341–342
Habte, M., 18
Hagström, A., 10–12, 15, 144–145
Hairston, N. G., 359
Halobacterium, 234
Halophilic bacteria, 231
Hamilton, R. D., 134
Hansen, P. A., 33
Hanson, R. B., 10, 135
Hansen, S. R., 302
Harder, W., 302
Hardin, G., 299
Hartleb, R., 254
Hazen, L. D., 36
Heat flux, 138
Hedrick, D., 193

Heliobacterium chlorum, 256
Hemin procedures and biomass, 109
Heterotrophic activity
  in aquatic ecosystems, 135–138
  in microorganisms, 343–344
  protein synthesis, and $^{14}CO_2$ technique, 147–148
  and succession of microbial communities, 360
^{3}H-labeled organic compounds, *see also* [^{3}H-methyl]thymidine; Radioisotopic assay
  and active substrate uptake, 21
  assimilation of, and metabolic activity, 136–138, 194
  as isotopic tracers for carbon, 97–98
  and microautoradiography, 26, 29
[^{3}H-methyl]thymidine, *see also* ^{3}H-labeled organic compounds
  and FDC method, 146
  and impurities, 96
  and metabolic activity, 194
  and microhabitats, 184
  and rate of nucleic acid synthesis, 150–153
Hobbie, J. E., 3, 10, 23, 77, 112, 134
Hobson, P. N., 31
Hodson, R. E., 99, 135
Hoechst dye No. 33258, 9, 103
Holm-Hansen, O., 107, 136
Holoxenique animals: *see* Conventional animals
Homeostasis, and diversification of function, 358–360
Homogenization, in AODC method, 4–5, 7
Hoppe, H.-G., 21, 119
Horner, S. M. J., 98
Horstmann, U., 21
HPLC-fluorometric method, 128
Hubbel, S. P., 302
Hughes, T. A., 35
Hungate, R. E., 306, 309–310
Huston, M., 357
Hutchinson, G. E., 303
Hydrogen: *see* ^{3}H-Labeled organic compounds; [^{3}H-methyl]thymidine
Hydrogen sulfide
  and anaerobic production, 350
  in photosynthesis, 342
Hyphomicrobium, 254, 298

Iannoti, E. L., 311
IL techniques: *see* Immunolabeling
Immunofluorescence: *see* Fluorescent antibody technique
Immunolabeling, 31–37

Impedins, 180
Incubation
  and activity detection in biofilms, 76
  and estimation of microbial biomass,
    114–116
  and *in situ* study, 93
  length of
    and detection limit of radiorespirometric
      assay, 80
    and *in situ* study, 93
    and INT-reduction method, 25, 27
    and microautoradiography, 25
    and nalidixic acid method, 27
  and size fractionation procedures, 99
Indigenous microflora, and GF versus CV
    animals, 209–213, 218–222
Indirect parasitism, 306
Industrial systems, and biofilms, 73–74, 79
Infrared (IR) spectroscopy, and
    microhabitats, 188–189
In situ measurement, versus *in toto*
    measurement, 88
Internal metalization technique, with SEM,
    59
INT labeling, 121–122, 126–127
INT-reduction method
  applications of, 20
  in Chesapeake Bay comparative study,
    27–30
  and electron transport activity, 20, 23, 25
  and FAINT technique, 36
  modifications of Zimmerman method, 27
Irgalon black-stained filters, and UFOs in IL
    procedures, 34–35
Isotopic tracers, 93–99
  impurities in, 96
  isotope-dilution method, and metabolic
    activity, 136
  and isotope discrimination, 97
  and labeling concerns, 97
  and mathematical formulations, 98–99
  specific activity of, 98
  theoretical bases, 98–99
Iturriaga, R., 20

Jannasch, H. W., 14, 140
Jassby, A. D., 147
Jenkinson, D. S., 114–115
Johnson, K. M., 132
Johnson, P. W., 6
Jones, G. W., 215
Jordon, T. L., 359
Jorgensen, N. O. G., 137
J parameter, 283, 301–304

Kalff, J., 121
Karl, D. M., 128, 149, 153–156
Kim, Y. B., 210
King, F. D., 28
King, G. M., 135, 138
Kingma-Boltjes, T. Y., 254
Kirchman, D., 6, 8, 92
Klebsiella aerogenes, 278
Klug, M. J., 135, 138
Kluyver, A. J., 237, 239
Knoechel, R., 121
Koch, R., 213
Kogure, K., 22, 123
Konigs, W. N., 310
Kormondy, G. J., 346
Krambeck, C., 6, 105
Kriss, A. E., 2
Kronenberg, L. H., 119
Kuparinen, J., 136
Kurath, G., 20

Laanbroek, H. J., 287
Lachnospira multiparus, 308
Lactate, in rumen fermentation, 309
Lactobacillus, 296–298
Ladd, T. I., 76–77, 80
Lalonde, M., 32, 35
LAL test: *see Limulus* amoebocyte Lysate test
Larsson, U., 11
Laser-based flow, and cell counts, 105
Latham, M. J., 308
Laws, E. A., 115, 131, 136, 142, 148
Lectins, as TEM stabilizers, 54
Lee, C., 114
Leeuwenhoek, Antonie van, 261
Legionella pneumophila, 35–36, 122, 252–253
Levenson, S. M., 212
Levin, G. V., 107
Lewis, D. H., 295
Liebig's *law of the minimum*, 351, 355
Lignin, 344
Limiting resources, and coexistence of
    competitors, 303
Limulus amoebocyte lysate (LAL) test,
    111–112
Linear alkyl benzenesulphonates (LAS), 326
Lineweaver-Burk equation, and metabolic
    activity, 134
Lipid analysis
  versus ATP analysis, 110
  and biomass, 109–110, 114
  phospholipids, and detection of microbial
    cells, 190
  and population structure, 109–110

Lipid phosphate: *see* Lipid analysis
Lipopolysaccharide (LPS) measurement,
      111–112
London, J., 255
Low-shear systems, and antibody stabilization
      technique, 60
Luckey, T. D., 207
Luminous bacteria, 231
Lundgren, B., 118

MacArthur, R. H., 358
Mah, R. A., 312, 318
Maki, J. S., 123
Malachite green staining, and INT labeling,
      121
Mann, S. O., 31
Marine bacteria, *see also* Aquatic ecosystems
   and assimilation of organic compounds,
      134–135
   predation on, 18
Markovitz, A. J., 32
Marr, A. G., 14
Marsh, D. H., 9
Mateles, R. I., 274, 277–278, 287
Matsuzawa, T., 207
MBC method, 71–72
MC: *see* Microautoradiography
McCambridge, J., 18
McInerney, M. J., 318
McMeekin, T. A., 18
Mean cell volume, as productivity measure,
      14
Meers, J. L., 296
Megasphaera elsdenii, 307–309
Megee, R. D., 296
Metabolic activity
   of animals, and indigenous microflora,
      210–212
   and cell division, 88
   and growth, 88
   in situ rates of, incubation, 93
   microbial, and habitat physicochemistry,
      193–194
   in natural microbial samples, 90, 132
   potentials for, and enrichment technique,
      232
   and signature compounds, 189
   status determination, 123–132
Metal coating technique, in SEM, 57–59
Methane fermentation
   in rumen, 307, 310
   signature lipids of bacteria, 192
   sulfate inhibition of, 324–325
   three-stage scheme for, 314–315

Methanobacillus omelianskii, 317
Methanobacterium thermoautotrophicum, 312
Methanobrevibacterium ruminantium, 308, 310
Methanogenic bacteria, 231, *see also specific*
      genera
Methanogenic environments, microbial
      interactions in, 314–316
Methanogens
   competition with sulfate reducers,
      324–326
   and idealized reactions of bacteria, 312
   in methane fermentation, 314–316
Methanomicrobium mobilis, 310
Methanosarcina, 234, 310, 315, 318
Methanospirillum hungatei, 319
Methanothrix, 315
Methanotrophic bacteria, 231
Methyl-[^{3}H]thymidine: *see* [^{3}H-
      methyl]thymidine
Methylotrophic bacteria, 231
Meyer-Reil, L. A., 4, 183–184
Meyer-Reil technique, 3, 119
Michaelis–Menton kinetics, 134–137
Microautoradiography
   and biofilm activity detection, 75–76
   in Chesapeake Bay comparative study, 23,
      25–30
   and estimation of viability, 118–121
   and individual cell activity, 20–21
   and rates of primary production,
      120–121
Microbial infallibility, principle of, 347
Microbial production: *see* Primary
      productivity; Productivity,
      determination of
Microcalorimetry, and metabolic activity,
      138–139
Micrococcus luteus, 305
Microelectrodes, and measure of habitat
      parameters, 182–183, 193
Microeukaryotes: *see* Eukaryotes
Microhabitats
   and ecological interaction principles, 352
   recognition of, and enrichment techniques,
      232
Miller, T. L., 310
Mills, A. L., 20, 36
Minimum bactericidal concentration (MBC)
      method, 71–72
Minimum uptake rate, as quantitation
      method, 29
Mitotic index, and specific growth rate (μ),
      144
Mitskevich, I. N., 2

Mixed cultures
  and competition for substrates, 267–273
  on mixed substrates, 272–273, 277–285
  stability on methane of, 298
Mixed substrates
  chemostat enrichments with, 285–287
  limitations in, and growth, 266
  mixo- and autotroph competition on,
      279–282
  mixo- and heterotroph competition on,
      279–282
Modified Robbins device (MRD), and biofilm
      bacteria, 71–72, *see also* Robbins device
Monheimer, R. H., 146
Monks, R., 98
Monoassociated animals, 206, 214
Monoclonal antipilus antibodies, 57
Monod, J., 263
Monod-type kinetics, 265–267, 300
Moriarty, D. J. W., 7, 151–152
Morita, R. Y., 20, 136
Most popular number (MPN) method,
      251–252
Mucus-covered tissues, and microbial
      populations, 59–60, 62–67
Multiple substrate concentration technique,
      and intact biofilms, 77–79
Muramic acid, 110–111
Mutualism, 295, 297–299
Mycobacterium vaccae, 296
Myxobacterales, 305–306

Nalidixic acid method, *see also* Yeast extract/
      nalidixic acid technique
  and biofilm activity, 75
  in Chesapeake Bay comparative study,
      22–25
  and determination of cell activity, 21–22
Nedwell, D. B., 324
Neihof, R. A., 28–29, 119, 121
Neutralism, 295–296
Newell, R. C., 15
Newell, S. Y., 7–8, 10, 12–13, 18, 105–106,
      121, 145–146
Niche sharing, 356–357
Nicholas, J. L., 221
Nichols, P. D., 188
Nickel, J. C., 79
Nitrate respiration, 243, *see also* Anaerobic
      respiration
Nitrification, 348–349
Nitrobacter, 34, 234, 304
Nitrogen cycling, 125–126, 347–349
N-limited autotrophic populations, 131

Nonaxenic cultures, 32
Nonbacterial microorganisms, 87–88
Nonincubation methods, and *in situ* rate
      processes, 93
Nonspecific staining, in FA procedures, 34
Novitsky, J. A., 119–120
Nucleic acid synthesis, 148–155
Nucleosides, 127
Nucleotide ratios, 127–128
Nutrient concentrations, 344, 352–353
Nutritional status, 90, 194–196

Obligatory interactions, definitions of, 297
Odum, E. P., 295, 352
Odum, H., 345
Odum, W. E., 9
Odum's combined law, 352
Oligotrophic bacteria
  enrichment cultures for, 247
  heterotrophic activity of, 344, 346
  and survival strategies, 353
Oligotrophic ecosystems
  cell activity in, 21
  metabolic activity in, 135–136
  photosynthetic rates in, 141
Olsen, R. A., 105–106
Onderdonk, A. B., 214, 222
Open ecosystems, steady-state theory in,
      139–140
Opportunistic species, and survival strategies,
      354
Oquist, G., 129
Oremland, R. S., 326
Organic compounds, uptake of, 134–138, *see
      also* ^{14}C-Radiolabeled organic
      compounds; ^{3}H-Labeled organic
      compounds
Osmophilic bacteria, 231
Owens, T. G., 28

Packard, T. T., 126–127
Paerl, H. W., 121–122
Paracoccus denitrificans, 234
Parasitism, 295, 305–306, *see also* Indirect
      parasitism; Predation
Parathion, degradation of, 326–327
Parsons, T. R., 1, 134
Particulate matter, and FDC method, 13
Pathogenic mechanisms, and gnotobiotic
      animals, 213–216
Paul, E. A., 8–9, 118
Paul, J. H., 9
Pectinase treatment, and autofluorescence, 35
Pelobacter, 318, 321

"Pelochromatium roseoviridie," 323
"Pelochromatium roseum," 323
Petzold, H., 255
Pfennig, N., 310, 315, 321–322, 324
PHA: *see* Poly-β-hydroxy alkanoate
PHB: *see* Poly-β-hydroxybutyrate
Phagocytosis, bacterial resistance to, 180–181
Philips, O. M., 280
Phospholipids: *see* Lipid analysis
Phosphorylation state, 127–128
Phosphosphingolipids, 193
Photoautotrophic microorganisms, 131–132, 341
Photoautotrophy, 341–343
Photometric techniques, and IL procedures, 36–37
Photosynthesis
    anoxygenic, 342
    in microorganisms, 341–343
    oxygenic, 140–142
    and phylogenesis, 239
    rates of, and DCMU-enhanced fluorescence, 129–130
Phototrophic bacteria
    and enrichment cultures, 242
    sources of samples, 231
    and symbiotic associations, 323–324
    and synotrophic associations, 322–324
Physiological effects, and gnotobiotic animals, 216–219
Physiologic potentials
    and enrichment techniques, 232
    and inference of habitat properties, 189–196
    and *in vivo* activity, 88
    methods for estimation of, 90, 123–132
Physiologic traits, and classification, 235
Physiotypes, bacterial, and sources for samples, 231
Phytohemagglutinin (PHA), in mucus layer, 67–68
Phytoplankton
    and autofluorescence in IL studies, 34–35
    estimation of standing stock, 112–113
    growth rate of, 131–132
    paradox of, 303
Plant polysaccharides, in rumen fermentation, 307–308
Plasmalogen phospholipids, and anaerobic bacteria, 192–193
Plastic substratum, and bacterial biofilm, 52–53
Polcin, S., 326
Pollard, M., 207
Pollard, P. C., 151–152

Poly-β-hydroxy alkanoate (PHA), 128–129, 195
Poly-β-hydroxybutyrate (PHB), 128–129
Polyanions, stain specificity and, 54
Polyassociated animals, 206
Polymer degradative bacteria, 231
Polymers, 128–129, 180–181
Populations
    adaptation in, 353–354
    consumer, distribution of, 342–343
    interactions of, 295, 351–358
    oscillations of, and niche sharing, 356–357
Powell, E. O., 267
Predation, 295, 305–306, *see also* Grazing; Parasitism
    and extracellular polysaccharides, 181
    and *in situ* growth rate, 156–158
Primary productivity, *see also* Productivity, determination of
    aphotic, 140
    and availability of minerals, 351
    of organic matter, 140–142
    rates of, by photoautotrophs, 341
Primuline: *see* Epifluorescent microscopy; Exogenous fluorochromes
Principle of competitive exclusion, 268
Prochlorales, 341, *see also* Photosynthesis
Productivity, determination of, *see also* Primary productivity
    bacterial versus microbial, 87, 91
    and estimation of grazing output, 17–19
    and FDC method, 13
    and [^{3}H]thymidine uptake, 150–153
    with incubation, 14–19
    without incubation, 10–14
    by light microscopic methods, 10–19
    by prevention of grazing output, 16–17
    and use of $^{35}SO_4^{2-}$, 146–148
    using isolates at natural substrate concentration, 15
Proflavine, 9, 104, *see also* Exogenous fluorochromes
Prokaryotes
    activity of, compared with microeukaryotes, 193–194
    biomass of, 110–112
    phototrophic, 99
    phylogenetic classification of, 235–240
Propionate degradation, 318–319
Propionibacterium shermanii, 297
Proteus rettgeri, 193
Proteus vulgaris, 298
Protocooperation, 295, 297–299
Protozoa, 185, *see also* Eukaryotes

Prouty, C. C., 254
Pseudomonas, 296, 298
Pseudomonas aeruginosa, 71–74, 278,
    326–327
Pseudomonas fluorescens, 274
Pseudomonas stutzeri, 326–327
Psychrophilic bacteria, sources of samples of,
    231
Purcell, E. M., 179
Pure cultures
    growth of, on mixed substrates, 273–277
    instability on methane of, 298
    and planktonic versus sessile cell
        differentiation, 50
    problems in, 49–50, 57

Quispel, A., 32, 35

Rabbit serum, and IL methods, 34
Radioactive adenine, 153–155
Radioactive nuclides, 94–95, *see also* Isotopic
    tracers
Radioisotopic assay, *see also* Isotopic tracers
    and biofilm activity, 79–80
    compared to FDC method, 13
Raibaud, P., 206, 219–220
Ramsay, A. J., 119
Redalje, D. G., 115–116, 142
Redfield ratio, and growth rate, 131
Red-orange fluorescent markers, and
    autofluorescence, 35
Redox potential (Eh)
    bacterial modification of, 180
    microelectrode measurement of, 183
    of respiratory quinones, 193
Remsen, C. C., 123
Renger, E. H., 129
Resource, defined, 299
Respiration rate, and ETS activity,
    19–20
Revsbech, N. P., 133, 183
Rhodamine isothiocyanate (RITC)-
    hydrolyzed gelatin counterstain
    technique, 34–35
Rhodopseudomonas, 234, 285
Riemann, B., 152
RITC technique, 34–35
Rittenberg, S. C., 255, 328
RNA : DNA ratios, 130
RNA synthesis: *see* Nucleic acid synthesis
Robarts, R. D., 9
Robbins device, 71–74
Roberts, R. B., 146
Romanenko, V. I., 105, 143

Romanenko technique, and dark CO_2
    uptake, 143–144
Root tissue, and autofluorescence, 35
rRNA analysis, as phylogenetic approach,
    235–240
Roser, D. J., 9
Rozee, K. R., 64
Rublee, P. A., 7, 10, 106
Ruby, E. G., 328
Rullman, W., 254
Rumen ecosystem
    bacterial competition in, 308–309
    energy sources in, 308
    fermentation interactions in, 307–310
    interspecies hydrogen transfer in, 310–312
    microbial interactions in, 306–316
    nutritional interactions in, 313–314
    and population niches, 359
    and protozoan populations, 306–307
    and TEM of biofilm from, 51
Ruminococcus, 308, 311, 313
Russell, J. B., 309
Ruthenium red stain, with bacteria biofilm,
    51–52, 54–56, 58
Rutter, J. M., 215
Ryder, D. N., 266

Saccharomyces, 274, 296
Salmonella typhimurium, 216–217
Samuelsson, G., 129
Sasser, Myron, 192
Scanning electron microscopy (SEM)
    in bacterial ecology, 32
    and biofilms, 57–64
    and cell size distributions, 105
    and microhabitats, 185–187
    photographs, artifacts of, 69
    and shrinkage problem, 105, 185
Schink, B., 310, 321
Schleyer, M. H., 7
Schmidt, E. L., 32–33, 35–36
Scintillation autoradiography, 120
Secondary ion mass spectroscopy (SIMS),
    187–188
Secondary productivity, 344, 346, *see also*
    Productivity, determination of
Sediments, *see also* Aquatic ecosystems
    anaerobic, 191
    AODC method with, 5, 7
    compared with rumen environments, 314
    and epifluorescent counting, 184
    and FDC method for bacterial productivity
        in, 13
    fermentative bacteria in, 315–316

Sediments (*Cont.*)
  growth conditions in, 195
  individual cell activity in, 20
  metabolic activity in, 138–139, 192–193
  methanogenesis in, 324
  O$_2$ flux in, 133–134
  respiratory quinones in, 193
  total microbial biomass in, 155
Selenomonas ruminantium, 308–309
Sephton, L. M., 9
Sequence analysis, and bacterial classification, 235–237
Sessile bacteria: *see* Biofilms
Shannon index, 359–360
Shehata, T. E., 14
Shelford's *law of tolerance*, 351–352, 355
Shields, J. A., 115
Shingella flexneri, 219–220
Sieburth, J. McN., 6, 185
Signature compounds, assay of, 189–190, 192–193
Silver, R. S., 274
Sinclair, C. G., 266
Single time-multiple substrate concentration technique, with biofilms, 77
Size fractionation procedures, 99
Slime layers, and micro habitat modification, 180–182
Smith, D. F., 98, 136
Smith, M. E., 274
Solar energy: *see* Energy, capture of
Sondergaard, M., 137
Specific-activity indices, 20, 22
Specific growth rate (μ), and concentration of limiting substrate, 301
Specific pathogen free (SPF) animals, 211
Specificity, and IL studies, 33–34
Spread plate technique: *see* Culture techniques
Spirillum strain G7, 304
Sporocytophaga myxococcoides, 234
Stable isotopes, 94, 96, *see also* Isotopic tracers
Staley, J. T., 21, 359
Standing crop, 2–10, 87–88, 112–113
Stanley, P. M., 21
Staphylococcus aureus, 35, 62
Statzell-Tallman, A., 105–106
Steady-state theory, in open ecosystems, 139–140, 284–285
Stephanopoulos, G. N., 301, 303
Sterility testing, 206–207
Stolp, H., 255
Strayer, R. I., 32

Streptococcus, 217, 298, 309
Streptovertcillium, as predator, 305
Stress-tolerant species, 354
Strickland, J. D. H., 134
Stripping film technique, and estimation of viability, 119
Strugger, S., 103
Stull, E. A., 121
Stutzer, A., 254
Subjectivity, 3, 13, 101
Subsampling, 92
Substrate(s)
  comparison of, in microautoradiography method, 26
  choice of
    and kinetic approach to heterotrophic activity, 135–136
    and viability measures, 119–120
  concentration of, and growth rate, 267
  mixed, growth of cultures on, 272–285
  removal of, and biofilm bacteria, 77
  sequential utilization of, 273
  uptake of
    and individual bacterial cells, 20–21
    and terminal fermentation process, 325–326
Successions
  autotrophic, 360
  and climax community, 359–360
  and enrichment cultures, 249–250
  of habitat colonizers, 191
Succinate, in rumen fermentation, 309–310
Succinovibrio dextrinosolvens, 308
Sulfate
  and aerobic production, 350
  reduction of, and microelectrode measurement, 183
  $^{35}SO_4^{2-}$ uptake, 146–148
Sulfate-reducing bacteria, 192, 324–326
Sulfolobus acidocalcarius, 31
Sulfur cycling, 350
Sulfur metabolism, and growth, 146–148
Sulfur oxidizers, 343
Surfaces, nature of, 196
Survival strategies, 353–354
Symbiosis, as term, 299, *see also* Mutualism
Synthetically active bacteria (SAB), 22–30
Synthetics, man-made, microbial degradation of, 344
Syntrophic associations, 322–323
Syntrophobacter wolinii, 318–320
Syntrophomonas wolfei, 318, 320
Syntrophus buswellii, 320
Syntrophy, 298, *see also* Protocooperation

Tabor, P. S., 28–29, 119, 121
Tabor-Neihof method, for electron-transport activity, 23, 28–29
Tamminen, T., 136
Taxonomic classification, techniques for, 233–240
Taylor, P. A., 271
Teichoic acids, 181
TEM: *see* Transmission electron microscopy
Temporal niche, and competitive exclusion, 356–357
Terrestrial ecosystems
  AODC method with, 3
  dominant grazers in, 345
  intertidal beach sand, biomass in, 115
  nitrification in, 348–349
  primary producers in, 345
  and productivity determination, 16–17
  soils
    bacterial count in, 4, 7, 9
    IL method in, 34
    metabolic activity in, 138
    microhabitat modification in, 181
    O_2 flux in, 133–134
    zymogenous populations in, 353
  stratification patterns in, 342
Tetrazolium salts, and ETS activity, 19
Thermal rift ecosystems, 343
Thermoanaerobium brockii, 318
Thermophilic bacteria, 231
Thermus, 245
Thicapsa pfennigii, 234
Thingstad, T. F., 272, 280
Thiobacillus
  acidification of environment by, 179
  and amensalism, 304
  and coexistence, 304
  and enrichment procedures, 234
  and FAINT technique, 36
  growth of, 275–276, 279, 304
Thiocarbohydrazide (TCH) compound, 59
Tiedje, J. M., 32
Tilman, D., 270, 283–284
Tobramycin sulfate, and biofilm bacteria, 80
Tolerance range, 351
Total adenine nucleotide concentration, and microbial biomass, 108–109
Toxins, secretion of, 179–180
Track autoradiography, and viability, 121

Transmission electron microscopy (TEM), 32
  and biofilm bacteria, 50–57
  and microhabitats, 185–187
Treponema hyodysenteriae, 214
Trexler, P. C., 205
Tritium: *see* ^{3}H-Labeled organic compounds
Trolldenier, G., 3
Trophic levels, 345

UFOs: *see* Unidentified fluorescent objects
Ultrasonication techniques, with soils, 7
Unidentified fluorescent objects (UFOs) in IL studies, 34–35
Universal acceptors, 35
UV epifluorescence counting: *see* Epifluorescence techniques

Van Cleve, K., 132
Van Gemerden, H., 283, 322
Van Niel, C. B., 86, 232, 235, 237, 239
Van Verseveld, H. W., 14
Veillonella alcalescens, 309
Veldkamp, H., 302, 310
Viability
  and biogeochemical transformations, 117
  estimation of, 90
  and microbial biomass, 116–123
  and potential for reproduction, 117
Vibrio succinogenes, 311, 313
Viral contamination of germfree animals, 207

Wagner, M., 207, 215
Walker, H. H., 143
Wand, M., 96
Waste digestors, 314–316
Water column, distribution of microorganisms in, 342
Watson, S. W., 111–112
Wedemyer, G., 327
Weiler, C. S., 144
Weimer, P. J., 311–312
Wheeler, P., 99
White, D. C., 128–129, 195
Widdel, F., 324
Wiebe, W. J., 135
Wilkins, J. R., 116
Williams, P. J. leB., 15, 127, 134–135, 271
Winkler titration, 133
Winogradsky, S., 229, 232, 240, 286
Wolin, M. J., 304, 310, 312–313
Wostmann, B. S., 211, 213
Wright, R. T., 1, 77, 135–136

Xenobiotic compounds, degradation of, 326–327, 347

Yeast extract/nalidixic acid technique, 23–25, 123, *see also* Nalidixic acid method
Yoon, H., 272
Young, L. Y., 320

Zeikus, J. G., 312
Zimmerman, R., 3, 28, 183
Zonation
    nitrogen cycling, 349
    and photosynthesis, 342
Zoogleal flocs, 182